**Maker Innovations Series**

Jump start your path to discovery with the Apress Maker Innovations series! From the basics of electricity and components through to the most advanced options in robotics and Machine Learning, you'll forge a path to building ingenious hardware and controlling it with cutting-edge software. All while gaining new skills and experience with common toolsets you can take to new projects or even into a whole new career.

The Apress Maker Innovations series offers projects-based learning, while keeping theory and best processes front and center. So you get hands-on experience while also learning the terms of the trade and how entrepreneurs, inventors, and engineers think through creating and executing hardware projects. You can learn to design circuits, program AI, create IoT systems for your home or even city, and so much more!

Whether you're a beginning hobbyist or a seasoned entrepreneur working out of your basement or garage, you'll scale up your skillset to become a hardware design and engineering pro. And often using low-cost and open-source software such as the Raspberry Pi, Arduino, PIC microcontroller, and Robot Operating System (ROS). Programmers and software engineers have great opportunities to learn, too, as many projects and control environments are based in popular languages and operating systems, such as Python and Linux.

If you want to build a robot, set up a smart home, tackle assembling a weather-ready meteorology system, or create a brand-new circuit using breadboards and circuit design software, this series has all that and more! Written by creative and seasoned Makers, every book in the series tackles both tested and leading-edge approaches and technologies for bringing your visions and projects to life.

More information about this series at `https://www.apress.com/series/17311`

# Mastering Analog Electronics

## Unlocking the Power of Circuits and Semiconductors

**Hubert Henry Ward**

Apress®

***Mastering Analog Electronics: Unlocking the Power of Circuits and Semiconductors***

Hubert Henry Ward
Leigh, UK

ISBN-13 (pbk): 979-8-8688-0247-8
ISBN-13 (electronic): 979-8-8688-0245-4
https://doi.org/10.1007/979-8-8688-0245-4



Managing Director, Apress Media LLC: Welmoed Spahr
Acquisitions Editor: Miriam Haidara
Development Editor: James Markham
Project manager: Jessica Vakili

Cover designed by eStudioCalamar

Distributed to the book trade worldwide by Springer Science+Business Media New York, 1 New York Plaza, Suite 4600, New York, NY 10004-1562, USA. Phone 1-800-SPRINGER, fax (201) 348-4505, e-mail orders-ny@springer-sbm.com, or visit www.springeronline.com. Apress Media, LLC is a California LLC and the sole member (owner) is Springer Science + Business Media Finance Inc (SSBM Finance Inc). SSBM Finance Inc is a **Delaware** corporation.

For information on translations, please e-mail booktranslations@springernature.com; for reprint, paperback, or audio rights, please e-mail bookpermissions@springernature.com.

Apress titles may be purchased in bulk for academic, corporate, or promotional use. eBook versions and licenses are also available for most titles. For more information, reference our Print and eBook Bulk Sales web page at http://www.apress.com/bulk-sales.

Any source code or other supplementary material referenced by the author in this book is available to readers on GitHub. For more detailed information, please visit https://www.apress.com/gp/services/source-code.

If disposing of this product, please recycle the paper

*Dedicated to my wife Ann.*

*Love always.*

*Pinch! Pinch! Pinch!*

*I also dedicate these books to my grandchildren in the hope they may use them in their studies just like their dad did.*

# Table of Contents

# About the Author

**Hubert Henry Ward** has nearly 25 years of experience as a college lecturer delivering the BTEC, and now Pearson's, Higher National Certificate and Higher Diploma in Electrical and Electronic Engineering. Hubert has a 2.1 Honors Bachelor's Degree in Electrical and Electronic Engineering. He has also worked as a consultant in embedded programming. His work has established his expertise in the assembler and C programming languages, within the MPLAB X IDE from Microchip, as well as designing electronic circuits, and PCBs, using ECAD software. Hubert was also the UK technical expert in Mechatronics for three years, training the UK team and taking them to enter in the Skills Olympics in Seoul 2001, resulting in one of the best outcomes to date for the UK in Mechatronics.

# About the Technical Reviewer

**Sai Yamanoor** is an embedded engineer based in Oakland, CA. He has over ten years of experience as an embedded systems expert, working on hardware and software design and implementations. He is a coauthor of three books on using Raspberry Pi to execute DIY projects, and he has also presented a Personal Health Dashboard at Maker Faires across the country. Sai is also working on projects to improve the quality of life (QoL) for people with chronic health conditions. Check out his projects at `https://saiyamanoor.com`.

# Introduction

## The Aims and Objectives of the Book

My main aim in writing this book is to introduce you to the exciting and challenging field of analog electronics. I want to develop your desire and ability to understand how analog circuits work.

The Objectives of the Book

After reading this book, you should be able to do some or all of the following:

- You will appreciate what a semiconductor is.
- You should understand what a PN junction is and how we can use it as a diode.
- You should be able to appreciate how an SCR works and where we might use them as well as a Diac and a Triac.
- You will understand how the PN junction led to the NPN and PNP transistors.
- You will have learned how to design a simple BJT amplifier and a more stabilized amplifier.
- You will have some appreciation of what FETs are and the JFET and MOSFET devices.
- You will have a good understanding of Opamps and how they can be configured in a variety of modes.

- You will appreciate what a multivibrator is, and we can create a monostable and an astable multivibrator.
- You will learn about oscillators and how we can design the phase shift and Wien Bridge oscillators using Opamps.
- You will appreciate some of the ways we can use the 555 timer IC.
- You will understand passive and active filters and how we can use complex numbers to help analyze them.
- You should be able to use the ECAD software Tina 12. We will use the Tina ECAD software to simulate the circuits in the book and thus prove they work as expected.

## Prerequisites

There are no real prerequisites for you except a desire to learn about this exciting and challenging field of analog electronics. An appreciation of Ohm's Law and the main Kirchhoff's laws would be an advantage. However, we will recap them as we use them in the book.

## Engineering Numbers and How I Express Them

There will be numerous times when we need to write numbers in their engineering format. For example, 2200 can be written as 2.2k. We can also use scientific notation, and this will write 2200 as 2.2 but with an indication that the decimal point has been moved from 2200.0 three places to the left to be at 2.2. As a college lecturer, there were times when writing this in scientific notation caused some issues. Therefore, to try and avoid confusion, I would write the scientific notation in the following manner:

2200=2.2E3

In this format, the capital letter "E" means the EXP button, for Exponent, on your calculator. So, you would type 2.2 on your calculator, then press the EXP button, then type 3 for the number of places you have moved the decimal point. Some calculators, like the CASIO fx-83GT PLUS shown in Figure 1, have the x10 button instead of the EXP.

Another example would be 3500000.0. This would be written as

$3.5E^{6}$

The number 6 is because we move the decimal point six places to the left.

Another example:

0.0000045

We would write this as

$4.5E^{-6}$

The number 6 again because the decimal point has been moved six places. However, this time it has been moved six places to the right, and so we give it the minus or negative sign. To put this into your calculator, you would type 4.5. Then press the EXP button, then –6, or press 6 and use the +– button to make it –6.

***Figure 1.*** *The Basic Calculator*

## The Natural Number "e"

You must not mistake my capital "E" for the letter "e" on your calculator. The letter "e" is a representation of a real number that occurs in nature. Almost everything that happens in nature has some relationship to this number. Well, what is the number? When you look at your calculator, like the one shown in Figure 1, you will see a button with "In". This is really Ln for natural or Napierian Logs. The shift function button above it is "$e^n$" which is the antilog for Ln. If you press this button and put a 1 in the brackets that appear, then after pressing enter the display will give the value of "2.71821828r". This is the value of the natural number "e". I hope we are already happy to use the symbol "π" to represent the number of 3.14159r, then really, as an engineer, we should be happy to use the symbol "e". However, you must not mistake it with my use of the capital letter "E".

I hope these examples, and my explanation, will help you understand how I will write these scientific numbers in all the chapters of the book. In this introduction, I have tried to give you a feel for what you will learn when you read this book.

I hope I have got a good balance between explaining how things work and giving you enough examples that you understand and find them useful. I know I cannot cover everything there is about this and any subject, as it is a growing area of electronics and there is always something new and exciting happening. However, I hope there is enough in the book for most of your needs and that you find it a useful resource to help in a career as an analog electronics engineer. Happy reading.

# CHAPTER 1

# Electrical Current

In this chapter, we will learn what electrical current is. We will delve deep inside the atom to understand the workhorse of electrical engineering, the electron. We will study the principle on which all electrical circuits work.

We will examine the structure of the semiconductor and how we, as engineers, make use of it. This will lead us, through the PN junction, to consider the major semiconductor devices we use in our everyday life.

## The Main Operation of All Electrical Circuits

Before we delve into what electrical current is, it might be useful to try and understand the principle on which all electrical circuits work. The main concept of what engineers are trying to do, with their electrical circuits, is to move the energy from the source, be it a DC battery or an ac power source, to a load. They will convert that energy into something useful such as sound, light, or movement. This chapter is concerned with how we move that energy from the source to the load, that is, the medium which carries the energy around the circuit.

I feel I should say that the following is my interpretation of what I have read about the process. It is based on years of studying, reading a lot of material on the subject, and teaching the subject to my students. I hope you will find it illuminating and useful and hopefully a good read. However, I cannot guarantee that it is 100% correct; well, who can? Even Bohr's writings, on which some of this text is based, are termed Bohr's


H. H. Ward, *Mastering Analog Electronics*, Maker Innovations Series,
https://doi.org/10.1007/979-8-8688-0245-4_1

postulates, that is, theories. However, my interpretation has served me well over the years, and I think it does follow some common-sense approach to the subject.

## The Mighty Electron

The ELECTron is at the very heart of ELECTricity. It is the workhorse of all electrical circuits, and I think that is why we have given the subject the name "Electricity," hence the capital letters in bold. However, the electron is so small; it is invisible to the naked eye.

The importance of the electron is not confined to electrical circuits. Indeed, the electron is the flexible glue, a glue that is flexible (what?) that binds all things, including ourselves, together – a pretty strong glue with a lot of force behind it. I compare the glue of the electron to the glue that binds our universe together. Indeed, the way the electron orbits the nucleus is similar to the way the planets of our universe orbit the sun. However, the electron is very small; it's miniscule in fact.

So, for something that has such a tiny mass that it takes approximately 1 million, million, million, million, million, or 1.098 $E^{30}$, to make 1 gram of mass, that's not bad going. Yes, the mass of an electron has been measured, and the method scientists used to measure its mass is quite extraordinary, but that's not for this book.

So how does this little electron do it? It is all down to force "F" and the interaction of two or more forces. The basic expression for force F is

$$F = ma, \text{ or Force} = \text{mass x acceleration.}$$

The mass of an electron, given the symbol "$m_e$", has been measured to be

$$\text{Mass of an electron } (m_e) = 9.10939 \text{ E}^{-31} \text{ g}$$

This means that to get any level of force, there must be a very high degree of acceleration. Well, the acceleration involved is another study entirely and not for me to try and discuss; I will leave that for a far better scientist than me.

## The Charge "C" of an Electron

Charles Coulomb, conducting some experiments back in 1888, discovered that the electron had some form of energy which he termed electrical charge. He named the unit for charge, the Coulomb, after himself as he found it, and he measured the charge on the electron at

Charge on the electron given the symbol "e" = $1.6022\ E^{-19}$ C

Note "C" stands for Coulomb. During these experiments, he determined that the electron was attracted to the nucleus by an invisible force pulling the electron toward the nucleus. If this attraction is compared with the attraction of magnets, then it would seem that, as with magnets, like poles repel and unlike poles attract, then the electron must be attracted by an unlike or opposite charge on the nucleus. He used the convention of positive and negative charge and stated that the electron had a negative charge, while the nucleus, or rather something inside the nucleus, had a positive charge. The difference in these charges brings the electron close to the nucleus and allows it to settle down into its orbits around the nucleus. When there are enough electrons orbiting the nucleus, then the atom is said to be in a stable state, as the positive attractive force of the nucleus is canceled out by the negative charge of the electrons orbiting the nucleus.

## Bohr's Postulates

It is inside the atom that the electron resides. An atom is made up of a collection of electrons which are continually moving around a nucleus. In the early days, scientists thought the electrons were moving around the nucleus in the same way that the Earth orbits the Sun. Indeed, just as the Earth spins on its own axis, thus causing the period of the day, while it orbits the Sun, thus causing the period of the year, then the electron is spinning around on its own axis as it moves around the nucleus in its orbit.

The Danish physicist, Niels Bohr, came up with his theory that the electrons orbit the nucleus in a series of shells that are at a set distance from the nucleus while spinning around on their own axis at the same time. This means that the arrangement of the electrons orbiting the nucleus can be as shown in Figure 1-1.

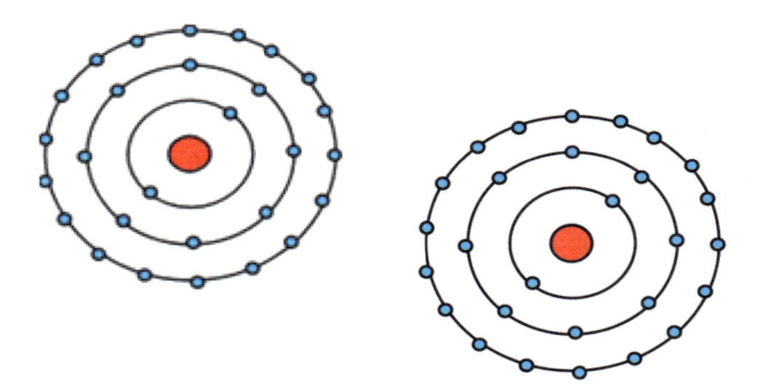

***Figure 1-1.*** *The Simplified Orbits of the Electron*

Figure 1-1 is a simplified representation of how the electrons orbit the nucleus of an atom. It really only shows the shells that Bohr's postulates put forward. The nucleus, at the center, shown in red, is a collection of protons that have a positive charge and neutrons that are neutral, that is, no charge. It is the number of protons, in the nucleus, that gives the atom its atomic number in the periodic table. For an atom to be in its stable state, the number of electrons orbiting the nucleus must be the same as the number of protons in the nucleus. This would mean that the overall charge on the atom would be neutral, that is, no charge.

The shells that the electrons exist in are given a number starting at 1, where 1 is the shell closest to the nucleus. An alternative way of identifying the shells would be to use letters starting at "K." Bohr's theory has given us an expression for the number of electrons in each shell as

$$Number\ of\ electrons = 2n^2$$

where "n" is the number of the shell.

Using this expression, we can list the number of electrons in each shell, as shown in Table 1-1.

***Table 1-1.*** *The Number of Electrons in Each Shell*

| Shell Letter | Shell Number | Number of Electrons $2n^2$ |
|---|---|---|
| K | 1 | 2 |
| L | 2 | 8 |
| M | 3 | 18 |
| N | 4 | 32 |
| 0 | 5 | 50 |

The table lists the maximum number of electrons in each shell; this is the maximum number, and there will be atoms with less electrons in a particular shell.

# The Valence Shell

The valence shell is the outer shell of the orbits. It is in this shell in which there may be less electrons than predicted by Table 1-1. As an example, we will look at the copper atom. It has an atomic number of 29. This means there will be 29 protons in the nucleus, but more importantly, well I think it's more important, there will be 29 electrons orbiting the nucleus. Figure 1-2 shows a simplified representation of the copper atom.

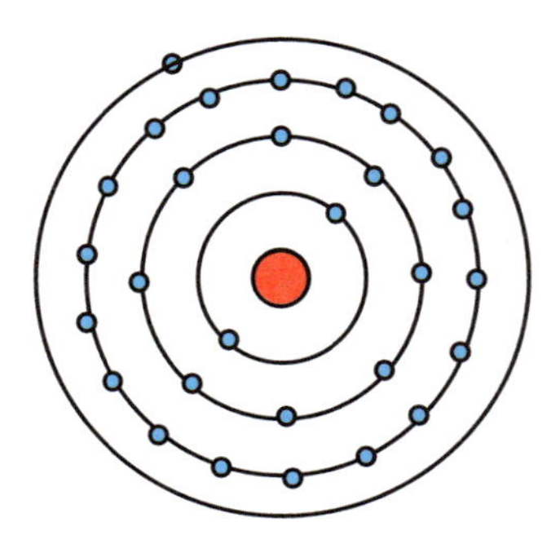

***Figure 1-2.*** *The Simplified Representation of the Copper Atom*

Using Bohr's theory, there will be 2 electrons in the first shell, 8 in the next, and 18 in the third. This must leave 1 in the outer or valence shell. This is shown in Figure 1-2. I don't think it is too difficult to see that the electron in the outer shell has only a weak bond to the nucleus, as it has to overcome the repelling forces of the electrons between it and the nucleus. However, this does not mean the electron in the valence shell has no energy. Indeed, according to Bohr's theory, the outer electrons have more energy than the electrons in the lower shells, that is, the closer the electrons get to the nucleus, the less energy levels they have.

We should be able to see that as the electron tries to move toward the nucleus, because of the attractive force between the positive protons inside the nucleus and the negative charge of the electron, the electrons in the lower shells actually try to repel it, stopping it from getting any closer to the nucleus. It is this same attracting force of the protons in the nucleus and the repelling force of the electrons close by that keep the electrons in their respective orbits.

This electron in the valence shell is the weak link in this state of equilibrium as the copper atoms that are close by also have a single electron in their valence shell. As these electrons are orbiting their respective nuclei, then at some moment in time, these two electrons, each in their respective valence shells, will come close to each other. It is at this instant in time that the like charges of the two negatively charged electrons cause a repelling force that is strong enough to force the two electrons to spin out of their orbits and away from their respective nuclei. This has some very important consequences:

1. The two electrons accelerate at a tremendous rate causing huge forces.
2. The two atoms they have left are now in an unstable state.
3. The atoms that have lost this outer electron will do all they can to get an electron back so that they can return to their natural stable state of a neutral charge.

## The Sea of Free Electrons

We have atoms in which the electrons have just been forced out of the valence shell, which means that the atoms are in an unstable state, which is intolerable for nature and these atoms. The nucleus of these two atoms will exert great forces to bring an electron back to them. They may even bring back the electron that has just left it, but they are more likely to pull another electron out of the valence shell of a nearby copper atom. In this way then, there is a continual movement of electrons accelerating away from, and to, the nucleus of nearby copper atoms. This is termed a sea of free electrons all jumping, or being pushed, from one copper atom to another. However, we must realize that this movement of electrons is totally random as they are jumping about in all different directions. The fact that, in copper, there are electrons freely moving around it makes copper a good conductor. This is because, as we will see, it does not need much effort to make these electrons move in a uniform direction and so create a flow of current. Indeed, one of the definitions of current flow is as follows:

> Current flow can be defined as the flow of electrons moving in a uniform direction.

## The Octal Rule

To fully understand the movement of electrons around an atom, we would have to delve deep into quantum mechanics. That is far beyond my expertise and the context of this book. However, to understand conductors and insulators we need to cover some of the basic aspects of atomic theory. We have already looked at some of Bohr's theories, and it was only a glimpse. Figure 1-1 is a simplified diagram to try and explain the orbits of the electron. It has, I hope, helped us appreciate the importance of the valence shell. However, to move on and learn about conductors, insulators, and the usage of the semiconductor, we need to consider the octal rule. An American chemist called Gilbert Lewis came up with the theory that if a compound of atoms could create the situation whereby their valence shell had the magic number of eight electrons in them, then they could mimic the stable state of the inert gases. He went on to say that most of the elements, especially those in groups 1, 2, and 14 to 18 of the periodic table will try their best to form a valence shell with eight electrons in them. This is what he termed the "octal rule or octet rule," and some materials will create huge forces as their atoms try to comply with this octal rule.

If a material has eight electrons in its valence shell, then, according to the octal rule, it can be said to have a full valence shell. It will make a good insulator, as opposed to a good conductor, that has only one electron in its valence shell.

We can summarize the important points of what we have discussed as follows:

- The atomic number of an atom is based on the number of protons in the nucleus.
- The nucleus has protons with a positive charge and neutrons with a neutral charge.
- Electrons have a negative charge.
- Electrons orbit the nucleus in a series of shells.
- The number of electrons in an atom is the same as the number of protons in the nucleus.
- It is the valence or outer shell that determines the action of the atom.
- Materials or compounds that have atoms with a full eight electrons in their valence shell make good insulators.
- Materials or compounds that have atoms with only one electron in their valence shell make good conductors.
- The octal rule states that atoms within the groups 1, 2, and 14 to 18 exert large forces to maintain eight electrons in their valence shell

## The Semiconductor

In our study so far, we have learned what makes a good conductor and what makes a good insulator. We have seen that it is the number of electrons in the atom's valence shell that controls this aspect, that is, a full valence shell for an insulator and a valence shell with only 1 electron for a good conductor. This is in accordance with the octal rule. Now we are going to look at elements whose atoms have valence shells that are half full, that is, those that have only four electrons in their valence shell. As the number of four is halfway between eight and one, we could say that elements that have these four electrons in their valence shell could be called semiconductors as they are half way between conductors and insulators. Using Bohr's expression, we can create a table that determines the required atomic number for a semiconductor element related to the number of shells used in the atom. This table, shown as Table 1-2, can then allow us to determine if such an element exists and resides in the groups 1, 2, and 14 to 18.

***Table 1-2.*** *The Shell Arrangement of Electrons Required for a Semiconductor Atom*

| Shell | Atomic Period | Shell Arrangement of Electrons | Atomic Number |
|---|---|---|---|
| 1 + valence shell | 2 | 2 **+ 4** | 6 |
| 2 + valence shell | 3 | 2 + 8 **+ 4** | 14 |
| 3 + valence shell | 4 | 2 + 8 + 18 **+ 4** | 32 |
| 4 + valence shell | 5 | 2 + 8 + 18 + 32 **+ 4** | 64 |
| 5 + valence shell | 6 | 2 + 8 + 18 + 32 + 50 **+ 4** | 114 |

The first column indicates the shells of the atoms of the element with the valence shell added. Therefore, the first row in this column indicates we are looking for an atom that has one shell not including the valence shell. This means that an atom will actually have two shells when we include the valence shell.

The second column in Table 1-2, headed Atom Period, relates to the row number on the periodic table.

Therefore, using this table we can see that, if we consider the elements within the atomic row 2, for the element to be termed a semiconductor having four electrons in its valence shell, it must have an atomic number of "6." Similarly, if we consider the next row in Table 1-2, we can see that the element must have an atomic number of 14. The next row states that the element should have an atomic number of 32. Similarly, we can see that the next two would have atomic numbers of 64 and 114.

The middle column of Table 1-2 states the required arrangement of electrons to provide four electrons in the valence shell. The last column states the required atomic number of the element, which is the same as the total number of electrons stated in the middle column. We can use this last column to determine if such an element exists and if it is within the groups that follow the octal rule.

If we now look at the periodic table (see Figure 1-3), we can see that only the first three elements, that is, those with atomic numbers of 6, 14, and 32, actually reside in the groups that follow the octal rule. Indeed, all three exist in group 14. The elements are

- Carbon with the number 6
- Silicon with the number 14
- Germanium with the number 32

This means that these three elements are suitable for use as a semiconductor. However, carbon is not suitable as its electrons require too much energy to jump between levels. This means we are left with just silicon or germanium as our semiconductor elements. The most widely used of the two is silicon, although germanium can be used in the manufacture of diodes, which are sometimes referred to as signal diodes.

| Atomic Period | Group | 1 | 2 | 3 | 4 | 5 | 6 | 7 | 8 | 9 | 10 | 11 | 12 | 13 | 14 | 15 | 16 | 17 | 18 |
|---|---|---|---|---|---|---|---|---|---|---|---|---|---|---|---|---|---|---|---|
| 1 | | 1 H | | | | | | | | | | | | | | | | | 2 He |
| 2 | | 3 Li | 4 Be | | | | | | | | | | | 5 B | 6 C | 7 N | 8 O | 9 F | 10 Ne |
| 3 | | 11 Na | 12 Mg | | | | | | | | | | | 13 Al | 14 Si | 15 P | 16 S | 17 Ci | 18 Ar |
| 4 | | 19 K | 20 Ca | 21 Sc | 22 Ti | 23 V | 24 Cr | 25 Mn | 26 Fe | 27 Co | 28 Ni | 29 Cu | 30 Zn | 31 Ga | 32 Ge | 33 As | 34 Se | 35 Br | 36 Kr |
| 5 | | 37 Rb | 38 Sr | 39 Y | 40 Zr | 41 Nb | 42 Mo | 43 Tc | 44 Ru | 45 Rh | 46 Pd | 47 Ag | 48 Cd | 49 In | 50 Sn | 51 Sb | 52 Te | 53 I | 54 Xe |
| 6 | | 55 Cs | 56 Ba | | 72 Hf | 73 Ta | 74 W | 75 Re | 76 Os | 77 Ir | 78 Pt | 79 Au | 80 Hg | 81 Tl | 82 Pb | 83 Bi | 84 Po | 85 At | 86 Rn |
| 7 | | 87 Fr | 88 Ra | | 104 Rf | 105 Db | 106 Sg | 107 Bh | 108 Hs | 109 Mt | 110 Ds | 111 Rg | 112 Cn | 113 Uut | 114 Fl | 115 Uup | 116 Lv | 117 Uus | 118 Uuo |

***Figure 1-3.*** *The Periodic Table*

# Covalent Bonding

Now that we know we can use silicon as a semiconductor, we need to examine how we use it. The first aspect is how the pure silicon atom, in its natural or intrinsic state, reacts when multiple atoms come together. As silicon, being in group 14, is one of the elements that complies with the octal rule, the atom will exhibit huge forces to try and increase the number of electrons in its valence shell to the magic 8. If it can do this, then it will be able to act in a similar fashion to one of the noble gases, that is, it will become very stable and not interact with other atoms. Figure 1-4 is an attempt to show how five silicon atoms can come together to share one of their four valence shells to enable one of them, in this case the center atom, to think it has a full valence shell of 8 electrons.

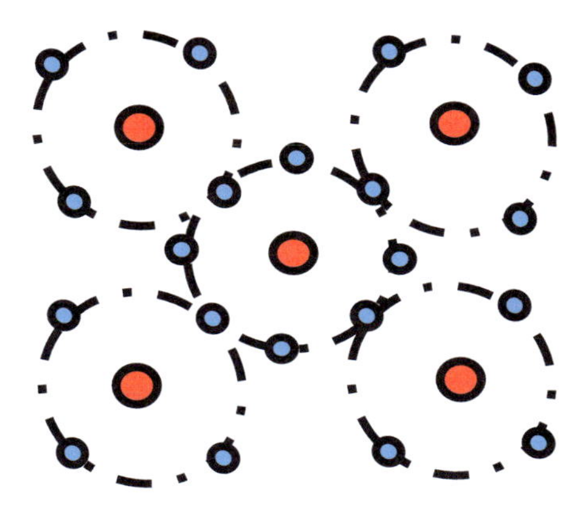

***Figure 1-4.*** *Covalent Bonding*

The center atom appears to have eight electrons orbiting it. We need to appreciate that this is happening in a three-dimensional arrangement. This actually means that all five of the atoms would have the eight electrons orbiting their respective nucleus. This may not be the best two-dimensional drawing to show this, but I hope it does help you appreciate the principle behind what I am trying to show you here. This covalent bonding is what happens with silicon, and it makes a lattice type structure for silicon that is very stable. However, we must still appreciate that the actual individual atoms of silicon still have only four electrons in their valence shell. It is only when they come together and form this covalent bonding that the atoms appear to have a full valence shell. This is called covalent bonding because it is a bonding formed when multiple atoms cooperate with each other, using electrons in their respective valence shell, to create a valence shell that has eight electrons in it.

## The Extrinsic Semiconductor Introducing Some Impurities

The covalent bonding we have looked at so far uses just pure silicon atoms and so is said to be in its intrinsic state, that is, with no impurities. We have achieved this full valence shell because we were using just silicon atoms which have four electrons in their valence shell. Now we will see what happens if we add an element that has atoms that has only three electrons in its valence shell. These types of elements are called trivalent elements. If we look at the periodic table, we can see that the element Boron, which is in group 13, has five electrons, and so it has three electrons in its valence shell. Therefore, if we used Boron as one of the five atoms shown in Figure 1-4, we would create the covalent bonding as shown in Figure 1-5.

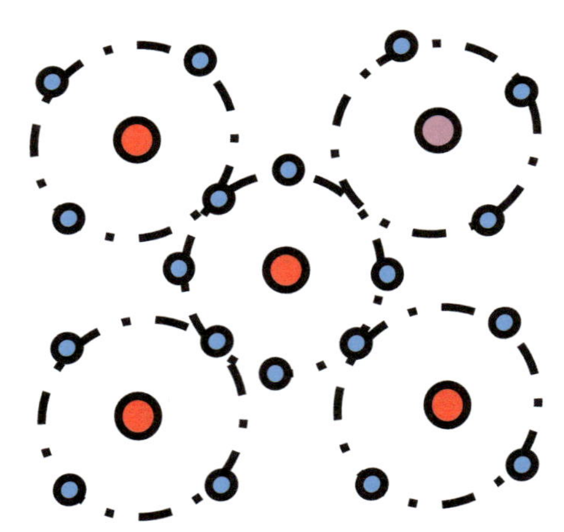

***Figure 1-5.*** *The Bonding of Four Silicon Atoms and One Boron Atom*

This arrangement makes the center atom appear as though it is short an electron. Indeed, to be able to comply with the octal rule, it is short an electron. This means this central atom will be trying its best to attract an electron to it and so fulfill its desire to comply with the octal rule. If we realize this is done throughout the three-dimensional lattice of the piece of silicon, we should be able to understand how we have turned this stable inert piece of silicon into a substance that is doing all it can to attract electrons to it. As this piece of silicon is now attracting electrons to it, we can say it works as if it has a positive charge on it, as electrons are attracted to a positive charge or potential. That is why when we add a trivalent atom into the mix with silicon atoms, we create what can be termed a P-type material.

## The Pentavalent Impurity

We have seen what happens if we introduce some Boron atoms into the mix; now, we will see what happens if we introduce some antimony atoms into the mix. Antimony has 51 electrons, and it has 5 electrons in its valence shell. Figure 1-6 shows what could happen.

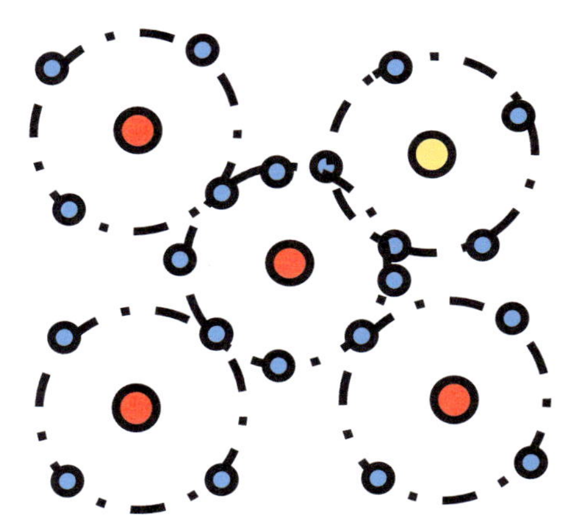

***Figure 1-6.*** *Adding a Pentavalent Atom*

In this situation, the central atom now appears to have nine electrons in its valence shell. Now, to comply with the octal rule, it has too many electrons, and so it will try to get rid of it. Again, if we apply this to the three-dimensional lattice bonding, the silicon semiconductor material can act like it has a negative charge on it, as it has too many electrons. Hence, using a pentavalent addition to the mix, we have created a material with an apparent negative charge, that is, an N-type semiconductor.

This act of adding impurities to the covalent bonding of silicon is termed "doping," and the intrinsic semiconductor has now become an extrinsic semiconductor.

# The PN Junction

Now that engineers have found that they can create a P-type and an N-type semiconductor, by doping silicon compounds in this way, they had to find a way of using the two types of semiconductors. What those engineers were about to discover actually revolutionized electronic engineering. What we had to use up to this point was valve technology. Just to give you some idea of what this was, we should consider the simple diode manufactured using valve technology. A schematic of one is shown in Figure 1-7.

***Figure 1-7.*** *The Schematic of the 54RA Rectifier Valve*

These are still manufactured today, as some people use them in hi-fi equipment; however, I don't quite understand why. The schematic does not show you much about them, but they are glass tubes that have a vacuum inside them to help the electrons move from the cathode, the long connection close to the bottom of the valve, to the anode, the thicker connection at the top of the valve. The last connection at the bottom of the valve is actually a heater element. The way in which the diode works is that a low voltage, around 6V ac, is applied to the heater element. This heating heats the cathode and gets the electrons excited by giving them extra energy. This gets them close to moving out from the cathode to the anode, which will be connected to a large positive voltage. If an ac signal is then applied to the cathode, this will, when going positive, inject more electrons into the valve, and these are attracted by the positive potential at the anode.

This is a very awkward process and expensive, as it requires a heating voltage to be applied to the heater. Also, valves are expensive to produce and can easily be broken. That being the case, it was not a very efficient component. If engineers could come up with a much more efficient method of rectifying a source, then it would be a great advancement. Well, that is exactly what they did with the PN junction.

What engineers tried was to bring a P-type semiconductor and an N-type semiconductor together and watch what happened. Figure 1-8 shows the PN materials brought together.

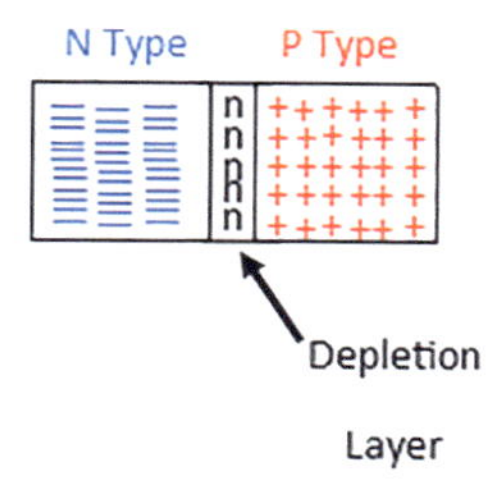


***Figure 1-8.*** *A PN Junction*

Figure 1-8 shows the PN junction in its stable state with no voltage applied across it. It is called a junction because the important action happens at the point where the two materials meet, that is, at the junction of the two materials. We should appreciate that the electrons in the N-type material are surplus, as there are nine electrons in the valence shell of the atoms, as we have explained earlier. Also, the P-type material is short of electrons, as we have explained earlier. This shortage of electrons can be viewed as there being holes in the P-type material that want electrons, from anywhere, to move into and fill them. This is what happens as soon as the two materials are brought together. The surplus electrons in the N-type material start to combine with the holes in the P-type material. This sets up a layer of atoms between the N-type and P-type materials that are now neutral in charge and are quite happy to stay that way as they are in a very stable state. This creates a layer which is termed the "depletion layer," as it is depleted of charge. This depletion layer makes it too difficult for any more electrons to move across from the N-type material into the P-type material and so combine with the holes there. This means that any further movement of electrons in the PN junction stops.

Even in this state, the PN junction has some electrical properties. The first is that there are two conductive plates, that is, the N-type and P-type plates, separated by some medium. This is the basic requirement of a capacitor, that is, two conductive materials separated by some medium. This is the inherent capacitance of the PN junction. We will see how this affects performance when we look at the bandwidth of a transistor amplifier in Chapter 6. Also and slightly more obvious is the fact that the PN junction has a positive plate and a negative plate.

## Biasing the PN Junction

We now need to see what happens when we apply a voltage across the PN junction; this is called "biasing." In the first instance, we will apply the biasing with a negative potential applied to the P-type material with respect to the N-type material. This is shown in Figure 1-9.

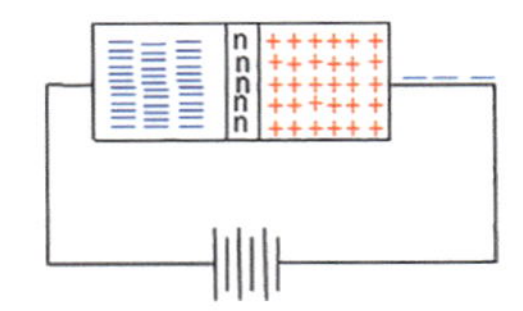

***Figure 1-9.*** *Applying a Negative Voltage to the P-Type Material*

To appreciate what happens, it is best to consider electron flow. Current actually flows in a circuit, because the excess electrons, on the negative plate of the voltage source, are trying to flow around the circuit and enter the holes in the positive plate of the voltage source, thus returning the source to its natural neutral state. This is termed electron flow, as opposed to conventional current flow, which flows from positive to negative. We use conventional current flow in our analysis of most electrical circuits. However, in this instance, and some others, we will use electron flow.

Any positive potential will attract electrons to move toward it, as it wants the electrons to recombine with those atoms that are short of an electron, thus creating a neutral charge. This movement of electrons is not the same electron making the whole trip from negative to positive; it is one electron knocking into another and making that electron move in the same direction. Note current flow is defined as electrons moving in a uniform direction.

With this in mind, we can see that there are electrons that are trying to enter into the P-type material as shown in Figure 1-9. Also, the electrons in the N-type material are attracted to the positive potential of the voltage source and so leave the N-type material. This means we do get movement of electrons. However, as electrons leave the N-type material, they leave behind atoms that have a neutral charge. Also, electrons moving into the P-type material combine with the holes, which also results in atoms that now have a neutral charge. This results in two more depletion areas developing within the PN junction. This results in an increase of the depletion area of the PN junction. This means that the electrons that are trying to move through the PN junction have to overcome a more dominant depletion area. This means that the electrons eventually find it too difficult to move through the PN junction and so current flow stops.

This type of biasing is termed reverse biasing where the N-type material is connected to a positive voltage supply and the P-type material is connected to the negative terminal of the voltage source, as shown in Figure 1-9.

Now let's see what happens when the voltage is applied in the other polarity. This is shown in Figure 1-10.

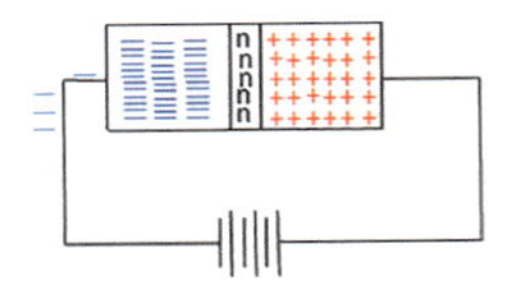

***Figure 1-10.*** *Forward Biasing of the PN Junction*

First of all, it can be seen that more electrons are trying to enter the N-type material. These electrons will increase the number of excess electrons in this region. Also, there is now a positive potential on the P-type material, and this has the effect of increasing the attractive force on the excess electrons in the N-type material. This encourages these electrons to try and cross over the depletion layer. When the force is strong enough, that is, the voltage across the PN junction is high enough, these electrons will break through the depletion layer and rush through the P-type material, heading toward the positive potential applied to the P-type terminal. This rush through the PN junction is assisted by the electrons entering the N-type material, and so an avalanche effect occurs, and we have current flow through the PN junction.

The amount of force required to overcome the depletion layer depends upon the semiconductor material, and for silicon it is around 0.7V, and for germanium it is around 0.2V. We should appreciate that voltage is the force that makes electrons move. Indeed, it was termed the "EMF" (ElectroMotive Force) as it was the force that made electrons move in a uniform direction. Now, thanks to Alessandro Volta, we call it a voltage.

This then means that when the voltage is applied as in Figure 1-9, it is termed reverse biasing, and in Figure 1-10 it is termed forward biasing. I should point out that this really means current will only flow one way through the PN junction. This is the action of a diode, and so we can see that the PN junction actually creates the component that is the diode. The symbol of the diode is related to the PN junction as shown in Figure 1-11.

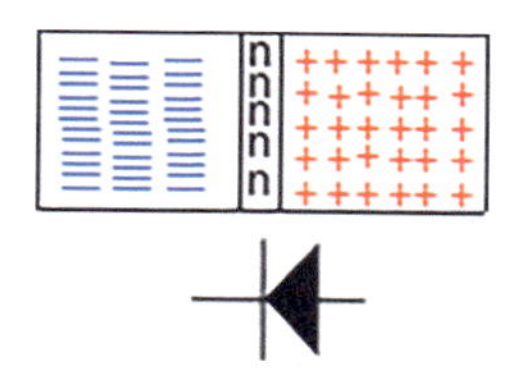

***Figure 1-11.*** *The PN Junction Related to the Diode*

This means that the P-type material is the anode, and the N-type material is the cathode. Note the arrow symbol, in the symbol for the diode, points in the direction of conventional current flow through the diode, that is, from positive to negative.

This then shows that the semiconductor can be put to good use in that we have made a diode from the simple PN junction. This was actually a bigger breakthrough than it might at first seem as engineers have gone on to create a wide range of semiconductor devices, and we will look at some of them in this book.

# The NPN Sandwich

We will look at the diode and some useful variations of them in the next chapter. However, for now we will look at what would have been the natural extension of the PN junction, that is, the NPN sandwich, as shown in Figure 1-12.

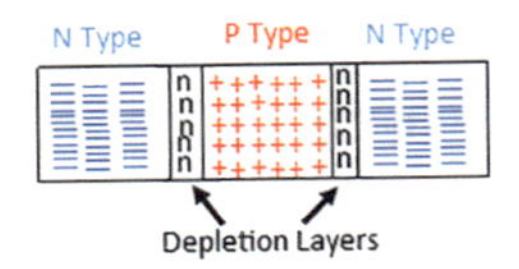


***Figure 1-12.*** *The NPN Sandwich*

We have seen from our analysis of the PN junction that we could view the PN junction as a diode and that we would have a depletion layer between the P- and N-type materials. Therefore, it should be no surprise that we will now have two depletion layers, but also, we will have two diodes. This is because the two diodes would have their anodes sharing the P-type material. This concept is shown in Figure 1-13.

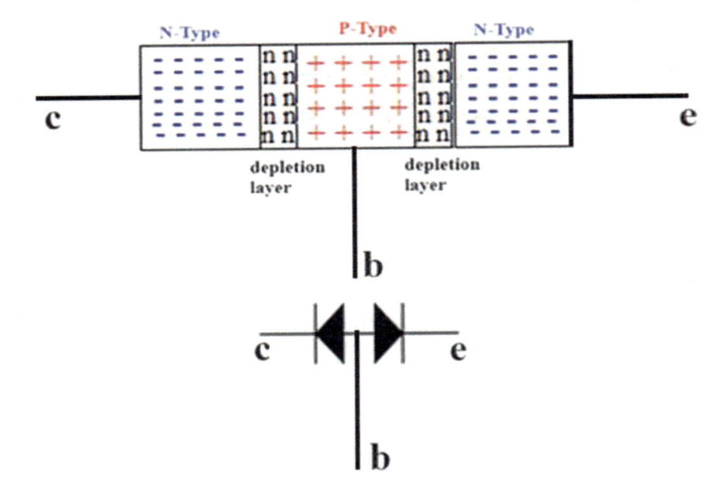


***Figure 1-13.*** *The NPN Sandwich Represented As Two Diodes Back to Back*

With the PN junction shown in Figure 1-10, we have shown that we can forward bias that junction by applying a positive potential to the P-type material and a negative potential to the N-type. So, this would suggest that if we apply a positive potential to the "b" terminal and a negative potential to the "c" and "e" terminals, the two diodes would conduct. However, we do not connect the device to the power supply in that way. The correct manner in which we connect this device is shown in Figure 1-14.

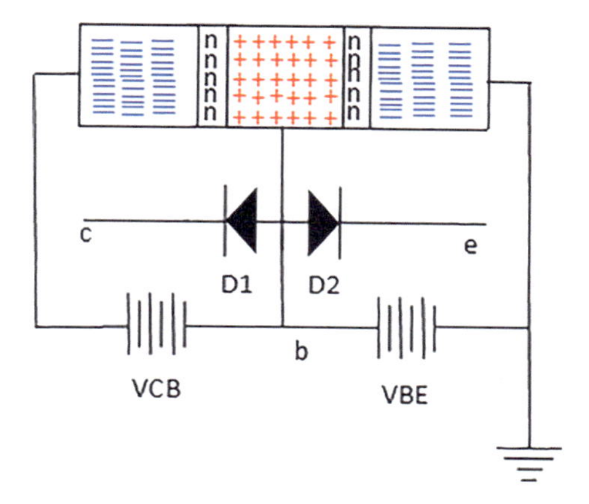


***Figure 1-14.*** *The NPN Sandwich Connected to the Power Supplies*

The two diodes have been left in the diagram to try and remind us that the NPN sandwich is two diodes connected as shown, that is, back to back. This arrangement of connecting the batteries will try to force electrons into the two N-type materials.

The letters at the connection have the following meanings:

- b is for base as we call this the base terminal.
- c is for collector as this terminal collects the electrons that are forced to flow through the NPN sandwich.
- e is for emitter as this is the terminal from which the electrons are emitted.
- VBE is the DC voltage between the base "b" and the emitter "e."
- VCB is the DC voltage between the collector "c" and the base "b."

If we compare this figure to Figure 1-9, we should see that the NP junction on the left-hand side is reverse biased. Indeed, if we look upon it as a diode, we see that the diode D1 is reverse biased. However, voltage is a force that is used to make current flow. We will see later that if we apply enough force, then a diode that is reverse biased can be made to break down and conduct. Indeed, that is what we do with Zener diodes that are manufactured so that they work in this breakdown region. We will look at Zener diodes in Chapter 2.

If we compare Figure 1-14 to Figure 1-10, we will see that the NP junction on the right-hand side is actually forward biased. Indeed, if we consider the diode D2 representation of the junction, we can see that the anode is connected to the positive terminal, and the cathode is connected to the negative terminal. This will apply enough forward biasing to make that diode conduct.

In the configuration shown in Figure 1-14, if the voltage between the base "b" and the emitter "e," the VBE voltage, is enough to overcome the depletion layer between the right-hand PN junction, then that diode, D2, will be conducting. This will allow the emitter terminal to take electrons into the PN junction. The electrons would normally flow out at the base terminal. However, as they enter the PN junction, they are attracted by the positive potential of the VCB + VBE terminal. So, they would attempt to flow toward that terminal, but that would mean they would have to flow the wrong way through diode D1. Remember, we are talking about electron flow which flows in the opposite direction of conventional current flow. Also, we said that the arrowhead in the symbol of the diode points in the direction conventional current would flow through the diode. This means the electrons would have to flow through a diode that has broken down. Well, surprise, surprise, that is exactly what these electrons do. That is because diode D1 in the NPN sandwich is a very, very soft diode, and it soon breaks down. Remember, we said voltage was a force, and diode D1 does not need much of a force to break it down.

This means that of the electrons that enter the PN junction at the emitter, some will flow out through the base, but the majority, which would be a very large number, of electrons flow through the now broken-down diode D1 and are collected by the collector terminal, being attracted by the positive voltage VCB + VBE.

This was another breakthrough as engineers have now used the PN junction, really two PN junctions, to create the transistor. It is called the BJT, bipolar junction transistor, as it uses two PN junctions. These semiconductor diodes and transistors are much more efficient than their valve equivalent, as they don't need the heater voltage or a vacuum tube and they are not easily broken. Also, they are much, much smaller than the valve equivalents. We will look at transistors in detail in Chapter 3.

# The PNP Sandwich

Just as engineers created an NPN sandwich, they also created a PNP sandwich. This is shown in Figure 1-15.

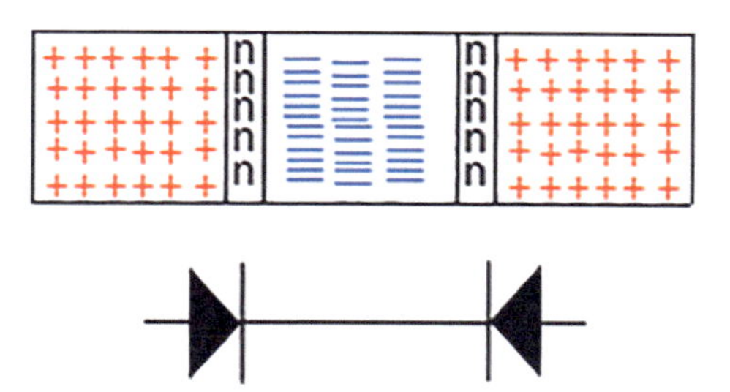

***Figure 1-15.*** *The PNP Sandwich*

This semiconductor sandwich works in the same way as the NPN sandwich except the polarities are reversed.

These two semiconductor sandwiches have given engineers the two types of BJTs, that is, the more common NPN transistor and the less used PNP transistors. They both have their uses, and we will look at them in Chapter 3.

## More Common Semiconductor Devices

This revolutionary semiconductor material, as that was what it was at the time, has been used to create a whole range of semiconductor devices, some of which are listed here:

- The rectifying diode
- The voltage regulating Zener diode
- The switching device and amplifying device of the BJT and FET (field effect transistor)
- The SCR (silicon-controlled rectifier) or thyristor
- The DIAC and TRIAC

We will look at these devices in the various chapters in the book.

## Current Flow, DC and ac, Voltage, PD, Amp, etc.

Now that we have studied the atomic structure of conductors, insulators, and semiconductors, we need to finish this chapter with some of the common definitions that we will come across in electrical and electronic engineering.

## Current Flow

In electrical and electronic engineering, we use two types of current flow, and they are

- Conventional current flow
- Electron flow

I tend to think, and I am not saying it is correct, that engineers first decided that current flows from positive to negative. However, later it has been shown that electrons flow from the negative plate of a source to the positive plate. So, in respect to the early scientists, engineers decided to call the first idea of current flow "conventional current" as it has been the convention to apply this direction of current flow in circuit analysis. An example of how this conventional current flow has been adopted is shown in Figure 1-16.

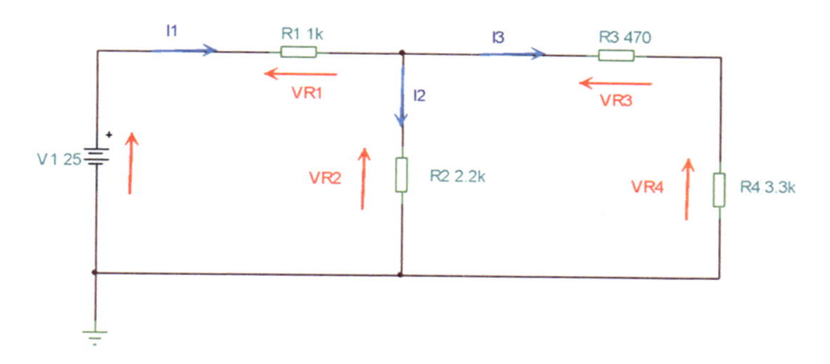


***Figure 1-16.** A Typical DC Circuit Using Conventional Current Flow*

The idea of adding the voltage arrows, in red, and the current arrows, in blue, to the circuit drawing, is to help in analyzing the circuit. The voltage arrows follow the convention that the positive end of the volt drop is at the end with the arrowhead. We can use the voltage arrows to help us apply Kirchhoff's Voltage Law (KVL) in that in a closed loop the voltage arrows pointing in one direction add up to equal the sum of the voltage arrows pointing in the other direction. With respect to the circuit in Figure 1-16, the voltage source $V_1$ and the resistors, $R_1$ and $R_2$, are in a closed loop. Therefore, applying KVL, we can say

$$V_{G1} = V_{R1} + V_{R2}$$

The voltage arrow for $V_1$ points one way, while the voltage arrows for $V_{R1}$ and $V_{R2}$ point the other way.

Using the convention that current flows from positive to negative, we can place the current arrows in blue, as shown in Figure 1-16. The convention also states that conventional current will always flow into the positive end of a volt drop. Note, $V_1$ is not a volt drop, it is a voltage source, and conventional current will always flow out of the positive end of a voltage source and flow into the negative end.

The current arrows will help us apply Kirchhoff's Current Law (KCL) as the law states that the algebraic sum of the currents at a node must equal zero. A more helpful statement might be that the sum of the currents flowing into a node must equal the sum of the currents flowing out of a node. With respect to the circuit shown in Figure 1-16, we can say that $I_1$ flows into the node and both $I_2$ and $I_3$ flow out. Therefore, using KCL, we can say

$$I_1 = I_2 + I_3$$

We will use these conventions later in the book when we analyze some of the analog circuits.

When scientists realized that electrons actually flow in the opposite direction, they simply called this "electron flow" and kept using conventional current flow in their analysis. However, there are some who do adopt the electron flow in their analysis. This basically has the effect of changing the use of the hands in applying Fleming's left-hand and right-hand rules. In this book, I will use conventional current flow in my analysis, unless I state otherwise.

# DC

This stands for direct current which is a phrase that I don't really think explains what it is. A slightly better name might be unidirectional current, UDC, as DC current flows in one direction only.

DC sources are normally batteries, although we can use power supplies which usually convert the mains ac supply to DC using rectifying circuits that use diodes and other components.

## PD and Voltage

Batteries are used as a source in DC applications, and as such they can create what is termed a "PD," or a "potential difference," across the battery or across the components in the circuit. This potential difference is measured in volts.

A simple way of defining voltage is to say that

> **Voltage is the force that makes the electrons move in the circuit.**

It used to be termed EMF, which stands for ElectroMotive Force. This describes it as the force that makes electrons move.

A more technical definition of the volt is

> **1 Volt is the usage of 1 joule of energy per coulomb of charge**

The potential difference "pd" or voltage is different in batteries for different types of simple or wet cells, that is, zinc/copper in sulfuric acid = 0.75v, nickel/cadmium = 1.3v, and the lead/acid cells as in car batteries = 1.8v.

## What Is Current Flow

To start with, we must first state "The Law of Conservation of Energy"; this law states

> Energy cannot be destroyed, it can only be converted into another *format.*

Originally, we had, in respect of a battery, a **chemical energy** in the chemical solution of the battery. This energy was used to strip electrons from the positive plate and deposit them on the negative plate. This meant that the chemical energy in the battery was changed into potential energy in the electron when it gets to the negative plate as it has the potential to move around the circuit to the copper plate. The chemical energy of the solution prevents the electron from moving through the battery back to the positive plate, unless the battery is flat.

Once the circuit is closed, the potential energy on the electron forces the electron to move. The electron then gains kinetic energy (KE), as the electron has a mass and it is moving. Note kinetic energy can be expressed as $KE = \frac{1}{2}mv^2$. The electron then moves around the circuit until it reaches the components in the circuit. Note, it is not the same electron moving all the way around the circuit, it is one electron bumping into another and passing the KE onto the next. If the component is a resistor, then the electron will slow down. This slowing down of the electron means the kinetic energy is reduced. However, using the Law of Conservation of Energy, energy cannot be destroyed; it must be converted into another form. Indeed, the energy is converted into heat energy, which is why resistors get hot.

Other components store the energy as with inductors which store the energy as a magnetic field and capacitors which store it as a charge. The energy can also be used to create movement as with motors or sound as with speakers.

## The Amp or Ampere

An electrical circuit must have a *circ*ular path along which *elec*trons are forced to move by applying a voltage to the circuit. As the electrons move, they transfer energy from the source to a load. Note, current flow is defined as

**The movement of electrons in a uniform direction.**

The unit of current is the Ampere or Amp, and the unit 1 Amp is defined as

**1 Amp of current is 1 Coulomb of charge passing a point in 1 second.**

It is the electron that carries the charge, and the charge on an electron has been measured to be $1.602E^{-19}$ Coulombs. Therefore, it takes $6.241E^{18}$ electrons passing a point to constitute a current flow of 1 Amp. That is a lot of electrons, but they are very, very small.

## Alternating Current (ac)

We have discussed DC; now we should consider "ac." I am using the convention of using capital letters for DC quantities and lowercase letters for "ac" quantities. The term "ac" stands for alternating current, and the current does alternate. This actually means the current changes direction in that at one point in time it is being forced to move around the circuit in one direction and that at another point in time it is being forced to move in the opposite direction. However, conventional current can be, and is, applied to ac circuits. This is because the current still flows from the positive terminal toward the negative terminal. This happens because the polarity of the supply simply swaps over to push the current in the different direction. It's not as complex as it sounds. We generate an ac type voltage by rotating a coil inside a magnetic field. This naturally produces a voltage that is of a sine wave form as shown in Figure 1-17.

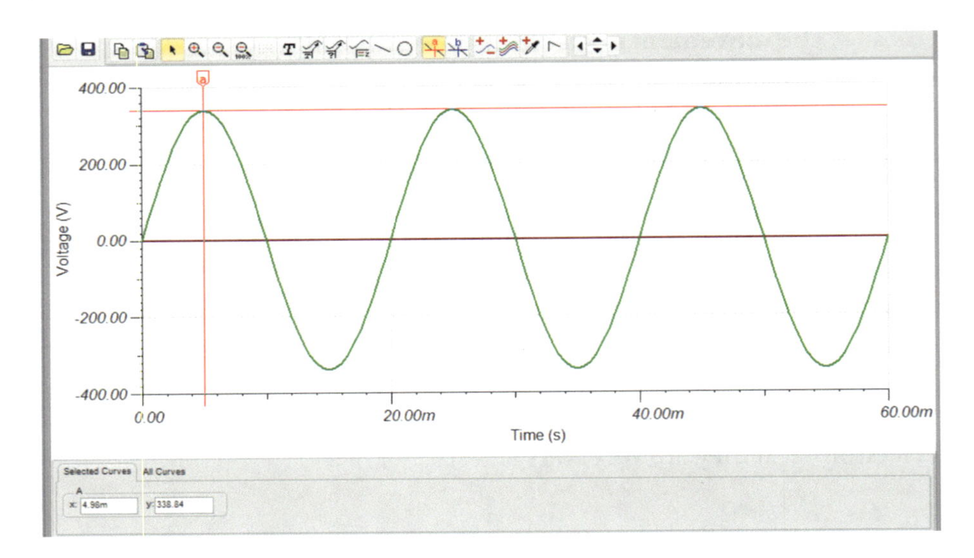


***Figure 1-17.*** *A Typical ac Voltage*

Figure 1-17 shows what the UK mains voltage would look like if you displayed it on an oscilloscope. I am assuming the rms voltage of the UK mains is 240v. This is what it was when I studied engineering, but now it is more likely to be around 220v. For an rms of 240, the peak voltage would be 339v as shown in Figure 1-17.

In the UK, the turbines that rotate the coil inside the magnetic field rotate at a speed of 3000 rpm. This would produce the frequency of 50Hz, that is, 50 cycles per second. We can calculate the periodic time "T" for the UK mains using

$$f = \frac{1}{T} therefore\, T = \frac{1}{f} = \frac{1}{50} = 20ms$$

Figure 1-17 shows that the time to complete one cycle is 20ms. We can see from Figure 1-17 that the voltage starts off at zero and then rises positively until it reaches the peak of 339. It then reduces to zero volts. From then on, it starts to go negative until it reaches –339v. Then it starts to rise until it reaches zero. At that point, it starts to repeat the process. I hope you can see how this would come about by rotating a coil through 360° or $2\pi$ radians inside that magnetic field. Figure 1-17 shows clearly that the polarity of the voltage does change as it completes one cycle. That is why the current will alternate, that is, change direction during each cycle.

## The rms or Root Mean Square

One thing I would like to mention is that if you tried to use the average to give one value that would describe the mains voltage, you would get 0v, which is nonsense. This is because there are as many positive values as there are negative, and so when you add the values up, before dividing to get the average, the sum would equal zero. This is one reason why we use the rms, which stands for root mean square, to express the voltage for the mains. With the rms, we square the expression for the voltage, which gets rid of any negative values. This means we can then determine the mean or average of that squared expression. This would give us a nonzero value. Finally, we need to determine the square root of that result to cancel out the squaring of the expression at the beginning. This method does give us a value that we can use to describe the ac voltage or quantity.

This method may seem a cheat at first, but the rms value of the ac voltage does have a realistic value in that the rms is the same value of the DC voltage that would be required to get the same power in the load. This suggests that the 240v mains is like having a 240V DC battery in your home. Daft as that might seem, there is a simple experiment that might help you appreciate this concept. If we simulate the circuit shown in Figure 1-18, we could measure the current flowing through the 1kΩ load.

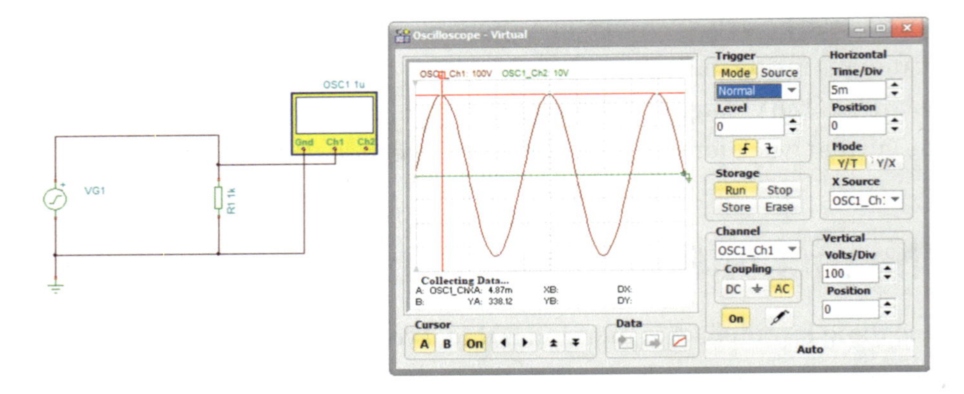


***Figure 1-18.*** *Test Simulation to Measure the Current Through R1*

An oscilloscope cannot measure current; it can only measure voltage and time. However, we can use those measurements to calculate the frequency of the waveforms measured and, as in this case, the current flowing through the resistor R1. To calculate the frequency, we must use the expression

$$f = \frac{1}{T}$$

The term "f" is for frequency and "T" is for the periodic time. This is the time the waveform takes to go through one complete cycle that covers all the values the waveform has. This waveform is a simple sinusoidal wave, and from the oscilloscope, we can see it takes four divisions to complete one cycle. As each division is set to 5ms, then we can see it takes 20ms to complete one cycle. Therefore, the value for "T" is 20ms. Now we can calculate the frequency as follows:

$$f = \frac{1}{20E^{-3}} = 50Hz$$

To calculate the current flowing through R1, we simply need to divide the peak voltage of the waveform by the value of the resistor. This would give the peak value of the current as

$$i_{Peak} = \frac{338.12}{1E^3} = 338.12mA$$

Knowing that, the rms of a sine wave can be calculated as

$$rms = \frac{Peak}{\sqrt{2}}$$

Then the rms current flowing through R1 is

$$i_{rms} = \frac{338.12mA}{\sqrt{2}} = 239.07mA$$

If a DC voltage of 240V was applied across a 1kΩ resistor, then the current flowing through the resistor can be calculated using Ohm's law as

$$I = \frac{V}{R} = \frac{240}{1E^3} = 240mA$$

This does confirm that the rms, of the ac voltage, does represent the DC voltage required to get the same power in the load. Note, the power can be calculated using

$$P = i^2 R = I^2 R = VI = \frac{V^2}{R}$$

To be more correct when dealing with ac circuits, the expression for power becomes

$$P = viCos(\theta)$$

However, when looking at DC and purely resistive loads, we can ignore the Cos(θ) as θ = 0.

# Summary

In this chapter, we have looked at the atomic structure of the main elements that are used in electrical engineering. We have studied what makes a good conductor and what makes a good insulator. We then went on to study what a semiconductor is and how we make a P-type and an N-type semiconductor.

Finally, we looked at the principle of what current flow is and how we make it flow around a circuit. We have looked at some definitions for DC, ac, voltage, current, etc.

In the next chapter, we will start our investigation of semiconductor devices. We will look at the diode and the Zener diode and their usage in power supply designs.

I hope you have found this chapter useful, and it has given you some background understanding of this exciting and challenging field of electronic engineering.

# CHAPTER 2

# Semiconductor Devices

In this chapter, we will learn how engineers tried to make use of the semiconductor materials we looked at in Chapter 1. In this chapter, we will concentrate on the diode and how we make use of it.

## The Diode

This is the first device engineers created with the P- and N-type semiconductor material. Indeed, it was actually the PN junction that we looked at in Chapter 1. A brief recap is covered here. Figure 2-1 shows the PN junction that is created when an N-type material is brought into contact with a P-type material.

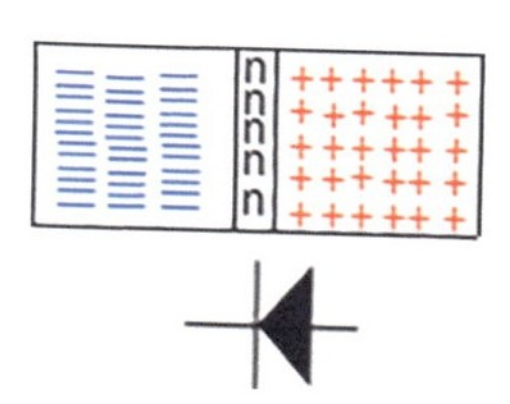

***Figure 2-1.*** *The PN Junction Related to the Diode*


H. H. Ward, *Mastering Analog Electronics*, Maker Innovations Series,
https://doi.org/10.1007/979-8-8688-0245-4_2

The PN junction basically has two conductor materials, one N-type and one P-type semiconductor material, that are separated by an area that is free from charge. Whenever we have two conductors separated by a medium, we have a capacitor, and so there is this capacitance effect within the PN junction. It has to be said that it is a very small value of capacitance, but, as we will see when we study the transistor, in Chapter 6, it does have an effect that we need to be aware of.

However, the basic use of the diode is as an electronic switch. It's a switch, which is normally open, that we can close by applying a voltage across it. This is called biasing the diode, and when we are forward biasing the diode, we force the switch to close and so allow current to flow through the diode. When we are reverse biasing the diode, we keep the switch open and don't allow current to flow through it. We will go through some experiments to study this biasing, but before we do, I feel I should remind you that the arrow shape of the symbol for the diode points in the direction of conventional current flow through the diode. Also, the diode has two terminals, one being the anode and the other being the cathode. Figure 2-2 shows these terminals of the diode.

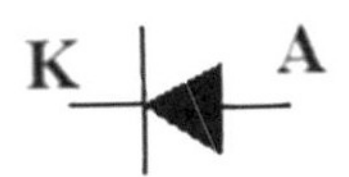


**2-2.** *The Symbol for the Diode*

e the letter "A" for anode, pretty obvious, but "K" for cathode, not This is because the letter "C" is used for capacitance. We can ure 2-2, that conventional current flows from the anode to the

# Forward Biasing the Diode

This is when we apply a positive voltage to the anode of the diode with respect to the cathode. The test circuit to examine what happens with forward biasing is shown in Figure 2-3.

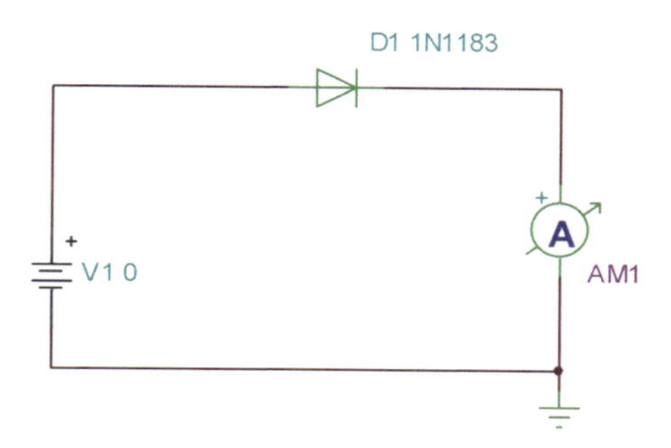


***Figure 2-3.*** *Experiment for Forward Biasing the Diode*

In this simple experiment, we will increase the voltage applied to the anode in steps of 100mV starting from 0V and record the current flowing through the diode. We can then record the results and complete the table of results as shown in Table 2-1.

***Table 2-1.*** *The Results for Forward Biasing the Diode*

| V1 | AM1 |
| --- | --- |
| 0 | 0 |
| 100mV | 367.42E-9 |
| 200mV | 4.48E-6 |
| 300mV | 50.63E-6 |
| 400mV | 567.67E-6 |
| 500mV | 6.63E-3 |

(*continued*)

*Table 2-1.* (*continued*)

| V1 | AM1 |
|---|---|
| 600mV | 71.05E-3 |
| 700mV | 769.77E-3 |
| 800mV | 6.53 |
| 900mV | 27.09 |
| 1V | 60.47 |
| 1.1V | 100.06 |
| 1.2V | 142.71 |

# Reverse Biasing

We will simulate the circuit shown in Figure 2-4 so that we can examine the reverse biasing characteristics for the diode.

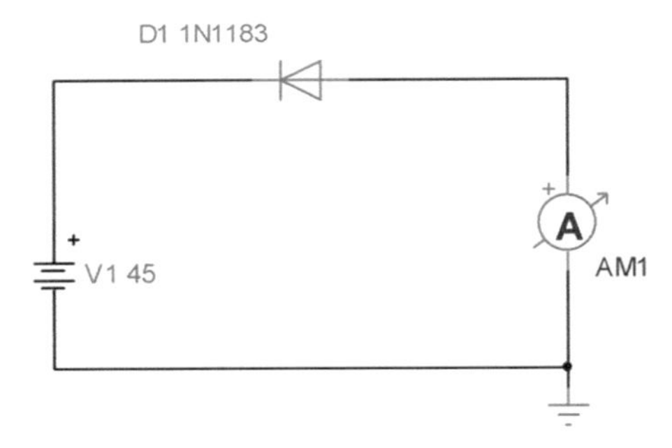


***Figure 2-4.*** *The Reverse Biasing Experiment*

In this experiment, we will increase the voltage applied to the cathode in steps of 1V starting from 45V and record the current flowing through the diode. We can then record the results and complete the table of results as shown in Table 2-2.

*Table 2-2. The Results for Reverse Biasing the Diode*

| V1 (Volts) | AM1 |
|---|---|
| 45 | 36.04E-9 |
| 46 | 36.05E-9 |
| 47 | 36.05E-9 |
| 48 | 36.05E-9 |
| 49 | 36.05E-9 |
| 50 | 4.96E-3 |
| 51 | 231.33 |
| 51.5 | 463.93 |
| 52 | 703.53 |

We should be able to use the results of these two experiments to draw up the characteristics of the diode. These are shown in Figure 2-5.

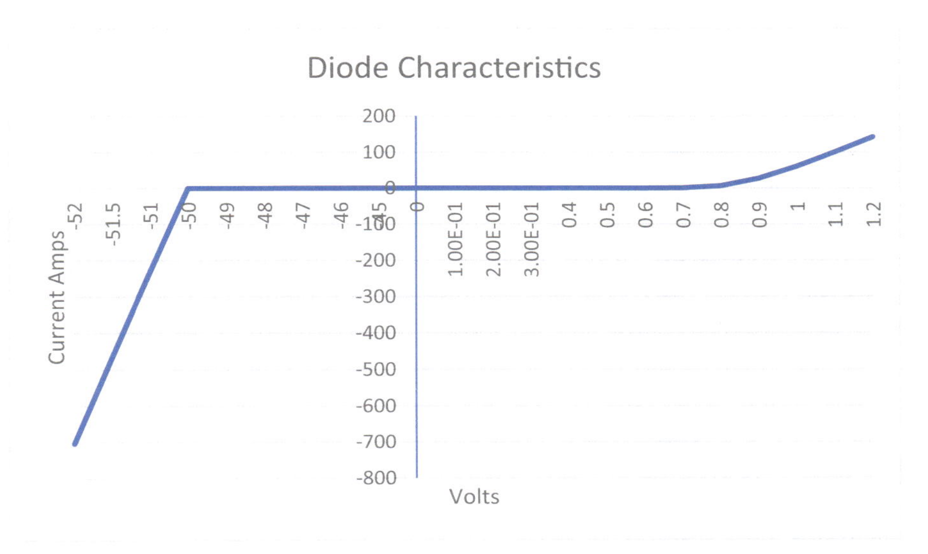


***Figure 2-5.*** *The Diode Characteristics*

The scaling in Figure 2-5 is not the best as with the negative voltages, to the left of the 0 crossing, the voltages jump from 0 to -45 and then increase in steps of 1 up to -51. After the graph reaches the -51V, the voltages increase in steps of 0.5V. This means that the scaling of the graphs is not great. Going positive, to the right of the 0 crossing line, the voltage increases in steps of 100mV. However, the graph does give us the impression I want to put across.

The graph to the right of the 0 crossing is when we were forward biasing the diode by applying a positive voltage to the anode. We can see that there is little or no current flowing through the diode until the voltage at the anode reaches about 0.7V. This is because we have to apply enough force, in the form of a voltage, at the anode to overcome the depletion layer at the PN junction, discussed in Chapter 1. From that point on, the switch is closed, and current can flow through the diode. We can see that with little increase in the voltage applied to the anode, the current soon reaches such high values of 200A. That is only a mathematical value as in real life, if we had allowed the diode to pass such a high value of current through it, the diode would soon burn out and destroy itself. That is why we would need to insert some resistance or impedance to limit the current to a safe value, usually around the hundred milliamps or so.

The graph to the left of the 0 crossing line shows what happens when we reverse bias the diode by applying a negative voltage to the anode. We can see that there is virtually no current flowing through the diode until we reach a voltage of around -50V. At this voltage, the force being applied across the diode is too much for the diode to withstand, and it actually gives up trying to prevent current flow through the diode, even though it is in the wrong direction as it would be flowing from the cathode through to the anode. This is indicated by the fact that the current is shown as being negative, which means it is flowing in the opposite direction. What is interesting with this area of the reverse bias characteristics is that the current flowing through the diode can change by a large amount even though the voltage only changes by a small amount. If we changed the current flowing through a resistor in that way, the voltage across the resistor would have had to change in proportion to the current change. This large current change for a small voltage change is made use of in voltage regulator circuits as we will see when we look at the Zener diode later in this chapter.

The value of -50V is not random. If we examine the datasheet for the 1N1183 diode, we will see that there is a maximum reverse voltage rating for the diode, and this is -50V for this diode. This means that the manufacturer will only guarantee the diode will not conduct while reverse biased as long as the user does not exceed a reverse voltage or -50V. Each diode will have a list of ratings that the user must adhere to when using a diode. We will look at these when we make use of the diode later in this chapter.

## The Forward Diode Voltage Drop

In this next experiment, we are going to see how the voltage dropped across the diode, measured by VM1, varies as we increase the voltage applied to the anode of the diode. We will simulate the circuit shown in Figure 2-6 to see what happens.

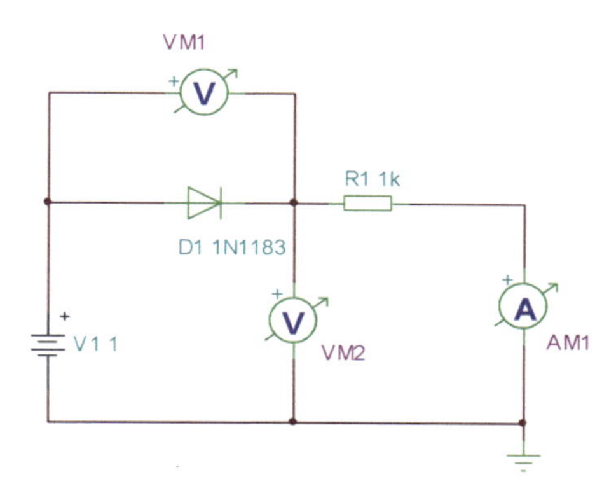


***Figure 2-6.*** *The Forward Diode Voltage Drop Experiment*

The procedure is to increase the voltage applied to the anode in steps of 5V starting at 1V while recording the voltage dropped across the diode, using VM1, and the current flowing through the diode, measured using AM1. We should then be able to complete the table of results as shown in Table 2-3.

***Table 2-3.*** *The Table of Results for the Circuit Shown in Figure 2-6*

| V1 (Volts) V | VM1 | V1 – VM1 Or VM2 | AM1 |
|---|---|---|---|
| 1 | 402.14mV | 597.86mV | 597.86μA |
| 5 | 485.81mV | 4.51V | 4.51mA |
| 10 | 516.54mV | 9.48V | 9.48mA |
| 15 | 534.02mV | 14.47V | 14.47mA |
| 20 | 546.29mV | 19.45V | 19.45mA |
| 25 | 555.75mV | 24.44V | 24.44mA |
| 30 | 563.45mV | 29.44V | 29.44mA |
| 35 | 569.95mV | 34.43V | 34.43mA |
| 40 | 575.56mV | 39.42V | 39.42mA |

The circuit shown in Figure 2-6 is similar to the forward biasing circuit shown in Figure 2-3. The main difference is that there is a resistor, $R_1$, placed in series after the diode. This resistor will limit the current flowing through the diode as we don't want a level of 10s of amps flowing through it. When we look at the table of results, in Table 2-3, we see that once the diode switch has been closed, at some voltage between 1V and 5V, the current flowing through the diode has been limited by the resistor. That is the real use of the resistor in that it is used to resist current flow, hence its name, resistor.

Table 2-3 shows us that once the diode starts conducting, the voltage across the diode, measured by VM1, stays pretty constant at around 0.5 to 0.6V. This is an important characteristic of the diode and one that we can make use of as we will see later in the book.

One thing I think I should point out is that the reverse breakdown voltage of -50V and the forward voltage drop of around 550mV are values for the 1N1183 diode that has been used in these simulations. There are a wide range of diodes manufactured particularly to give a whole range of different characteristic values, and you need to be aware of them when you choose a diode for a particular use. We will discuss some of them when we move on to rectification next in this chapter.

## Rectification

This was one of the applications that engineers first used the diode for. Rectification is the act of converting an ac voltage to a DC voltage. The term "ac" stands for alternating current, and an ac voltage is one that changes polarity, in that it goes from positive to negative with a regular frequency. DC stands for Direct Current, and a DC voltage is one that never changes polarity; it is always positive or it is always negative.

If we consider a basic ac voltage, as shown in Figure 2-7, then to convert this to DC we simply need to stop the negative part of the voltage from being applied to the load, in this case the resistor R1. We could stop the positive part, but it is easier to stop the negative part of the ac voltage.

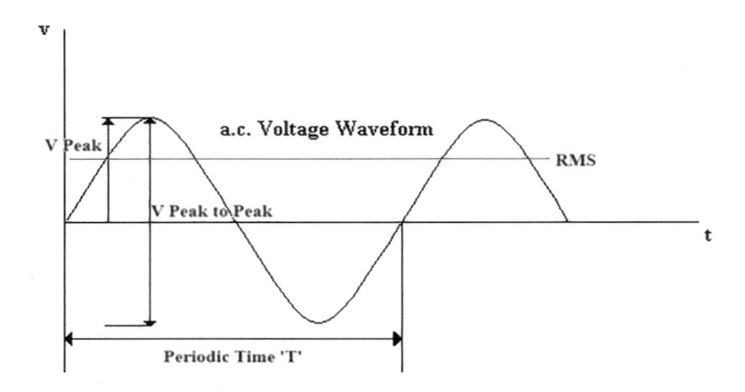


***Figure 2-7.*** *A Basic ac Voltage Waveform*

With respect to Figure 2-7, it might be useful to explain what some of the terms in the figure mean. To do that, I will relate the voltage waveform to the UK mains. The UK mains is just one example as there are a wide variety of ac voltages, all with their own individual values for the terms discussed here.

The UK mains voltage has the following parameters.

The frequency is 50Hz; this is because it is created by rotating a coil inside a magnetic field at a speed of 3000rpm, that is, 50 revs/sec. This means that the periodic time, which can be calculated using $T = \frac{1}{f} = \frac{1}{50} = 20ms\ or\ 0.02sec$

1. The V peak is approximately 339v.
2. The RMS, which stands for root mean square, is approximately 240v. However, nowadays it is said to be 220v. If we use 220v, then the V peak would be 312v.
3. The V peak to peak is simply twice that of the V peak voltage.

The simplest way to change the ac voltage to a DC voltage, that is, rectify the voltage, would be to stop the negative half from being passed onto the load. We could try putting a switch between the voltage source,

VG1, and the load $R_1$ so that we could turn off the source every time the voltage went negative. However, we would have to physically operate the switch at precisely the right time and at the required frequency, which, in the case of the UK mains, would be 50 times a second. I think you should realize that this would be impossible to do manually. Luckily though, we have an electronic switch in the form of a diode.

We have shown the diode acts like a switch in that it will automatically close and allow current to flow through it, when the voltage at the anode becomes approximately 0.7V greater than the voltage at the cathode. This will happen during the positive half of the ac voltage input. When the voltage at the anode falls below this 0.7V greater than the cathode voltage, then the switch will open and the diode prevents current from flowing through it. This will happen during the negative half of the input voltage. In this way, we can see that we could prevent the negative half cycle of the input voltage being passed onto the load, by using the diode as the required switch; the diode will open and close automatically without any action from us. This is shown in Figure 2-8.

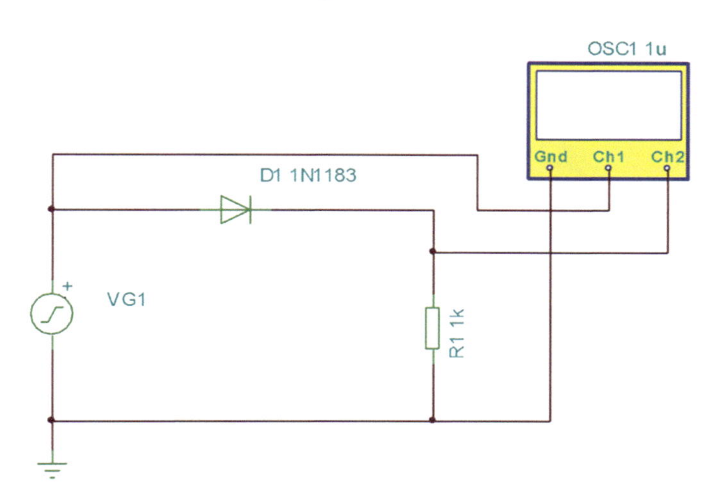


***Figure 2-8.*** *The Simple Diode Used As an Electronic Switch to Remove the Negative Half Cycle*

Figure 2-8 shows a simple circuit we can use in TINA to test the operation of a diode in removing the negative half cycle of the ac voltage source.

Figure 2-9 shows the oscilloscope display showing both the input voltage source and the voltage after the diode, that is, the voltage across the load $R_1$.

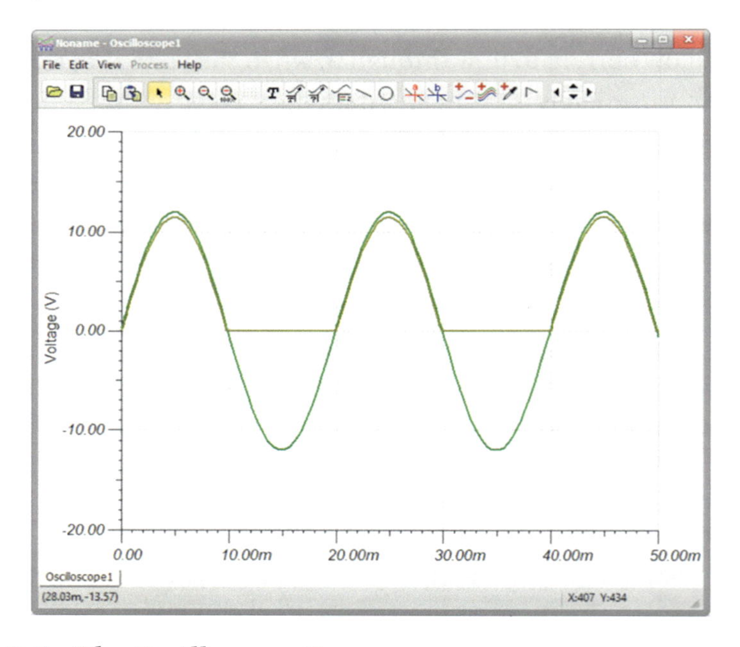


***Figure 2-9.*** *The Oscilloscope Traces*

If you look very closely at the two waveforms, shown in Figure 2-9, you can just make out that one trace is slightly lower in peak voltage than the other. This is the voltage across $R_1$, that is, the voltage after the diode. The slight difference is due to the volt drop across the diode which is approximately 0.7V. The peak voltage of the ac input has been kept low at 12v so that we could see the voltage difference between the two traces. If we had set the peak voltage at 339v, you would not be able to see the difference. Also, the 1N1183 diode would have a problem coping with a voltage level of 339v peak, but we will look at that later in this chapter.

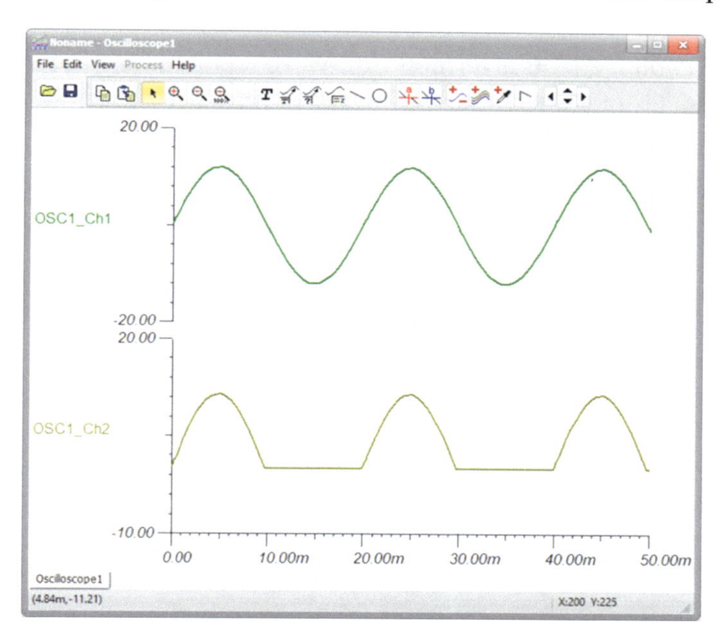


***Figure 2-10.*** *The Traces of the Oscilloscope Separated*

Figure 2-10 shows more clearly that the diode has prevented the negative half of the input being passed onto the output. It also shows why this is called half-wave rectification as only half of the input voltage is passed onto the output; see trace OSC1_Ch2. This is actually a drawback of this type of rectification as only half of the input voltage is used, which is an obvious waste. However, we have created a DC voltage from an ac source, and this rather lumpy voltage waveform can be used with some simple DC loads. The equipment would respond to the average voltage of the waveform. We will look more closely at this later in this chapter.

# Full-Wave Rectification

Now we will look at how we can use both the positive and negative half of the input voltage. This is termed "full-wave rectification" as we will use all of the input voltage waveform. One of the early solutions was to use a split transformer. This is a transformer that had two sets of coils on the secondary. A simple circuit showing the arrangement is shown in Figure 2-11.

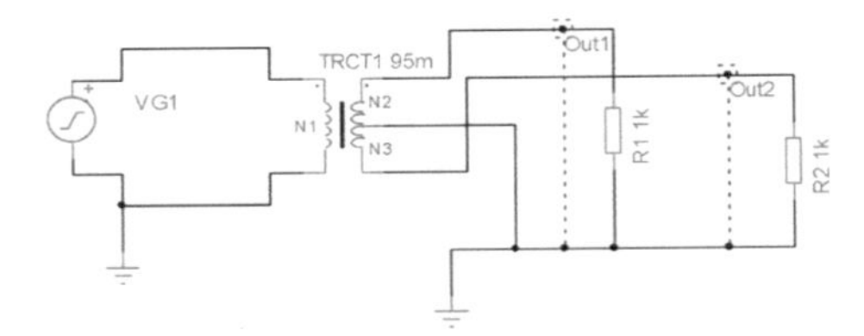


***Figure 2-11.** A Transformer with a Center Tap on the Secondary*

If, as shown in Figure 2-11, we connect the center tap to be used as the reference to all voltages by connecting it to ground, then the secondary will act as two coils on the output of the transformer. This will halve the voltage of the complete secondary, but it will allow us to use two voltage outputs from the transformer. Also, there will be a complete 180° phase difference between the two outputs. This means that when one voltage waveform is going through its positive half cycle, the other voltage waveform will be going through its negative half cycle. Figure 2-12 shows the two separate output waveforms from the split secondary. This does confirm that the two outputs, Out1 and Out2, are in antiphase, that is, 180° out of phase with each other.

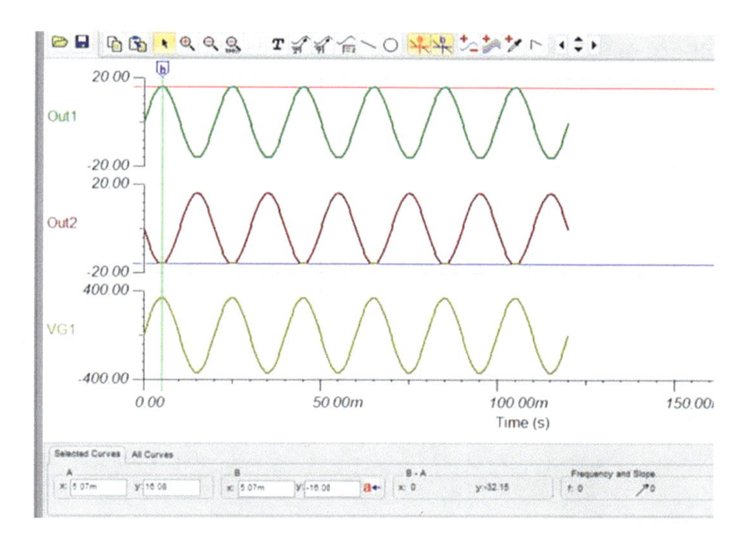


***Figure 2-12.*** *The Two Voltage Outputs from the Center Tapped Transformer*

Figure 2-12 shows that when Out1 is at +16.08v, Out2 is at –16.08v. This shows they are in antiphase with each other.

If we now add two diodes to the circuit, as shown in Figure 2-13, we should be able to use the output from both halves of the secondaries.

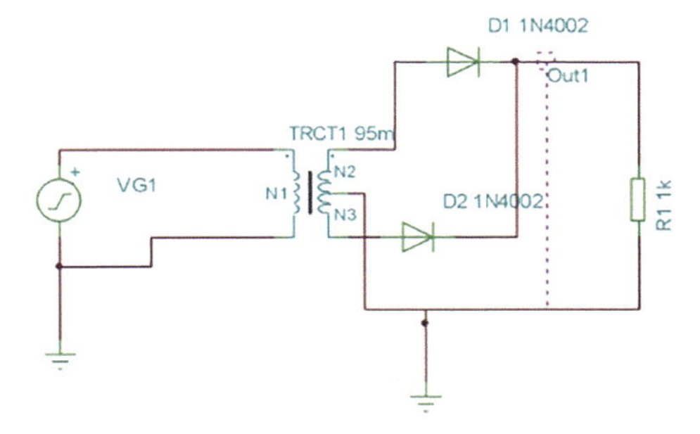


***Figure 2-13.*** *Full-Wave Rectification Using a Center Tapped Transformer*

The voltage across the load is displayed in Figure 2-14 as the Out1 trace.

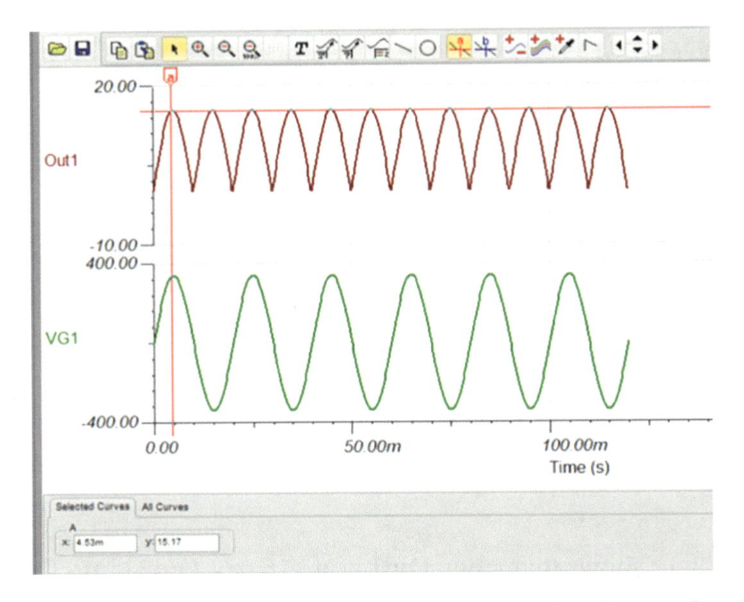


***Figure 2-14.*** *The Output of the Full-Wave Rectifier Using the Center Tapped Transformer*

We can see that the trace OUT1, which is the voltage across the load, does not go negative. It is a DC voltage even though it is not a constant voltage, as it is always positive.

This circuit has two major drawbacks which are

- The maximum voltage is half that what could be achieved from using the complete secondary output.
- The manufacture of a true center tapped transformer is very difficult.

An easier and better alternative would be to use the "bridge rectifier."

# Full-Wave Rectification Using the Bridge Rectifier

The circuit for the bridge rectifier is shown in Figure 2-15.

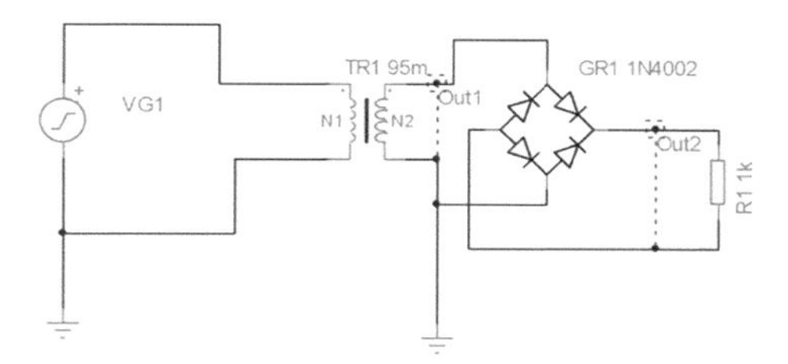


***Figure 2-15.*** *The Typical Bridge Rectifier Circuit*

Figure 2-15 shows the more conventional layout of the bridge rectifier. Indeed, the four diodes can be bought as a single IC. The VSIB440 IC is a bridge rectifier as shown in Figure 2-16. There is a whole range of such ICs available to us.

***Figure 2-16.*** *The VSIB440 Bridge Rectifier from Multicomp*

However, Figure 2-17 might be a better circuit that we can use to analyze how the bridge rectifier works. I have added some red and blue voltage arrows to try and help with the analysis of how the circuit works.

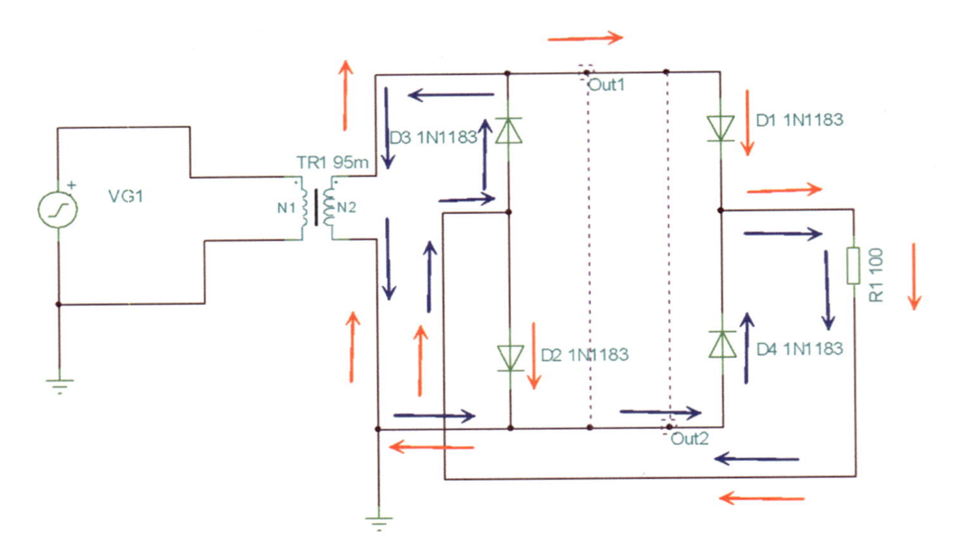


***Figure 2-17.** The Full-Wave Rectifier Circuit with No Smoothing*

It is really the same circuit as that shown in Figure 2-15, but the four diodes have been separated so that we can see how the circuit works.

To explain how the circuit works, we need to examine the waveforms shown in Figure 2-18.

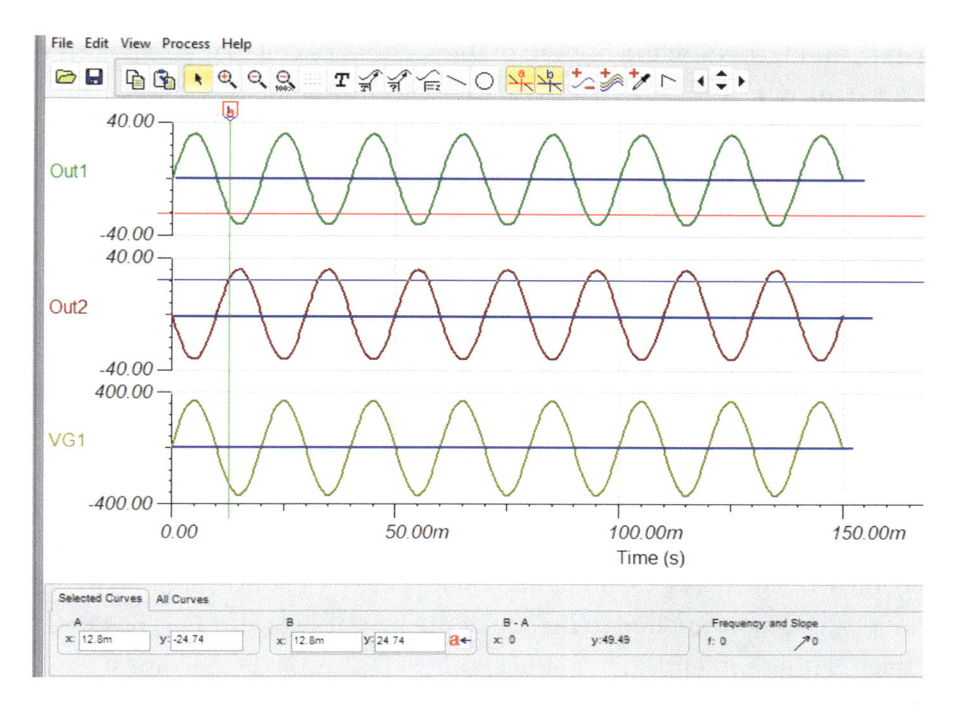


***Figure 2-18.*** *The Traces Out1 and Out2*

Trace Out1 shows the voltage at the anode of $D_1$ and cathode of $D_3$. Trace Out2 shows the voltage at the anode of $D_4$ and cathode of $D_2$. VG1 is the ac input voltage. We can see that when the ac input voltage starts to go positive, the voltage at the anode of $D_1$ goes positive and so starts to forward biases $D_1$. The same positive voltage is at the cathode of $D_3$, and so that diode is reverse biased. As soon as this positive voltage goes above 0.7v, then the diode $D_1$ closes, and this positive going voltage is applied to the load resistor $R_1$.

At the same time, if we look at trace Out2, we can see that the anode of diode $D_4$ goes negative, and so it is reverse biased. Also, the cathode of $D_2$ starts to go negative, which helps to forward bias that diode. Its anode has just started to go positive, as $D_1$ is now closed. The diode $D_2$ will quickly become forward biased and so close, allowing current to flow through $D_1$, then $R_1$ and $D_2$ back to the bottom of the secondary coil in the transformer. The red arrows try to show this path that the current follows. This will continue until the ac input voltage goes below the 1.4V required to forward bias the diodes $D_1$ and $D_2$.

When the ac input starts to go negative, the trace Out1 also goes negative, which turns $D_1$ off, as its anode has gone negative. However, at the same time, it drives the cathode of $D_3$ negative, which will try to forward bias $D_3$.

If we consider the trace Out2, we can see that when the ac input starts to go negative, the trace Out2 voltage starts to go positive. This means the cathode of $D_2$ goes positive, which turns it off, but the anode of $D_4$ goes positive, which starts to forward bias $D_4$. The cathode of $D_4$ has just gone to 0V, and so when the voltage at the anode of $D_4$ rises to around 0.7V, $D_4$ becomes forward biased and so allows the positive voltage, applied to the anode of D4, to be applied to the top of the load resistor $R_1$. This positive voltage is now applied via $R_1$ to the anode of $D_3$, and so $D_3$ becomes forward biased. This then allows current to flow out of the bottom of the secondary coil through $D_4$, then $R_1$ and then $D_3$ and back into the top of the secondary coil of the transformer. The blue arrows try to show this path that the current follows.

The arrows placed in Figure 2-17 are an attempt to show you how the current flows around the circuit during both halves of the ac input voltage. The red arrows show the direction the current flows during the positive half cycle of the input voltage. The blue arrows show the direction of current flow during the negative half cycle. We can see that the current flow through the load resistor $R_1$ does not change direction, and so we can confirm it is a DC current. However, when we look at the secondary of the transformer, we can see that the red arrow flows in the opposite direction to the blue arrows. This confirms that the current through the transformer does change direction, and so it is an ac current.

We are now using both halves of the ac input voltage, but the voltage across the load is still lumpy. We can improve this by adding a smoothing capacitor.

# Designing a Power Supply Unit (PSU)

To complete our look at the diode and move it on to another use, we will look at the design of a power supply unit, or PSU. We will start by designing a simple power supply that will connect to the UK mains voltage and produce a lower DC voltage.

The simplest power supply would use a transformer to reduce the peak of the mains voltage from around 339v to a lower voltage. The circuit for this basic operation is shown in Figure 2-19. This is the circuit for half-wave rectification with no smoothing.

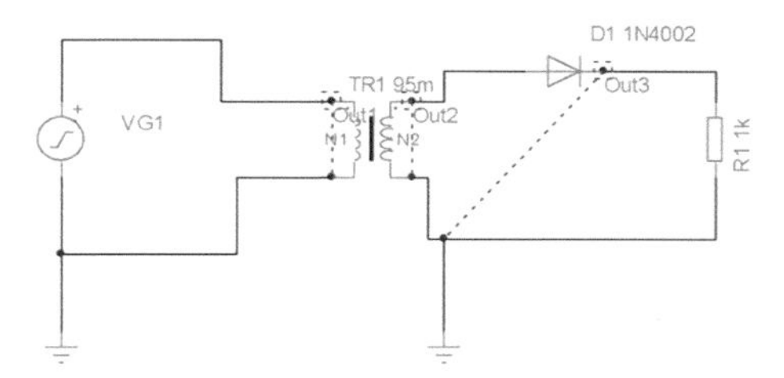


***Figure 2-19.*** *The Basic Transformer with Half-Wave Rectification*

The input and output waveforms are shown in Figure 2-20.

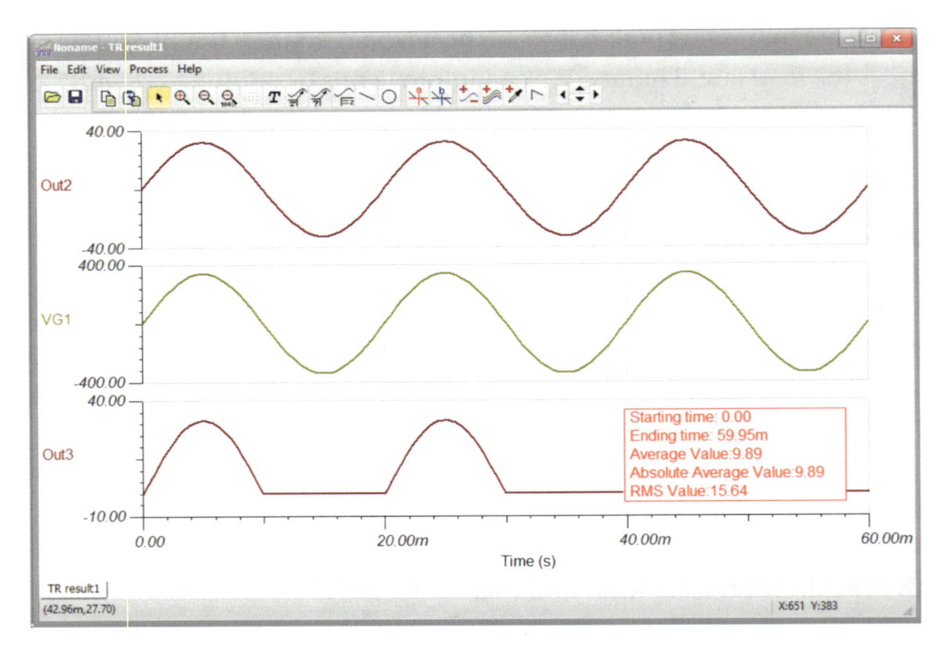


***Figure 2-20.*** *The Waveforms for the Half-Wave Rectifier in Figure 2-19*

The trace OUT1 is the UK mains voltage on the primary of the transformer. This peaks at 339v. Out2 is the voltage on the secondary of the transformer, and this peaks at 32.11v. The turns ratio of a transformer can be calculated using

$$Turns\ Ratio'NT' = \frac{V_{primary}}{V_{secondary}} = \frac{339}{32.11} = 10.56:1$$

We will be using the ECAD software TINA, and TINA defines the turns ratio in the opposite way, which means we would need a turns ratio of 1/10.56 = 0.0947 or 95m. This is what we have used in the simulation for the circuit shown in Figure 2-19.

The OUT3 trace, shown in Figure 2-20, shows the output of the half-wave rectifier. We can see that, without any smoothing, the waveform is lumpy. The load would respond to the average of the output waveform. We can use integration to determine the average of this waveform, and it can be shown to be

$$V_{Ave} = \frac{V_{max}}{\pi}$$

The integration to determine this expression is shown in the appendix.

Using the OUT3 trace, we can see that the $V_{Max}$ is 31.36v. This means that the average voltage would be 9.98v. This agrees with the average value shown in Figure 2-20.

# Smoothing

The output of the half-wave and full-wave rectifier could be improved by adding a capacitor across the output. This would act like a temporary source storing some charge while the diode or diodes are conducting and supplying the load and feeding the capacitor. Then, when the diode or diodes are turned off, the capacitor then feeds the load which will discharge some or all of the charge in the capacitor. Then, when the diodes turn back on, the capacitor can again charge up while the diode feeds the load.

It would be useful to be able to determine the output voltage of both the half-wave and full-wave rectifier circuits that incorporate a smoothing capacitor to the rectification circuit. The addition of the smoothing capacitor means that the output voltage across the load would be similar to that as shown in Figure 2-22. However, to help calculate the average voltage of that output, it would be easier if we used the approximated output waveform shown in Figure 2-21, to calculate the expression for the average.

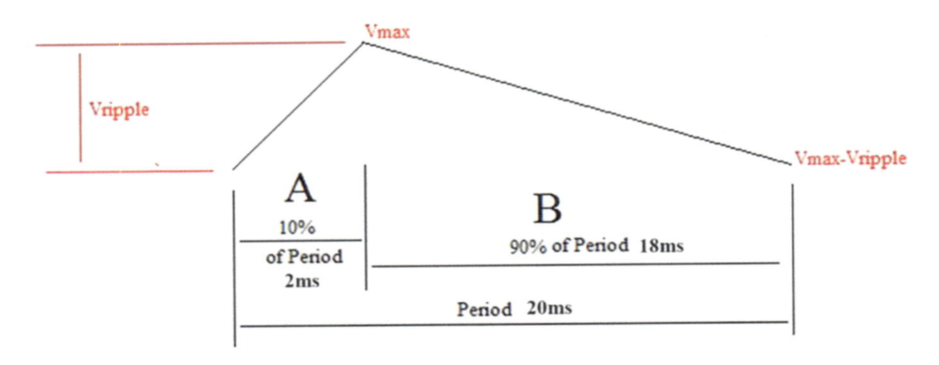


***Figure 2-21.*** *An Approximation of the Ripple Voltage at the Output of the PSU*

The output voltage would be the average, and this can be calculated using integration as

$$Vave = \frac{1}{T}\int_{0}^{T} f(x)\ dx$$

Note, the period "T," shown in Figure 2-21, for a half-wave rectifier is "2π," and for the full-wave rectifier, it is simply "π". This is because the full-wave waveform is cyclic over the period "π".

We should appreciate that the rectifier uses the ac voltage from the secondary of the transformer, which I will call "Vsecpk." The capacitor will charge up to the peak of this secondary voltage but only after it has passed through the diode. This means that the Vmax will actually be related to the peak voltage at the secondary of the transformer according to the two relationships.

$$\text{For half wave Vmax} = \text{Vsecpk} - 0.7$$
$$\text{For half wave Vmax} = \text{Vsecpk} - 1.4$$

This understanding will be used later when calculating the turns ratio of the transformer.

# The Average Voltage Output of the Half-Wave Rectifier

We will use the following example to show how the average can be calculated using the preceding expression.

Using the following values

- Vsecpk = 12v, then Vmax = 11.3v
- Vripple = 0.5V, the period = 2π

The waveform shown in Figure 2-21 has been split into two parts as the average can be calculated using

$$Vave = \frac{1}{2\pi}\left[\int_{0}^{10\%} A\ dx + \int_{10\%}^{2\pi} B\ dx\right]$$

The first thing to do is develop the expression for the waveforms for part "A" and part "B" as follows.

The approximation shows that both parts have linear expressions which must follow the general expression:

$$y = mx + C$$

Part "A" is over the period of 10% of the 2π. Therefore, this is $\frac{2\pi}{10}$.

The gradient "m" is the change in y divided by the change in x. Note, instead of "y," we have "v," and instead of "x," we have "θ". The change in "v" is simply the ripple voltage, which in this case is 0.5v. The change in "θ" is the 10% of the period 2π, which is $\frac{2\pi}{10}$. Therefore, we have

$$m = 0.5 \div \frac{2\pi}{10} \therefore m = 0.5 \times \frac{10}{2\pi} = \frac{5.0}{2\pi} \therefore m = \frac{2.5}{\pi}$$

The term "C" is the value of "v" at the start of the waveform, that is, the value of "v" when "θ" = 0. In this waveform, this is Vmax - Vripple = 11.3 - 0.5 = 10.8.

Therefore, C = 10.8.

Therefore, the expression for part "A" is

$$v = \frac{2.5}{\pi}\theta + 10.8$$

For part B, the change in "v" is 10.8 – 11.3 = –0.5v.

The change in "θ" is $2\pi - \frac{2\pi}{10} = \frac{18\pi}{10}$.

Therefore, the gradient "m" is

$$m = \frac{-0.5}{\frac{18\pi}{10}} = \frac{-5}{18\pi}$$

The value of C is the value of "v" when "θ" = 0.

We know that when "θ" was $\frac{2\pi}{10}$, v was 11.3.

Knowing this, the expression for "v" over part B is

$$v = \frac{-5}{18\pi}\theta + C$$

We can say

$$11.3 = \frac{-5}{18\pi} \times \frac{2\pi}{10} + C$$

We can say

$$11.3 = \frac{-1}{18} + C$$

Therefore

$$C = 11.3 + \frac{1}{18} = 11.356$$

This means that the expression for "v" over part "B" is

$$v = -\frac{5}{18\pi}\theta + 11.356$$

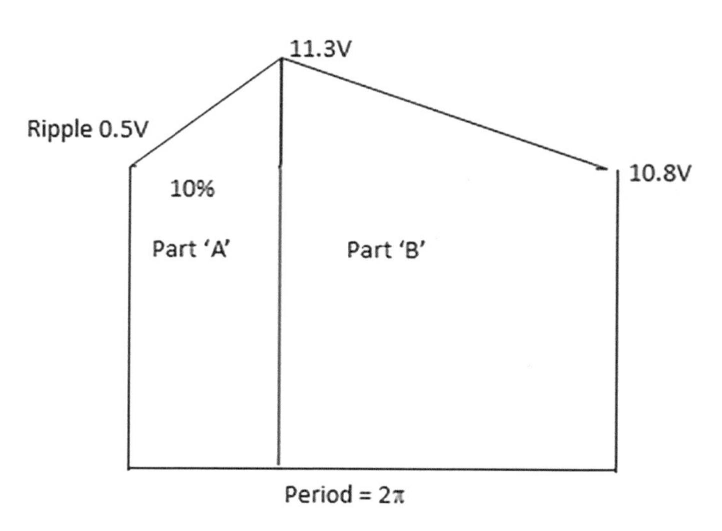


***Figure 2-22.*** *The Half-Wave Voltage with Smoothing*

Figure 2-22 should give us some idea of the shape of the voltage that we are integrating to determine the average voltage the load would respond to.

Therefore, to calculate the average output voltage, we have

$$V_{ave} = \frac{1}{2\pi}\left[\int_{0}^{10\%} Adx + \int_{10\%}^{2\pi} Bdx\right]$$

$$V_{ave} = \frac{1}{2\pi}\left[\int_{0}^{\frac{2\pi}{10}} \frac{25}{\pi}\theta + 10.8d\theta + \int_{\frac{2\pi}{10}}^{2\pi} -\frac{5}{18\pi}\theta + 11.356d\theta\right]$$

Therefore

$$V_{ave} = \frac{1}{2\pi}\left[\left[\frac{25}{2\pi}\theta^2 + 10.8\theta\right]_0^{\frac{2\pi}{10}} + \left[-\frac{5}{36\pi}\theta^2 + 11.356\theta\right]_{\frac{2\pi}{10}}^{2\pi}\right]$$

Therefore

$$V_{ave} = \frac{1}{2\pi}\left[\left(\frac{25}{2\pi}\times\frac{4\pi^2}{100} + 10.8\times\frac{2\pi}{10}\right) - (0)\right]$$
$$+\left[\left(-\frac{5}{36\pi}\times 4\pi^2 + 11.356\times 2\pi\right) - \left(-\frac{5}{36\pi}\times\frac{4\pi^2}{100} + 11.356\times\frac{2\pi}{10}\right)\right]$$

Therefore

$$V_{ave} = \frac{1}{2\pi}\left[\left(\frac{\pi}{2} + 2.16\pi\right) + \left[\left(-\frac{20\pi}{36} + 22.712\pi\right) - \left(\frac{20}{3600}\pi + 2.2712\pi\right)\right]\right]$$

Therefore

$$V_{ave} = \frac{1}{2\pi}\left[(2.56\pi) + (19.8908\pi)\right]$$

Therefore

$$V_{ave} = \frac{1}{2\pi}\left(\frac{22.4508\pi}{2\pi}\right) = 11.2254v$$

A simpler approximation that can be used to calculate the average of a waveform like that shown in Figure 2-22 is

$$Vave = \text{Vmax} - \frac{\text{Vripple}}{2}$$

$$\therefore Vave = 11.3 - \frac{0.5}{2} = 11.05v$$

As both approximations have errors, then for the rest of the text, we will use the second approximation, that is, for half wave and full wave:

$$Vave = \text{Vmax} - \frac{\text{Vripple}}{2}$$

## The Turns Ratio for the Transformer

This is an important aspect of the power supply design. You need to know what the turns ratio of the transformer needs to be to get the required DC output voltage. As stated earlier, the capacitor actually charges up to the peak voltage of the secondary side of the transformer. However, there will be a voltage drop across any diodes used. Note this text will assume the volt drop for a diode is 0.7v.

To calculate the turns ratio of the transformer, we need to know what the peak secondary voltage needs to be. This can be related to the required average voltage of the PSU using the following.

For the half wave, we have

$$\text{Vave} = \text{Vmax} - \frac{\text{Vripple}}{2}$$

But

$$\text{Vmax} = \text{Vsecpk} - 0.7$$

$$\therefore \text{Vave} = (\text{Vsecpk} - 0.7) - \frac{\text{Vripple}}{2}$$

$$\therefore Vave + \frac{\text{Vripple}}{2} = \text{Vsecpk} - 0.7$$

$$\therefore \text{Vsecpk} = Vave + \frac{\text{Vripple}}{2} + 0.7$$

For the full wave, we have

$$\text{Vave} = \text{Vmax} - \frac{\text{Vripple}}{2}$$

But

$$\text{Vmax} = \text{Vsecpk} - 1.4$$

$$\therefore \text{Vave} = (\text{Vsecpk} - 1.4) - \frac{\text{Vripple}}{2}$$

$$\therefore Vave + \frac{\text{Vripple}}{2} = \text{Vsecpk} - 1.4$$

$$\therefore \text{Vsecpk} = Vave + \frac{\text{Vripple}}{2} + 1.4$$

The turns ratio of the transformer can be set as follows:

$$\text{TurnsRatio} = \frac{\text{NP}}{\text{NS}} = \frac{\text{Vpripk}}{\text{Vsecpk}}$$

where Vpripk is the peak voltage on the primary, and Vsecpk is the peak voltage on the secondary.

We will continue the design exercise using a full-wave rectifier circuit, as most PSUs will use full-wave rectification, and I want to cover this aspect here. The specification is that the PSU supplies an average voltage of 10V with an allowable ripple of 1v at a maximum current of 1A.

To calculate the turns ratio of the transformer, we need to know the peak voltage of the secondary, that is, the Vsecpk. This can be calculated as follows:

Using

$$Vsecpk = Vave + \frac{Vripple}{2} + 1.4$$

Therefore

$$Vsecpk = 10 + \frac{1}{2} + 1.4 = 11.9$$

Knowing the primary voltage will be the UK mains, the Vpripk = 339. Therefore, the turns ratio can be calculated as follows:

$$Turns\ Ratio = \frac{Vpripk}{Vsecpk} = \frac{339}{11.9} = 28.49$$

Using TINA, we have

$$Turns\ Ratio = \frac{Vsecpk}{Vpripk} = \frac{11.9}{339} = 35.1m$$

## Choosing the Value for the Smoothing Capacitor

The capacitor is commonly called a "smoothing" capacitor, as it smooths the output voltage. The capacitor works like a battery in that it can supply energy that can be delivered to load later. However, it is a battery that soon discharges and goes flat. The concept is that while the secondary voltage is switched off from the output, by the action of the diode, and so supplying no energy to the load, the load can take the energy from the capacitor. Of course, the capacitor must be charged up at some time to be able to supply the load. This charging takes place when the diode is switched on and so conducting, allowing current to be passed into the load and the capacitor. When the diode is not conducting, the capacitor then takes over and supplies the current to the load. We should realize that when current flows into a capacitor, that is, when the diode is conducting, the capacitor will charge up. When current flows out of a capacitor, that is, when it is supplying current to the load, the capacitor will discharge. The state of the current charge on a capacitor can be determined by measuring the voltage across it. When the capacitor is fully charged up, the voltage will be at a maximum. When the capacitor discharges, the voltage will reduce. This discharging of the capacitor will result in a drop in the voltage at the output of the rectifier; however, hopefully the voltage will not go down to zero as it did without the smoothing capacitor. Figure 2-21 shows how the addition of a smoothing capacitor improves the output voltage of the rectifier circuit.

There has to be some method for choosing the value of this capacitor. Indeed, there are two relationships we can use to help us in this matter. They are

$$1. \quad Q = CV_c \qquad 2. \quad Q = \int i\,dt$$

Note, Q is the charge in Coulombs.

C is the capacitor value.

$V_C$ is the voltage across the capacitor.

"i" is the current flowing through the capacitor.

The two expressions can be combined as follows:

$$CV_C = \int i\,dt$$

If we now differentiate both sides, we get

$$i = C\frac{dV_C}{dt}$$

We should remember that differentiation is the opposite of integration.

The $\frac{dV_C}{dt}$ is the change in the capacitor voltage over time. This is the ripple voltage at the output. The dt is the time in which this takes place. The current is the current being drawn from the capacitor during this time.

To put the preceding relationship into place, the designer of the PSU should know these three values:

1. The maximum current that the user will demand from the PSU, "i"

2. The allowable ripple voltage they can tolerate at the output of the PSU, "dv"

3. The time the capacitor will be discharging over, "dt"

If the designer knows these values, then they can determine the value of the smoothing capacitor.

We will carry on with the design of our full-wave rectifier, and we will use a dt of 8ms. This can be confirmed from Figure 2-24. To calculate the value of the smoothing capacitor, we will use the following:

$$C = i\frac{dt}{dV_C}$$

Therefore

$$C = 1 \times \frac{8m}{1} = 8mF$$

This is a large capacitor, but we are taking a lot of current from the power supply. The capacitor of this size would be an electrolytic type. The circuit for the full-wave rectifier with the smoothing capacitor is shown in Figure 2-23.

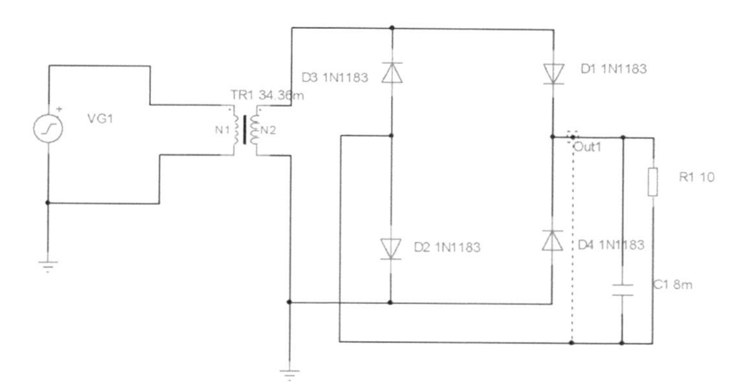


***Figure 2-23.** The Full-Wave Rectifier with Smoothing Capacitor*

We are using a transient analysis to measure the output voltage, and the display is shown in Figure 2-24. The resistor value for $R_1$ has been calculated using Ohm's Law as follows, knowing the voltage was the 10V average and the current was the maximum of 1Amp:

$$I = \frac{V}{R} \therefore R = \frac{V}{I} = \frac{10}{1} = 10\Omega$$

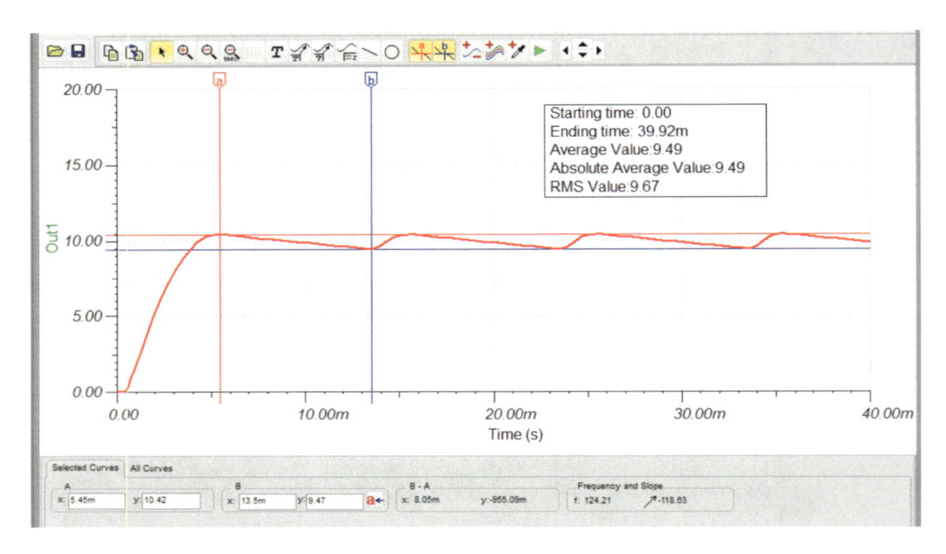


***Figure 2-24.*** *The Output Voltage of the Rectifier Circuit*

The display in Figure 2-24 shows that the dt time is 8.05ms and that the ripple is 955.09mV. It also shows that the average voltage, as calculated by the software, is 9.49V. These results are close enough to confirm that the method of calculations we have used for these parameters is fairly accurate.

# The Zener Diode

It is virtually impossible, using the basic smoothing capacitor, to produce a power supply that does not have any drop in voltage at its output. Also, it would be difficult to produce one that gives out the same voltage over a wide range of load currents drawn from the supply. However, some modern-day electronic equipment rely on the fact that they have a constant unchanging supply voltage. This has to be provided even if the load demand changes and the supply to the PSU varies. The simplest way of providing this type of stabilized power supply is with the use of a Zener diode.

Figure 2-25 shows the characteristics of a typical diode. The right-hand quadrant shows the normal forward biased conditions, and the left-hand quadrant shows the normal reverse biased conditions.

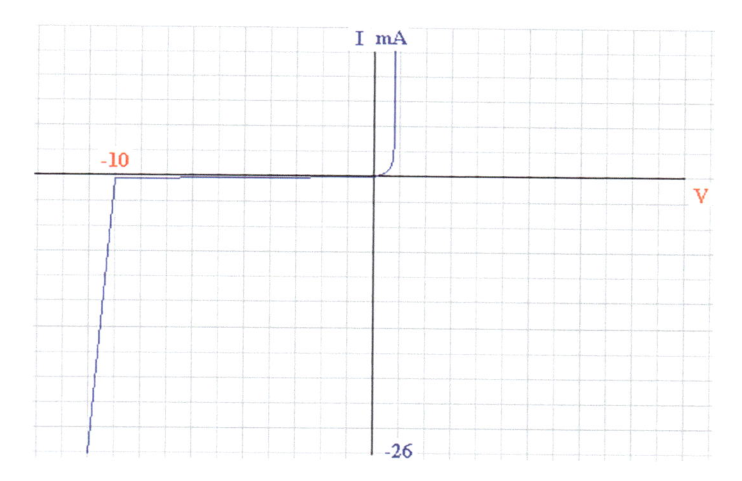


***Figure 2-25.*** *The Basic Diode Characteristics*

The characteristics show what happens if we apply enough voltage in the reverse biased region. We should remember voltage is a force that will try to make current flow. If we apply enough voltage, then the diode will break down as shown in Figure 2-25.

From the characteristics, it can be seen that once the reverse voltage has been taken past a point termed the breakdown voltage, or the knee voltage, there can be a very large change in current for a very small change in voltage. This aspect of the diode characteristics is termed the Zener region. Special diodes can be made to work in this region and put it to good use. These Zener diodes can be made for a whole range of different voltages, and the concept is that the voltage across them will remain virtually constant no matter what current is made to flow through them. Note, this is the ideal situation, and as can be seen from Figure 2-25, the Zener is not an ideal component, as the voltage across it does actually change as the current through it changes from 0 to 26mA. However, the change is very small. Figure 2-25 is not the characteristics of a real Zener diode. Figure 2-25 is really only to show the main idea of what a Zener diode can do.

These Zener diodes can be used to provide a constant voltage under various conditions. It might be useful to state the conditions that would make the output of the PSU change. There are basically two such conditions which are

1. The input voltage to the PSU could increase or decrease.

2. The current taken from the PSU could increase or decrease.

# The Shunt Regulator

Consider the circuit shown in Figure 2-26.

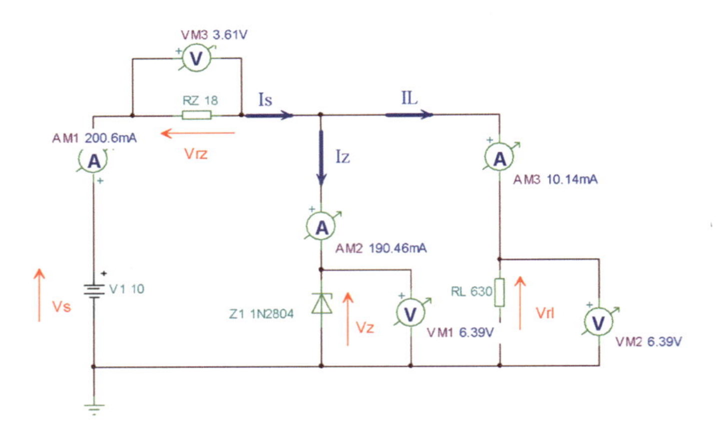


***Figure 2-26.*** *The Test Circuit for the Zener Diode*

We will use this circuit to show what a Zener can be used for and how this simple application of using a Zener diode works.

The arrows in the circuit are voltage arrows which are used to identify the particular volt drop, as well as which is the positive end of the volt drop. The arrow head indicates the positive end of the volt drop. The conventional current flows into the positive end of the volt drop.

The voltages around the circuit are identified as follows:

- Vs is the supply voltage.
- Vrz is the volt drop of the Zener resistor.
- Vrl is the volt drop across the load resistor.
- Vz is the Zener voltage.

The ammeters are used to measure the following currents:

- AM1 measures the supply current.
- AM2 measures the Zener current.
- AM3 measures the load current.

The voltmeters are used to measure the following voltages:

- VM1 measures the Zener voltage.
- VM2 measures the load voltage.
- VM3 measures the volt drop across $R_Z$.

Kirchhoff's Voltage Law (KVL) is a useful law which allows us to analyze electrical circuits. The law simply states that in a closed loop, the volt drops around the loop must add up to equal the supply to that loop. We can use the voltage arrows to help apply KVL in that all arrows pointing in one direction must add up to equal arrows pointing in the opposite direction.

# The Main Principle of Operation

The main principle of how the shunt regulator works is that we drive the Zener diode at close to its maximum wattage rating. This means the current flowing through the Zener would be close to its maximum. This is so that when the load wants more current, it can take it from the Zener diode instead of taking it from the source. The Zener has a lot of "spare" current as it doesn't need much to make the Zener work.

Obviously, nothing is ideal, and there will be limits on what the circuit can deliver. When designing any circuit, there will be a specification that states the limits of what the circuit can do. With a regulated power supply, we should specify the expected output with a maximum voltage and current. We should also specify the limits of which the circuit can cope with a varying input voltage.

To all intents and purposes, the output voltage, which is $V_Z$, will be constant at a value depending upon its manufacture. In this example, let $V_Z = 6.3V$.

We will specify the maximum input voltage would be 10V.

We now need to think about the current limits that we can work with. The customer, in this simple example, wants a nominal current of 10mA. We can simply use Ohm's Law to calculate the resistance, $R_L$, of our test circuit for the regulator circuit. Therefore, we have

$$R_L = \frac{V_Z}{I_L} = \frac{6.3}{10E^{-3}} = 630\Omega$$

Now we need to determine the maximum current the Zener diode can run at. The limiting factor as to how much current the Zener diode can take is the amount of power in watts the diode can safely dissipate. It is usual to specify Zener diodes in terms of their Zener voltage and the power they can dissipate. For example, if the Zener diode in the circuit is a 6.3V

Zener rated at 1.3 watts, we can determine the maximum current, $I_{ZMax}$, that can flow through the Zener as follows:

$$I_{ZMax} = \frac{Z_{Power}}{V_Z} = \frac{1.3}{6.3} = 206.35mA$$

This means we can now determine the maximum current that the circuit will use from the source; this is $I_S$. Using the test circuit shown in Figure 2-25, we can see that

$$I_{SMax} = I_{Zmax} + I_L = 206.35 + 10 = 216.35mA$$

# The Zener Resistor $R_Z$

The purpose of this resistor is to limit the current flowing out of the source to the $I_{SMax}$. This would be when the source voltage was at its maximum allowed value. In this example, $I_{SMax}$ is 216.35mA, and the maximum source voltage is set at 10V. We can now calculate the value of this resistor using

$$R_Z = \frac{V_{max} - V_Z}{I_{SMax}} = \frac{10 - 6.3}{216.35E^{-3}} = 17.09\Omega$$

The closest standard value of resistance using the E12 series is 18Ω. Note, when choosing a standard value of resistor, we must always go higher, not lower, as going lower would run the risk of allowing the current to rise above the maximum allowed.

So, we now have the values of the components to simulate the test circuit as shown in Figure 2-26. Figure 2-26 shows the value of the voltages and currents in the test circuit, and they agree very closely to the calculated values. These values can be compared in Table 2-4.

***Table 2-4.*** *The Comparison of the Calculated and Simulated Values*

| | $V_S$ | $V_{RL}$ | $V_Z$ | $V_L$ | $I_S$ | $I_Z$ | $I_L$ |
|---|---|---|---|---|---|---|---|
| Calculated | 10 | 3.7 | 6.3 | 6.3 | 216.35mA | 206.35mA | 10mA |
| Simulated | 10 | 3.61 | 6.39 | 6.39 | 200.6mA | 190.46mA | 10.14mA |

## The Minimum Working Source Voltage

The specification of the regulator would state the minimum source voltage as if the user tried to use it with a lower source voltage, then they can be said to be out of specification. We know from Ohm's Law that the current is proportional to the voltage. This means the minimum voltage would supply the minimum current. Therefore, we need to determine the minimum current that the circuit would demand from the supply. This would be the 10mA required by the load and the minimum Zener current. The minimum Zener current is the current required by the Zener to make it operate in the Zener region. This would be typically around 3mA. This means that the minimum current the circuit could demand from the source would be 13mA. As this minimum current must flow through $R_Z$, then some of the source voltage would be dropped across this $R_Z$. This volt drop would be equal to

$$V_{RZMin} = I_{SMin} \times R_Z = 13E^{-3} \times 18 = 234mV$$

This means the minimum source voltage can be calculated as follows:

$$V_{SMin} = V_{RZMin} + V_Z = 234E^{-3} + 6.3 = 6.534$$

We can test this value by simulating the circuit as shown in Figure 2-27, and the results are displayed on the circuit.

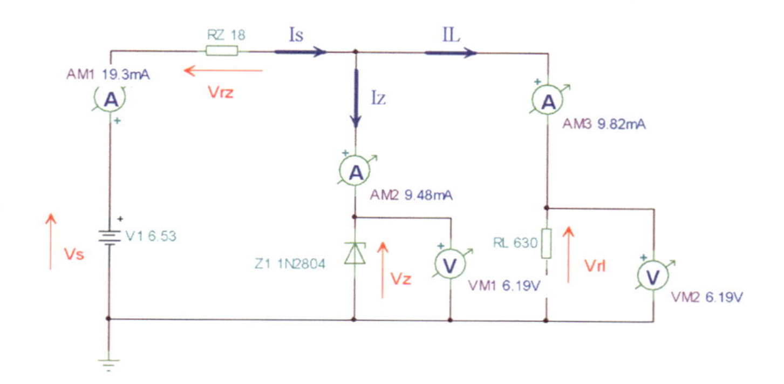


***Figure 2-27.*** *The Test Circuit with the Source Voltage Set to 6.534V*

The meter readings do agree closely to the value we have calculated.

# Testing the Regulation of the Circuit

We can test how well the circuit operates, and so regulates the voltage across the load, by changing the load resistance, while keeping the supply voltage at 10V, and recording the measured load values to complete Table 2-5.

***Table 2-5.*** *The Test Results for the Voltage Regulation*

| Row | $R_L$ | $V_{RZ}$ | $V_Z$ | $V_{RL}$ | AM1 | AM2 | AM3 |
|---|---|---|---|---|---|---|---|
| 1 | 1k | 10–6.39 | 6.39 | 6.39 | 200.51mA | 194.12mA | 6.39mA |
| 2 | 500 | 10–6.37 | 6.39 | 6.39 | 200.67mA | 187.89mA | 12.78mA |
| 3 | 100 | 10–6.36 | 6.36 | 6.36 | 202.11mA | 138.49mA | 63.62mA |
| 4 | 40 | 10–6.22 | 6.29 | 6.29 | 206.25mA | 49.06mA | 157.19mA |
| 5 | 35 | 10–6.25 | 6.25 | 6.25 | 208.07mA | 29.37mA | 178.71mA |
| 6 | 30 | 10–6.17 | 6.17 | 6.17 | 212.79mA | 7.13mA | 205.66mA |
| 7 | 20 | 10–5.26 | 5.26 | 5.26 | 263.16mA | 15.01nA | 263.16mA |

The results shown in Table 2-5 show that the Zener diode does maintain the output voltage at around 6.37V even when the load current increases from 6.39mA to 63.62mA, an increase of ten times. The load voltage only fell to 6.36V, a fall of 30mV from 6.39V. This is virtually no real change. Even when the load current increased to 178.71mA, an increase of 2800%, the voltage only fell by 104mV, a change of 1.63%; see row 5 in Table 2-5.

This does show that a Zener diode can be used as a voltage regulator in that it can maintain a set voltage across a load even if the current flowing through the load does change. There will be a limit to the variation of the load current; see row 7 in Table 2-5.

There is a wide range of Zener voltage that Zener diodes can be manufactured to regulate at and a range of different power ratings for them. The basic shunt regulator is quite a simple regulator circuit; however, it is not very efficient, and it is only usable with small load currents.

# Summary

In this chapter, we have looked at the PN junction and what type of component it creates. We have looked at the basic diode and one of its main uses as a rectifier. We then looked at the Zener diode and how it can be used as a basic voltage regulator. I hope you have found this chapter interesting and informative.

In the next chapter, we will look at some more semiconductor devices, such as the diac, the triac, and the SCR.

CHAPTER 3

# More Semiconductor Devices

In Chapter 2, we studied how the PN junction was used to create the first semiconductor device, the simple diode. In this chapter, we will look at some of the other initial semiconductor devices, such as the SCR, the DIAC, and the Triac.

## The SCR or Thyristor

SCR stands for silicon-controlled rectifier. It is also known as a thyristor. Really, it is a diode, hence the reference to the rectifier, but it has an extra terminal called the gate terminal. The symbol for the SCR is shown in Figure 3-1.

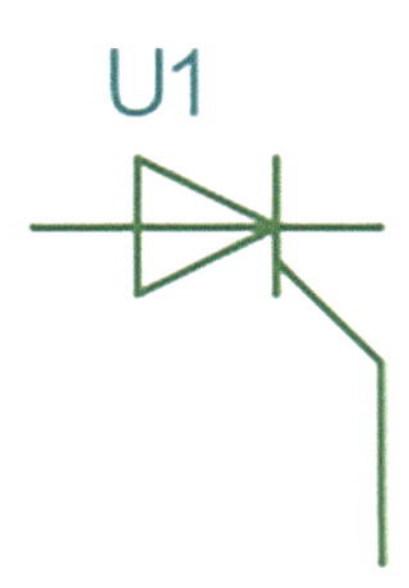


***Figure 3-1.*** *The Basic Symbol for the SCR*


H. H. Ward, *Mastering Analog Electronics*, Maker Innovations Series,
https://doi.org/10.1007/979-8-8688-0245-4_3

We can see from the symbol why it can be viewed as a controlled diode, that is, a controlled rectifier.

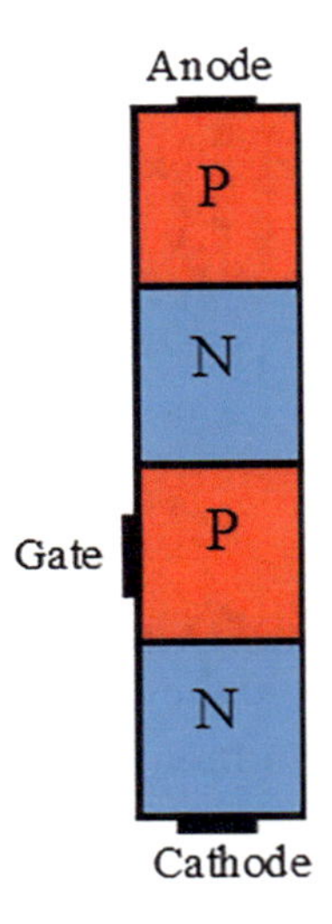


***Figure 3-2.*** *The Makeup of an SCR*

I am not going into a deep atomic analysis of the SCR as really there is no need to look that deep into the workings of the SCR. We really only need to know how to use them. In showing this diagram in Figure 3-2, I just want to show you that the SCR is still a semiconductor device. It is made up of segments of silicon that has been doped to create "P-type" and "N-type" materials. It is an amazing process by which they create the various levels of doping to create these different devices. It is a whole world of chemical engineering that is best left to the specialist. If we look at the PNPN arrangement as shown in Figure 3-2, we could see that the SCR could be viewed as two transistors connected in series. This would be a PNP in series with an NPN. However, we will leave that analysis for another book.

## The One-Pulse Converter

This is one of the simplest uses of the SCR. The circuit for this one-pulse converter is shown in Figure 3-3. We will use it to study how we can make the SCR work.

There is an anode, a cathode, and the extra terminal called the gate terminal. To make the SCR conduct, we must do the following:

- It needs the voltage at the anode terminal to go more positive than the voltage at the cathode terminal.
- It also needs a positive going pulse voltage, known as the $V_{GT}$, gate triggering voltage, to force current into the gate terminal.

However, turning the SCR off is not as simple as with the normal diode. To turn the SCR off, we must ensure the following happens:

- It needs the voltage at the anode terminal to go lower than the voltage at the cathode terminal.
- But more importantly the current flowing through the SCR must fall below a certain value known as the "holding current."

Different thyristors will have different values of the holding current, $I_H$, and the gate triggering voltage $V_{GT}$. Also, as the voltage we apply across the SCR is a force which could cause the SCR to break down, then different SCRs will have different values of the voltage across them that they can withstand before they break down. This is known as the "peak reverse blocking voltage." Therefore, before we use an SCR, we should check with the datasheet to ensure it will be fit for our purpose. The 2n159* is a range of SCRs, and the 2n1595 can block up to 50v, and the 2n1599 will block up to 400v. This means if we want to use the SCR to control rectifying the UK mains, then we should use the 2n1599 SCR. The holding current for the full range of the 2n159* SCRs is stated as 5mA. Of course, you should also make sure the maximum forward current, the IT, for the SCR is sufficient for your requirements. So, there is a lot to consider before you choose the correct SCR, but that is true for all your devices you want to use. In all our test circuits, we will use the 2n1599 SCR.

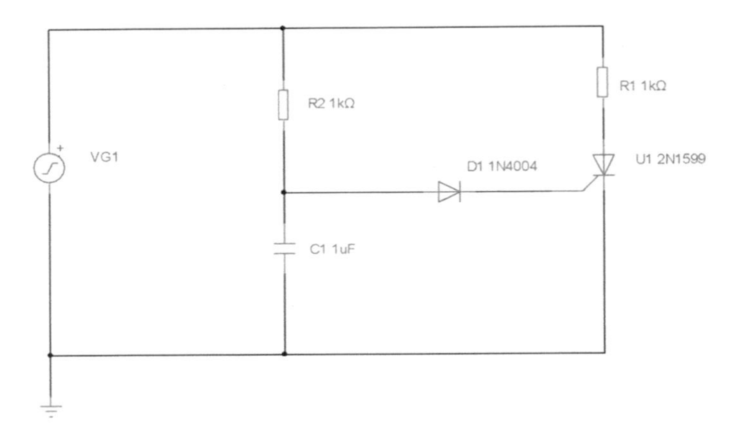


***Figure 3-3.*** *The One-Pulse Converter*

The circuit shown in Figure 3-3 is the basic one-pulse converter. It is using a CR circuit to control when the SCR turns on and delivers the supply voltage to the load. The CR circuit shown in Figure 3-3 is there to introduce some phase shift, or time difference, between the input voltage and the voltage across the capacitor, $v_{out}$. This will ultimately delay when the SCR turns on. At a frequency of 50Hz, the waveform will take 20ms to move through 360°. That means in 1ms the waveform would have moved through 18°. We can use this relationship to calculate the phase shift when we examine the waveforms in this chapter.

We can use the voltage divider rule to derive the expression for the transfer function. The expression for $v_{out}$ is

$$v_{out} = v_{in} \frac{-j\frac{1}{\omega C}}{R - j\frac{1}{\omega C}}$$

Therefore, we can show that the transfer function for this CR circuit is

$$TF = \frac{1}{1 + j\omega CR}$$

It can be shown that the magnitude of the transfer function is

$$|TF| = \frac{1}{\sqrt{1^2 + (\omega CR)^2}}$$

It can also be shown that the phase for the voltage across the capacitor will be

$$Phase = -\left\langle Tan^{-1}(\omega CR)\right\rangle$$

The proof of these expressions will be gone through in Chapter 11, as we will do a lot more work with complex numbers in that chapter.

We can use this transfer function to derive an expression for the voltage across the capacitor. We need to know what this expression is because this voltage will be used to trigger the SCR. The transfer function shows how the circuit will take the input and transfer it to become the output. The relation between the input and output voltages and the transfer function is

$$TF = \frac{v_{out}}{v_{in}}$$

Using this, we can say

$$v_{out} = v_{in} \times TF$$

With respect to the CR phase circuit shown in Figure 3-3, the input voltage is simply the UK mains voltage, and the output voltage would be the voltage across the capacitor. This means we can use the transfer function to develop an expression for the voltage across the capacitor as follows:

$$v_{out} = 339Sin(2\pi ft) \times \left(\frac{1}{1 + j\omega CR}\right)$$

It would be easier to multiply the two terms together if they were expressed in their magnitude and phase. We will use the peak value of the mains, $v_{in}$, so that we will derive the peak value of the output voltage.

The vin is

$$339 Sin(2\pi ft) = 339\langle 0 \rangle$$

This assumes the rms of the UK mains is 240v, but nowadays it has been reduced to 230v; however, I will use a 240v rms value as I am old-fashioned. We are using an angle of 0° as the voltage is the reference.

The transfer function in magnitude and phase is

$$\frac{1}{\sqrt{1^2 + (\omega CR)^2}} - \langle Tan^{-1}(\omega CR) \rangle$$

If we multiply the two expressions together, we get

$$v_{out} = (339\langle 0 \rangle) \times \left( \frac{1}{\sqrt{1^2 + (\omega CR)^2}} - \langle Tan^{-1}(\omega CR) \rangle \right)$$

To carry this multiplication out, we simply multiply the two magnitudes and add the two angles. The result is

$$v_{out} = \frac{339}{\sqrt{1^2 + (\omega CR)^2}} \langle 0 - Tan^{-1}(\omega CR) \rangle$$

Therefore, the expression for Vout is

$$v_{out} = \frac{339}{\sqrt{1^2 + (\omega CR)^2}} - \langle Tan^{-1}(\omega CR) \rangle$$

We can use this expression for different values of frequency and component values. With respect to the circuit shown in Figure 3-3, the frequency is 50Hz, the UK mains, the capacitor value is 1uF, and we will start with a resistor value of 1kΩ. Putting these values into the expression, we get

$$\omega CR = 2\pi \times 50 \times 1E^{-6} \times 1E^{3} = 0.31416$$

We can calculate the expression for the peak voltage at the output as

$$v_{out} = \frac{339}{\sqrt{1^2 + (0.31416)^2}} - \langle Tan^{-1}(0.31416) \rangle$$

Therefore

$$v_{out} = 323.416(-0.3044)$$

This means that the expression for Vout with respect to time is

$$v_{out} = 323.416Sin(100\pi t - 0.3044)$$

Note the phase shift and so the angle are expressed in radians. We can test this expression out with the simple test circuit as shown in Figure 3-4.

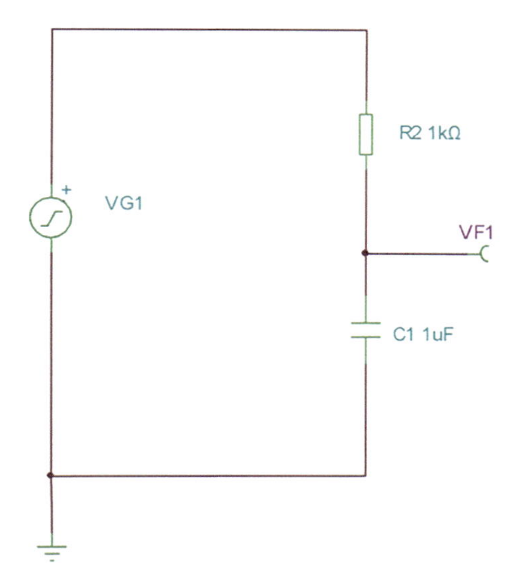


***Figure 3-4.*** *The Test Circuit for the CR Phase Shift Circuit*

If we simulate this circuit, the waveform for the voltage across the capacitor, the output voltage, can be displayed as shown in Figure 3-5.

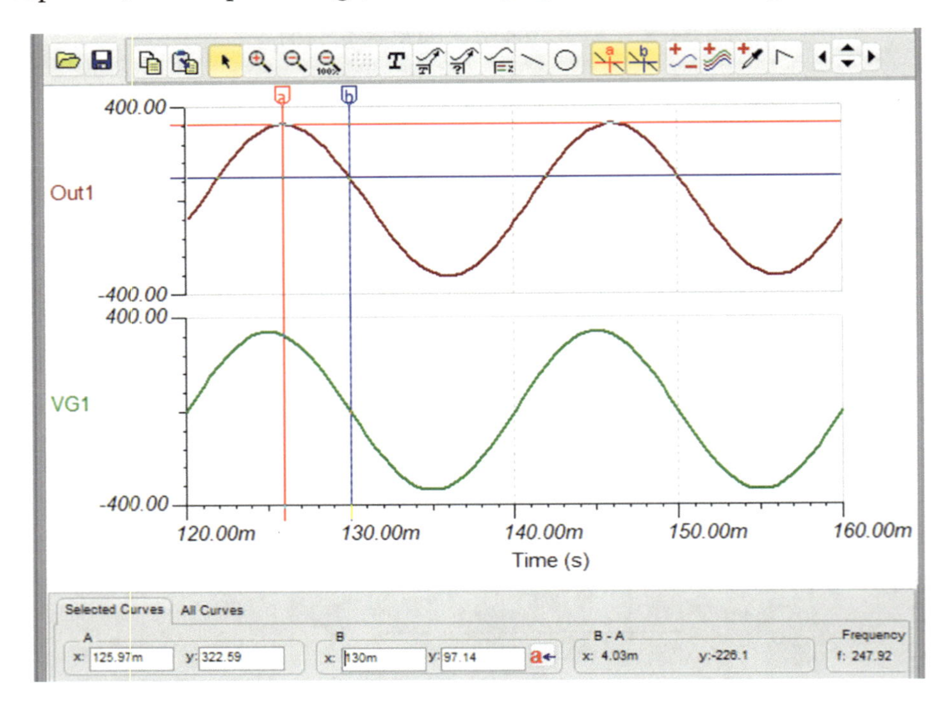


***Figure 3-5.** The Output Voltage Across the Capacitor*

Figure 3-5 shows the output voltage when C = 1uF and R = 1kΩ. We can see from cursor "a" that the peak voltage is 322.59v, and with cursor "b," when time was 130ms, the voltage was 97.14v. If we use the expression for Vout, we can calculate the output voltage when t = 130ms as

$$v_{out} = 323.416Sin\left(100\pi \times 130E^{-3} - 0.3044\right) = 96.934v$$

The simulated results compare very closely to the calculated results. We will calculate the expression for Vout when the resistance value was increased to 10kΩ. This would give

$$\omega CR = 2\pi \times 50 \times 1E^{-6} \times 10E^{3} = 3.1416$$

Therefore, we have

$$v_{out} = \frac{339}{\sqrt{1^2 + (3.1416)^2}} \langle -Tan^{-1}(3.1416) \rangle$$

which gives

$$v_{out} = 102.824 \langle -1.2627 \rangle$$

With respect to time, the expression for Vout would be

$$v_{out} = 102.824 Sin(100\pi t - 1.2627)$$

We can use this to determine the voltage across the capacitor when t = 130ms as

$$v_{out} = 102.824 Sin\left(100\pi \times 130^{-3} - 1.2627\right) = 97.98v$$

Simulating the circuit with $R_1$ set to 10kΩ produced the output voltage as shown in Figure 3-6.

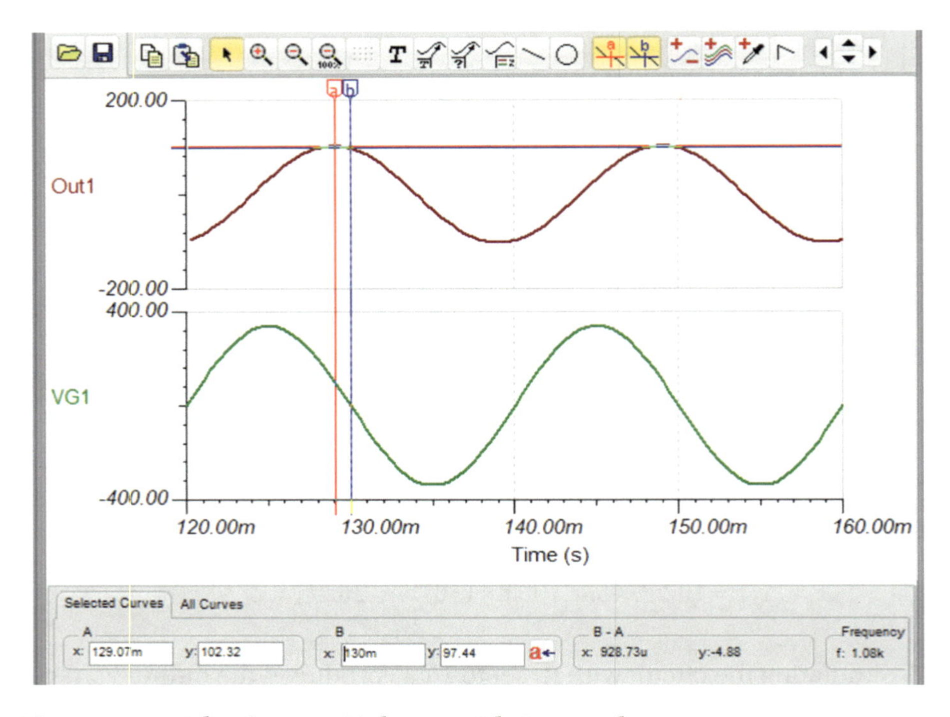


***Figure 3-6.*** *The Output Voltage with R1 = 10k*

Cursor "a" shows that the peak voltage is 102.32v, and cursor "b" shows that at 130ms the voltage was 97.44v. These two readings do again agree closely to the calculated values. This means that we can use the expression for the voltage across the capacitor with some confidence:

$$v_{out} = \frac{339}{\sqrt{1^2 + (\omega CR)^2}} \langle -Tan^{-1}(\omega CR) \rangle$$

This is for the peak voltage; if we wanted to calculate the rms voltage, we would simply replace the 339 with 240. Of course, this assumes that the input voltage was the UK mains.

# The Loading Effect of the SCR

We will now connect the SCR to the phase shift circuit and see what happens. The circuit to test this is shown in Figure 3-7.

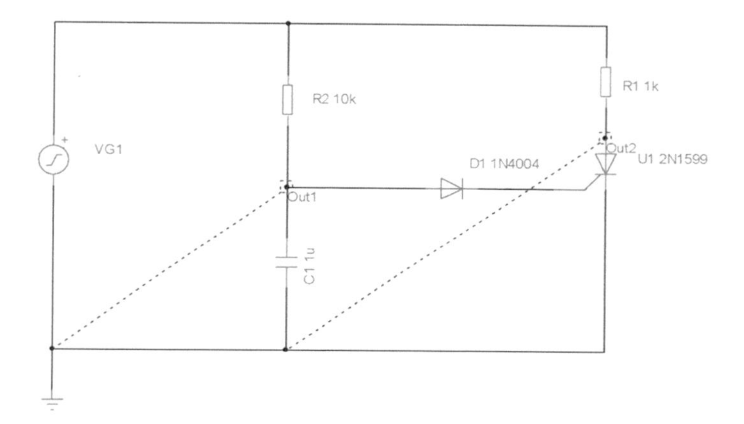


***Figure 3-7.*** *The Test Circuit for the Loading Effect of the SCR*

The SCR is the 2n1599 which can accommodate a forward voltage of 400V. We are also using a diode, D1, which is a 1n4004, which has a peak reverse voltage rating of 400V. This is because both components must be able to cope with the peak of the mains voltage which is set to 339v. When we simulate the circuit, we can display the waveforms as shown in Figure 3-8.

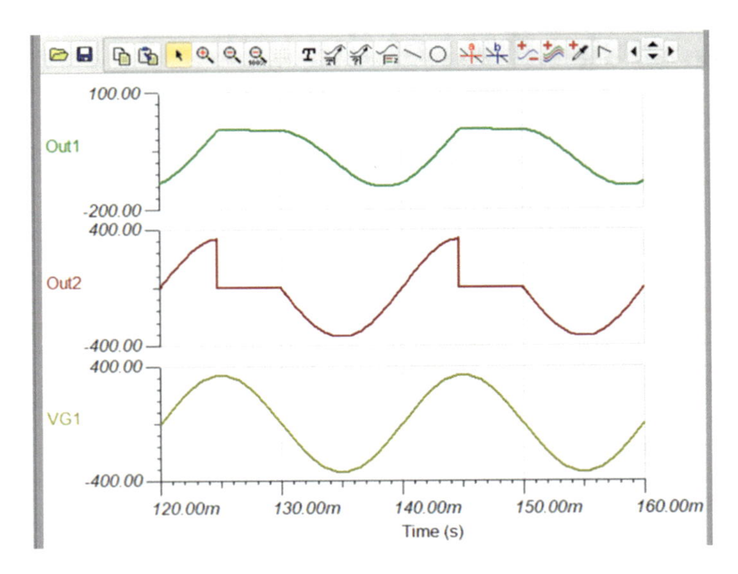


***Figure 3-8.*** *The Waveforms of the Test Circuit Shown in Figure 3-7*

The waveform out1 is the voltage across the capacitor. The positive half of the waveform is clamped to around 2.5v due to the action of D1 and the SCR itself. The diode, D1, is there to disconnect the SCR from the phase shift CR circuit when the voltage across the capacitor goes negative. However, when we measured the negative peak, it was around –141v. This does suggest that the SCR is still loading the CR circuit.

The waveform out2 is the voltage at the anode of the SCR. We can use this to study when the SCR turns on. When the SCR turns on, it will drive the voltage at the anode down toward 0V as it will connect the load to the ground, giving the load current a path to flow through. Therefore, when the out2 voltage waveform goes to 0V, the SCR has turned on. However, this does not easily show us the voltage across the load. We can do that with the waveforms shown in Figure 3-9.

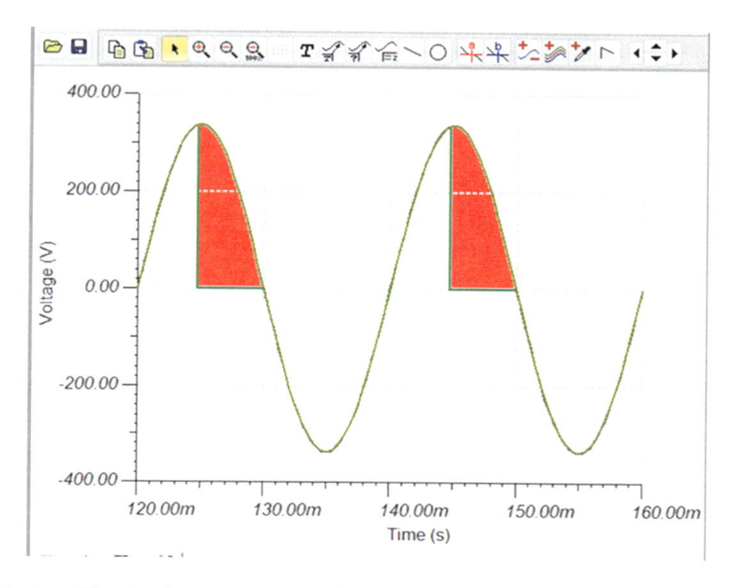


***Figure 3-9.** The Voltage Across the Load in Figure 3-7, R1 = 10k*

The voltage across the load is shaded in red. We can see we have allowed only a small amount of the positive half cycle of the input voltage to be passed onto the load. This will reduce the average voltage across the load, which is what we want to do with this type of circuit. So, the circuit seems to work. We can test this concept more by reducing the value of the resistor, $R_1$, in the CR phase shift circuit and see what happens.

After reducing the value of $R_1$ to 5kΩ, the circuit was simulated, and the display of the waveforms was as shown in Figure 3-10.

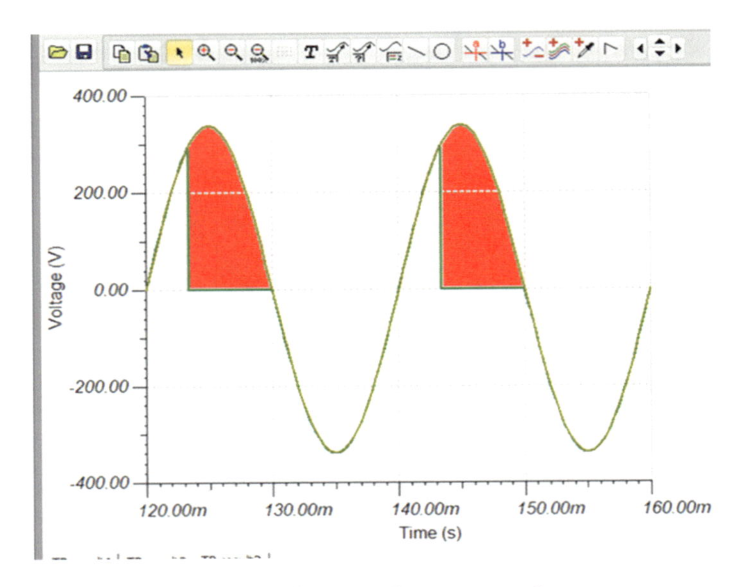


***Figure 3-10.*** *The Output Voltage When R1 = 5k*

We can see that the voltage across the load has been increased. This means there must be a relationship between the CR values of the phase shift circuit and the voltage across the load. However, the loading effect of the SCR has distorted the waveform of the voltage across the capacitor, which makes it difficult to determine this relationship. However, we can see that if we vary the value of the resistor $R_1$, we can vary the voltage across the load. There is nothing really wrong with just using a variable resistor for $R_1$ and just say that by varying the value of this resistor we can change the average voltage across the load, that is, dim a light if the load is a lamp or vary the speed of a motor if the load was a motor. However, I believe that a good engineer can design circuits with which they can predict what will happen.

# Buffering the Phase Shift Circuit

One way to reduce or even remove the effect of interstage loading, something we will look at in Chapter 9, is to place a buffer between the two stages. In this dimmer circuit, as that is what we are trying to create with the SCR, the first stage is the CR phase shift circuit, and the second stage is the SCR itself. So, what we will try to do is put a buffer between the two circuits. The buffer circuit we will use is the unity gain Opamp; we will study this buffer circuit later in Chapter 9. The test circuit using this buffer is shown in Figure 3-11.

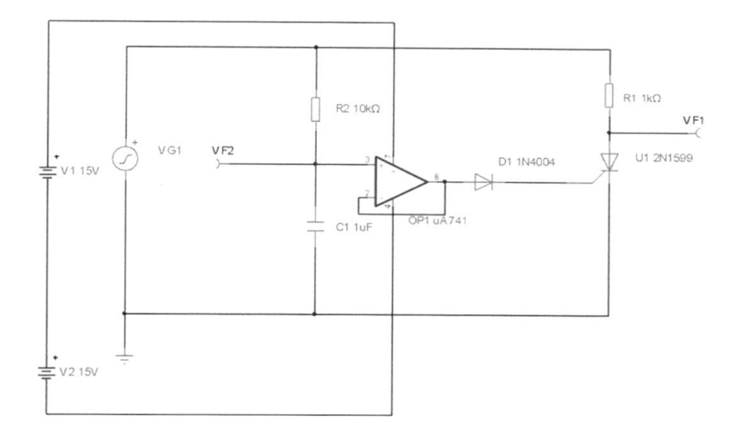


***Figure 3-11.*** *The Dimmer Circuit with a Buffer Circuit*

We are using voltage pins instead of the out pins to allow us to monitor the voltage at points around the circuit. We will look more closely at how to use these when we look at using the TINA software in Chapter 12.

The circuit is a bit more complex as we will need a DC power supply to supply the Opamp with. We will use a dual supply with +15V and -15V. When we simulate the circuit, the output waveforms we get are shown in Figure 3-12.

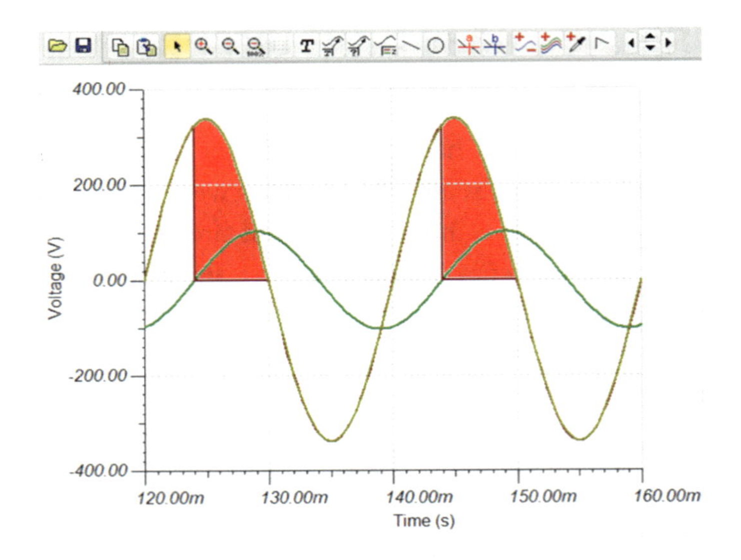


***Figure 3-12.*** *The Output Voltage R1 = 10k but with the Buffer*

The output voltage is now slightly less than when the buffer circuit was not added. See Figure 3-9. However, if we look at the waveform shown in Figure 3-13, we might see the advantage of adding this buffer circuit.

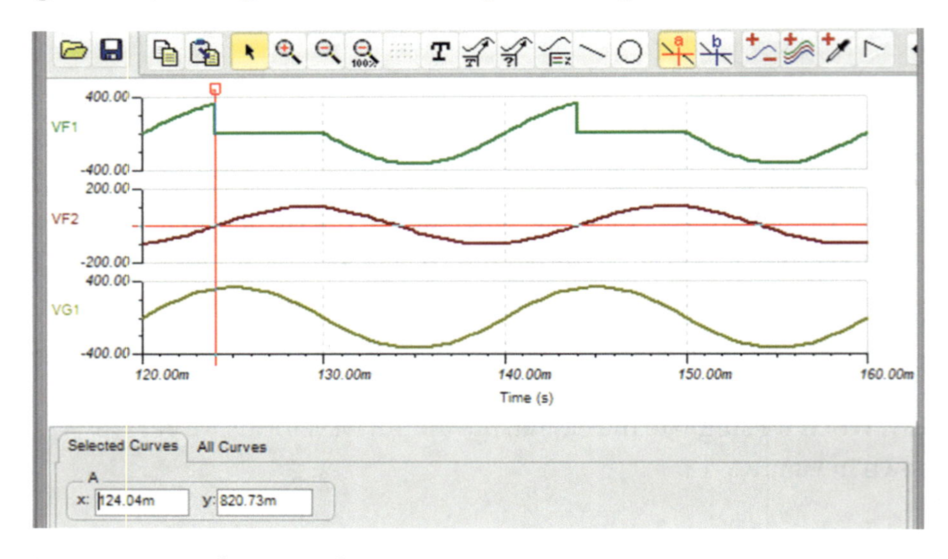


***Figure 3-13.*** *The Waveforms with the Buffer Circuit Added*

The trace VF2 is the voltage across the capacitor of the CR phase shift circuit. We can see that it is the same as when the SCR had not been included; see Figure 3-6. This means the buffer circuit has done its job. This means we may be able to predict when the SCR turns on. The ECAD software TINA states that the gate triggering voltage for the 2n1599 SCR is 700mv. We have shown that when $R_1$ was set to 10kΩ, the expression for the voltage across the capacitor, that is, Vout, of the phase shift circuit is

$$v_{out} = 102.824 Sin(100\pi t - 1.2627)$$

We can transpose this for an expression for time "t" as follows. Divide both sides by 102.824, which gives

$$\frac{v_{out}}{102.824} = Sin(100\pi t - 1.2627)$$

Now take the inverse sin of both sides, which gives

$$Sin^{-1}\left(\frac{v_{out}}{102.824}\right) = 100\pi t - 1.2627$$

Now add 1.2627 to both sides:

$$Sin^{-1}\left(\frac{v_{out}}{102.824}\right) + 1.2627 = 100\pi t$$

Finally, divide both sides by 100π:

$$t = \frac{Sin^{-1}\left(\frac{v_{out}}{102.824}\right) + 1.2627}{100\pi}$$

We can use this expression to determine the time when the Vout would be around 700mv:

$$t = \frac{Sin^{-1}\left(\frac{700E^{-3}}{102.824}\right) + 1.2627}{100\pi} = 4.041ms$$

This would be 4.041ms after the start of any new cycle. The waveforms shown in Figure 3-13 start at the beginning of the sixth cycle, that is, time is 120ms. Therefore, if we examine the VF2 trace at a time of 124.041ms, we should see a voltage close to 700mv. Cursor "a" is showing that when time is 124.041ms, the voltage across the capacitor, that is, VF2 trace, is 820.73mv. This is pretty close to what we expect. Also, we can see that it coincides with the time that the SCR turns on and the VF1 trace, the voltage at the anode of the SCR, falls to 0v. This time delay of 4.041ms relates to a phase delay of

$$Phase\ delay = 4.041 \times 18 = 72.738^{\circ} or\ 1.27\ Rads$$

If we reduce the value of $R_1$ to 5k, the expression for the voltage across the capacitor changes to

$$v_{out} = 182.053\,Sin(100\pi t - 1.004)$$

Therefore, the expression for time "t" becomes

$$t = \frac{Sin^{-1}\left(\frac{v_{out}}{182.053}\right) + 1.004}{100\pi}$$

Using this expression, we can calculate the time the voltage got to 700mv and so the time the SCR turned on as

$$t = \frac{Sin^{-1}\left(\frac{700^{-3}}{182.053}\right) + 1.004}{100\pi} = 3.208ms$$

Changing the resistor value to 5k and simulating the circuit gives the display as shown in Figure 3-14.

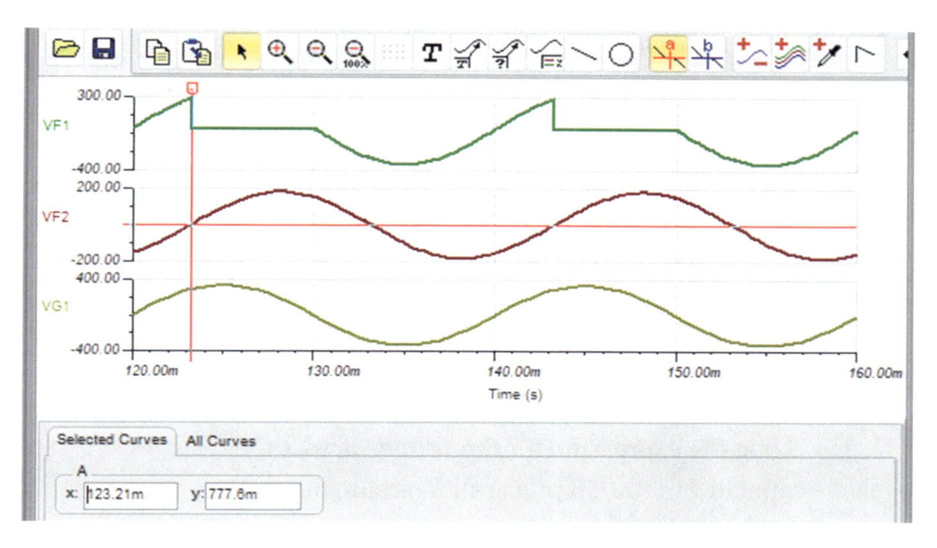


***Figure 3-14.*** *The Waveforms When R1 = 5k*

Cursor "a" shows that when t = 123.21ms, the voltage across the capacitor is 777.6mv, which is close to the expected 700mv. Also, it is the time when the SCR turns on. I hope this work does show you that by adding the buffer circuit and doing some maths work, we can predict when the SCR will turn on. I believe this is something that engineers should be able to do. However, I am not saying that the circuit shown in Figure 3-11 is the actual circuit of a dimmer. I am only trying to show you how we can apply some concepts and maths to analyze analog circuits.

We can generalize the expression for time "t" to

$$t = \frac{Sin^{-1}\left(\frac{v_{out}\sqrt{1^2 + (100\pi CR)^2}}{339}\right) + Tan^{-1}(100\pi CR)}{100\pi}$$

We can also generalize the expression for Vout as

$$v_{out} = \frac{339}{\sqrt{1^2 + (100\pi CR)^2}} Sin\left(100\pi t - Tan^{-1}(100\pi CR)\right)$$

This assumes we are dealing with the UK mains with an rms of 240v, and the angle is described in radians.

## Exercise 3.1

1. Using the expression for the voltage across the capacitor in the CR phase shift circuit, determine the time when the SCR would turn on if the resistor $R_1$ was changed to 15kΩ. Determine the delay in degrees and radians as well as time. Calculate the peak of the voltage across the capacitor. This is with the Opamp buffer included in the circuit.
2. Repeat the calculation but now with $R_1$ changed to 22k.

## Controlling an Inductive Load with an SCR

It is an unusual load that is purely resistive; even a bulb will have some inductance, and a motor will definitely have some inductance. Therefore, we need to investigate what happens if we have an RL load as shown in the test circuit in Figure 3-15.

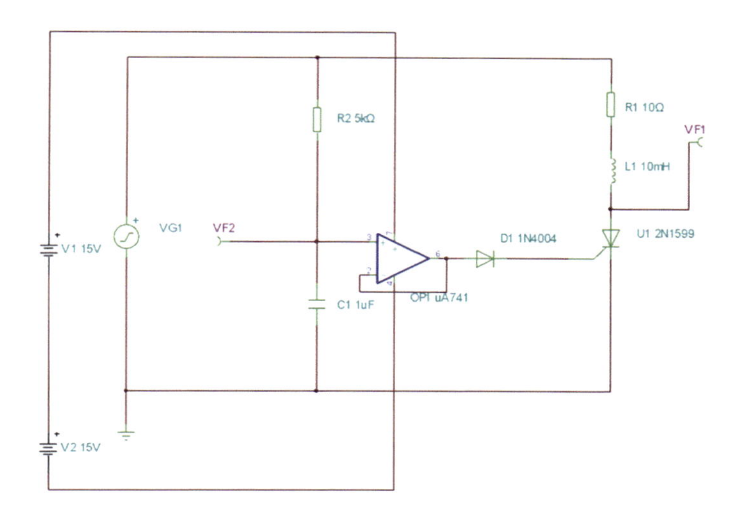


***Figure 3-15.** The Test Circuit for the RL Load*

We have kept the value of $R_2$ at 5k, and we are still using the buffer circuit. After simulating the circuit, the waveforms as shown in Figure 3-16 were obtained.

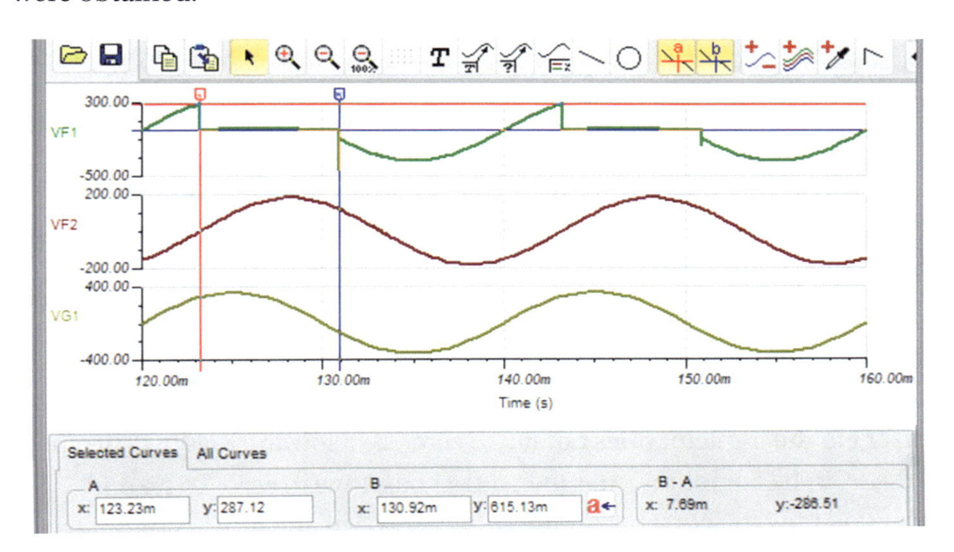


***Figure 3-16.** The Output Waveforms for the Test Circuit with the RL load*

The SCR turns on, as expected when time is 123.23ms. However, I hope we can see that the SCR turns off later than it did with just the resistive load. It actually turns off when the time is 130.92ms. This is actually 10.92ms after the time when the input voltage started to go negative. If we increase the inductance value to 100mH, the delay in the turn-off time is greater as shown in Figure 3-17.

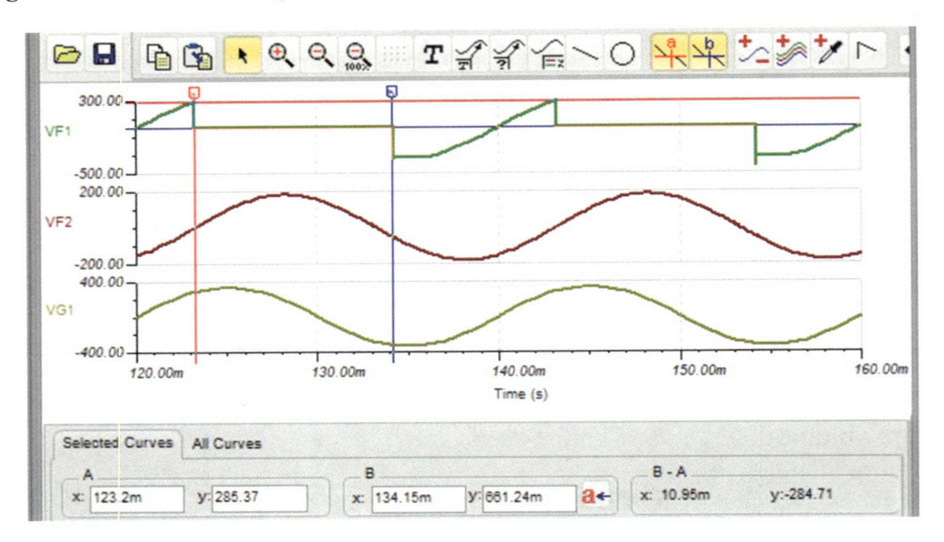


***Figure 3-17.** The Waveforms When L Was Increased to 100mH*

The turn-on time is unchanged, but the turn-off time is now around 134.15ms, which is 14.15ms after the input voltage has gone negative. The issue that this creates is that some of the negative half of the input voltage has been allowed to be passed onto the load. This would actually reduce the average voltage across the load. The reason this happens is because the current in an inductor lags behind the voltage across it. This means the current in the inductor does not fall to zero, or below the holding current of the SCR, till sometime after the voltage has gone to zero. To try and show you how the current lags the voltage, we can add a very small value resistance as shown in Figure 3-18.

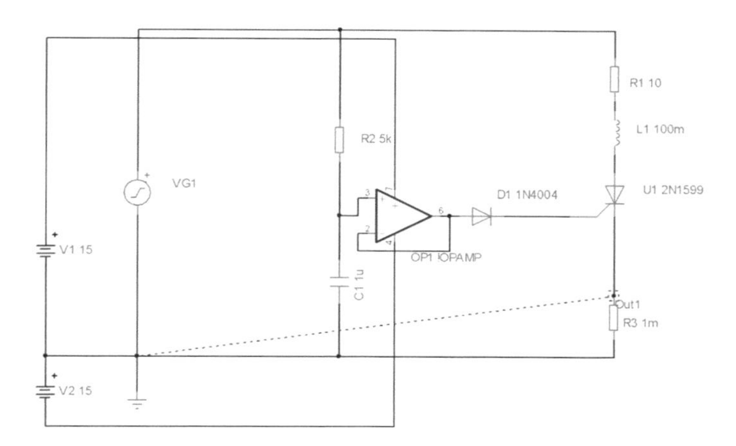


***Figure 3-18.** Adding the Resistor R3 to Measure the Current*

Adding this small resistor allows us to display the voltage across it as shown in Figure 3-19. If we divide the value of the voltage by the value of the resistor, we can convert that waveform to a current waveform.

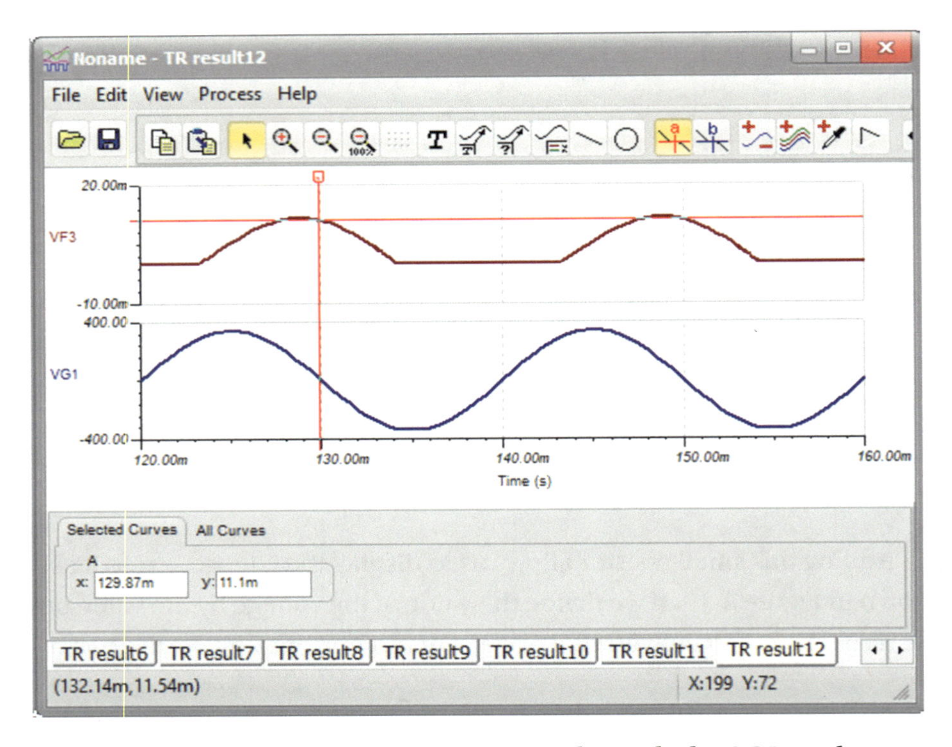


***Figure 3-19.*** *The Display of the Current Through the SCR and the RL Load*

The trace VF3 shows that the current is still positive after the input voltage has gone negative. This means it is above the holding current value, and the SCR stays closed, allowing some of the negative half of the input to be passed onto the load.

# Adding a Flywheel Diode

The fact that due to the inductive load, the SCR turns off later than expected reduces the average voltage and also makes it difficult to determine the average voltage across the load. It would be better if we could make the SCR turn off when the input goes negative even with an RL load. The act of turning the SCR off is termed "commutation." One solution we can use is shown in the circuit in Figure 3-20.

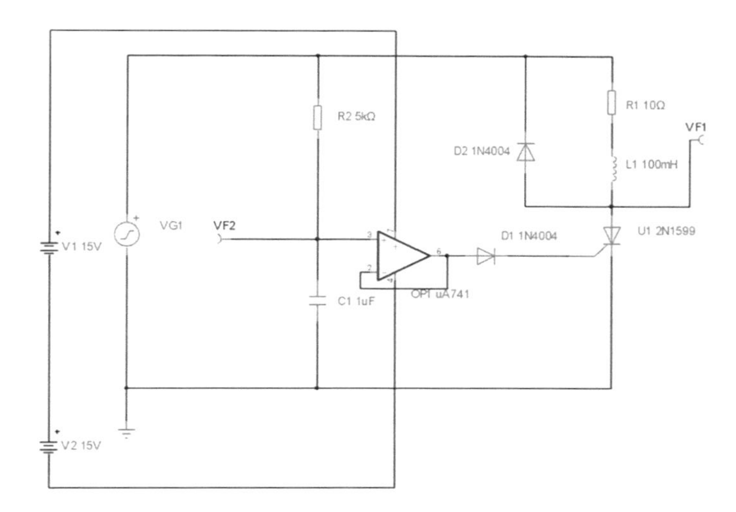


***Figure 3-20.*** *The Circuit with the Added Flywheel Diode*

When we simulate the circuit, the waveforms produced are as shown in Figure 3-21.

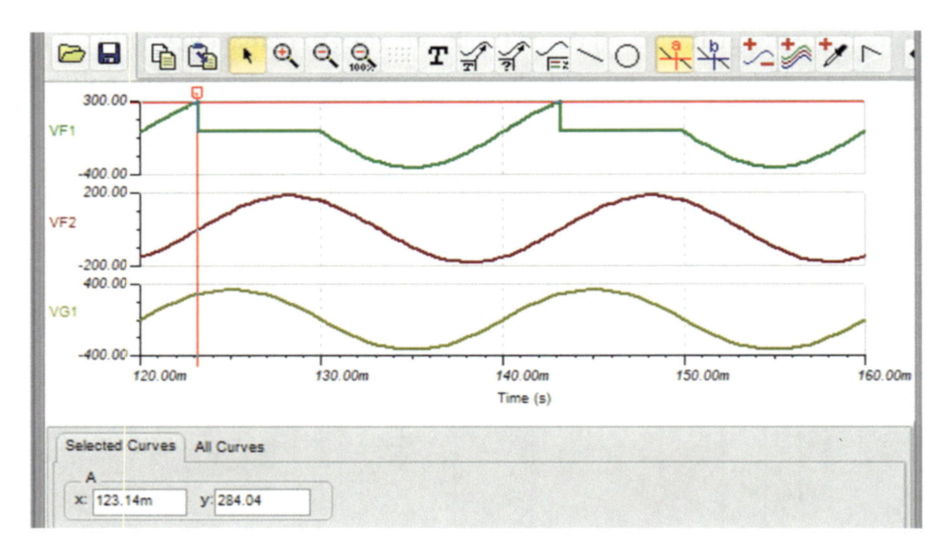


***Figure 3-21.*** *The Waveforms for the Circuit with the Flywheel Diode*

The SCR turns on as expected, but now it also turns off when the input voltage starts to go negative. This means none of the negative half cycle of the input is passed onto the load. The reason this works is that when the input voltage starts to go negative, the cathode of the flywheel diode goes negative, and the inductor will drive the anode more positive as it tries to keep the current flowing through the SCR. However, this situation means that the flywheel diode becomes forward biased and so closes. This provides an alternative path for the current that would have been forced to flow through the SCR, to flow through and so not through the SCR. This means that no current flows through the SCR, and so it can turn off when the input voltage goes negative. This means the RL circuit can operate just as if it was resistive.

# The Average Voltage Across the Load

Now that we can control when the SCR turns on and off, we can determine the average voltage across the load. If we consider the voltage across the load as shown in Figure 3-22, we should be able to calculate the average voltage across the load.

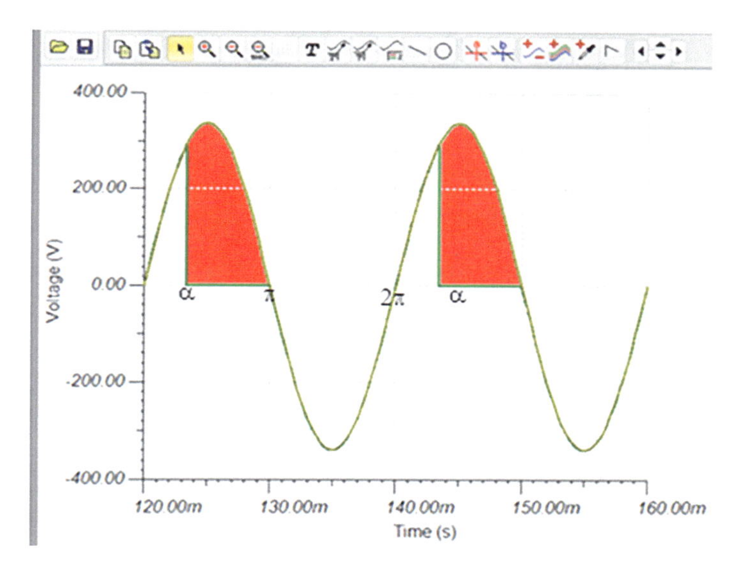


***Figure 3-22.** The Voltage Across the Load with a Flywheel Diode*

As average is a summing action with the result divided by the number in the sum, then, as integration is a summing action, we can use integration to calculate the average as

$$ave = \frac{1}{2\pi}\int_{\propto}^{\pi} v_{Peak}\, Sin(\theta)\, d\theta$$

Therefore, taking the constant of $v_{Peak}$ out of the integral, we have

$$ave = \frac{v_{Peak}}{2\pi}\int_{\propto}^{\pi} Sin(\theta)\, d\theta$$

Using the standard integrals, sin integrates to -cos. This then gives

$$ave = \frac{v_{Peak}}{2\pi} = \left[-Cos(\theta)\right]_{\alpha}^{\pi}$$

Putting the limits in, we get

$$ave = \frac{v_{Peak}}{2\pi}\left[\left(-Cos(\pi)\right) - \left(-Cos(\propto)\right)\right]$$

Knowing Cos (π) = −1, we get

$$ave = \frac{v_{Peak}}{2\pi}\left(1 + Cos(\propto)\right)$$

This is the general expression we can use to calculate the average voltage when the SCR turns off when the ac input starts to go negative.

This means that if we can calculate the value of "α" in radians, we can calculate the average voltage across the load. If we are using the circuit shown in Figure 3-20, we know that the value of "α" can be calculated using the expression from the phase shift circuit shown in Figure 3-4 as the Opamp in Figure 3-20 prevents the phase circuit from being loaded by the SCR. The expression we will use is

$$v_{out} = \frac{339}{\sqrt{1^2 + (\omega CR)^2}} - Tan^{-1}(\omega CR)$$

where the phase part is

$$Phase = -Tan^{-1}(\omega CR)$$

Therefore, the value of "α" depends upon the value of $\omega CR$. This can be calculated using

$$\omega CR = 2\pi fCR$$

As we are using the UK mains, then the frequency f is 50Hz. This means that

$$\omega CR = 100\pi CR$$

If we keep the capacitor value set at 1uF, then we can say

$$\omega CR = 100\pi \times 1E^{-6} R = 3.1416E^{-4} R$$

This means that

$$\alpha = Tan^{-1}\left(3.1416E^{-4} R\right)$$

We can transpose this expression for R to show that

$$R = 3183.1 Tan(\alpha)$$

This is for the phase circuit being supplied from the UK mains and using a 1uF capacitor. We can test this by using it to determine the resistor value to produce a value of 1.2 radians:

$$R = 3183.1 Tan(\alpha) = 3183.1 \times Tan(1.2) = 8.187 k\Omega$$

We can relate 1.2 Rads to degrees as 1 Rad = 57.296°; therefore, 1.2 Rads = 68.755°. We can then relate this angle to a time delay knowing that at 50Hz 1ms = 18°. Therefore, 68.755° will equate to a delay of 3.82ms. If we change the value of $R_1$ to 8.187k and simulate the circuit, then the waveforms produced would be as shown in Figure 3-23.

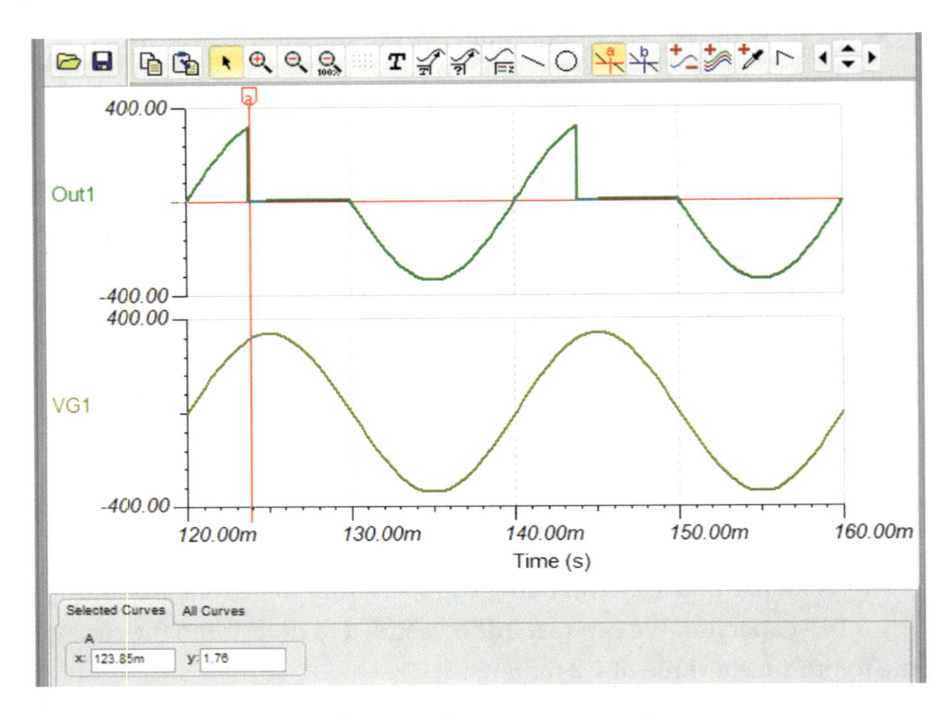


***Figure 3-23.*** *The Waveforms When R1 = 8.187k*

Cursor "a" shows that at a time of 123.85m, the SCR has just turned on. This is 3.85ms after the input voltage has started to go positive. We will try one more angle to test the relationship. This time, let α = 1.5 Rads.

Therefore

$$R = 3183.1Tan(\alpha) = 3183.1 \times Tan(1.5) = 44.886k\Omega$$

We can calculate the time delay using

$$time\ delay = \frac{Radsvalue \times 10}{\pi} = \frac{1.5 \times 10}{\pi} = 4.775ms$$

This can again be confirmed by changing the resistor value and simulating the circuit.

We can now use the expression for the average voltage to determine the average voltage when the resistor $R_1$ in the phase shift circuit was set to 5kΩ. This would produce the voltage waveform across the load as shown in Figure 3-22.

We can use the expression

$$ave = \frac{v_{Peak}}{2\pi}\left(1 + Cos(\propto)\right)$$

The vpeak is the peak of the UK mains at 339v, assuming an rms of 240v. The value of "α" can be calculated using

$$\alpha = Tan^{-1}\left(3.1416E^{-4}R\right)$$

Therefore

$$\alpha = Tan^{-1}\left(3.1416E^{-4} \times 5E^{3}\right) = 1.507\ Rads$$

We can now calculate the average voltage across the load as

$$ave = \frac{339}{2\pi}\left(1 + Cos(1.507)\right) = 57.393v$$

Note, your calculator must be placed into rads as we are using radians to describe the angle.

## Exercise 3.2

Determine the average voltage across the load if the resistor $R_1$ in the phase shift circuit was changed to 10k and then 20k, if the capacitor was set at 1μF.

# The Average Voltage with RL Load and No Flywheel Diode

For the sake of completeness, we should look at how we can calculate the average voltage if the SCR stays turned on and allows some of the negative voltage to be passed onto the load. This would happen if we did not place a flywheel diode across the load. This means that the SCR stays closed past the half cycle which is when the angle is 180° or π radians. The amount by which the SCR stays open pass π radians is expressed as "β" radians or degrees. We will use radians in our calculations. This means that the expression for calculating the average voltage applied across the load of an SCR when the SCR stays closed passed π radians is

$$ave = \frac{1}{2\pi}\int_{\propto}^{\pi+\beta} v_{Peak}\, Sin(\theta)\, d\theta$$

$$ave = \frac{v_{Peak}}{2\pi}\int_{\propto}^{\pi+\beta} Sin(\theta)\, d\theta$$

where "α" is the delay before the SCR turns on and "β" is the delay from "π" when the SCR turns off. If we are using the test circuit shown in Figure 3-15, we can see the load is made up with a resistor in series with an inductor. With this inductive load, the current flowing through the load will lag behind the voltage across the load by some angle "β." If we had an expression for the current flowing through the RL load, then we could work out the angle "β" that the inductor has introduced.

We can, as always, use Ohm's Law to derive the expression for any current. This would give

$$i = \frac{v}{z_T}$$

The term $z_T$ is simply the series sum of R plus $X_L$, but this is a phasor sum. The term $X_L$ is the impedance of the inductor, and it can be calculated as

$$X_L = 2\pi fL = \omega L$$

As a phasor, this impedance goes on the +j axis, so we can say

$$X_L = 0 + j\omega L$$

The "0" is there because this is a pure inductance with no internal resistance.

The resistance has no imaginary or j term, so we can say that as a phasor the resistance is

$$R + j0$$

If we now add the resistor which is in series with the inductor, we can say

$$z_T = (R + j0) + (0 + j\omega L) = R + j\omega L$$

When adding phasor numbers, we add all real parts together and then add all j terms together. There is a complete analysis of how we use complex numbers to represent impedances in Chapter 11.

This means we can calculate the current flowing through the load as

$$i = \frac{v}{R + j\omega L}$$

Expressing this expression in polar format, we have

$$i = \frac{v\langle 0 \rangle}{\sqrt{R^2 + (\omega_L)^2}\left\langle Tan^{-1}\left(\frac{\omega_L}{R}\right)\right\rangle}$$

This results in

$$i = \frac{v}{\sqrt{R^2 + (\omega_L)^2}} \left\langle -Tan^{-1}\left(\frac{\omega_L}{R}\right)\right\rangle$$

This can be split into two parts, which are as follows.

The magnitude is

$$\frac{v}{\sqrt{R^2 + (\omega_L)^2}}$$

The angle is

$$\left\langle -Tan^{-1}\left(\frac{\omega_L}{R}\right)\right\rangle$$

In the calculation for the average voltage across the load, we are only interested in the angle as this is what sets the delay angle "β." Therefore, we can say

$$\beta = -Tan^{-1}\left(\frac{\omega_L}{R}\right)$$

If we put in the values from the test circuit shown in Figure 3-15, we get

$$\beta = -Tan^{-1}\left(\frac{2\pi \times 50 \times 10E^{-3}}{10}\right) = 0.3044\ Rads$$

We need to calculate the delay angle "α." This delay angle is created using the CR phase shift circuit, and so it can be calculated using

$$\propto = Tan^{-1}(\omega CR)$$

Putting the values in from the test circuit shown in Figure 3-15, we get

$$\propto = Tan^{-1}\left(2\pi \times 50 \times 1E^{-6} \times 5E^{3}\right) = 1\ Rad$$

Putting these values into the expression for the average voltage, we get

$$ave = \frac{v_{Peak}}{2\pi} \int_{\propto}^{\pi+\beta} Sin(\theta)d\theta = \frac{339}{2\pi} \int_{1}^{\pi+0.3044} Sin(\theta)d\theta$$

Integrating, we get

$$ave = \frac{v_{Peak}}{2\pi} \left[-Cos(\theta)\right]_{1}^{\pi+0.3044}$$

This gives

$$ave = \frac{v_{Peak}}{2\pi} \left[\left(-Cos(\pi+0.3044)\right) - \left(-Cos(1)\right)\right]$$

This gives

$$ave = \frac{339}{2\pi} \left[\left(--0.954\right) - \left(-0.5403\right)\right]$$

$$ave = \frac{339}{2\pi} \left[1.4943\right] = 128.87v$$

If we calculate the average voltage when the flywheel is in place, we should be able to compare the two values. The average voltage with the flywheel diode in place can be calculated using the expression

$$ave = \frac{v_{Peak}}{2\pi} \left(1 + Cos(\propto)\right)$$

Putting the values in, we get

$$ave = \frac{339}{2\pi} \left(1 + Cos(1)\right) = 166.21v$$

This shows that with the inductive load without the flywheel diode, the average voltage across the load has been reduced.

# Exercise 3.3

1. If the inductance value in the test circuit shown in Figure 3-15 was increased to 50mH, calculate the new angle "β" and the average voltage across the RL load.

2. Repeat the calculations if the resistor in the CR phase shift circuit was increased to 20k and the inductance value in the load was reduced to 40mH.

# The Triac

So far, we have just used half of the input waveform. This is probably OK if we were just interested in using DC devices. However, if we wanted to control an ac motor, then the simple SCR would not do. We will now look at the Triac, and as its name does suggest, as it has the "ac" in it again, the Triac has something to do with ac currents. Also, the term "Tri" does suggest that it has three terminals. The symbol for the Triac is shown in Figure 3-24.

***Figure 3-24.*** *The Symbol for the Triac*

The symbol looks like two SCRs connected in parallel but in opposite directions. That is because that is basically what they are. Their semiconductor makeup is shown in Figure 3-25.

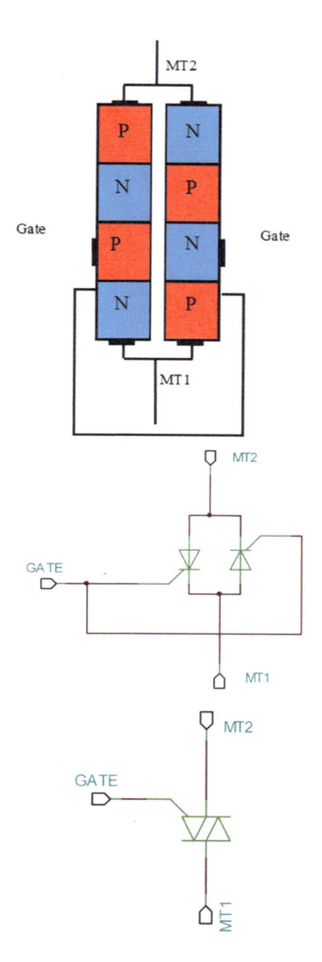


***Figure 3-25.*** *The Makeup of a Triac*

Figure 3-25 shows how the semiconductor structure of the Triac relates to two SCRs and finally the Triac. The Triac is an attempt to give the engineer the ability to chop the mains input in both ways. The voltage across the Triac will be forward biased when the ac input goes both negative as well as positive, whereas with the SCR it was only forward biased when the ac input went positive. Just as with the SCR, we need to provide a triggering pulse to turn the Triac on. Due to the work we have done with the SCR, it is hoped that we will be able to trigger the Triac with the CR phase shift. However, we need to investigate what is the requirement for triggering the Triac. The test circuit we are going to use to investigate the Triac is shown in Figure 3-26.

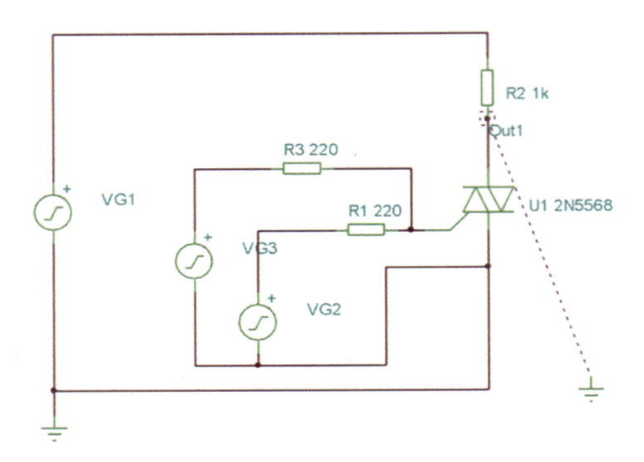


*Figure 3-26. The Test Circuit for the Triac*

We can examine the datasheet and the TINA model we will be using in our investigations. The parameters for the TINA model are shown in Figure 3-27.

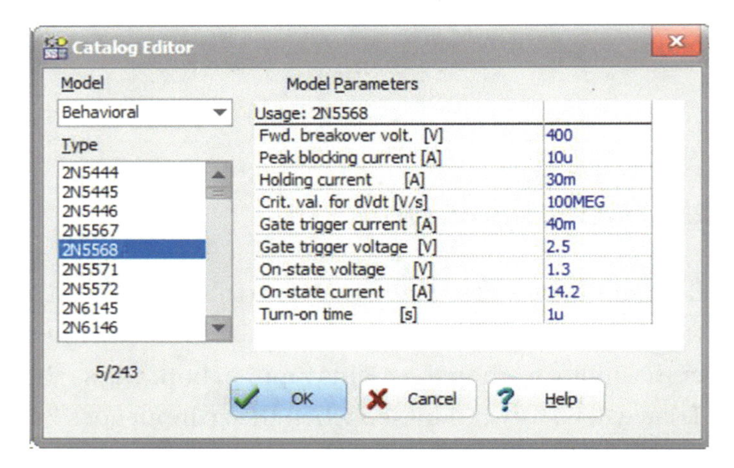


*Figure 3-27. The Model Parameters for the 2N5568 Triac*

The main parameters are

- The forward breakdown voltage, 400v, enough for the UK mains
- The holding current, 30mA.
- The gate triggering voltage, 2.5v. This is the minimum voltage.
- The gate triggering current, 30mA. This is the minimum current.

We are using a voltage source to produce the triggering pulses as this is the easiest to alter and so determine the requirements of the triggering circuit.

After quite a lot of trials and changing the parameters, the test circuit was simulated, and the waveforms as shown in Figure 3-28 were displayed.

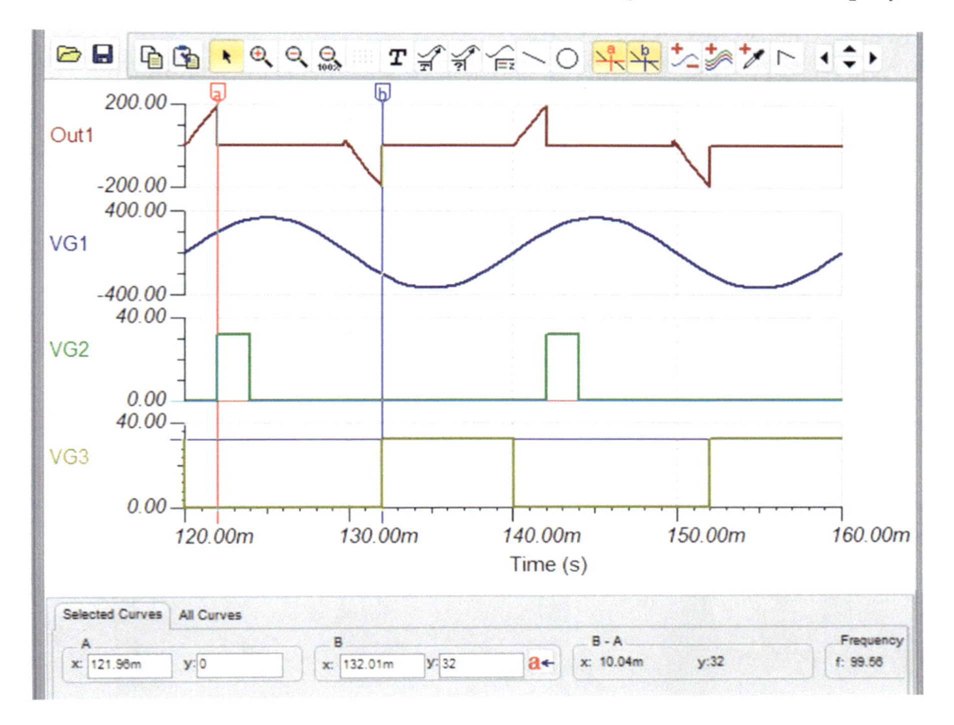


***Figure 3-28.** The Waveforms from the Test Circuit*

The circuit uses two voltage sources to produce the triggering as we need to trigger the Triac during both the positive and negative half cycles of the input voltage. VG2 triggers the Triac during the positive half cycle of the input, and VG3 triggers the Triac during the negative half cycle. The idea is to delay the triggering of the Triac during both half cycles the same. VG2 is a 2ms pulse rising to 32v that is delayed from rising by 2ms. Note, at 50Hz the periodic time for one cycle is 20ms. Therefore, VG2 goes high for 2ms and then stays low for the remaining 18ms. VG3 controls the triggering of the negative half cycle. The negative half cycle starts when time is 10ms into the full cycle. Therefore, VG3 stays low for the first 12ms and then goes high for the remaining 8ms. This ensures that the Triac triggers the negative half cycle with the same 2ms delay. I have set the pulses to rise to 32V because I am thinking of using a Diac to assist the triggering circuit, and they do not conduct until the forward biasing is around this 30V level. However, this is later in the chapter.

The trace Out1 is the voltage at the anode of the Triac. When this voltage goes down to 0V, we know the Triac has been turned on. Using the cursors "a" and "b," we can see that this happens at the same time the respective triggering pulses go high. I have tried a variety of changes for the triggering waveforms, and I have determined that this arrangement when the two pulses are not both high at the same time is the better approach.

Figure 3-29 shows the actual voltage across the load.

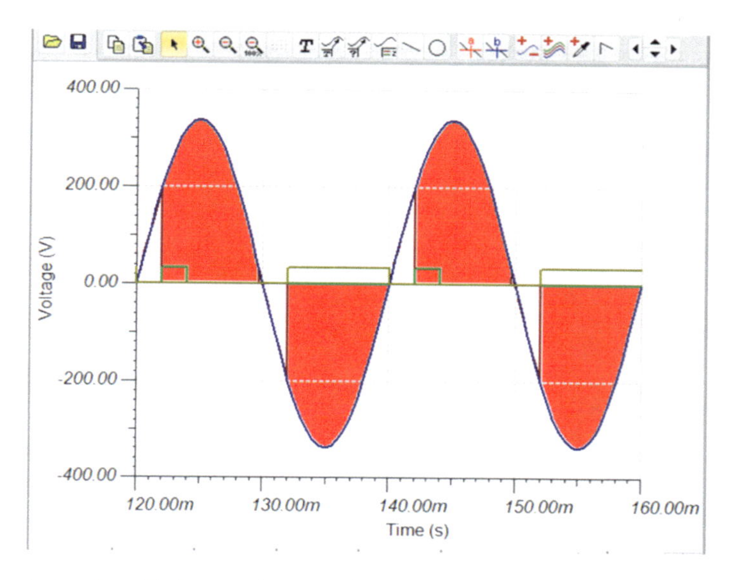


***Figure 3-29.*** *The Voltage Across the Load Is Shown in Red*

Just to show that we can reduce the voltage across the load to a very low value, we can re-simulate the circuit with a delay of 8ms. The result is shown in Figure 3-30.

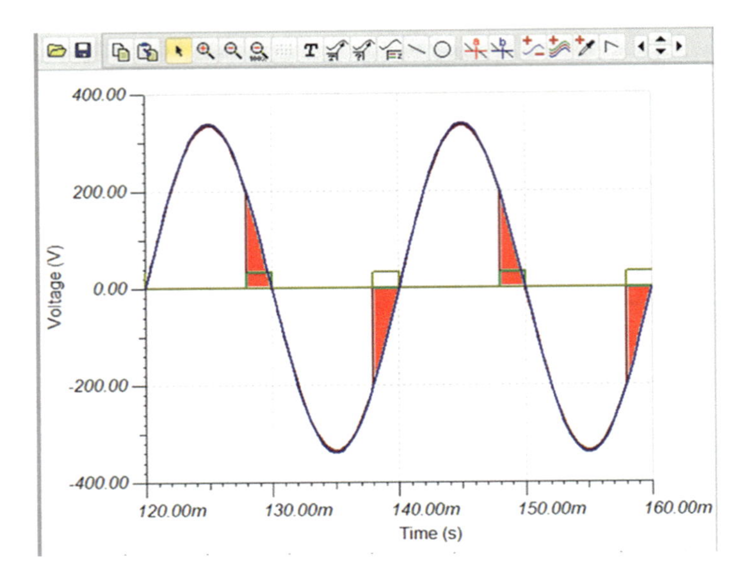


***Figure 3-30.** The Voltage Across the Load Is Shown in Red*

I hope it is quite obvious that the voltage across the load has been reduced.

Using the test circuit has shown us that we should try and create two pulses to drive the Triac. This could be achieved to some extent if we fed the output of the CR phase shift circuit via a Diac to trigger the Triac. Therefore, it would be good to look at what a Diac is.

# The Diac

We have described the diode as an electronic switch for a DC circuit, as it closes when the anode voltage is around 0.7v greater than the cathode. The switch can then be opened when the anode is less than 0.7v more than the cathode. In this way, the diode is used to control when current passes through it in just one direction, that is, conventional current flows from the anode to the cathode.

The Diac, as the "ac" seems to suggest, is an electronic switch which controls when it will pass current in both directions. Well, it is easy to see how the diode can be used in rectifying an ac input, but why would we want to control the switching of current in both directions? Really, the main use of the Diac is in controlling the triggering of a Triac.

## The Operation of the Diac

Just as the diode has to be forward biased, that is, its anode voltage has to be approximately 0.7v greater than the cathode, then to make the Diac conduct in either direction, the voltage at the anode of the Diac must become greater than the specified breakdown voltage, for it to conduct in either direction. We can look at that concept with the test circuit shown in Figure 3-31.

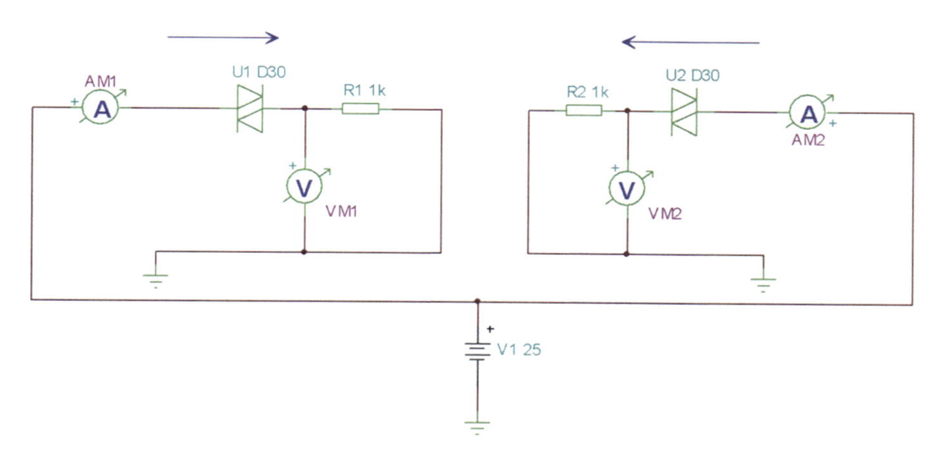


***Figure 3-31.*** *Test Circuit for Biasing the Diac*

TINA states that the forward breakover voltage for this D30 diac is 32V. Therefore, we will apply the biasing voltage from 25 to 37 in steps of 1v to see how the diac conducts. Simulating this test circuit allows us to complete a table of results as shown in Table 3-1.

***Table 3-1.*** *The Results of the Test Circuit Shown in Figure 3-31*

| Biasing Voltage V1 (Volts) | AM1 | VM1 | AM2 | VM2 |
|---|---|---|---|---|
| 25 | 76.72u | 76.72m | 76.72u | 76.72m |
| 26 | 79.83u | 79.83m | 79.83m | 79.83u |
| 27 | 82.94u | 82.94m | 82.94m | 82.94u |
| 28 | 86.05u | 86.05m | 86.05m | 86.05u |
| 29 | 89.16u | 89.16m | 89.16m | 89.16u |
| 30 | 92.28u | 92.28m | 92.28m | 92.28u |
| 31 | 95.39u | 95.39m | 95.39m | 95.39u |
| 32 | 98.5u | 98.5m | 98.5m | 98.5u |
| 33 | 101.62u | 101.62m | 101.62m | 101.62u |
| 34 | 177.63u | 177.63m | 177.63m | 177.63u |
| 35 | 9.5m | 9.5 | 9.5 | 9.5m |
| 36 | 9.78m | 9.78 | 9.78 | 9.78m |
| 37 | 10.06m | 10.06 | 10.06 | 10.06m |

The results from the table show quite clearly that the Diac is passing only a small current until the biasing voltage reaches a value of 34V. This does confirm that the Diac acts as an electronic switch, but unlike the diode, it works in both directions. Also, the voltage at which the switch turns on is a higher voltage than the basic diode, and it is different for different Diacs.

# Using the Diac

One of the main uses of the Diac is to assist in the switching on and off of a Triac. Indeed, we are hoping that this will produce the triggering pulses like the ones we used in Figure 3-28. A typical Diac-Triac circuit is shown in Figure 3-32.

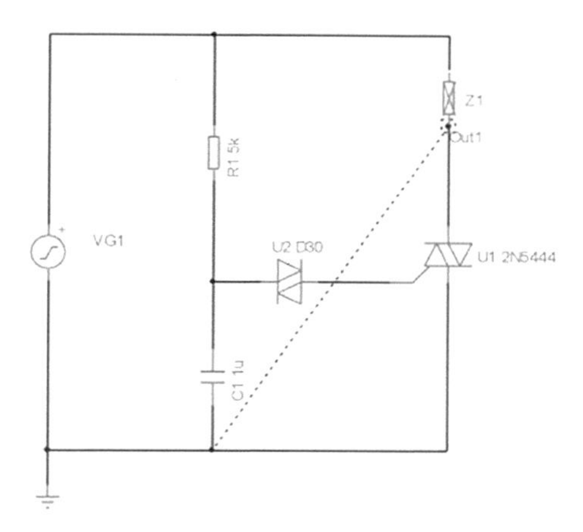


***Figure 3-32.*** *A Typical Diac-Triac Triggering Circuit*

This is one method of triggering the Triac. It is using the CR phase shift as we did with the simple SCR circuit. With this simulation, we are using an impedance "$Z_1$" to represent the load. This is a useful facility that TINA provides us with. This is because all ac loads are impedances, even the simple resistor, which is what is being used here. We can change the load to make it an RL load by modifying the impedance. The process is shown in Figure 3-33.

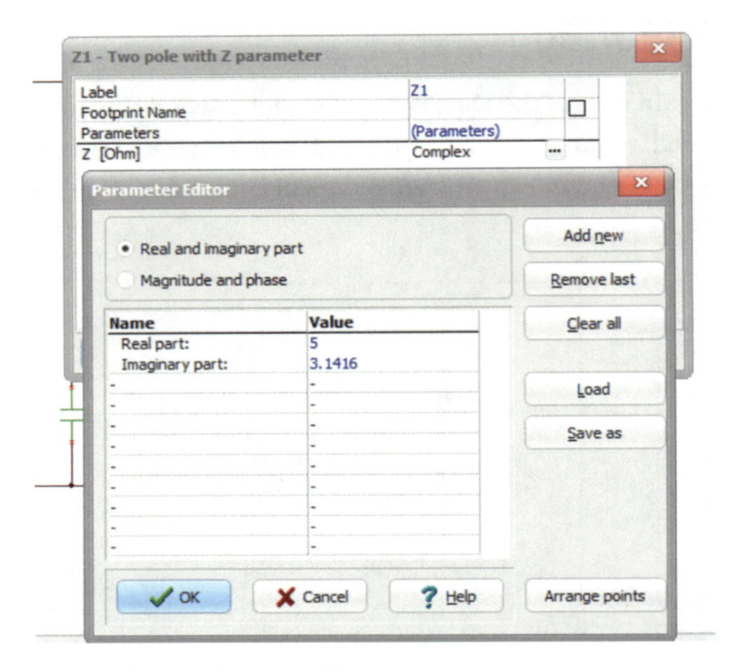


***Figure 3-33.*** *Editing the Impedance*

The real part is simply the resistance value you want for the load. The imaginary part is the actual impedance of the inductor we want for the load. If the inductor we want to use is a 10mH coil, then the impedance can be calculated as

$$X_L = 2\pi fL = 2\pi \times 50 \times 10E^{-3} = 3.1416\Omega$$

If we change the load to just a simple resistance of 5Ω and simulate the circuit, the waveforms we get are as shown in Figure 3-34.

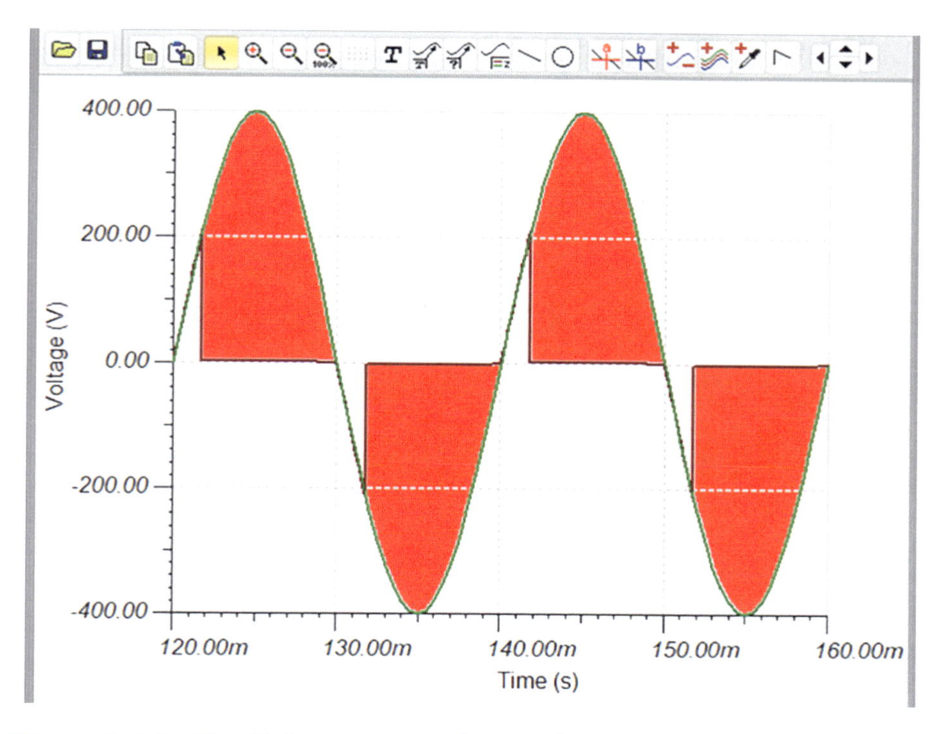


***Figure 3-34.** The Voltage Across the Load*

The voltage across the load is shown in red.

The CR phase circuit is one way of triggering the Triac. However, the loading effect of the Triac will make it difficult to produce a relationship between the phase shift across the capacitor and the output voltage. To go any further would take up too much analysis, and really power electronics is a book on its own. I will leave the analysis for that book if I ever write one.

Before I leave the work on Triac, etc., we will look at the test circuit shown in Figure 3-35.

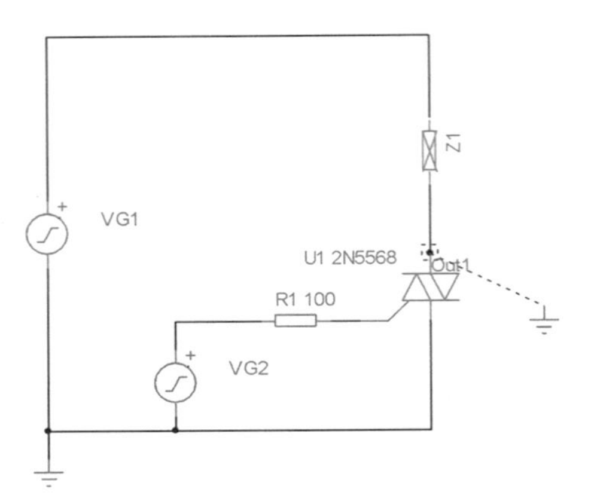


***Figure 3-35.*** *The Test Circuit for the Final Look at a Triggering Circuit*

This does look a bit like the triggering circuit shown in Figure 3-26, except that there is only one source for the triggering pulse, VG2. This is because the idea of the circuit in Figure 3-26 was to supply two pulses in the 20ms period of the 50Hz ac input. This used two sources, each producing one of the two pulses in their own 20ms period. In this final circuit, the trigger source produced one pulse but at a frequency of 100Hz, not 50Hz. This would mean that there would be two pulses in the 20ms period of the 50Hz ac input. The Triac used in the test circuit is the 2N5568, which has a breakdown voltage of 400v, so it is suitable for use with the UK mains voltage. The waveforms obtained when the circuit was simulated are shown in Figure 3-36.

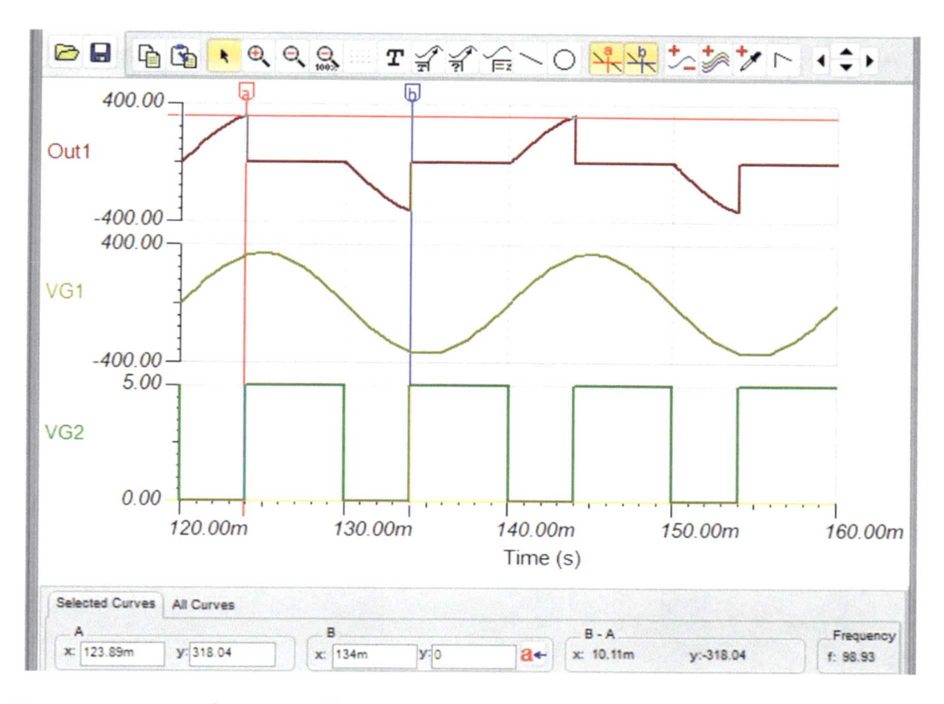


***Figure 3-36.*** *The Waveforms from the Test Circuit Shown in Figure 3-35*

The VG2 trace is the 100Hz square wave used to trigger the Triac to conduct during both the positive and negative halves of the ac input. As the triggering source is a simple square wave, then all we need to do is lengthen the space time of the square wave to reduce the voltage across the load or shorten the space time to increase the voltage. This would not be too difficult, and the relationship between the triggering and the output voltage would be quite simple as, for example, a 50/50 duty cycle triggering square wave would produce half of the output voltage. If the space time was set at 25%, then the output voltage would be at 75%. Sounds good, well do not get carried away. In the test circuit shown in Figure 3-35, the load was purely resistive at 5Ω, that is, the load was 5 +j0. The space time was set to 4ms which is simply 40%, and the output voltage was also 60%. The output voltage is shown in Figure 3-37.

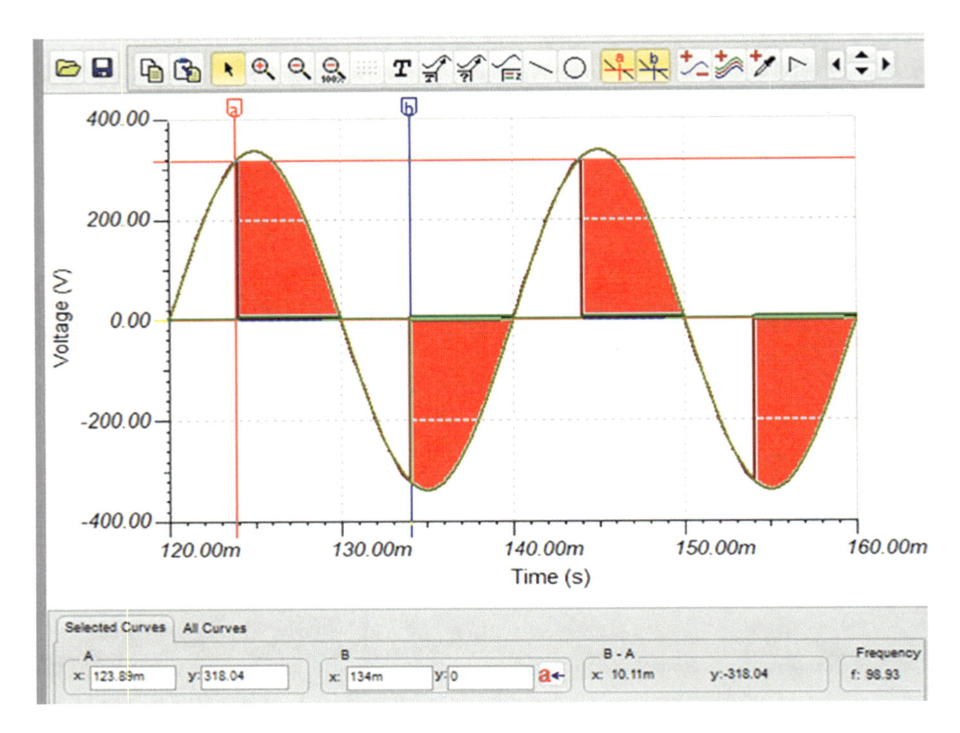


***Figure 3-37.** The Output Voltage with 40% Triggering*

Figure 3-37 shows that the output voltage is at 60%. When we changed the triggering to 60% space time, the average voltage reduced to 40%. This does suggest that this form of triggering could work and give you better control over the amount of voltage that could be developed across the load. It would not be too difficult to produce this type of triggering source; we might find we could use the 555 timer that we will look at in Chapter 10. If we were to add some inductance to the load, then we would probably use a flywheel diode across the load to make it work as though it was a purely resistive load.

Please note I am not saying this is the best triggering circuit. The design of all these types of power electronic circuits will require a lot more work. However, I hope I have given you some insight into what we can do with these analog circuits.

# Summary

I feel I must say that I am not trying to design actual circuits that you could use, as there are many variations of the circuits we could use for all types of electronics we will come across. I am really only trying to introduce you to the major concepts of the different analog circuits and devices we could come across.

In this chapter, we have introduced some of the main semiconductor devices used in power electronics. We have looked at how the SCR, or thyristor as it is sometimes known as, can be used to chop up the ac supply to provide a variable DC voltage for use as a dimmer circuit. We then went on to look at how the Triac could be used to produce a similar variation with an ac voltage. Finally, we looked at how we could use the Diac.

I hope you have found this chapter useful. The way I set some of it out was an attempt to show how you could investigate some concepts and then produce a solution to an idea you have behind the design of a circuit.

In the next chapter, we will move on from the PN junction to the NPN and start our investigation into the bipolar junction transistor, the BJT.

CHAPTER 4

# The NPN and PNP Junctions

In this chapter, we will learn how engineers tried to make use of the PN junction we have looked at in Chapter 2. We will study the NPN and PNP construction and discover how engineers make use of them.

## The NPN Junction

The makeup of the diode, studied in Chapter 2, led engineers to create a semiconductor device that was made up of a P-type material sandwiched between two N-type materials as shown in Figure 4-1.

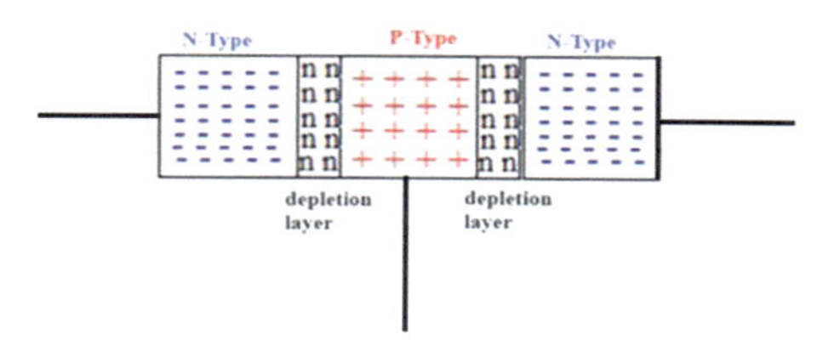


***Figure 4-1.*** *The NPN Sandwich*

Figure 4-1 is a very much simplified representation of the NPN sandwich; however, it does give us the idea behind the analysis in this chapter. The manufacturing process that is used in making these


H. H. Ward, *Mastering Analog Electronics*, Maker Innovations Series,
https://doi.org/10.1007/979-8-8688-0245-4_4

sandwiches is quite remarkable, and it is way beyond me. However, they succeed in making the emitter region a heavily doped region, which, in the case of the NPN sandwich, means there are a lot of atoms that have an excess of electrons in it. The base, which is quite thin, is lightly doped, which, in the case of the NPN sandwich, means there are not a lot of atoms that are short of electrons. Finally, the collector is moderately doped, which means there are atoms that have an excess of electrons but not as many as with the emitter.

The PNP transistor will be doped in the same manner, but the emitter and collector have atoms that are short of electrons, while the base has atoms that have an excess of electrons.

As with the PN junction, when a P-type and an N-type semiconductor material are brought together, there is a depletion layer, where some of the electrons that are surplus to the atoms in the N-type material combine with the atoms that are short an electron in the P-type material. This then creates a layer that has no charge, that is, it is depleted of charge, between the two types of semiconductor material. Now, because we have a sandwich-type arrangement, there will be two such depletion layers created as shown in Figure 4-1.

From our work in Chapter 2, we know that just one PN junction will create a basic diode, with the P-type material acting as the anode and the N-type acting as the cathode; see Figure 2-1 in Chapter 2. Therefore, it shouldn't be too difficult to see that the NPN sandwich shown in Figure 4-1 can be viewed as two diodes back to back as shown in Figure 4-2.

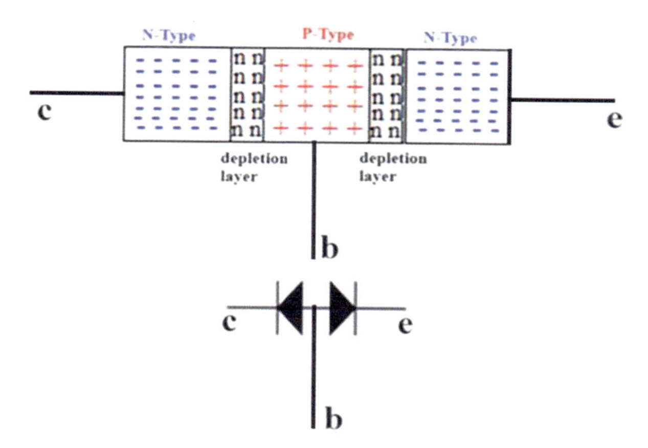


***Figure 4-2.*** *The Two Diodes Back to Back*

Figure 4-2 shows three terminals connected to the NPN sandwich arrangement, and they are

The base "b" terminal

The collector "c" terminal

The emitter "e" terminal

We will learn what they are used for later in this chapter.

We know, from our work in Chapter 2, that if you made the anode of a diode more positive, by about 0.7V, than the cathode, we would be forward biasing the diode. However, if you made the anode more negative than the cathode, you would be reverse biasing it. This would suggest that we could do something similar with this NPN sandwich. A typical biasing arrangement for the NPN sandwich is shown in Figure 4-3.

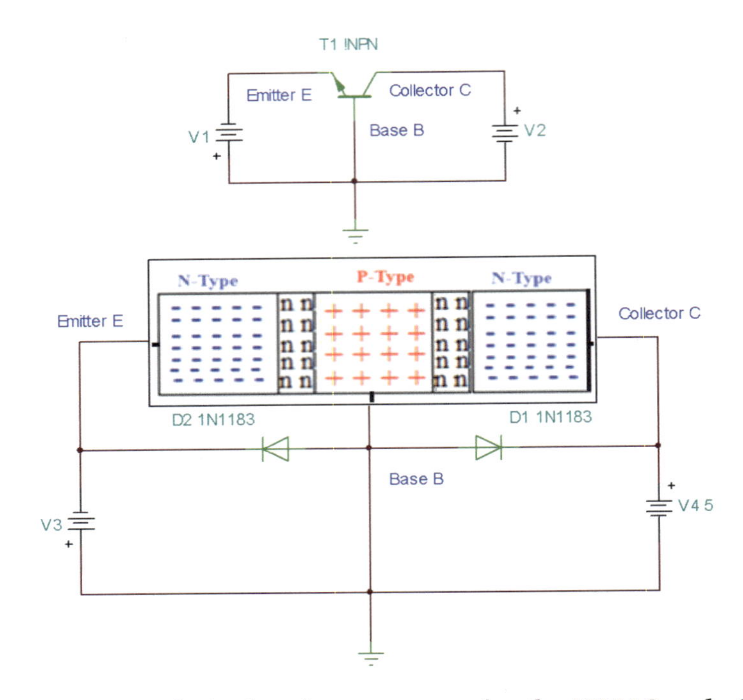


***Figure 4-3.*** *Typical Biasing Arrangement for the NPN Sandwich*

The two diagrams in Figure 4-3 are trying to show how the doping arrangement creates the N-type and P-type regions of the NPN transistor and how it relates to the typical symbol for an NPN transistor. The symbol for the transistor has an arrow between the base and the emitter. The arrow head is used to indicate the direction of conventional current flow through the transistor. With the NPN transistor, the conventional current flows into the collector and out of the emitter as shown in Figure 4-3. However, with a PNP transistor, which we will look at later in this chapter, the conventional current flows into the emitter and out of the collector. Also, we should remember that conventional current flow goes from positive to negative, whereas electron flow goes from negative to positive. In the analysis of the biasing of the transistor, we will use electron flow.

If we look at the NPN sandwich shown in the lower diagram in Figure 4-3, we can see that the voltage source $V_{4,}$ in Figure 4-3 is applying a positive potential to the cathode of the internal diode $D_1$. This would mean that the internal diode, $D_1$, between the base and collector region is reverse biased. Also, the positive terminal of the voltage source $V_3$ is applied to the anode of the internal diode, $D_2$, between the base and emitter region, then that diode, $D_2$, will be forward biased. With the ground, which is normally considered as 0V, attached to the base connection, then the only way the base can be more positive than the emitter is with the emitter being a negative voltage, which this arrangement with the sources V3 and V4 gives us.

This biasing arrangement means the base emitter diode, $D_2$, is closed and allows the surplus electrons in the N-type material, at the emitter, to flow into the depletion region around the base area. However, as the emitter is heavily doped compared to the base region, more electrons flow out of the emitter toward the base than can be absorbed into the base depletion layer. This means there is an excess of electrons that need somewhere to go, but the diode, $D_1$, is reverse biased. Indeed, it is placed in the wrong direction for conventional current to flow through it.

However, we should not forget there is a positive voltage at the collector, provided by V4. Also, this voltage is greater than the voltage at the base, provided by V3. These excess electrons that do not flow into the base region would be highly attracted to this positive voltage at the collector, and so they are trying very hard to flow through $D_1$. But $D_1$ is reverse biased! That is true, but with some clever doping of the NPN sandwich, the diode $D_1$ is a very, very soft diode, and it soon breaks down. This soft diode, $D_1$, soon gives up any resistance to the electrons trying to flow through it and out to the positive potential at the collector. This means this diode, $D_1$, allows this excess of electrons that have been sent out from the emitter region to flow out to the collector region. Indeed, this positive potential at the collector creates an avalanche effect, whereby many more electrons are forced to leave the emitter and flow through the collector. Of course, the biasing voltage, $V_3$, is happy to push more electrons into the NPN sandwich via the emitter terminal. This then sets up the NPN sandwich to allow current to flow through the NPN transistor. If we consider conventional current, we can say it flows down from the collector to the emitter and around the complete biasing circuit to reenter at the collector.

This analysis tries to show you how, by forward biasing the base emitter junction, we have allowed current to flow through the NPN transistor even though the base collector junction was reverse biased. We will see later that this arrangement is a typical one when using the NPN transistor.

## The Symbols and Acronyms of a Transistor

Before we go too far down the line of analyzing these transistors, it may be useful to discuss the various symbols we use for transistors. There are two main ways of symbolizing transistors which are shown in Figure 4-4.

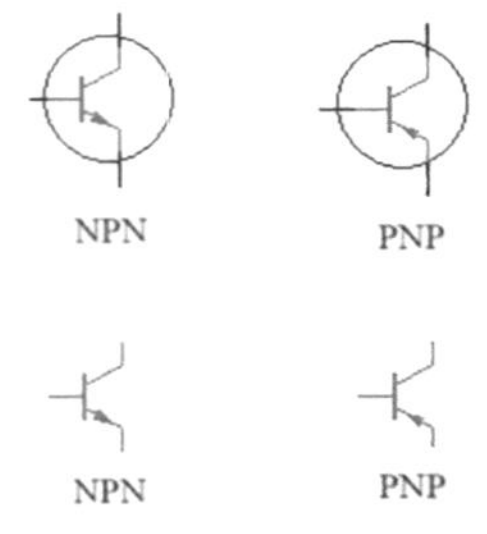


***Figure 4-4.*** *The Symbols for Transistors*

The top two transistors are shown enclosed within a circle. We should always use these symbols when we are discussing circuits that have individual transistors or sections of circuits with just individual transistors in them. The symbols without the circle are used when discussing integrated circuits that use many transistors in them. However, because most ECAD software, and TINA is one of them, don't put circles around their transistors in their schematics, we will only use the lower symbols in our circuits, even though we are discussing individual transistors. So, I apologize in advance for this small error in my representation of circuits with just one transistor in them. However, I do hope you will forgive me for this little discrepancy.

Also, I would like to point out the difference between the NPN and the PNP symbol. The NPN has the arrow pointing "Noutwards," a phrase I was introduced to when I was a student, or outwards, if we are using English, as with the NPN the arrow head is pointing outward from the base to come out at the emitter end of the transistor, whereas the PNP has the arrow head pointing inward; see Figure 4-4.

We can also use the acronym FEN and FEP to explain how we forward bias the base emitter diode inside the transistors. The term FEN stands for Forward Emitter Negative, and it is used with the NPN transistor, meaning to turn the NPN transistor on we must make the emitter around 0.7V more negative than the base.

The term FEP stands for Forward Emitter Positive, and it is used with the PNP transistor, meaning to turn the PNP transistor on we must make the emitter around 0.7V more positive than the base. I hope this helps in analyzing how we use these transistors.

# The Characteristics of a Transistor

To be able to use any device, we need to appreciate the main characteristics of the device. With respect to a transistor, the most important characteristics are

- The input characteristic
- The output characteristic
- The transfer characteristic

We can carry out a series of experiments to determine these characteristics. However, before we do that, we need to appreciate there are three main configurations we can use the basic transistor in, and these are

- Common base
- Common collector
- Common emitter

# The Common Base Configuration

We will look at the common base configuration first. The test circuit we will use to look at the characteristics is shown in Figure 4-5.

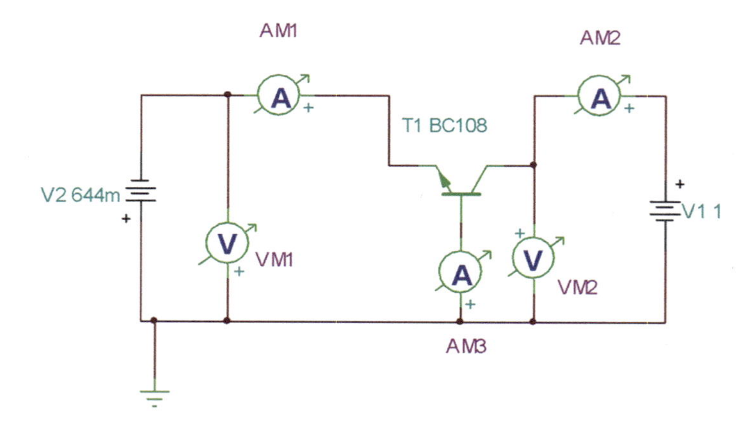


***Figure 4-5.*** *The Common Base Configuration*

This is called the common base configuration because the base terminal is common to both the input, which will be $V_2$, the base-to-base emitter $V_{BE}$ measured by VM1, and the output side, $V_1$, the collector-to-base voltage $V_{CB}$ measured by VM2.

Note, when symbolizing a voltage such as $V_{BE}$, we use the V for voltage, then follow it with the notation for the positive end of the voltage followed by the notation for the negative end. In this way, the term $V_{BE}$ is the voltage between the base and emitter of a transistor with the base being the positive end of the voltage. Also, we use capital letters when referring to DC quantities and lowercase letters when referring to ac quantities.

We can produce a graph of the input characteristics by performing the following experiment:

1. Set the $V_{CB}$ voltage, that is, $V_1$, to 1V and maintain it at that throughout the procedure.
2. Set the $V_{BE}$ voltage, $V_2$, to -250mV.
3. Turn the circuit on and record the emitter current $I_E$ using AM1, the collector current $I_C$ using AM2, and the base current $I_B$ measured using AM3.
4. Then increase the $V_{BE}$ voltage by 50mV and take the same measurements again so that we can complete the table of results shown in Table 4-1.
5. Repeat the procedure until you have used a $V_{BE}$ of 950mV.
6. Now change the $V_{CB}$ voltage to 20V and repeat steps 2 to 5.

Therefore, simulating the circuit to carry out the experiment, we can complete the table of results shown in Table 4-1.

*Table 4-1. Results for Common Base Experiment*

| | $V_{CB}$ = 1v | | | $V_{CB}$ = 20V | | |
|---|---|---|---|---|---|---|
| $V_{BE}$ mV V2 | $I_E$ | $I_B$ | $I_C$ | $I_E$ | $I_B$ | $I_C$ |
| –250 | 747.62f | –250.02f | 997.65f | 27.5p | –4.05p | 31.55p |
| –200 | 797.58f | –249.92f | 1.05p | 27.53p | –4.05p | 31.58p |
| –150 | 847.43f | –249.83f | 1.1p | 27.56p | –4.05p | 31.61p |
| –100 | 899.11f | –247.95f | 1.15p | 27.59p | –4.05p | 31.64p |
| –50 | 954.79f | –243.16f | 1.2p | 27.63p | –4.04p | 31.68p |
| 0 | 1.03p | –226.02f | 1.26p | 27.69p | –4.03p | 31.72p |
| 50 | 1.22p | –164.02f | 1.38p | 27.89p | –3.96p | 31.86p |
| 100 | 2.04p | 61.2f | 1.98p | 28.89p | –3.74p | 32.62p |
| 150 | 6.72p | 881.57f | 5.84p | 34.92p | –2.92p | 37.84p |
| 200 | 36.52p | 3.88p | 32.64p | 74.35p | 82.13f | 74.27p |
| 250 | 235.45p | 14.94p | 220.51p | 341.13p | 11.14p | 329.99p |
| 300 | 1.6n | 56.26p | 1.54n | 2.18n | 52.46p | 2.13n |
| 350 | 11.01n | 214.6p | 10.8n | 14.95n | 210.8p | 14.74n |
| 400 | 76.63n | 848,39p | 75.79n | 104.19n | 844.59p | 103.35n |
| 450 | 535.6n | 3.56n | 532.04n | 729.27n | 3.56n | 725.71n |
| 500 | 3.75u | 16.35n | 3.73u | 5.11u | 16.35n | 5.1u |
| 550 | 26.29u | 83.36n | 26.21u | 35.87u | 83.35n | 35.78u |
| 600 | 184.04u | 470.65n | 183.57u | 251.03u | 470.3n | 250.56u |
| 650 | 1.27m | 2.86u | 1.27m | 1.73m | 2.84u | 1.72m |
| 700 | 8.04m | 17.11u | 8.03m | 10.67m | 16.6u | 10.65m |
| 750 | 36.63m | 80.48u | 36.55m | 45.53m | 72.74u | 45.46m |

(*continued*)

***Table 4-1.*** *(continued)*

| | $V_{CB}$ = 1v | | | $V_{CB}$ = 20V | | |
|---|---|---|---|---|---|---|
| $V_{BE}$ mV V2 | $I_E$ | $I_B$ | $I_C$ | $I_E$ | $I_B$ | $I_C$ |
| 800 | 99.53m | 246.69u | 99.29m | 116.48m | 205.4u | 116.27m |
| 850 | 186.97m | 541.64u | 186.42m | 211.59m | 424.1u | 211.16m |
| 900 | 287.16m | 972.46u | 286.19m | 319.46m | 729.69u | 318.73m |
| 950 | 393.59m | 1.54m | 392.05m | 434.07m | 1.12m | 432.94m |
| 1000 | 502.82m | 2.24m | 500.58m | 552.26m | 1.6m | 550.66m |

Using the table of results shown in Table 4-1, a graph of the input characteristics can be drawn as shown in Figure 4-6.

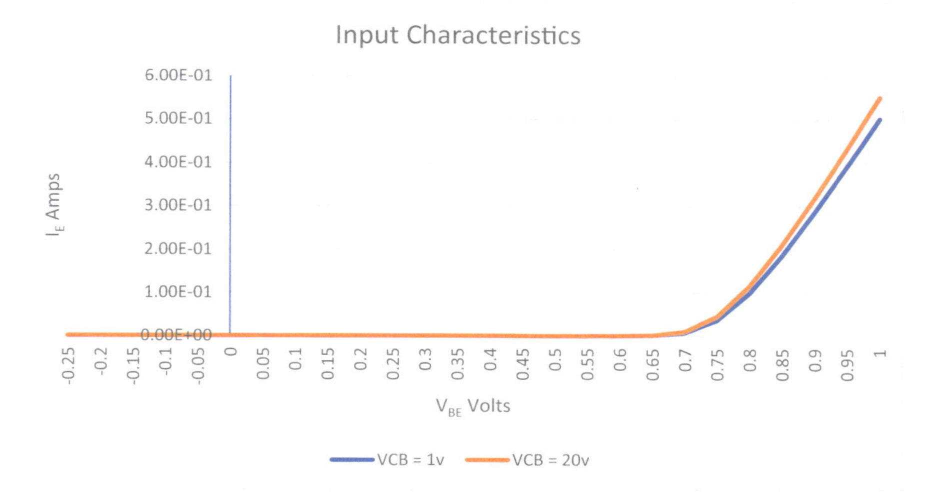


***Figure 4-6.*** *The Input Characteristics for the Common Base Configuration*

The gradient of the graph shown in Figure 4-6 can be determined using

$$Gradient = \frac{Change\ in\ I_E}{Change\ in\ V_{BE}}$$

This has two problems in that the gradient is not constant, and it would calculate a value for the input conductance or admittance, not the input resistance or impedance. This is because, transposing Ohm's law for resistance, we have

$$\frac{I}{V} = \frac{1}{R}$$

The term 1/R is the inverse of resistance, which is conductance or admittance.

This means that to calculate the input resistance, we must use the inverse of the gradient. Therefore, we can calculate the input resistance using

$$Input\ Resistance = \frac{change\ in\ V_{BE}}{change\ in\ I_E}$$

If we use an area where the gradient is very steep, we would get a very low value of resistance as follows:

$$Input\ Resistance = \frac{change\ in\ VEB}{change\ in\ IE} = \frac{0.9 - 0.8}{0.27884 - 0.09779} = 0.55$$

This is the value of the input resistance at that particular tangent on the characteristics. It should be clear that the input resistance does depend upon where we are on this curve and also other parameters such as the VCB and internal characteristics of the transistor. It does show that the input impedance of a transistor in common base configuration is low; a more realistic typical value is in the order of Ohms, say, 10 to 20Ω. We will look at this again in Chapter 6 when we use the Ebers-Moll model to determine this resistance more carefully.

# The Current Characteristics and the Current Gain "Alpha α"

The input resistance or impedance is the important characteristic we can determine from the input characteristics; however, another important consideration we can get from the results of the simulation is the relationship between the three currents measured during the simulation. If we consider the emitter current and collector current in Table 4-1, we can see that once the $V_{BE}$ voltage starts to go positive, then $I_E$ and $I_C$ are very close to each other in value. We can see that $I_E$ is slightly greater than $I_C$. The following expression can be used to express the relationship between the two currents:

$$I_C = \alpha I_E$$

where the term "alpha α" is the current gain between the collector and emitter currents. It is the ratio of the two currents. The expression for alpha is

$$\alpha = \frac{I_C}{I_E}$$

As the collector current, which is the output current, is always going to be less than the emitter current, which is the input current with the common base configuration, then alpha will always be less than one. Typical values for alpha in the common base configuration are 0.95 to 0.99.

If we consider all three currents that are recorded in Table 4-1, we should be able to confirm the following expression:

$$I_E = I_B + I_C$$

If we take the reading when $V_{BE}$ was 750mV, we can add the three currents as follows:

$$0.03663 = 0.00008046 + 0.03655$$

This does confirm the expression for the emitter current $I_E$.

# Analyzing the Test Results from the Circuit in Figure 4-5

The first thing we can see is that even when the base emitter junction is reverse biased, there is a leakage current flowing through the transistor. This internal diode is reverse biased when $V_2$ is negative, which takes a bit of thinking about, but if you simulate the circuit shown in Figure 4-7, you get the same results.

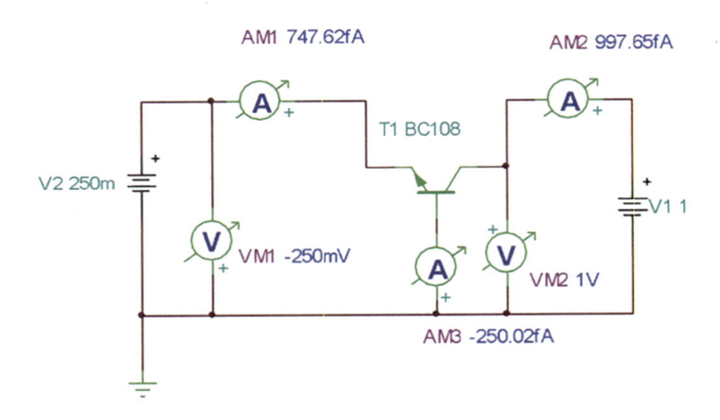


***Figure 4-7.*** *Reverse Biasing D2, the Base Emitter*

We can see that the three currents have the same reading as that shown in Table 4-1 when the VBE voltage was set at –250mV.

Using Table 4-1 again, we should appreciate when V1 is 0v and above; the base emitter diode is now becoming forward biased. The currents in the transistor become more measurable when $V_1$ reaches 400mV. If we now plot the change in the emitter current as we continue to increase the $V_1$ voltage, we will be able to plot the input characteristics for the common base configuration. This is shown in Figure 4-6.

The input to the common base configuration is simply the voltage across the base to emitter and is labeled $V_{BE}$. The other parameter of the input characteristics is the emitter current, $I_E$, that flows out of the emitter.

The graph, shown in Figure 4-6, shows that the current starts to increase when the $V_{BE}$ starts to go above 0.7V. The graph is very similar to the diode characteristics we looked at in Chapter 2. This confirms that the base emitter junction does have a real diode within it as expected.

# The Output Characteristics

The next important characteristic of the transistor is the output characteristic. To create a graph of the output characteristics, we need to carry out the following experiment. We will simulate the circuit shown in Figure 4-5 to gain three series of results so that we can fill in Table 4-2. To get the circuit ready for the three series of simulations, we should set up the circuit as follows:

- With $V_{CB}$ set to 1V, set $V_{BE}$ to 644mV to achieve a value of approximately 1mA for $I_E$. Then, while keeping VBE at 644mV, adjust $V_{CB}$ accordingly so that you can complete Table 4-2.
- Now reset $V_1$ to 1V and set $V_2$ to 673mV to produce an emitter current of 3.03mA. Then, while keeping VBE at 673mV, adjust $V_{CB}$ accordingly so that you can complete Table 4-2.
- Now reset V1 to 1V and set V2 to 707mV to produce an emitter current of 10.02mA. Then, while keeping VBE at 706.5mV, adjust $V_{CB}$ accordingly so that you can complete Table 4-2.

***Table 4-2.** The Table of Results for the Output Characteristics*

| | $I_E$ = 1.02mA | | $I_E$ = 3.03mA | | $I_E$ = 10.02mA | |
|---|---|---|---|---|---|---|
| $V_{CB}$ | $I_E$(mA) | $I_C$(mA) | $I_E$ | $I_C$ | $I_E$ | $I_C$ |
| −0.5 | 0.982 | 0.97 | 2.94 | 2.91 | 9.75 | 9.64 |
| 0 | 0.995 | 0.985 | 2.97 | 2.94 | 9.84 | 9.74 |
| 0.5 | 1 | 0.994 | 3 | 2.97 | 9.93 | 9.82 |
| 1.0 | 1.01 | 1 | 3.03 | 3 | 10.02 | 9.91 |
| 1.5 | 1.02 | 1.01 | 3.05 | 3.02 | 10.1 | 10 |
| 2.0 | 1.03 | 1.02 | 3.08 | 3.05 | 10.19 | 10.08 |
| 2.5 | 1.04 | 1.03 | 3.11 | 3.08 | 10.27 | 10.17 |
| 3.0 | 1.05 | 1.04 | 3.14 | 3.11 | 10.36 | 10.26 |
| 3.5 | 1.06 | 1.05 | 3.17 | 3.14 | 10.44 | 10.34 |
| 4.0 | 1.07 | 1.06 | 3.19 | 3.16 | 10.53 | 10.43 |
| 4.5 | 1.08 | 1.07 | 3.22 | 3.19 | 10.62 | 10.51 |
| 5.0 | 1.09 | 1.08 | 3.25 | 3.22 | 10.7 | 10.6 |
| 5.5 | 1.1 | 1.09 | 3.28 | 3.25 | 10.79 | 10.68 |
| 6.0 | 1.11 | 1.1 | 3.31 | 3.27 | 10.87 | 10.77 |
| 6.5 | 1.12 | 1.11 | 3.33 | 3.3 | 10.95 | 10.85 |
| 7.0 | 1.13 | 1.12 | 3.36 | 3.33 | 11.04 | 10.94 |
| 7.5 | 1.14 | 1.13 | 3.39 | 3.36 | 11.12 | 11.02 |
| 8.0 | 1.15 | 1.14 | 3.42 | 3.39 | 11.21 | 11.1 |
| 8.5 | 1.16 | 1.15 | 3.44 | 3.41 | 11.29 | 11.19 |
| 9.0 | 1.17 | 1.16 | 3.47 | 3.44 | 11.38 | 11.27 |

Using the results as shown in Table 4-2, we can complete the graph of the output characteristics for the common base configuration. This is shown in Figure 4-8.

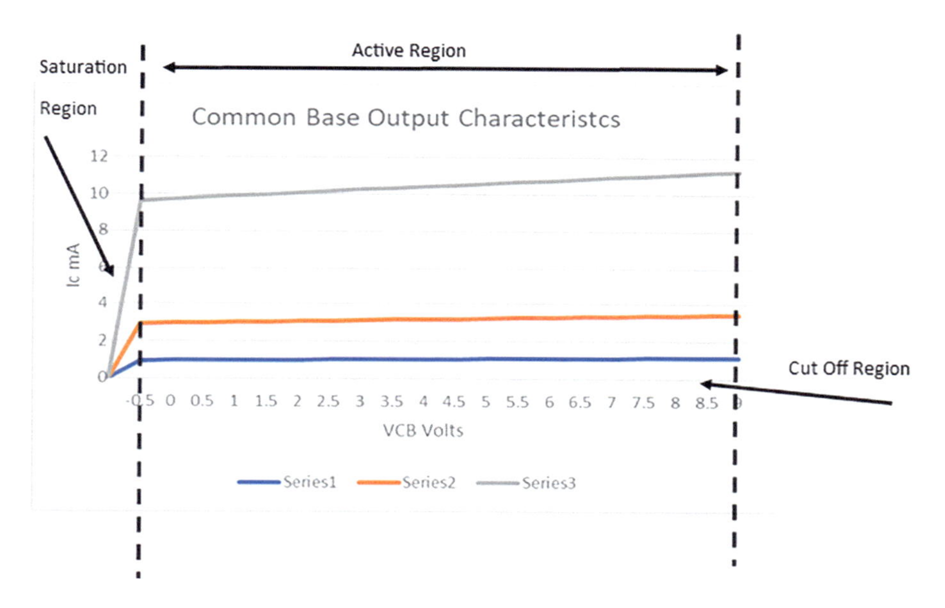


***Figure 4-8.*** *The Output Characteristics for the Common Base Configuration*

When we examine the characteristics, we see that there are three regions the transistors can work in, and they are

- Active region
- Cutoff region
- Saturation region

The active region on the graph is the area between the two dashed lines. In this active region, the collector current does not change much, and it is almost the same as the emitter current.

The cutoff region is the point where the collector current is zero or very close to zero. It is called the cutoff region as the transistor has switched off as no current is flowing through it.

The saturation region is where the collector current has reached a maximum, and so the transistor is said to be saturated in current flow.

# Current Characteristics

The current characteristics allow us to look at the relationships between the currents flowing through the BJT amplifiers. The BJT, in all its configurations, is a current device which normally amplifies its input current; however, that depends upon the type of configuration the BJT is used in. In this section of the text, we are looking at the common base configuration. In this configuration, the input current is the emitter current, $I_E$, while the output current is the collector current, $I_C$. We can state the relationship between the collector current and the emitter current as

$$I_C = \alpha I_E$$

With the output characteristics, we have seen that the collector current is very close to the value of the emitter current, but it is slightly less; examine the results shown in Table 4-2. This would suggest that the term "alpha" $\alpha$ is just less than 1, typically between 0.95 and 0.99.

We can rearrange the expression for $I_C$ to make "$\alpha$" the subject as follows:

$$\alpha = \frac{I_C}{I_E}$$

# The Common Base Amplifier

The most common use of a transistor is to use it as an amplifier. We have shown that the common base configuration does not act as a current amplifier, as the output current, $I_C$, is less than the input current, $I_E$. Therefore, we must investigate whether or not it can be used as a voltage amplifier. To test this concept, we can simulate the circuit shown in Figure 4-9.

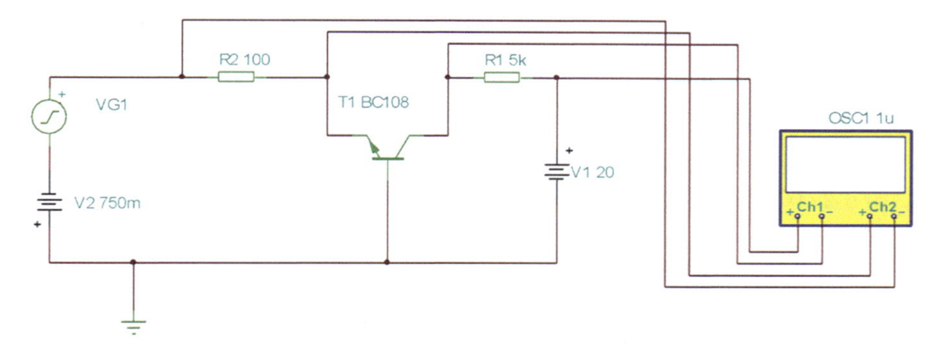


***Figure 4-9.*** *The Common Base Amplifier Circuit*

The input voltage to the circuit is VG1. This is set to give a 5mv peak ac voltage with a frequency of 1kHz; this is just an example; it could be any small voltage at any frequency. This is applied between the emitter and base as this is the input to the amplifier. I am using channel 1 of the oscilloscope to measure the voltage across $R_2$, the emitter resistor. This is because if we divide the voltage waveform by the 100Ω value of $R_2$, we can interpret the trace of channel 1 as the current flowing through the amplifier. In the same way, by using channel 2 to measure the voltage across $R_1$, the collector or load resistor, we can divide that trace voltage by the 5kΩ value and interpret that trace as the output current. The two traces of the oscilloscope are shown in Figure 4-10.

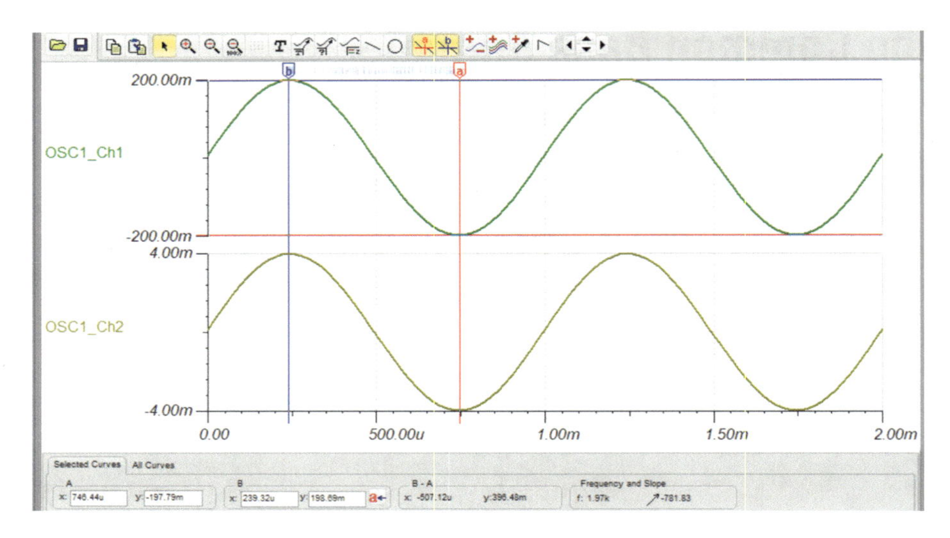


***Figure 4-10.*** *The Two Traces of the Oscilloscope*

The OSC1_Ch1 is the output voltage, and if we divide the peak of 200mV by the 5k value of $R_2$, we get a peak current of 40μA. The trace OSC1_Ch2 is the input voltage, and if we divide the peak of 4mV by the 100Ω value of $R_1$, we get a peak value of 40μA. This confirms what we expect: that the output current, $I_C$, is the same as the input current, $I_E$. It also confirms that in the common base configuration the value of alpha is very close to one.

If we now divide the peak to peak of the output voltage by the peak to peak of the input voltage, we can calculate the voltage gain of the common base amplifier:

$$Voltage\ gain = \frac{peak\ to\ peak\ Output\ Voltage}{peak\ to\ peak\ Input\ Voltage} = \frac{396.48E^{-3}}{7.9E^{-3}} = 49.93$$

Note, gain does not have any units as, with this expression, we have volts divided by volts, and therefore, the units cancel out.

We can change the value of $R_1$ to 1k. We can simulate the circuit again to see how the output changes.

The resultant traces from the oscilloscope are shown in Figure 4-11.

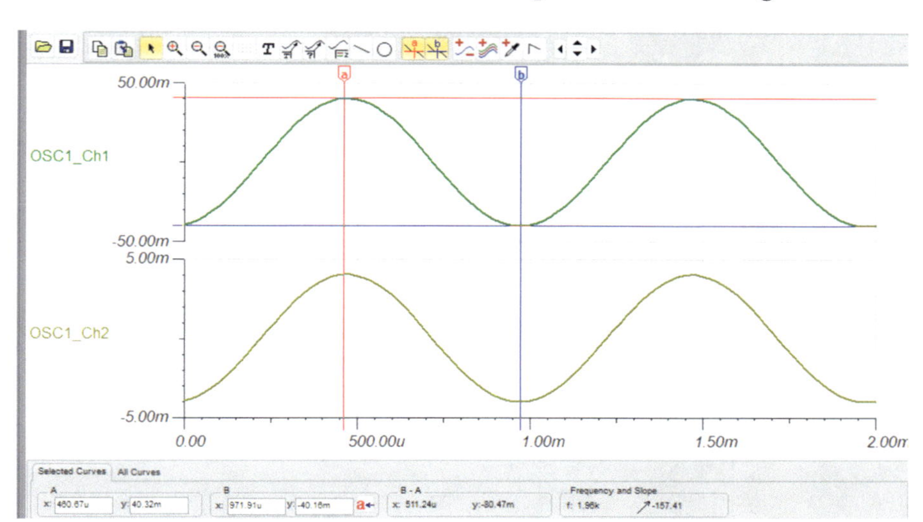


***Figure 4-11.*** *The Traces from the Second Test Simulation*

The peak input voltage is shown as the peak of the OSCI_Ch2 as 4.31mV. This means that the input current would be

$$I_E = \frac{4.04E^{-3}}{100} = 40.4\mu A$$

The peak of the output voltage is 40.24mV as shown as the peak of the Ch1 trace. We can use this to calculate the peak of the output current as follows:

$$I_C = \frac{40.24E^{-3}}{1^3} = 40.24\mu A$$

This agrees with the concept that $I_E = I_C$. We can use these two current values to calculate the current gain, alpha α, of the circuit as follows:

$$\alpha = \frac{I_C}{I_E} = \frac{40.24E^{-6}}{40.4E^{-6}} = 0.996$$

This is a typical value for alpha.

We can calculate the voltage gain as follows:

$$VGain = \frac{VoutPk - Pk}{VinPk - Pk} = \frac{80.47E^{-3}}{8.06E^{-3}} = 9.98$$

We will try one more simulation this time with the emitter resistor $R_2$ set to 5k and the load or collector resistor set to 100k. The traces of the output are shown in Figure 4-12.

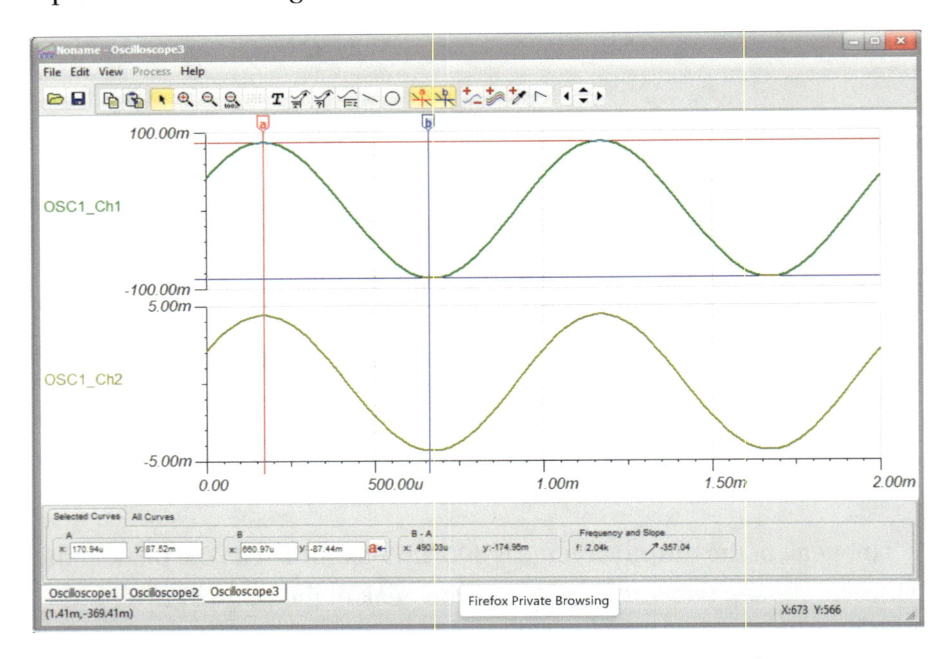


***Figure 4-12.*** *The Oscilloscope Traces of the Third Simulation*

The peak voltage at the input is 4.39v, and with $R_2$ set to 5k, the input current is

$$I_E = \frac{4.39E^{-3}}{5E^{-3}} = 878nA$$

The peak voltage at the output is 87.54mV, and with $R_1$ set to 100k, the output current is

$$I_C = \frac{87.54E^{-3}}{100E^{3}} = 874nA$$

The current gain α is

$$\alpha = \frac{I_C}{I_E} = \frac{874E^{-9}}{878E^{-9}} = 0.994$$

The voltage gain can be calculated as follows:

$$VGain = \frac{Vout}{Vin} = \frac{87.54E^{-3}}{4.39E^{-3}} = 19.9$$

These simulations do confirm the theory in that the emitter current $I_E$ is very nearly the same as the collector current $I_C$, and the current gain α is typically between 0.95 and 0.99. The voltage gain can be good; the examples put it between 5 and 40, but higher voltage gains can be achieved. If we look at the resistor values and the calculated gain, we could say that

$$Vgain \oplus \frac{The\ Load\ Resistor}{The\ Emitter\ Resistor}$$

There are always some limitations, and one that I have determined is that due to the low input resistance, the value of the emitter resistor should not be too low.

# The Common Base Leakage Current (ICBO)

This stands for the current that flows through the collector when the emitter is open circuited, where "I" stands for the DC current and the CB indicates it flows through the collector to base junction in that direction. This means it is the reverse leakage current flowing through $D_1$. If we are to fully understand the total current flow through the transistor, although it is not essential that you do, then we need to consider the $I_{CBO}$. This can be tested using TINA with the circuit shown in Figure 4-13. This, with respect to the output characteristics, is when we are working in the cutoff region with the emitter set to zero. This is when the input is open circuited as shown in Figure 4-13. In this arrangement, we would not expect any collector current to flow through the transistor. This is because the base collector diode, D1 in Figure 4-3, is reverse biased. However, as the simulation shows, there is a very small current that flows in at the collector and out of the base. The base current, measured by AM3 in Figure 4-13, has the same value as the collector current and, as it is shown as –251.39fA, then it is flowing out of the base terminal. The "f" in 251.39fA stands for femto, which stands for $10^{-15}$, a very, very small value.

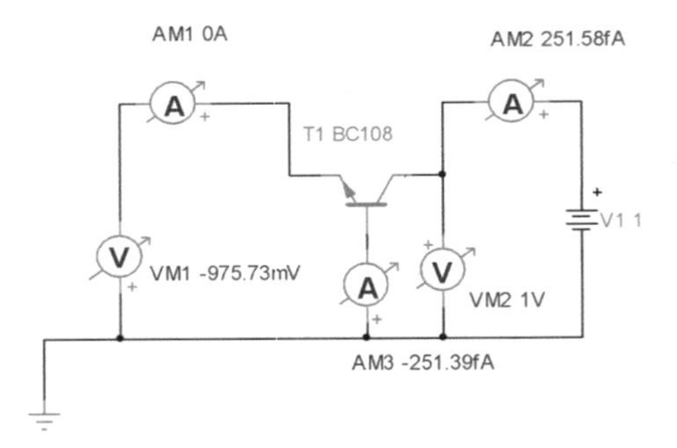


***Figure 4-13.*** *The Test Circuit for ICBO*

With $V_1$ set to 1V, the leakage current, measured by AM2, was 251.58fA. This current flows into the cathode of $D_1$ and out of the anode, then out of the base terminal. If $V_1$ was reduced to 0.1V, the values of the currents were all reduced by a factor of 10. If $V_1$ was increased to 10V, then all the currents were increased by a factor of 10, showing that this reverse leakage current, ICBO, has a linear relationship. This current is the reverse leakage current of D1, the internal base collector diode. It is this diode starting to break down due to the voltage at the collector terminal of the transistor.

If we now consider this leakage current, then a more complete expression for the total collector current, which we will label $I_{CT}$, will be

$$I_{CT} = I_C + I_{CBO}$$

However, because this current is very small, we can ignore it in our normal analysis of the BJT. I really just wanted to make you aware that it does exist.

## The Gain Terms "Alpha α" and "Beta β"

We should now consider these two gain terms so that we can show how the terms are related to each other. The term alpha $\alpha$ is the ratio of the collector current $I_C$ and the emitter current $I_E$ as we have already stated. The term $\beta$ is the ratio of the collector current $I_C$ and the base current $I_B$. This means that

$$\beta = \frac{I_C}{I_B}$$

We have stated earlier that the emitter current $I_E$ is the sum of the base current and the collector current:

$$I_E = I_C + I_B$$

We have also stated that

$$I_C = \alpha I_E$$

We can use these two expressions to develop a relationship between "$\alpha$" and "$\beta$" as follows:

If we substitute $I_C + I_B$ for $I_E$, we get

$$I_C = \alpha\left(I_C + I_B\right) = \alpha I_C + \alpha I_B$$

Therefore, we can say

$$I_C - \alpha I_C = \alpha I_B$$

From which we can say

$$I_C(1-\alpha) = \alpha I_B$$

Therefore

$$I_C = \frac{\alpha}{1-\alpha} I_B$$

Therefore

$$\frac{I_C}{I_B} = \frac{\alpha}{1-\alpha}$$

However, from before we know

$$\frac{I_C}{I_B} = \beta$$

Therefore, we can say

$$\beta = \frac{\alpha}{1-\alpha}$$

We can use three results from Table 4-1 to get an average value for β as follows:

First result when $V_{BE}$ = 700mv

| | | | |
|---|---|---|---|
| 700 | $I_E$ = 8.04m | $I_B$ = 17.11u | $I_C$ = 8.03m |

$$\beta = \frac{8.03E^{-3}}{17.11E^{-6}} = 469.32$$

Second result when $V_{BE}$ = 750mv

| 750 | $I_E$ = 36.63m | $I_B$ = 80.48u | $I_C$ = 36.55m |
|---|---|---|---|

$$\beta = \frac{36.55E^{-3}}{80.48E^{-6}} = 454.09$$

Third result when $V_{BE}$ = 800mv

| 800 | $I_E$ = 99.53m | $I_B$ = 246.69u | $I_C$ = 99.29m |
|---|---|---|---|

$$\beta = \frac{99.29E^{-3}}{246.69E^{-6}} = 402.489$$

Using these three values gives the average value of β as being

$$average\ \beta = \frac{469.32 + 454.09 + 402.489}{3} = 441.97$$

If we now look at the results when $V_{BE}$ = 850 mv

| 850 | $I_E$ = 186.97m | $I_B$ = 541.64u | $I_C$ = 186.42m |
|---|---|---|---|

$$\beta = \frac{186.42E^{-3}}{541.64E^{-6}} = 344.18$$

These results show that the value of β is not constant, and it does fall off as the collector current rises. If we look at the datasheet for the BC108, we will see that this is what the datasheet shows as there is a graph that shows how β varies when IC varies.

However, if we use the average value for β at 442, then we can try and calculate a value for α using

$$\beta = \frac{\alpha}{1 - \alpha}$$

This can be rearranged for "α" as

$$\alpha = \frac{\beta}{1 + \beta} = \frac{442}{1 + 442} = 0.998$$

This agrees closely to our previous calculations for α.

This work confirms that in the common base configuration we can show that

$$\beta = \frac{\alpha}{1-\alpha}$$

$$\alpha = \frac{\beta}{1+\beta}$$

Now we can look at the other configurations for the BJT.

# The Common Emitter Configuration

This configuration is the most commonly used configuration that transistors are used in. With this configuration, the emitter is common to both the input and the output. The input is applied to the base of the transistor, and the output is taken from the collector.

We will use the circuit shown in Figure 4-14 to look at the input and output characteristics.

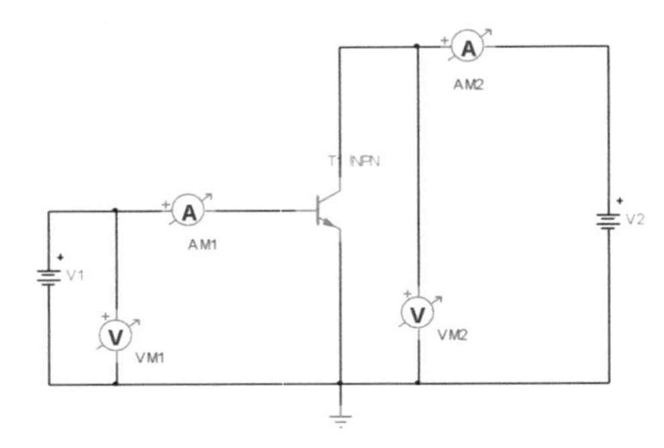


***Figure 4-14.*** *The Common Emitter Configuration*

This is called the common emitter because the emitter is common to the input, which is $V_{BE}$, V1, applied across the base to emitter junction, and the output $V_{CE}$, V2, which is taken from across the collector and emitter terminals. If we consider the two internal diodes that make up the NPN transistor, we can see, from Figure 4-14, that the base emitter junction diode is forward biased and the base collector diode is reverse biased. This is similar to the common base configuration we have just looked at.

Knowing that, the arrow pointing out of the emitter indicates the direction of conventional current flow through the transistor from the collector down through the transistor and out at the emitter. We normally use the collector current $I_C$ as the output current, even though this is in the reverse direction with respect to the internal diode between the base and collector. Also, as the base emitter diode is forward biased, the base current $I_B$ will be the input current. This means that the input parameters are $V_{BE}$ and $I_B$, while the output parameters are $V_{CE}$ and $I_C$.

## The Input Characteristics of the Common Emitter Configuration

To plot these, we will measure the input current $I_B$ while varying the input voltage $V_{BE}$ for a constant output voltage $V_{CE}$. We will run three separate simulations, the first when $V_{CE}$ was set to 1V, the second when $V_{CE}$ was set to 10V, and the third when $V_{CE}$ was set to 20V. This is so that we can see what effect the $V_{CE}$ voltage has on the current flowing through the transistor.

This is done by simulating the circuit shown in Figure 4-14 and completing the table of results shown in Table 4-3.

*Table 4-3.* *The Common Emitter Input Characteristics Results*

| $V_{CE}$ | 1V | | | 10V | | | 20V | | |
|---|---|---|---|---|---|---|---|---|---|
| $V_{BE}$ mV | $I_B$ | $I_E$ | $I_C$ | $I_B$ | $I_E$ | $I_C$ | $I_B$ | $I_E$ | $I_C$ |
| –100 | –267.95f | 1p | 1.27p | –2.07p | 11.96p | 14.02p | –4.07p | 27.77p | 31.84p |
| –50 | –253.16f | 1.01p | 1.26p | –2.05p | 11.93p | 13.99p | –4.05p | 27.72p | 31.78p |
| 0 | –226.02f | 1.03p | 1.26p | –2.03p | 11.93p | 13.96p | –4.03p | 27.69p | 31.72p |
| 50 | –154.02f | 1.17p | 1.32p | –1.95p | 12.06p | 14.01p | –3.95p | 27.8p | 31.76p |
| 100 | 81.2f | 1.93p | 1.85p | –1.72p | 12.89p | 14.61p | –3.72p | 28.71p | 32.43p |
| 150 | 911.57f | 6.55p | 5.64p | –888.43f | 18.13p | 19.92p | –2.89p | 34.64p | 37.53p |
| 200 | 3.92p | 36.19p | 32.27p | 2.12p | 52.32p | 50.2p | 122.13f | 73.88p | 73.76p |
| 250 | 14.99p | 224.15p | 219.16p | 13.19p | 282.4p | 269.21p | 11.19p | 339.65p | 328.46p |
| 300 | 56.32p | 1.59n | 1.53n | 54.52p | 1.86n | 1.81n | 52.52p | 2.17n | 2.12n |
| 350 | 214.67p | 10.94n | 10.72n | 212.87p | 12.8n | 12.59n | 210.87p | 14.88n | 14.67n |
| 400 | 848.47p | 76.05n | 75.21n | 846.67p | 89.11n | 88.26n | 844.67p | 103.61n | 102.77n |
| 450 | 3.56n | 531.02n | 527.45n | 3.56n | 622.75n | 619.19n | 3.56n | 724.69n | 721.13n |
| 500 | 16.35n | 3.72u | 3.7u | 16.35n | 4.36u | 4.34u | 16.35n | 5.08u | 5.06u |
| 550 | 83.36n | 26.02u | 25.93u | 83.35n | 30.55u | 30.47u | 83.35n | 35.59u | 35.51u |

| | | | | | | | | | |
|---|---|---|---|---|---|---|---|---|---|
| 600 | 470.66n | 181.92u | 181.45u | 470.49n | 213.67u | 213.2u | 470.31n | 248.92u | 248.45u |
| 650 | 2.86u | 1.25m | 1.25m | 2.85u | 1.47m | 1.47m | 2.84u | 1.71m | 1.71m |
| 700 | 17.13u | 7.94m | 7.93m | 16.88u | 9.21m | 9.19m | 16.62u | 10.57m | 10.55m |
| 750 | 80.83u | 36.24m | 36.16m | 76.87u | 40.67m | 40.6m | 73.01u | 45.21m | 45.13m |
| 800 | 248.85u | 98.72m | 98.47m | 226.71u | 107.37m | 107.15m | 206.82u | 115.84m | 115.64m |
| 850 | 548.53u | 185.67m | 185.12m | 483.73u | 198.46m | 197.98m | 428.21u | 210.63m | 210.2m |
| 900 | 987.93u | 285.31m | 284.33m | 851.87u | 302.32m | 301.46m | 738.46u | 318.16m | 317.42m |
| 950 | 1.57m | 391.1m | 389.53m | 1.33m | 412.65m | 411.32m | 1.14m | 432.38m | 431.24m |
| 1000 | 2.29m | 499.56m | 497.28m | 1.92m | 526.15m | 524.22m | 1.63m | 550.12m | 548.49m |

We can use these results to plot the input characteristics of the common emitter configuration. This is shown in Figure 4-15.

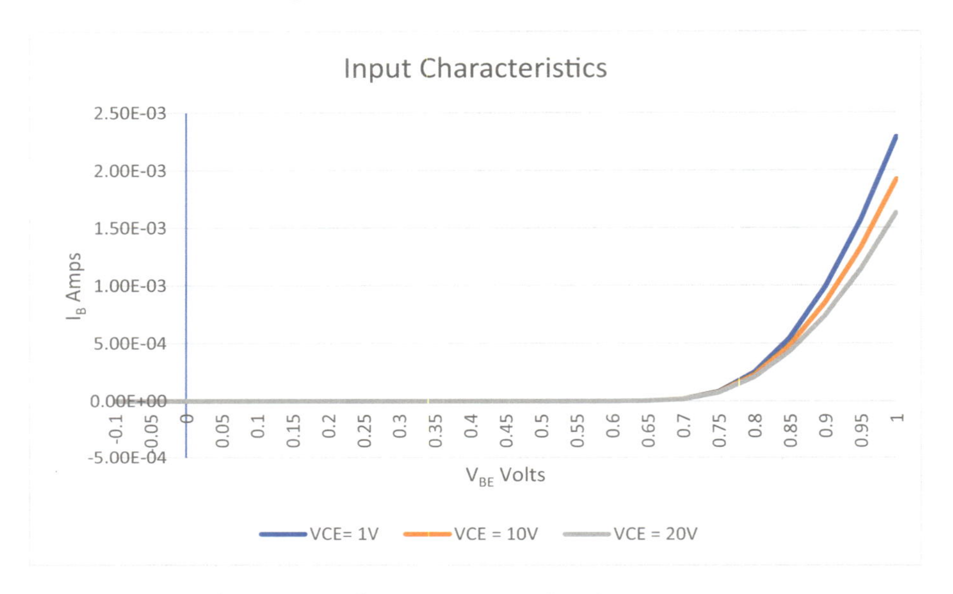


***Figure 4-15.*** *The Input Characteristics for the Common Emitter Configuration*

These characteristics are similar to that of the common base configuration. We can determine the input resistance or impedance by calculating the inverse of the gradient of the curve. Just as with the common base configuration, we can see that the gradient is not constant. We can also see that it will be quite low in value, as the gradient is steep and getting steeper as the voltage $V_{BE}$ rises.

The characteristics show that as the $V_{CE}$ voltage rises, the base current reduces, which is the opposite to what the emitter and collector currents do. There is a lot of discussion as to why this happens, some of which talk about minority and majority carriers and holes moving. I try to think of it in a more practical way. I am not an atomic scientist, and I don't live in the tiny world of electrons, so I can't say which theory is right or wrong; you must decide for yourself.

My explanation of why the base current reduces when the $V_{CE}$ voltage rises relies on the concept that a voltage is a force; indeed, it used to be called an EMF, which stands for ElectroMotive Force. Voltage is the force that makes electrons move through conductors and changes their random movement into what constitutes current flow, that is, electrons moving in a uniform direction. Also, if enough force is applied, then most things will break down. So, it is with the diode between the base and collector. I have stated that it is a weak diode, as the doping is very light. We have seen that there is already a breakdown current, ICBO, even when the emitter base is open circuited as shown in Figure 4-13. Also, we have seen in Chapter 2 that any diode can be forced to work in the reverse region. When this happens, the current flowing through the diode can vary a great deal, without a lot of change in the voltage across the diode. Therefore, I think that is what is happening now as when $V_{CE}$ is increased, we increase the breakdown force on the base collector diode. This then enables this diode to allow more current to flow through it. This means that some of the electrons that are pushing to flow out through the base find an easier path to leave this congested area via the collector. The area around the base area has a congestion of electrons because the emitter region is very highly doped, and so when the base emitter diode is forward biased, the emitter can flood this area with an excess of electrons more than the lightly doped area of the base can cope with. This leaves electrons around the base region struggling to find a path out. When we raise the VCE voltage, these electrons find they can leave the area easier through the base to the collector diode that has just been broken down.

I like to think that this explanation makes sense, but I am not saying it is the most correct. You should and will consider different explanations and decide which one you think fits it best. It is not essential for you to know exactly what is happening as you need to know how to use these transistors. We will look at how to use these transistors later in this chapter.

## The Output Characteristics of the Common Emitter Configuration

To plot the output characteristics, we will carry out a similar experiment as we did with the common base configuration.

- Firstly, set $V_{CE}$ to 1V and set $V_{BE}$ to 684.7mV to achieve a base current of 10.02μA. Then vary the output voltage $V_{CE}$ while keeping $V_{BE}$ set to 6484.7mV and record the measurements so that you can complete Table 4-4.
- Now reset $V_{CE}$ to 1V and set $V_{BE}$ to 704.6mV to set the base current to 20.04uA. Then while keeping $V_{BE}$ at 704.6mV, vary the $V_{CE}$ voltage and so complete the table.
- Now reset $V_{CE}$ to 1V and set $V_{BE}$ to 726.78mV to set the base current to 40.04uA. Then while keeping $V_{BE}$ at 726.78mV, vary the $V_{CE}$ voltage and so complete the table.

***Table 4-4.*** *The Common Emitter Output Characteristics*

| | $I_B$ Set to 10μA | | | $I_B$ Set to 20μA | | | $I_B$ Set to 40μA | | |
|---|---|---|---|---|---|---|---|---|---|
| $V_{CE}$ | $I_B$ | $I_E$ | $I_C$ | $I_B$ | $I_E$ | $I_C$ | $I_B$ | $I_E$ | $I_C$ |
| 0 | 743.23u | 93.1u | –650.13u | 1.29m | 242.21u | –1.04m | 2.14m | 566.85u | –1.58m |
| 0.5 | 10.02u | 4.56m | 4.55m | 20.06u | 9.23m | 9.21m | 41.35u | 18.89m | 18.85m |
| 1 | 10.01u | 4.65m | 4.64m | 20.03 | 9.39m | 9.37m | 41.28u | 19.05m | 19m |
| 1.5 | 10.01u | 4.69m | 4.68m | 20.01u | 9.47m | 9.45m | 41.21u | 19.2m | 19.16m |
| 2 | 10u | 4.73m | 4.72m | 19.99u | 9.55m | 9.53m | 41.13u | 19.35m | 19.31m |
| 2.5 | 10u | 4.78m | 4.77m | 19.97u | 9.64m | 9.62m | 41.06u | 19.5m | 19.46m |
| 3 | 9.99u | 4.82m | 4.81m | 19.95u | 9.72m | 9.7m | 40.99u | 19.66m | 19.62m |
| 3.5 | 9.99u | 4.86m | 4.85m | 19.93u | 9.8m | 9.78m | 40.92u | 19.81m | 19.77m |
| 4 | 9.98u | 4.9m | 4.89m | 19.91u | 9.88m | 9.86m | 40.85u | 19.96m | 19.92m |
| 4.5 | 9.98u | 4.95m | 4.94m | 19.89u | 9.96m | 9.94m | 40.78u | 20.11m | 20.07m |
| 5 | 9.97u | 4.99m | 4.98m | 19.88u | 10.04m | 10.02m | 40.71u | 20.26m | 20.22m |
| 5.5 | 9.97u | 5.03m | 5.02m | 19.86u | 10.12m | 10.1m | 40.64u | 20.41m | 20.37m |
| 6 | 9.96u | 5.07m | 5.06m | 19.84u | 10.2m | 10.18m | 40.57u | 20.56m | 20.53m |
| 6.5 | 9.96u | 5.12m | 5.11m | 19.82u | 10.29m | 10.27m | 40.5u | 20.71m | 20.67m |
| 7 | 9.95u | 5.16m | 5.15m | 19.8u | 10.37m | 10.35m | 40.43u | 20.86m | 20.82m |
| 7.5 | 9.95u | 5.2m | 5.19m | 19.78u | 10.45m | 10.43m | 40.36u | 21m | 20.96m |
| 8 | 9.94u | 5.24m | 5.23m | 19.77u | 10.53m | 10.51m | 40.29u | 21.15m | 21.11m |
| 8.4 | 9.94u | 5.28m | 5.27m | 19.75u | 10.61m | 10.59m | 40.23u | 21.3m | 21.26m |
| 9 | 9.93u | 5.33m | 5.32m | 19.73u | 10.69m | 10.67m | 40.17u | 21.45m | 21.41m |
| 9.5 | 9.93u | 5.37m | 5.36m | 19.71u | 10.77m | 10.75m | 40.09u | 21.59m | 21.55m |

Using the results shown in Table 4-4, we can plot a graph of the output characteristics as shown in Figure 4-16.

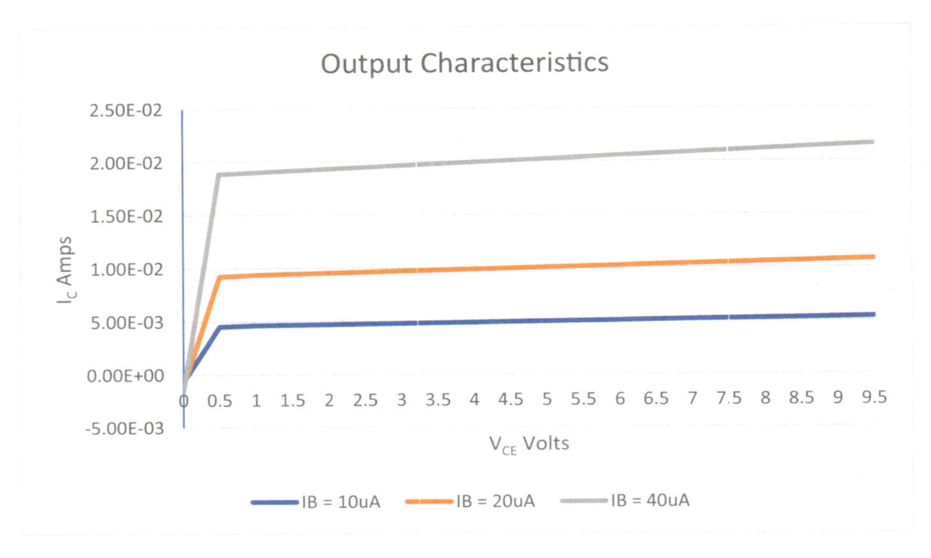


***Figure 4-16.*** *The Output Characteristics for the Common Emitter*

The main regions of this graph are the active and saturation regions. The active region is when the collector current has just passed the knee point with $V_{CE}$ around 0.6V and out to where $V_{CE}$ has reached 9.5V and beyond. Over this region, we can see that the collector current does increase slightly, which means the gradient is slightly more than that with the common base configuration. This means that the output impedance is less than that of the common base, but it is still very high. We can use the same type of expressions for the relationship between the three currents, that is

$$I_E = \alpha I_C$$

Therefore

$$\alpha = \frac{I_C}{I_E}$$

Also

$$I_C = \beta I_B$$

Therefore, we can say

$$\beta = \frac{\alpha}{1-\alpha}$$

Also

$$\alpha = \frac{\beta}{1+\beta}$$

Taking some values from Table 4-4, we have the following:

When $V_{CE}$ = 4.5V and IB = 10μA, we have

| 4.5 | 9.98u | 4.95m | 4.94m |
|---|---|---|---|

$$\beta = \frac{4.9E^{-3}}{9.98E^{-6}} = 490.98$$

When $V_{CE}$ = 4.5V and IB = 20μA, we have

| 19.89u | 9.96m | 9.94m |
|---|---|---|

$$\beta = \frac{9.94E^{-3}}{19.89E^{-6}} = 499.75$$

When $V_{CE}$ = 4.5V and IB = 40μA, we have

| 40.78u | 20.11m | 20.07m |
|---|---|---|

$$\beta = \frac{20.07E^{-3}}{40.78E^{-6}} = 492.15$$

The average of these three values gives an average value for β of

$$Average\ \beta = \frac{490.98 + 499.75 + 492.15}{3} = 494.29$$

Using this value, "b," the value for α can be calculated using

$$\alpha = \frac{494.29}{1 + 494.29} = 0.998$$

We can confirm this result using

| | | | |
|---|---|---|---|
| 4.5 | 9.98u | 4.95m | 4.94m |

$$\alpha = \frac{4.94E^{-3}}{4.95E^{-3}} = 0.998$$

## The Common Emitter Amplifier

The most common use for a transistor is as an amplifier, and so it is with the common emitter configuration; it can be used as an amplifier. We will look at a basic circuit to use the common emitter configuration in this way. This is shown in Figure 4-17.

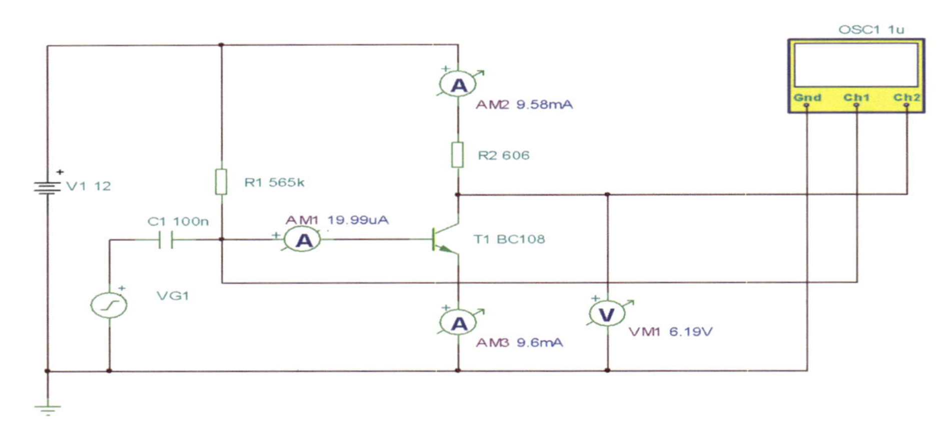


***Figure 4-17.*** *The Basic Common Emitter Amplifier*

When designing this circuit, there are basically two aspects we need to consider. The first aspect is designing the DC quiescent operating point. To design this, we need to appreciate that there are three main DC voltages we need to consider, and they are as follows.

**VCC is the common collector voltage.**

This is basically the supply voltage applied to the transistor. It is from this power supply that the transistor must get the power to amplify the signal applied to its input.

**VCE is the voltage between the collector and the emitter.**

In this basic arrangement, this $V_{CE}$ is the voltage at the collector with respect to the ground. If we want the output to be a true amplified reflection of the input, then ideally this $V_{CE}$ voltage should be half that of $V_{CC}$. This is to allow the output voltage to swing as far positive until it reaches $V_{CC}$ as far as it can swing negative until it reaches the ground voltage. This can be achieved by biasing the transistor correctly.

**VBE is the voltage at the base terminal.**

When the transistor is biased to operate the transistor in the active region, this voltage will be clamped to the forward diode volt drop of approximately 0.7V due to the internal base emitter diode inside the transistor. This assumes that the emitter is grounded as it is in the circuit shown in Figure 4-17.

We will use Ohm's law to calculate the values of $R_1$ and $R_2$ in Figure 4-17, assuming we want a base current of 20μA, and the transistor has an average β of 494.29 that we have obtained from the results in Table 4-4.

If we set $V_{CC}$ to 12v, then we can calculate the value of $R_1$ as follows.

We know the base voltage will be clamped to around 0.7V; this is the $V_{BE}$ in our calculations. We can use the following expression to determine the value of $R_1$:

$$R1 = \frac{V_{CC} - V_{BE}}{I_B} = \frac{12 - 0.7}{20E^{-6}} = 565k\Omega$$

$R_1$ is there to set this base current, and we know this should now produce a base current of around 20μA. We also know that the collector current will be β times greater than this base current $I_B$. Therefore, using

$$I_C = \beta I_B$$

Then

$$I_C = 494.29 \times 20E^{-6} = 9.886mA$$

Knowing that we want a $V_{CE}$ of around 6V, that is, half of $V_{CC}$, we can calculate the value of $R_2$ using

$$R2 = \frac{V_{CC} - V_{CE}}{I_C} = \frac{12 - 6}{9.886E^{-3}} = 606.92\Omega$$

The circuit shown in Figure 4-17 was simulated, and the DC currents and voltages were recorded. The capacitor $C_1$, which is placed in series with the signal source, is used to block any DC shift from the signal source, affecting the DC quiescent biasing voltage at the base of the transistor.

The base and collector currents were measured to be close to those calculated, and the $V_{CE}$ voltage was close to half $V_{CC}$ at 6.19V. This confirms that our calculations work well.

In the next part of the simulation, the signal was connected to the base to give the transistor a signal to amplify. The input and output waveforms were measured using the oscilloscope. The two waveforms obtained are shown in Figure 4-18.

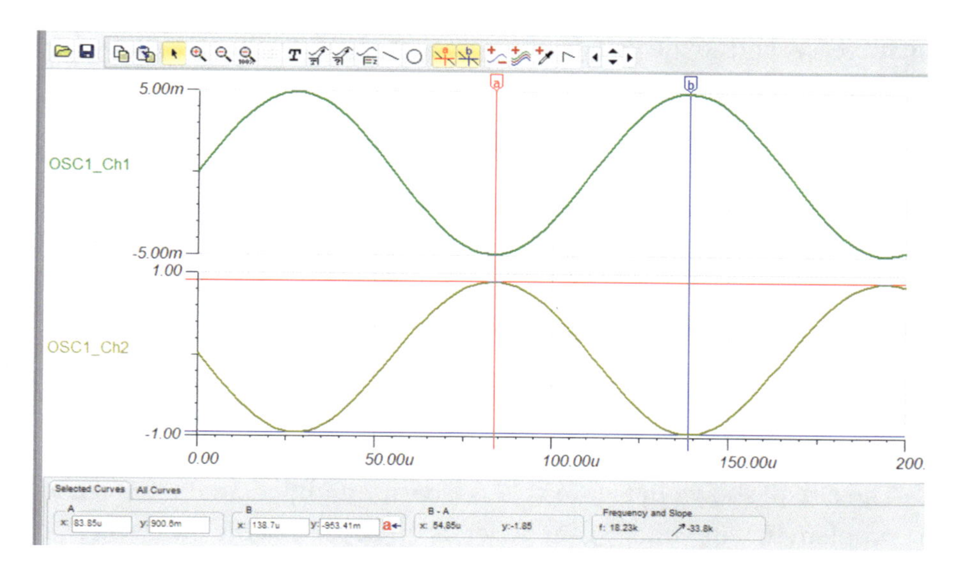


***Figure 4-18.*** *The Input and Output Traces for the Common Emitter Amplifier*

The input voltage is measured by the Ch1 trace. It has a peak-to-peak voltage of 10mV as expected. The output voltage is measured by the Ch2 trace. It has a peak-to-peak voltage of 1.85V. We can use these two values to calculate the voltage gain as follows:

$$Vgain = \frac{Vout\ Peak\ to\ Peak}{Vin\ PeakPeak\ to\ Peak} = \frac{1.85}{10E^{-3}} = 185$$

This is a good voltage gain. We can also see from the two traces that there is a phase difference between the input and the output in that when the input goes high the output voltage goes low and vice versa. This can be explained as follows:

When the base voltage at the input rises, then the input current $I_B$ must rise. As the output current $I_C$ is proportional to the base current, then $I_C$ will increase. This means that the current flowing through $R_2$, the load resistor, will increase, and so the volt drop across that resistor will increase. This, in turn, means the voltage $V_{CE}$ will decrease as

$$V_{CE} = V_{CC} - I_C R2$$

Then, when the voltage at the base reduces, the base current $I_B$ will reduce. This means that the collector current $I_C$ will reduce. This means that the volt drop across $R_2$ will decrease, and so $V_{CE}$ will increase. This inversion across the transistor is one reason that amplifiers may use two stage amplifiers in their circuitry. However, the phase shift does have a frequency dependency, and we need to take this into account when we design our amplifiers.

The circuit shown in Figure 4-17 is a basic amplifier circuit, but it has the problem that its voltage gain depends upon the β of the transistor. The problem is that if the transistor has to be changed, then the gain of the circuit could change as the β of a transistor can vary from 50 to 500, especially if we use a different transistor to the BC108 shown in Figure 4-17. The problem could still be a problem if we stayed with the BC108 as these transistors are manufactured in runs of hundreds if not more at a time. This means that the next BC108 could have a β that is 100, not the 494.29 we have used in our calculations. This would throw all our calculations out, and the amplifier would not work as we expected. For example, using this value of β at 100, we would get the following:

$$I_C = \beta I_B = 100 \times 20E^{-6} = 2mA$$

This means that the $V_{CE}$ voltage would be

$$V_{CE} = V_{CC} - I_C R2 = 12 - 2E^{-3} \times 606.92 = 10.786V$$

This would now push the output voltage too close to $V_{CC}$, which is the upper limit the output voltage could rise to. This may produce clipping if the input voltage rises to a higher level. To show what would happen if the beta for the BC108, which defaults to a figure of 529 with the TINA software, was changed to 100 we can simulate the circuit again with beta changed to 100, as shown in Figure 4-19. The DC voltage that this would produce is shown in Figure 4-19.

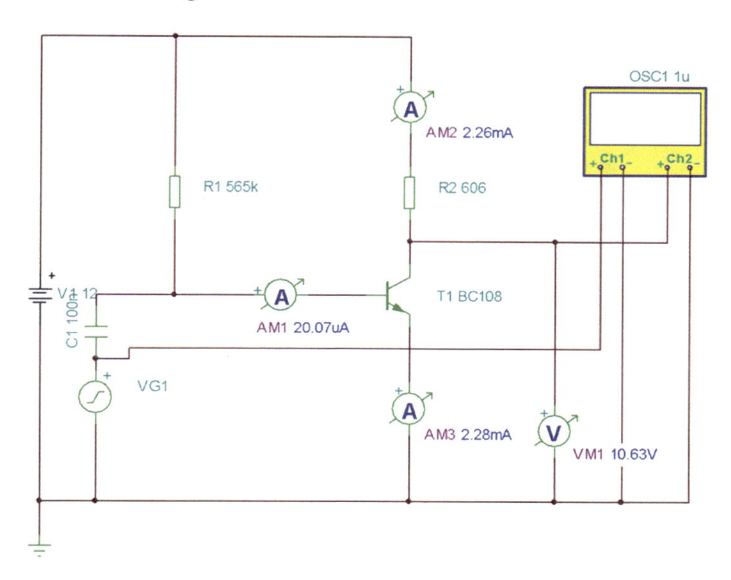


***Figure 4-19.*** *The Changed DC Quiescent Operating Point When b Was Changed to 100*

We can see that the $V_{CE}$ voltage has risen to 10.63V, and the collector current has dropped to 2.26mA. If we change the input signal to 20mV, the clipping of the output voltage can become clear if we look at Figure 4-20.

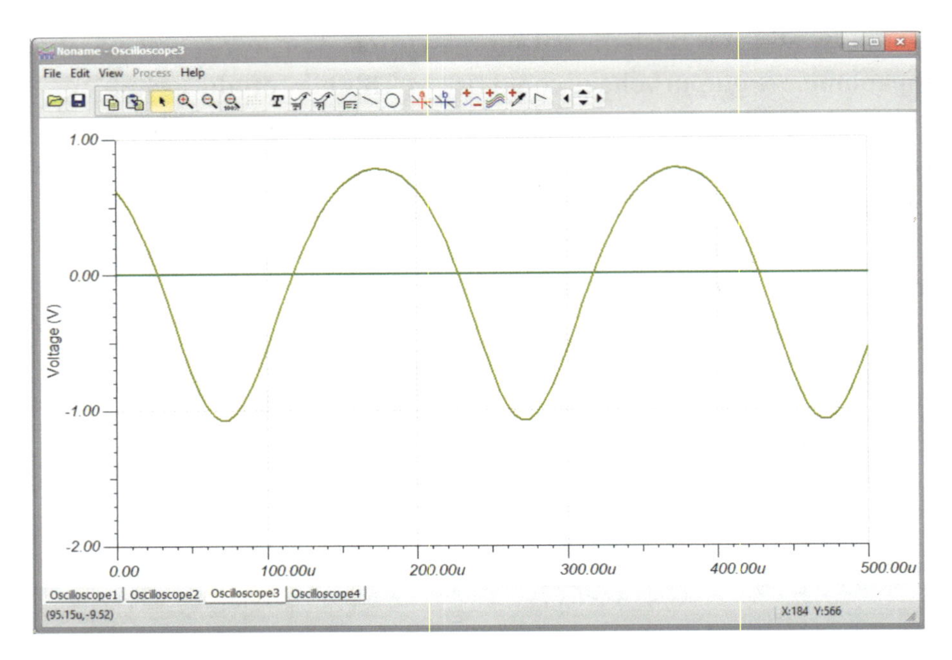


***Figure 4-20.*** *The Output Voltage with b Set to 100*

From Figure 4-20, we can see that the output does not go as far positive above 0V as it does go negative. This will produce a distortion of the signal at the output. When we look at designing amplifiers in more detail in Chapter 5, we will see how we can overcome this problem.

# The Current Characteristics of the Common Emitter Configuration

Using the circuit shown in Figure 4-19, we can see that the transistor creates a node into which the collector current $I_C$ and base current $I_B$ flow. The only current that flows out of it is the emitter current $I_E$. This means that we can see the expression for the emitter current is

$$I_E = I_C + I_B$$

From this, we can say

$$I_C = I_E - I_B$$

However, we know that this is a current amplifier and that “beta” is the current gain for this type of amplifier. Therefore, as the input current is the base current $I_B$ and the output current is the collector current $I_C$, we can say

$$I_C = \beta I_B$$

From this, we can say

$$\beta = \frac{I_C}{I_B}$$

Looking at the expression for $I_E$, we can say

$$I_E = \beta I_B + I_B = I_B(\beta + 1)$$

From this, we can say

$$\frac{I_E}{I_B} = 1 + \beta$$

Therefore, we can say

$$\beta = \frac{I_E}{I_B} - 1$$

If we consider the relationship between the collector current and the emitter current, we can say

$$I_C = \alpha I_E$$

From before, we can say

$$I_C = I_E - I_B$$

We can substitute $\alpha I_E$ for $I_C$ as follows:

$$I_E - I_B = \alpha I_E$$

Therefore, we have

$$I_B = I_E - \alpha I_E$$

Therefore

$$I_B = I_E(1 - \alpha)$$

This means

$$I_E = \frac{I_B}{1 - \alpha}$$

Therefore, we have

$$\frac{I_E}{I_B} = \frac{1}{1 - \alpha}$$

From before, we have stated

$$\frac{I_E}{I_B} = 1 + \beta$$

This then means that

$$1 + \beta = \frac{1}{1 - \alpha}$$

From this

$$\beta = \frac{1}{1 - \alpha} - 1 = \frac{1 - (1 - \alpha)}{1 - \alpha} = \frac{1 - 1 + \alpha}{1 - \alpha} = \frac{\alpha}{1 - \alpha}$$

This gives

$$\beta = \frac{\alpha}{1 - \alpha}$$

We can use this to create an expression for "α" in terms of "β" as follows:

$$\beta(1-\alpha)=\alpha$$

Therefore

$$\beta-\beta\alpha=\alpha$$

Therefore

$$\beta=\alpha+\beta\alpha=\alpha(1+\beta)$$

Therefore

$$\frac{\beta}{1+\beta}=\alpha$$

This means we now have the expressions for "α" and "β" for the common emitter configuration, and they are

$$\alpha=\frac{\beta}{1+\beta}\ and\ \beta=\frac{\alpha}{1-\alpha}$$

These are the same as the expressions for "α" and "β" for the common base configuration.

# The Common Collector Configuration

This is the third and last configuration that we will look at. It is sometimes, and more correctly, called the emitter follower. The normal circuit for this arrangement is shown in Figure 4-21.

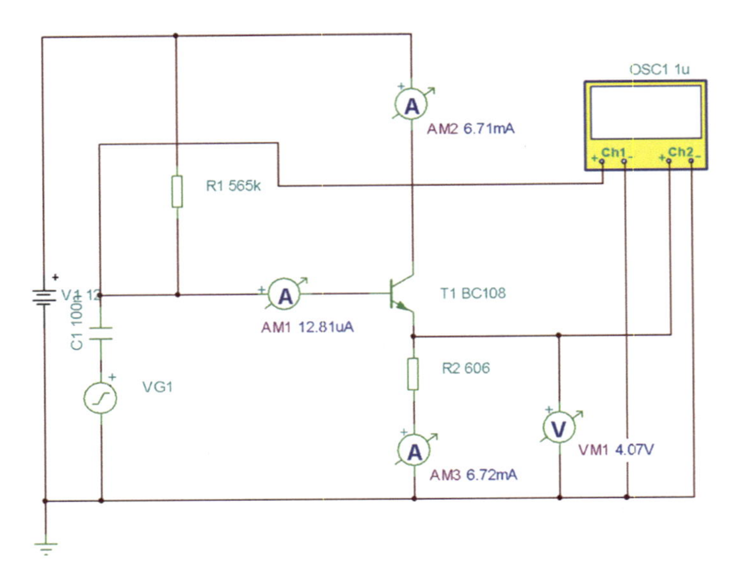


***Figure 4-21.*** *The Emitter Follower or Common Collector Configuration*

This circuit does look very much like the common emitter. The only difference is that the load resistor $R_2$ is now in the emitter leg of the transistor. To be called the common collector, the collector must be common to both the input and the output. However, we can see that the input is between the base and ground, while the output is between the emitter and ground. So how does the collector become common to them? Well, we are really considering the ac input and the ac output. If we do that, then, due to the capacitor that is across the power supply that provides the $V_{CC}$ voltage, the $V_{CC}$ rail and the ground rail are connected together. This may seem difficult to accept, but it is correct and it is used when we carry out an ac analysis of the amplifier.

The main use of the common collector configuration is with the emitter follower amplifier. With this circuit, the output current is $I_E$.

However, just as with the common emitter amplifier, the emitter current can be expressed as

$$I_E = I_C + I_B$$

Also, as with the common emitter, the relationship between the base current $I_B$ and the collector current $I_C$ is

$$I_C = \beta I_B$$

Therefore, substituting for $I_C$, we get

$$I_E = \beta I_B + I_B$$

Taking $I_B$ out as a common factor gives

$$I_E = I_B(1+\beta)$$

From this, we can say

$$\frac{I_E}{I_B} = 1+\beta$$

Knowing that, the current gain can be expressed as

$$Current\ Gain = \frac{Output\ Current}{Input\ Current}$$

With the common collector configuration, the output current is $I_E$ and the input current is $I_B$. This means the current gain can be expressed as

$$Current\ gain = \frac{I_E}{I_B}$$

This then means that in the common collector configuration, the current gain can be calculated using

$$Current\ Gain = 1+\beta$$

This means that the current gain in the common collector circuit is slightly more than in the common emitter configuration. This is because the output current in this configuration is the emitter current $I_E$, whereas in the common emitter configuration the output current was the collector current $I_C$.

We can use the oscilloscope, as shown in Figure 4-21, to compare the output voltage with the input voltage and determine what the voltage gain would be. The display of the two waveforms is shown in Figure 4-22.

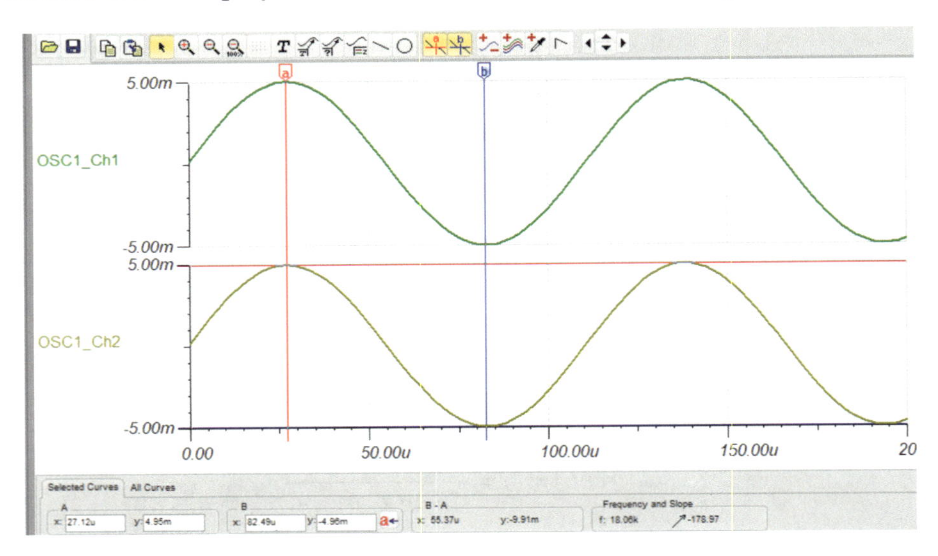


***Figure 4-22.** The Input and Output Voltage of the Emitter Follower Circuit*

However, if we look at the traces of the output voltage, that of Ch2 in Figure 4-22, and the input voltage, that of Ch1 in Figure 4-22, we can see that the voltage gain is less than one. If we take the peak-to-peak values of the two voltage traces, we get

$$Voltage\ Gain = \frac{Vout\ Peak\ to\ Peak}{Vin\ Peak\ to\ Peak} = \frac{9.91E^{-3}}{10E^{-3}} = 0.991$$

Also, we can see that the output voltage is in phase with input voltage.

So we can sum up our short investigation of the common collector configuration of the BJT in that.

The current gain is

$$Current\ Gain = 1 + \beta$$

The voltage gain is slightly less than 1.

# Summary

In this chapter, we have looked at the input and output characteristics of the three configurations of the BJT transistor. We have derived expressions for the current and voltage gain of all three configurations.

In the next chapter, we will concentrate on the common emitter amplifier, as this is the main configuration used with BJTs. We will look at what a load line is and how it can be used in the design of an amplifier. We will look at how the stabilized amplifier overcomes the drawbacks of the basic amplifier we have looked at in this chapter.

CHAPTER 5

# The Common Emitter Amplifier

In Chapter 4, we learned about the input and output characteristics of the BJT transistor. We also studied the basic transistor amplifier. With respect to that circuit, we mentioned there was a drawback with it.

In this chapter, we will learn about the load line and how it relates to the output characteristics of the BJT transistor. We will learn about the drawback with the basic amplifier and how we can correct it.

Finally, we will learn how we can use a load line to design a better basic amplifier and move on to designing the class "A" stabilized amplifier.

## The Basic Amplifier or Fixed Bias Amplifier

The circuit for the basic amplifier is shown in Figure 5-1.


H. H. Ward, *Mastering Analog Electronics*, Maker Innovations Series,
https://doi.org/10.1007/979-8-8688-0245-4_5

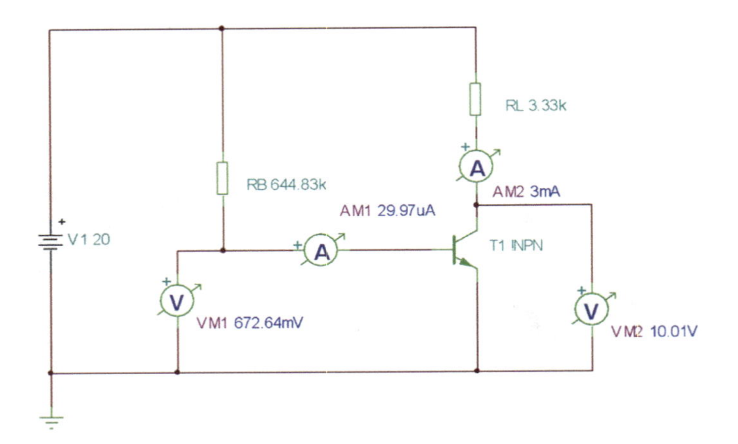


***Figure 5-1.*** *The Basic Amplifier Circuit*

This amplifier is sometimes referred to as the fixed bias amplifier, because once you have chosen the value of $V_{CC}$ and the $R_B$ resistor, the base current, $I_B$, is fixed and will not change.

When designing an amplifier circuit, there are three main operational considerations to take into account. They are as follows:

- The DC quiescent conditions which are the DC voltages and currents around the circuit when no ac signal is applied.
- The mid-frequency operating conditions: This is how the circuit is set up to work with an ac signal applied. As the response of the circuit will alter as the frequency of the signal changes, then it is normal to set up the circuit to operate, as you want, around the mid-frequency of your operating range of frequencies.
- The final part is what class of amplifier you are going to design. There are typically three classes of amplifiers, and they are

- Class A: This is the simplest one of the most common classes of amplifier. This is when the output is a true replica of the input but with some increase in voltage.
- Class B: This is when only one half, the positive or negative half, of the ac signal is amplified. This is an attempt to get more of a voltage swing at the output. However, the two halves have to be brought back together so as to reproduce the actual signal. When this is done with two class B amplifiers, known as push-pull amps, there is a risk of crossover distortion.
- Class AB: This is similar to class B, but some small part of the other half, be it negative or positive, is included in the amplification. This is an attempt to remove the crossover distortion when the two signals are brought back together.

In our first look at amplifiers, we will concern ourselves with just the class A amplifier. That being the case then, to get a true even reproduction of the input, at the output, we must set the DC quiescent voltage at the collector to half of $V_{CC}$. For example, if $V_{CC}$ was 20V as shown in Figure 5-1, then the DC voltage at $V_C$, the collector, should be 10V. This is to allow the output voltage to swing up to 20V, a 10v swing, when the transistor turns fully off. Then swing down to 0V, that is, another 10v swing, when the transistor turns fully on. That is when the emitter of the transistor is connected to ground, as it is with the basic amplifier shown in Figure 5-1. Also, we should note that only an ideal transistor could turn on so much that the collector voltage would fall to 0V. In real life, there is always some voltage, known as the saturation voltage, across the transistor, even when it is fully turned on. This is typically small, usually in the millivolt range.

Before we can determine any values for the resistors in the circuit, we must decide on the DC quiescent base current, $I_B$, flowing into the transistor. Once we have chosen that base current, we can calculate the value of the base resistor $R_B$. The sole purpose of the resistor $R_B$ is to set this base current to the level we want. This is because the expression for $I_B$ is

$$I_B = \frac{V_{CC} - V_{BE}}{R_B}$$

Using this expression, the value of $R_B$ can be calculated as follows:

$$R_B = \frac{V_{CC} - V_{BE}}{I_B}$$

Using the values of $V_{CC} = 20V$; $V_{BE} = 0.655V$, which is the typical diode drop that the ECAD software TINA uses; and $I_B = 30\mu A$, our chosen value of the base current, the value of $R_B$ is

$$R_B = \frac{20 - 0.655}{30E^{-6}} = 644.83k\Omega$$

Knowing that, the collector current $I_C$ is

$$I_C = \beta I_B = 100 \times 30E^{-6} = 3mA$$

This is using the typical value for Beta of 100.

# The Value of the Load Resistor $R_L$

Knowing that, the $V_{CE}$ voltage, when the transistor is correctly biased, should be half $V_{CC}$. That is the value of $V_{CE}$ at point VQ; see the output characteristics in Figure 5-5. If this point is projected up to the point where the load line crosses the graph of the chosen base current, we get the point "Q" on the graph. This point can then be projected to the vertical axis where we have the point IQ. This would be the value of the collector current when $V_{CE}$ is at half of $V_{CC}$, and it would be simply half the value of the current at point "A." These are the correct quiescent operating conditions for the amplifier circuit. We can use the values we know to determine the value of the resistors for the basic amplifier.

Firstly, knowing $V_{CC}$ is 20V, then the VQ voltage would be 10V. Knowing the chosen base current is 30µA, then the quiescent collector current $I_{CQ}$ can be calculated to be 3mA; this is indeed half of the maximum collector current, as expected. Using these values, we can calculate the gradient of the load line as follows:

$$Gradient = \frac{3E^{-3}}{10} = 300E^{-6}$$

As the gradient is the ratio of current divided by voltage, then the gradient is the inverse of the resistance, and it can be shown that

$$R_L = \frac{1}{Gradient} = \frac{1}{300E^{-6}} = 3.33k\Omega$$

This agrees with the resistor value for $R_L$ in the circuit shown in Figure 5-1. This shows that the expression for the load line is

$$V_{CC} = I_C R_L + V_{CE}$$

$$I_C = \frac{V_{CC} - V_{CE}}{R_L}$$

The expression for $I_C$ is the expression for the load line. However, this was calculated with a Beta of 100. The maximum collector current, $I_{CMAX}$, is set at

$$I_{CMax} = \frac{V_{CC}}{R_L} = \frac{20}{3.333E^3} = 6mA$$

This means we now have the DC quiescent values and settings for the basic amplifier. The circuit shown in Figure 5-1 shows the base current is 29.92µA, the collector current is 3mA, the base voltage is 0.675V, and the collector voltage is 10.1V. These are all very close to the quiescent values we wanted.

## Applying an "ac" Signal

Now we need to apply an ac signal. We will use a mid-frequency of 8kHz and a peak voltage of 10mV. I am setting the mid-frequency at 8kHz because I want to design an audio amplifier. The typical audio range of frequencies is 30Hz to 18kHz, so I might be a bit low at 8kHz, but then I am older now, and I have a rather limited range of frequencies that I can hear.

The oscilloscope is used to measure the input and output voltages, and Figure 5-2 shows the display of the two traces.

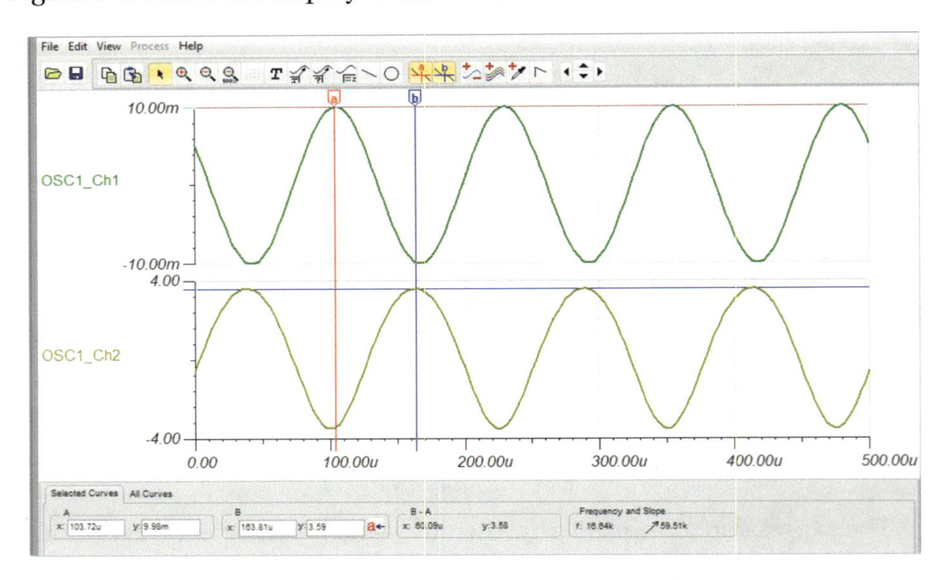


***Figure 5-2.*** *The Traces of the Input and Output Voltages*

Channel 1 is the input voltage with a peak of 9.98mv; really it is 10mV. Channel 2 is the output voltage with a peak of 3.59v. Using these two values, the voltage gain can be calculated using

$$V_{Gain} = \frac{PeakV_{OUT}}{PeakV_{IN}} = \frac{3.59}{10E^{-3}} = 359$$

# Changing the Value of Beta

However, if we now increase the Beta value of the transistor and simulate the circuit again with the same ac signal at the input, we get the traces on the oscilloscope as shown in Figure 5-3.

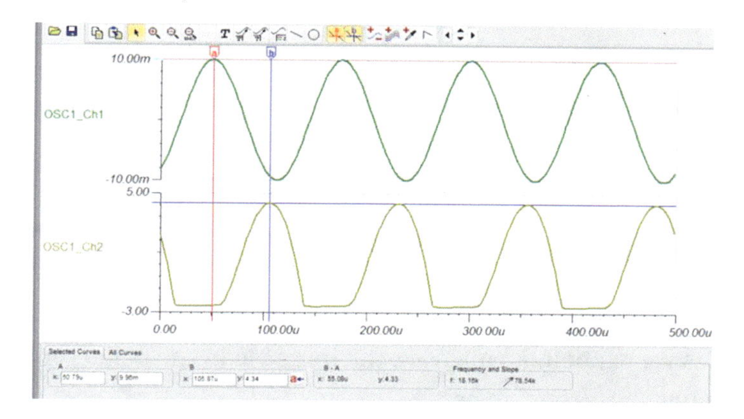


***Figure 5-3.*** *The Input and Output Voltages with a Beta of 300*

We can see that the peak voltage has only risen by a small amount, but the output voltage waveform is clipped as it goes negative, and so it is not a true representation of the input. If we look at the DC quiescent conditions, now that the Beta has been increased, we can see, from Figure 5-4, that the quiescent conditions are not now what we wanted.

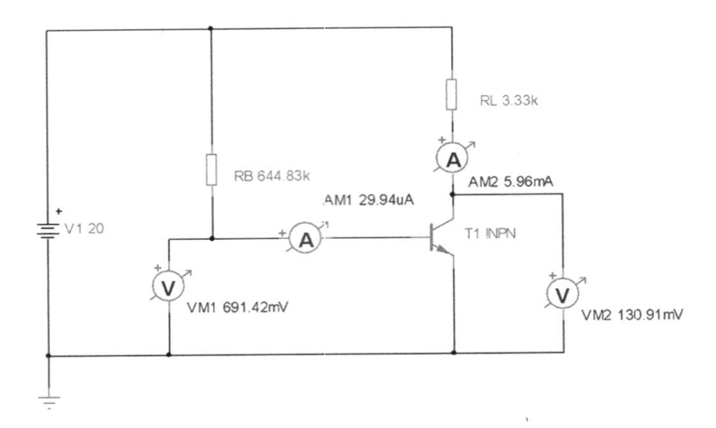


***Figure 5-4.** The Basic Amplifier with a Beta of 300*

The base current is still at 29.84μA, and the collector current has increased as we expected. However, the increase in the collector current is less than what was expected as

$$I_C = \beta I_B = 300 \times 29.94E^{-6} = 8.952mA$$

We expect the collector current to rise to 8.952mA but has only risen to 5.96mA. If we look at the output characteristics and the load line, we might be able to see why this has happened. The output characteristics with the load line are shown in Figure 5-5. The load line is shown as the straight line starting at point "A" when $V_{CE} = 0$ and ending at point "B" when $I_C = 0$. Point "A" is the collector current when $V_{CE} = 0$. Using the following expression for the collector current

$$I_C = \frac{V_{CC} - V_{CE}}{R_L}$$

Therefore, when $V_{CE} = 0$, the expression for $I_C$ becomes

$$I_{CMAX} = \frac{V_{CC}}{R_L}$$

Using the values of the circuit, we can calculate the maximum value the collector current can rise to as follows:

$$I_{CMAX} = \frac{V_{CC}}{R_L} = \frac{20}{3.3^3} = 6mA$$

This is the value of the collector current at point "A." This is assuming the transistor is ideal, and when it turns fully on, it would connect the collector to 0V, that is, the ground potential. This is not possible, and the maximum collector current is lower. However, we have shown that when the Beta value changes to 300, then the collector current should go up to 8.952mA. This is not possible as it is higher than the 6mA that is the maximum the collector current can rise to.

If we now consider the point "B" on the characteristics, then using the following expression for $V_{CE}$

$$V_{CE} = V_{CC} - I_C R_L$$

When $I_C = 0$ which is the current at point "B," then the expression for $V_{CE}$ becomes

$$V_{CE} = V_{CC}$$

This is the value of $V_{CE}$ at point "B."

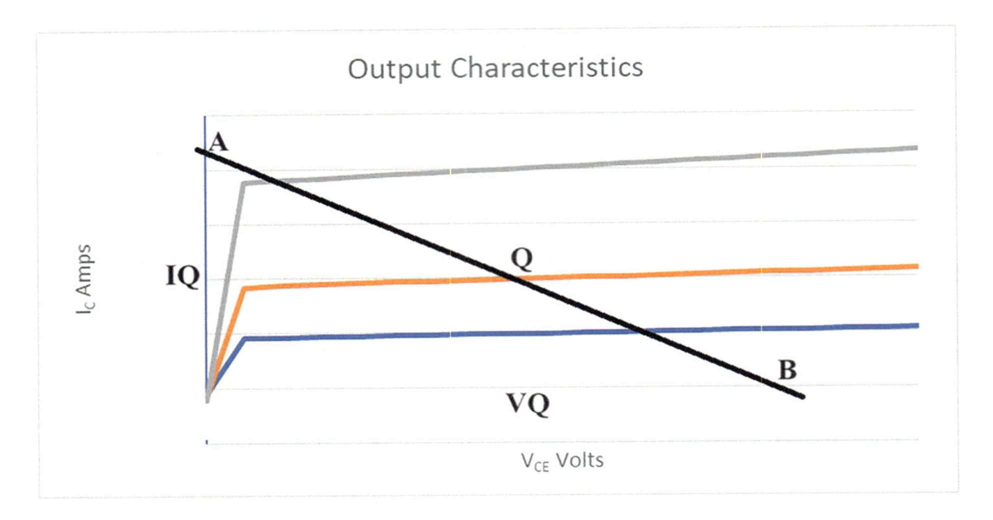


***Figure 5-5.*** *The Output Characteristics with the Load Line*

This means it can't reach the 8.952mA when Beta was increased to 300, and therefore the $V_{CE}$ voltage cannot reduce as much as is needed and it would be clipped.

Also, using the expression for $V_{CE}$, we have

$$V_{CE} = V_{CC} - I_C R_L$$

Knowing $I_C = \beta I_B$, then we get

$$V_{CE} = V_{CC} - \beta I_B R_L$$

When Beta was 100, then

$$V_{CE} = 20 - 100 \times 30E^{-6} \times 3.333E^{3} = 10.001$$

This is the correct value for VQ.

However, when Beta was increased to 300, then VCE becomes

$$V_{CE} = 20 - 300 \times 30E^{-6} \times 3.333E^{3} = -9.9999V$$

This would move the Q point too far up the load line and off the graph, and so it is an unusable quiescent operating point. Indeed, the further up the load line toward the vertical axis the Q point moves, the less the negative swing the output voltage could go and clipping would occur. The maximum value of Beta would be when $V_{CE} = 0$, which means

$$0 = V_{CC} - \beta I_B R_L$$

$$\beta I_B R_L = V_{CC}$$

Therefore, the maximum value of Beta would be

$$\beta_{MAX} = \frac{V_{CC}}{I_B R_L} = \frac{20}{30E^{-6} \times 3.333E^{3}} = 200.02$$

This would push the Q point up the load line to coincide with the "A" point on the graph. Moving the Q point up the line toward the maximum current value will reduce how higher the collector current can rise. This has the effect of reducing how low the output voltage can go, and so we get negative clipping of the output voltage.

If the Beta value was reduced, the opposite would happen, and the Q point would move down the line toward the horizontal axis. This would then limit how much lower the collector current could fall and so result in positive clipping of the output waveform.

This shows that a good amplifier is one that does not rely on the value of Beta. This is because changing the value of Beta could force the output voltage into clipping, either the positive peak, when Beta was reduced, or negative peak, when Beta was increased. We need to appreciate that the Beta could change with temperature, but that is probably small. However, more importantly, if we had to replace the transistor with another one, even if it was the same type, then the Beta value could change. This is because a typical transistor could have a Beta that was from 50 to 600. This means that just by replacing a broken transistor, even with one of the same type, could result in the amplifier no longer producing an exact replica of the input. It is something we cannot allow, and that is why this basic amplifier is not a common amplifier circuit that engineers use. A better circuit is that of the stabilized amplifier we will look at next in this chapter.

One final problem with the basic amplifier is the size of the base resistor $R_B$. This has to be a high value to ensure the base current is not too high. Large values of resistors can have some large percentage changes due to the tolerance of them.

## Designing a Class "A" Stabilized Amplifier

The stabilized amplifier is one that overcomes the problems with the basic amplifier we have just looked at. This analysis will look at how it does this.

The Class A amplifier reproduces an amplified version of the input without any distortion. To do this, the voltage at the collector of the transistor should be set at around half that of $V_{CC}$. This is to enable the output voltage to swing as low, in voltage, as it swings high.

To design any circuit, indeed anything, we need a specification, and we will start with the simple specification that gives the desired base current, $I_B$, and the $V_{CC}$ voltage. Then, we will use the standard output characteristics with a load line to determine the value of the load resistor, $R_L$, and the emitter resistor $R_E$. The process is as follows; the specification is that the base current, $I_B$, is 15μA, and the $V_{CC}$ voltage is 15V.

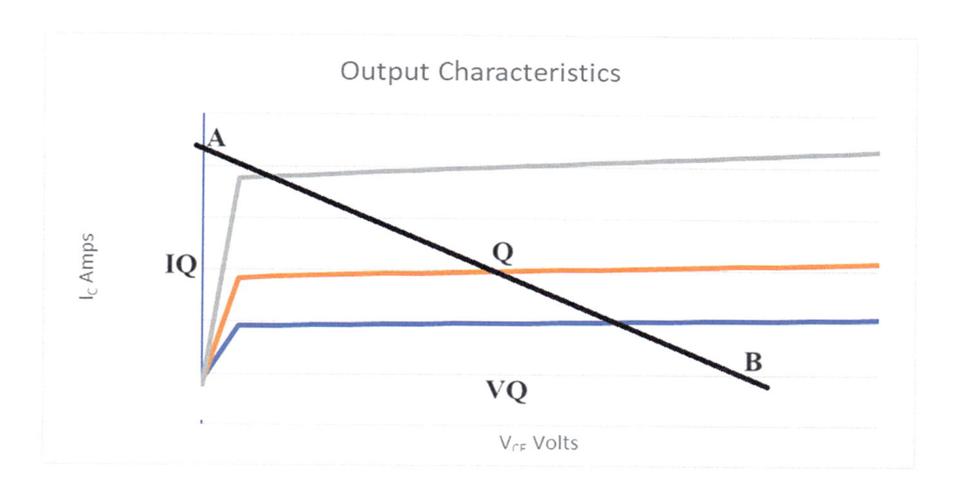


***Figure 5-6.*** *The Output Characteristics of the Common Emitter Transistor*

We do not need to use the particular output characteristics, as it is only the principle of the load line that we need to use.

Figure 5-6 shows the load line drawn on the output characteristics of the common emitter transistor.

Figure 5-7 shows the basic circuit of the transistor amplifier we will use as the class "A" stabilized amplifier.

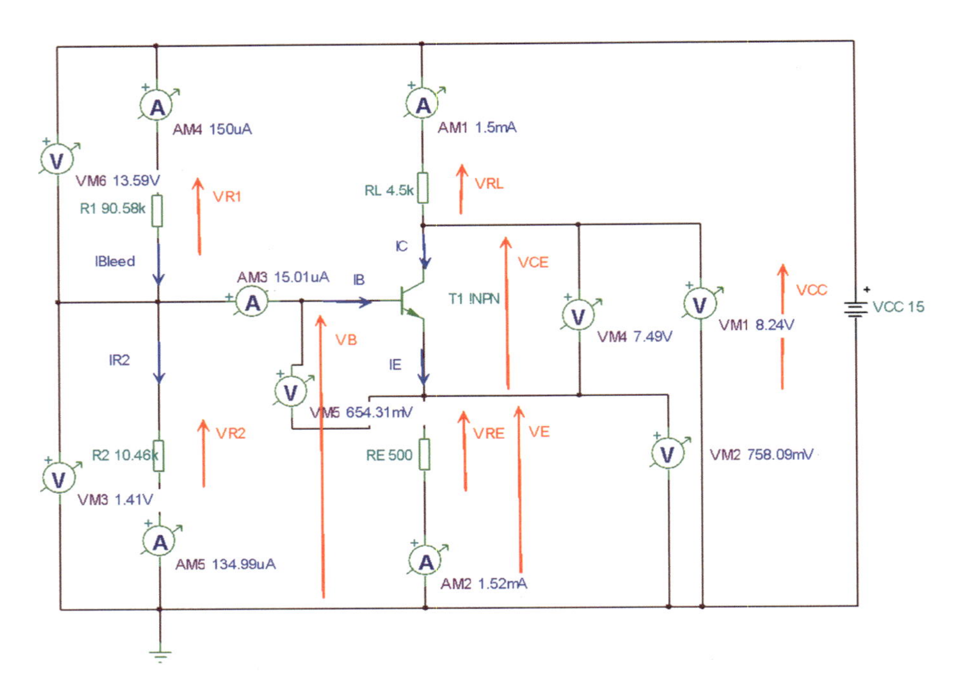


***Figure 5-7.*** *The Class A Stabilized Amplifier Basic Circuit*

I am including the voltage arrows, in red, and current arrows, in blue, on the circuit to try and help with the analysis. The convention of using the voltage arrows is that the arrow head indicates the more positive end of the volt drop. We can also use the voltage arrows to help apply Kirchhoff's Voltage Law (KVL) in that in a closed loop, the voltage arrows in one direction add up to equal voltage arrows in the opposite direction.

If we use Kirchhoff's Voltage Law on the output side of the circuit, we can say

$$V_{CC} = V_{RL} + V_{CE} + V_E$$

Using Ohm's law, we can say $V_{RL} = I_C R_L$, which means we can say

$$V_{CC} = I_C R_L + V_{CE} + V_E$$

We can see that $V_E$ is the same as the volt drop $V_{RE}$ and as $V_{RE} = I_E R_E$, we can say

$$V_E = I_E R_E$$

We know from Chapter 4 that $I_E \cong I_C$; we can say $V_E = I_C R_E$.
Therefore, we can say

$$V_{CC} = I_C R_L + V_{CE} + I_C R_E$$

Therefore, with a little bit of transposition, we can say

$$V_{CC} - V_{CE} = I_C \left( R_L + R_E \right)$$

Rearranging this for $I_C$, we get

$$I_C = \frac{V_{CC} - V_{CE}}{R_L + R_E}$$

This is actually the expression for the load line, shown in Figure 5-6. We can determine the value of the point "A" on the graph in Figure 5-6, as this will be when $V_{CE}$ is 0. This is when the transistor is fully turned on, and the voltage at the collector is the same as the voltage at the emitter. This is the ideal situation, as in real life there has to be a minimum voltage, $V_{CE}$, across the transistor. This is termed the saturation voltage, and it is typically in the millivolts. However, for the purpose of our design, we can assume this $V_{CE}$ can go to zero volts, and this will give us the expression for the maximum current, sometimes called the saturation current of the transistor. Using this assumption, we have

$$Saturation\ I_C = \frac{V_{CC}}{R_L + R_E}$$

This is point "A" on the graph.

If we now consider point "B," this will be the voltage when the current $I_C = 0$. Knowing this, we can say

$$0 = \frac{V_{CC} - V_{CE}}{R_L + R_E}$$

Therefore, multiply both sides by $R_L + R_E$; we have

$$0 = V_{CC} - V_{CE}$$

This means

$$V_{CE} = V_{CC}$$

This is the point "B" on the graph, that is, when $I_C = 0$, $V_{CE}$ will equal $V_{CC}$.

## The Gradient of the Load Line

This is the last aspect of the load line we need to consider. It is a useful parameter, and we need to determine the expression for the gradient. The expression for the load line describes how the collector current $I_C$ changes as the voltage $V_{CE}$ changes. We know that when $V_{CE} = 0$, the current $I_C$ is at a maximum of $\frac{V_{CC}}{R_L + R_E}$. Also, when $I_C = 0$, we know $V_{CE} = V_{CC}$. We can use these points to determine the gradient of the load line using

$$Gradient = \frac{Change\ in\ I_C}{Change\ in\ V_{CE}}$$

To determine the gradient, we take the last values, moving out along the horizontal from the origin, and subtract them from the initial values. The last values are those at point "B" on the graph, and the initial values are those at point "A." At point "B," the collector current $I_C = 0$, and at point "A" the current $I_C$ was $\frac{V_{CC}}{R_L + R_E}$. Using these two values, we can determine that the change in $I_C$ is $0 - \frac{V_{CC}}{R_L + R_E} = -\frac{V_{CC}}{R_L + R_E}$.

Similarly, the value of $V_{CE}$ at point "B" is $V_{CC}$, and at point "A" it is 0. Therefore, the change in $V_{CE}$ is $V_{CC} - 0 = V_{CC}$.

This means that the gradient can be expressed as

$$-\frac{V_{CC}}{R_L + R_E} \div V_{CC} = -\frac{V_{CC}}{R_L + R_E} \times \frac{1}{V_{CC}} = -\frac{1}{R_L + R_E}$$

This means that the expression for the load line, which is the same as the expression for the collector current as $V_{CE}$ changes, is

$$I_C = -\frac{1}{R_L + R_E} \times \left(V_{CC} - V_{CE}\right)$$

The current $I_C$ is the dependent variable, and $V_{CE}$ is the independent variable. The other terms are all constants. This fits the general expression for the straight-line graph of

$$y = mx + C$$

What we know, from the specification, is that $V_{CC} = 15V$ and $I_B = 15\mu A$. Knowing the collector current $I_C$ is related to the base current

$$I_C = \beta I_B$$

Then, assuming $\beta$ is the typical value of 100, we can say

$$I_C = 100 \times 15E^{-6} = 1.5mA$$

It should be said that the value of $\beta$ may be different from the 100 we have assumed here. Indeed, $\beta$ can vary from 50 to 600 and maybe more. This would normally cause a problem, but with this circuit we are designing, the problem is very much reduced as we will show later. Therefore, choosing a value of 100 for $\beta$ will be OK.

If we assume this is the collector current flowing at the "Q" point on the load line and that the $V_{CE}$ voltage would be at half $V_{CC}$, that is, at 7.5V in this case, then we can determine the value of the gradient as follows:

$$Gradient = \frac{0-1.5E^{-3}}{15-7.5} = \frac{-1.5E^{-3}}{7.5} = -200E^{-6}$$

Knowing that the gradient is $-\frac{1}{R_L + R_E}$, then we can say

$$-200E^{-6} = -\frac{1}{R_L + R_E}$$

Therefore

$$R_L + R_E = \frac{-1}{-200E^{-6}} = 5000$$

We can let $R_E = 500\Omega$ and $R_L = 4.5k$. It is fairly arbitrary as to how you split them up, but by creating the larger difference, you will be creating the greater voltage gain, as we will see later.

Now that we have a value for $R_E$, we can determine a value for $V_E$, knowing the current flowing through the emitter resistor would be

$$I_E = I_C + I_B = 1.5E^{-3} + 15E^{-6} = 1.515mA$$

This means we can determine a value for $V_E$ as

$$V_E = I_E R_E = 1.515E^{-3} \times 500 = 0.7575V$$

We can now carry on and calculate the voltage drops and voltage around the circuit. The volt drop across the resistor $R_L$, $V_{RL}$, can be calculated using

$$V_{RL} = I_C R_L = 1.5E^{-3} \times 4.5E^{3} = 6.75V$$

The voltage at the cathode, VC, can be calculated using

$$V_C = V_{CC} - V_{RL} = 15 - 6.75 = 8.25V$$

Now we can calculate the voltage across the transistor which is $V_{CE}$. This can be calculated as follows:

$$V_{CE} = V_C - V_E = 8.25 - 0.7575 = 7.4925V$$

Now we need to determine the value for the biasing resistors $R_1$ and $R_2$. The sole purpose of these two resistors is to provide a voltage divider network that will supply the voltage at the base with the correct voltage for $V_B$. We need to appreciate that, between the base and emitter, there is a real diode, and because the diode is conducting, the volt drop across this diode will be approximately 0.655V, which is the common volt drop shown using TINA; see Chapter 2. This means the base voltage will be 0.655V greater than the emitter voltage $V_E$. In this way, we can say

$$V_B = V_{BE} + V_E = 0.655 + 0.7575 = 1.4125V$$

Knowing the base current, $I_B$, is 15μA, then, if we assume this is 10% of the current flowing through $R_1$, known as the "bleed current," we know the current flowing through $R_1$ will be ten times bigger at 150μA. As 15μA of this 150μA flows into the base of the transistor, the current flowing through $R_2$ will be 135μA. We know the volt drop across $R_2$ will be 1.4125V, that is, $V_{R2} = V_B$ (see Figure 5-7), and so we can calculate the value of $R_2$ as follows:

$$R_2 = \frac{V_{R2}}{I_{R2}} \frac{1.4125}{135E^{-6}} = 10.463k$$

Also, we know the volt drop across $R_1$ can be expressed as

$$V_{R1} = V_{CC} - V_B = 15 - 1.4125 = 13.5875V$$

Therefore, the value of $R_1$ can be calculated using

$$R_1 = \frac{V_{R1}}{IBleed} = \frac{13.5875}{150E^{-6}} = 90.583k$$

This means the values of the resistors are

$R_1$= 90.583k, $R_2$ = 10.463k, $R_L$ = 4.5k, $R_E$ = 0.5k

Putting these values into the circuit and simulating it in TINA gives the results shown in Figure 5-7. Table 5-1 shows the comparison of the calculated and simulated values.

***Table 5-1.*** *The Calculated and Simulated Results*

| Item | Calculated | Simulated |
|---|---|---|
| R1 | 90.583k | Yes |
| R2 | 10.463k | Yes |
| RL (R3) | 4.5k | Yes |
| RE (R4) | 500Ω | Yes |
| VC | 8.25V | 8.24V |
| VE | 0.7575V | 0.758V |
| VCE | 7.4925V | 7.49V |
| VB | 1.4125V | 1.41V |
| VR1 | 13.5875V | 13.59V |
| IC | 1.5mA | 1.5mA |
| IE | 1.515mA | 1.52mA |
| IBleed | 150μA | 150μA |
| IR2 | 135μA | 134.99μA |

These results compare very well and so suggest the calculations are valid.

This means that to design a stabilized class A amplifier, we can use the following important expressions:

$$V_{CEQ} = \frac{V_{CC}}{2}$$

$$I_{CQ} = \beta I_B = 100 I_B$$

$$R_L + R_E = \frac{1}{Gradient} = \frac{V_{CEQ}}{I_{CQ}} = \frac{V_{CC}}{2I_{CQ}}$$

$$V_E = I_{CQ} \times R_E$$

$$V_B = 0.7 + V_E$$

$$I_B = \frac{I_{Bleed}}{10}$$

$$I_{Bleed} = 10 \times I_B$$

$$V_{R1} = V_{CC} - V_B$$

$$R_1 = \frac{V_{R1}}{I_{Bleed}}$$

$$R_2 = \frac{V_B}{I_{Bleed} - I_B}$$

Therefore, using all these expressions and just the specification of the $V_{CC}$ and $I_B$ for the circuit, you should be able to design a class A stabilized amplifier.

# Exercise 5.1

As an exercise, using the following specification of a $V_{CC}$ of 22V and a base current $I_B$ of 10μA, assuming the "β" was 100, design a class A stabilized amplifier using the preceding expressions. My circuit to meet this specification is shown in the appendix.

# The ac Gain of the Stabilized Class "A" Amplifier

The test circuit we will use to measure the ac voltage gain is shown in Figure 5-8.

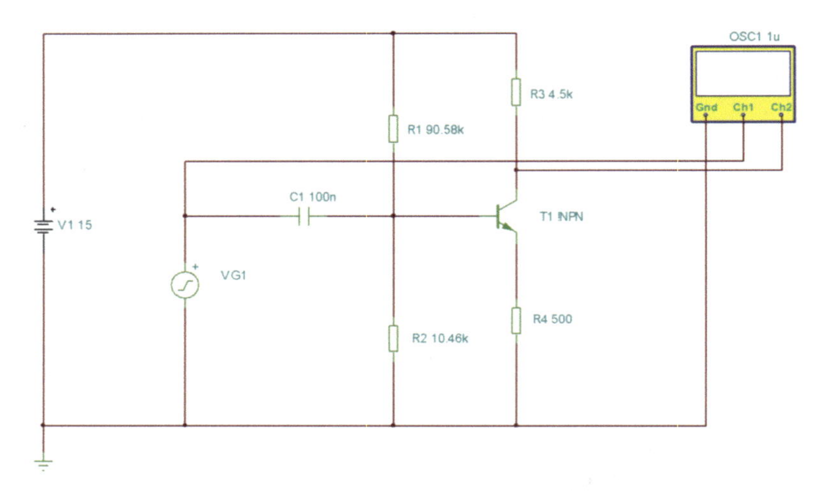


***Figure 5-8.*** *The ac Signal Applied to the Amplifier*

It is the same circuit as shown in Figure 5-7, but the arrows and meters have been removed. The oscilloscope is used to measure the input voltage with channel 1 and the output voltage with channel 2. The traces of the two signals are shown in Figure 5-9.

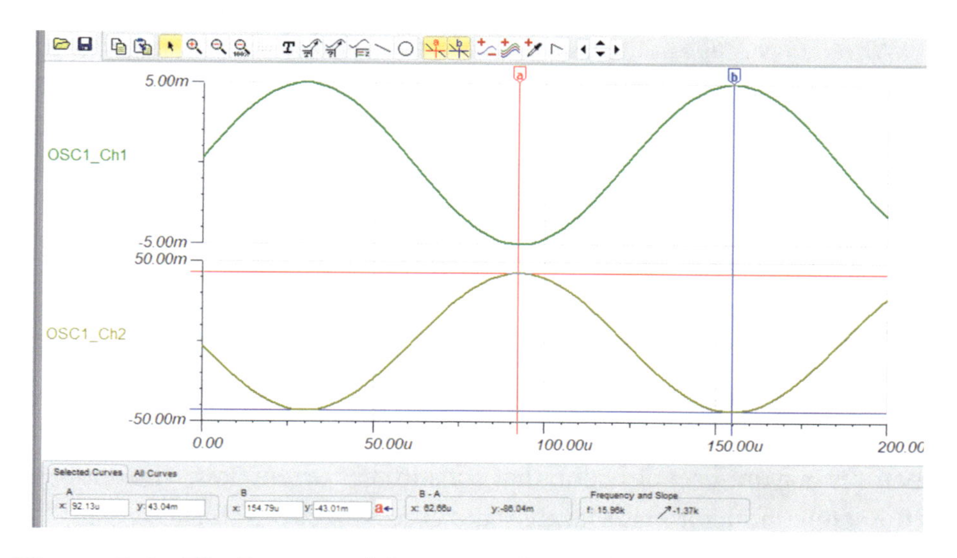


***Figure 5-9.*** *The Input and Output Voltage of the Amplifier*

We will use the peak-to-peak voltages to calculate the voltage gain as follows:

$$VGain = \frac{Vout\ Peak\ to\ Peak}{Vin\ Peak\ to\ Peak} = \frac{86.04m}{10m} = 8.604$$

We can also see that there is the 180° phase difference between the input voltage and the output voltage. This is because when the input voltage rises, the base current $I_B$ rises. This in turn means the collector current $I_C$ must also rise. This rise in the collector current means that the volt drop across $R_L$ must also rise. Knowing that, the voltage at the collector can be calculated as

$$V_C = V_{CC} - I_C R_L$$

Therefore, this rise in the volt drop across $R_L$ must mean that the voltage at collector must reduce.

When the voltage at the base reduces, the opposite occurs as the $I_B$ must reduce, which means $I_C$ must reduce. Therefore, the volt drop across $R_L$ must reduce, which explains why $V_C$ must increase. I hope this does explain why there is this 180° phase difference between the input and output or across the transistor.

# The Emitter Resistor $R_E$

This resistor has been added to the circuit of the basic amplifier to combat the effect of thermal runaway. We have learned how it is the movement of electrons in a uniform direction that constitutes current flow. The doping of the semiconductor material gives the excess electrons the potential energy to move toward the positive potential of the $V_{CC}$ supply, even overcoming the reverse biased diode between the base and collector. Well heat is another way of giving electrons some form of energy. We have all seen how water starts to bubble when it reaches boiling point. The heat has given the water molecules the energy to start bubbling. In the same way, heat can give electrons more energy. Indeed, when the current starts to flow through the transistor, it starts to generate its own heat. This heat then passes on some energy to the electrons in the transistor which allows the current to increase. This increase in current in turn increases the heat produced in the transistor, which in turn allows more current to flow. This could possibly cause a snowball effect which is called thermal runaway. The resistor $R_E$ is used to prevent this.

The way it works is that the two resistors, $R_1$ and $R_2$, fix the voltage at the base at its DC quiescent voltage, which in the test circuit in Figure 5-7 is 1.4125V. This voltage cannot change as the DC current through $R_1$ and $R_2$ is not affected by any increase in heat.

However, the voltage at the emitter, $V_E$, depends on the current flowing through the transistor as $V_E = I_E R_E$. As the current flowing through $R_E$ increases, due to the heat created inside the transistor, then this voltage

$V_E$ will increase. However, the voltage $V_{BE}$ must start to reduce when $V_E$ increases because the base voltage, $V_B$, cannot change. This can be shown as follows:

$$V_{BE} = V_B - V_E$$

This reduction in $V_{BE}$ has the effect of turning the base emitter diode off, which has the effect of reducing the flow of electrons across the base emitter junction. This then means that the current flowing through the transistor reduces, meaning $I_E$ must reduce. This reduction in the $I_E$ means the voltage $V_E$ must reduce, which then returns the $V_{BE}$ voltage back to what it needs to be, that is, between 0.655 and 0.7. Therefore, this counteracts the effect of the thermal runaway within the transistor.

This is an example of what is called negative feedback because as the heat causes the current to rise, the rise in the voltage $V_E$ causes the current to fall. In this way, the status quo is achieved.

# The Input Capacitor $C_1$

With this voltage divider network of $R_1$ and $R_2$, we have gone to a lot of effort to ensure the DC quiescent voltage at the base is around 0.655V to 0.7V higher than that of the DC voltage at the emitter. This is to ensure the transistor is biased such that the collector voltage sits at the mid-range of $V_{CC}$ to $V_E$. This would be at half $V_{CC}$ if the emitter was connected to ground. However, in this stabilized amplifier, we have inserted a resistor $R_E$ to prevent thermal runaway; the emitter voltage cannot go down to 0V. Indeed, it must sit at some voltage, due to $I_E R_E$, above 0V. In this case, with $R_E$ at 500Ω and $I_C$ at 1.5mA, then $V_E$ must be at 0.75V.

We now need to explain the purpose of the capacitor $C_1$. Capacitors have the ability to stop any DC voltage passing through them. This is because the impedance of a capacitor, XC, which is the ac equivalent to resistance, can be calculated as follows:

$$XC = \frac{1}{2\pi fC}$$

The frequency of DC is zero, which means to DC the impedance XC is

$$XC = \frac{1}{2\pi \times 0 \times C} = \infty$$

This is regardless of the value of the capacitor. In this way, the capacitor can be used to block any DC component that is attached to the input signal being passed onto the base of the transistor and upsetting the DC quiescent operating conditions of the transistor.

Well, that explains why we have the capacitor but not how we choose its value. The value of the expression for XC, the impedance of the capacitor, changes as the frequency changes. It will change from infinity when the frequency is zero, as with DC, and reduce in impedance as the frequency increases to a value that allows the ac signal to be passed onto the base. In this way, the capacitor can be used to select what frequencies are allowed to pass onto the next stage of the circuit. In this application, the next stage is the base of the transistor.

This is the situation whereby we are trying to create a high pass filter. That is a filter that will pass onto the next stage, in this case the base of the transistor, any signal that has a higher frequency above a chosen value. This means we need to specify what this low frequency value must be. We could say 0Hz, but that is almost impossible to achieve. Also, as the transistor is being used as an amplifier, then it would most likely be an audio amplifier. That being the case, it is unnecessary to set the low frequency to 0Hz, as the normal audio range for humans is between 50Hz to 18kHz. Some young people may go outside that range, but that is the normal audio range. As you get older, you lose the ability to hear the higher frequencies, which is why your granddad or grandma has difficulty in hearing a lot of words that end with "s" or "t." These are high-frequency letters. Therefore, we could set a low frequency of 30Hz, and in this example we will.

To explain how we choose the value of $C_1$, we must look at the transfer function, TF, of the high pass filter. The CR high pass filter, so called because it is made with a capacitor "C" in series with a resistor "R," is shown in Figure 5-10.

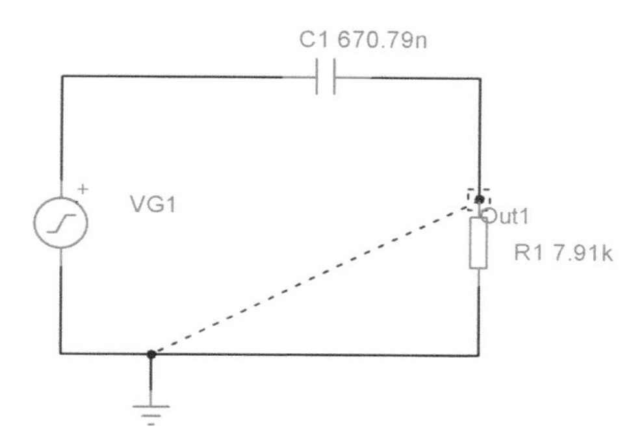


***Figure 5-10.*** *The Circuit of the Capacitive High Pass Filter*

The transfer function is the ratio of the input to the output. It can be used to show how the input is transferred, by the circuit, to become the output, hence the term transfer function (TF). The transfer function for the CR high pass filter is

$$TF = \frac{Vout}{Vin} = \frac{1}{1 - j\frac{1}{\omega CR}}$$

I say CR filter as we can create high and low pass filters using an inductor and resistor circuit. The proof for the expression for the transfer function is shown in the appendix. We will also look at these transfer functions in more detail in Chapter 11.

There will be a point, called the cutoff point, when the magnitude of the transfer function will be equal to $\frac{1}{\sqrt{2}}$ which is termed the half power point or the –3db point. This will be when the term $\omega CR = 1$. The half power point is the benchmark engineers use to say when the output of the filter, or many electronic circuits, will be good enough to do what is required of it. Any frequency that produces an output that is less than this half power point will not produce a usable output from the filter. This half power point is a common benchmark by which engineers measure many aspects of electronics. We will use it to measure the bandwidth of an amplifier in Chapter 6.

Using the expression $\omega CR = 1$ at the half power point or the –3db point, we can transpose the expression for "R" or for "C":

$$R = \frac{1}{\omega C} = \frac{1}{2\pi fC}$$

This means the cutoff point, or –3db point will be when the capacitor has a value that can be calculated using.

$$C = \frac{1}{2\pi fR}$$

We have decided that the cutoff frequency would be 30Hz as we are not interested in any frequencies lower than that. What we need now is the value of "R." This will be the actual input resistance of the base of the transistor to the ac signal. To determine that, we must appreciate that this will be different from the DC input resistance, as we are dealing with an ac signal. We need to remember that any capacitance can appear as a short circuit to an ac signal. Also, we need to realize that all power supplies, and there will be a power supply that delivers the $V_{CC}$ supply to the circuit, will have a high value capacitance across its output, used to smooth the output. This means that to ac, there is a short circuit between the $V_{CC}$ rail and the ground rail. This means that, with respect to the circuit shown in Figure 5-7, the top of the resistor $R_1$ is connected to the ground rail and so to the bottom of the resistor $R_2$. This means that to ac, $R_1$ and $R_2$ are in parallel.

There is one more resistance we need to consider and that is the emitter resistor $R_E$. We can see, from Figure 5-6, that the bottom of $R_E$ is connected to the ground rail as is the bottom of $R_2$ and now the top of $R_1$. However, what is slightly more difficult to see is that the top of $R_E$ is connected to the top of $R_2$ via the internal base to emitter diode inside the transistor. This diode is switched on and so closed. This is how the top of $R_E$ is connected to the top of $R_2$. However, due to the action of the transistor, the effective resistance of $R_E$ is multiplied by $\beta$+1. This means that the actual circuit of the capacitive high pass filter, at the input of the transistor, is as shown in Figure 5-11.

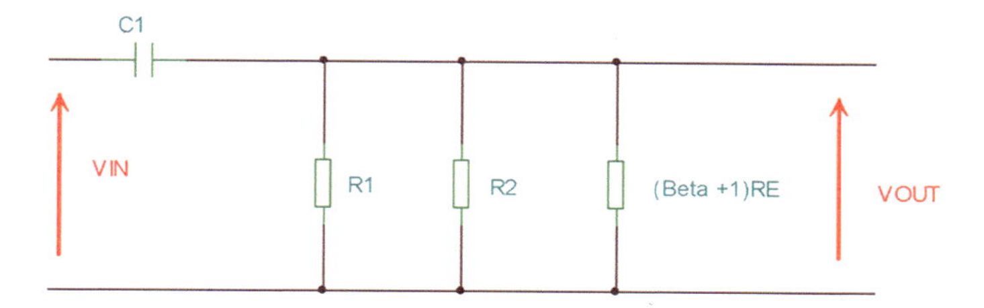


***Figure 5-11.** The High Pass Filter at the Base Input of the Stabilized Amplifier*

We can say R = $R_1$, $R_2$, and (Beta+1)$R_E$ in parallel. Putting in the value from the circuit, we calculate it as follows:

$$\frac{1}{R}=\frac{1}{R1}+\frac{1}{R2}+\frac{1}{(\beta+1)RE}=\frac{1}{90.58E^{3}}+\frac{1}{10.46E^{3}}+\frac{1}{101\times 500}$$

$$\frac{1}{R}=1.104E^{-5}+9.56E^{-5}+1.98E^{-5}=1.2644E^{-4}$$

Therefore

$$R=\frac{1}{1.2644E^{-4}}=7.908k$$

We can now calculate the value of the capacitance $C_1$ as follows:

$$C1 = \frac{1}{2\pi \times 30 \times 7908} = 670.79nF$$

Therefore, we now have a new value for $C_1$, but one which will stop any frequency below 30Hz from being passed onto the transistor. If we simulate the circuit again with the new value of $C_1$, we should be able to confirm our calculations are correct. The type of analysis we will simulate is called a Bode plot, after Hendrik Wade Bode, an American engineer who created a graph that shows how the gain of a system, measured in dBs, varies with frequency. The gain can be voltage or current gain and is the ratio of the output over input. That makes the Bode plot an ideal graph for displaying the output of this high pass filter and also the gain of the transistor as a whole.

The circuit we will use for the simulation is shown in Figure 5-12.

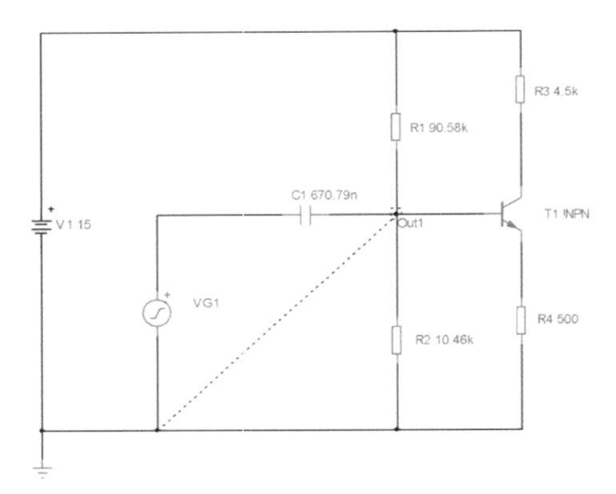


***Figure 5-12.** The Test Circuit for the Bode Plot*

We need to insert an output connection, and I have placed it at the base of the transistor. This will be the output of the high pass filter as shown in Figure 5-11.

To create the Bode plot, we must select the AC Transfer Characteristic option from the flyout menu that appears when you select AC Analysis from the drop-down menu that appears when you choose the Analysis option from the main menu bar, as shown in Figure 5-13.

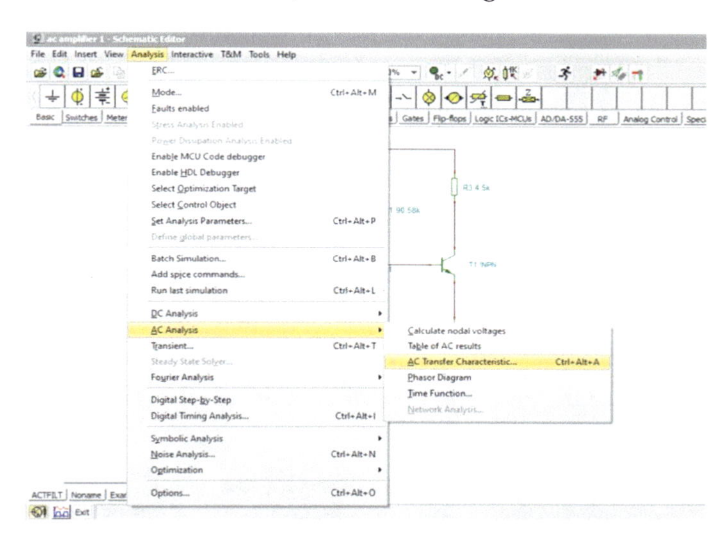


*Figure 5-13.* *Selecting the AC Transfer Characteristic Option*

When you select the AC transfer function, you will be presented with the window as shown in Figure 5-14.

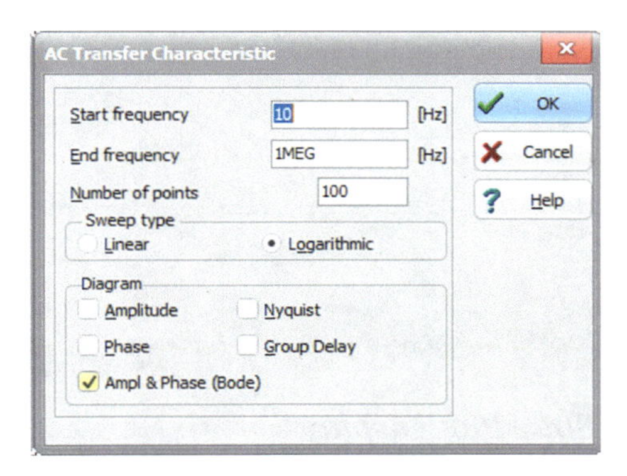


*Figure 5-14.* *The AC Transfer Characteristic Setup Window*

With this simulation, there is no need to change any of the parameters from their default settings. The start frequency at 10Hz is low enough, and the end frequency at 1MHz is high enough. The default diagram will display the amplitude and phase which is OK. We could just use the Amplitude only option as we are not concerned with the phase, but it is useful to see that the phase shift is a function of frequency as well. We should keep the sweep type set to logarithmic because if you choose linear, we would not have a page long enough to plot the frequency range of 10Hz to 1MHz. If you choose just 1mm to represent 1Hz, then you would need a sheet of paper nearly 1m wide to display the whole frequency range. The logarithmic scale reduces this to a workable length. You could change the number of points from 100 to 1000, if you so wish. This would produce a more accurate and finer graph.

Once you click OK, the software will run the analysis and the display of the results as shown in Figure 5-15.

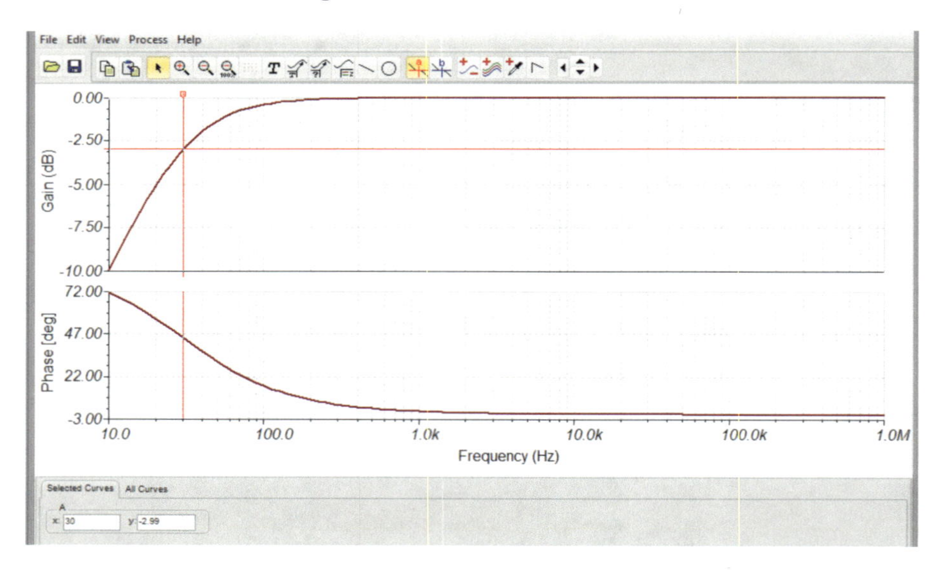


***Figure 5-15.*** *The Bode Plot Display*

We can use the cursors to measure the gain at high frequencies, where it will be a maximum; this is shown as 0dbs. This is a gain of 1, which in dBs is 0dbs. Cursor A is used to display the gain at a frequency of 30Hz. We can see that it shows the value of –3dbs which is a gain of $\frac{1}{\sqrt{2}}$ in dbs. This is what we expected the response to be. This indicates that our calculations are correct and that the circuit of the high pass filter, as shown in Figure 5-11, is also correct. This in turn confirms our thoughts that the emitter resistor can be imposed on the input at the base by multiplying it by $\beta+1$.

We can use the Bode plot to display how the gain of the transistor varies over frequency. To do this, we must move the output trace to connect it to the collector of the transistor, shown in Figure 5-16.

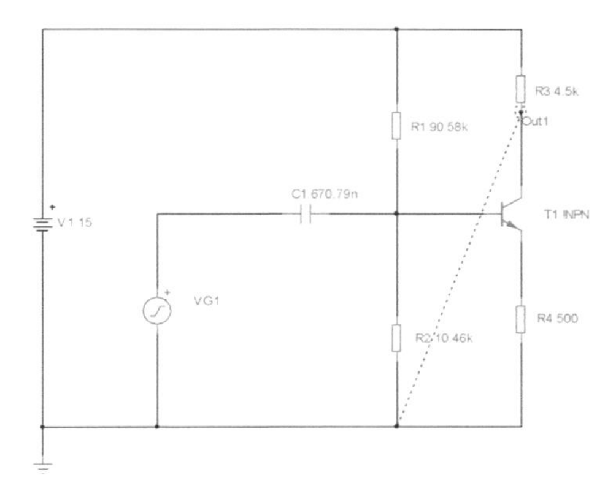


***Figure 5-16.*** *The Circuit with the Output Connection Moved to Display the Output of the Transistor*

When we run the simulation, the Bode plot is shown as that in Figure 5-17.

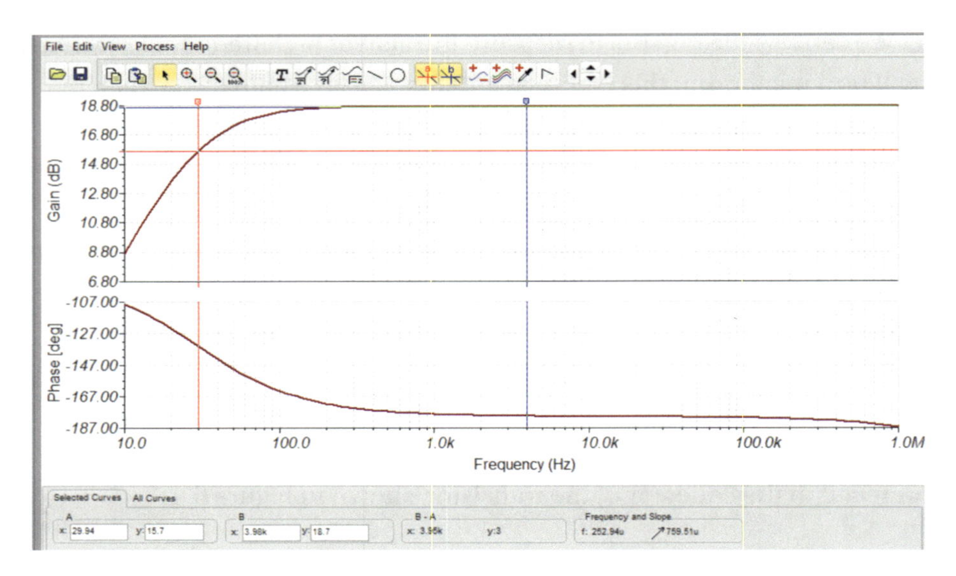


***Figure 5-17.** The Bode Plot to Show How the Gain of the Transistor Varies with Frequency*

The maximum gain, using cursor B, is shown as 18.7dbs. Cursor A is showing that at 29.94Hz the gain is 15.7dbs which is 3dbs down from the maximum. This is what is meant by the –3db point, that is, 3dbs down from the maximum. This Bode plot shows that at 30Hz the gain is at the half power point, and so it has met the required benchmark for when it will be deemed enough to be useful.

# Exercise 5.2

We can calculate the input resistance $R_{IN}$ and the filter capacitance value $C_1$ for the high pass filter, using the following expressions:

$$R_{IN} = \frac{1}{\frac{1}{R_L} + \frac{1}{R_E} + \frac{1}{101 \times R_E}}$$

$$C_1 = \frac{1}{2 \times \pi \times f_C \times R_{IN}}$$

Use the preceding expression to calculate the high pass filter capacitor "$C_1$" for the amplifier in the first exercise if the cutoff frequency was 30Hz. The answer is in the appendix.

The simulations have shown that using the calculations so far, we have designed an amplifier that does not suffer from thermal runaway. It does seem to have the frequency response we want. We now need to see if it has overcome the reliance of the gain of the circuit on the current gain Beta of the transistor. To show that this has been achieved, we need to look at the expression for the gain of the transistor. This will be the voltage gain, and this can be calculated using

$$VGain = \frac{V_{OUT}}{V_{IN}}$$

We can calculate the output voltage $V_{OUT}$ as

$$V_{OUT} = I_C R_L$$

We can calculate the input voltage $V_{IN}$ using

$$V_{IN} = I_B R_B$$

We know that the emitter resistor $R_E$ is reflected into the base by multiplying it by $\beta+1$. This means that $R_B$ can be expressed as

$$R_B = (\beta + 1) R_E$$

Using this, we can say that

$$V_{IN} = I_B (\beta + 1) R_E$$

We also know that the collector current $I_C$ is related to $I_B$ using

$$I_C = \beta I_B$$

This means that we can express $V_{OUT}$ as

$$V_{OUT} = \beta I_B R_L$$

We know the expression for the voltage gain becomes

$$V_{Gain} = \frac{\beta I_B R_L}{I_B(\beta + 1)R_E}$$

We know that β is very much higher than 1, as β can vary from 50 to 500. This means the 1 can be ignored. Therefore, the expression for the voltage gain becomes

$$V_{Gain} = \frac{\beta I_B R_L}{\beta I_B R_E} = \frac{R_L}{R_E}$$

This expression suggests that the voltage gain is not dependent upon the Beta of the transistor. Indeed, it is simply the ratio of the resistors $R_L$ divided by $R_E$. Using the values in the circuit shown in Figure 5-7, that would give the following:

$$V_{Gain} = \frac{R_L}{R_E} = \frac{4.5E^3}{500} = 9$$

Using the values obtained in Figure 5-9, the $V_{Gain}$ was calculated at 8.604. This is pretty close. Also, we can calculate the voltage gain in dbs using

$$V_{Gain} = 20\log\frac{R_L}{R_E} = 20\log 9 = 19.08\,dbs$$

This is again close to the measured value of 18.7dbs. However, it is still slightly out. However, if we added a small 25Ω value to the emitter resistor, the calculated gain would change to

$$V_{Gain} = \frac{4.5E^3}{525} = 8.57 = 20\log 8.57 = 18.66 dbs$$

In reality, we do need to add this 25Ω as this is the internal resistance of the transistor termed "$r_e$." It is dependent upon the heat around the transistor, and at a nominal room temperature, the value of $r_e$ is 25Ω.

Just to show that the voltage gain is independent of Beta, we will simulate the class A amplifier again, but this time we will change the value of Beta for the NPN transistor to 300, which is three times what it was. The result is shown in Figure 5-18.

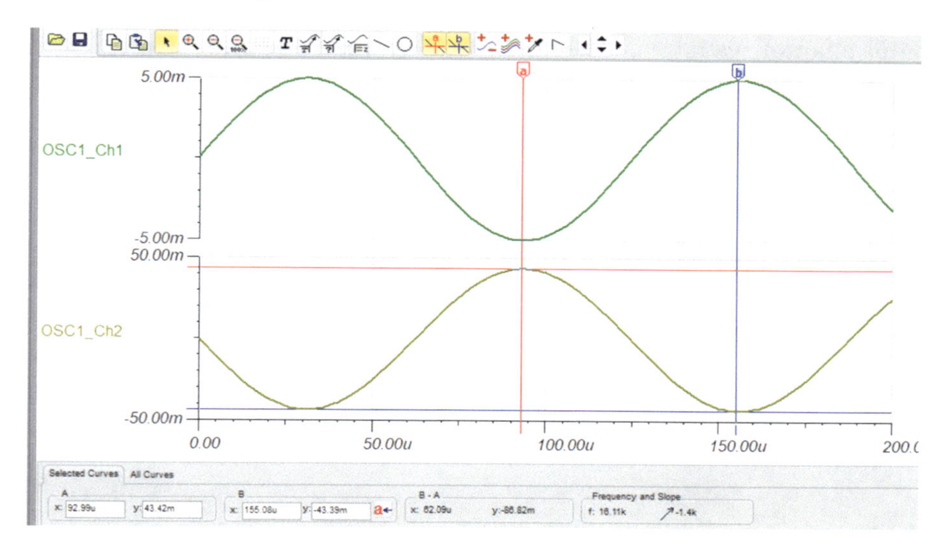


***Figure 5-18.*** *The Input and Output Voltages When Beta Was Set to 300*

If we take the peak-to-peak measurements from the two traces, we can see that the voltage gain is now

$$VGain = \frac{Vout\ Peak\ to\ Peak}{Vin\ Peak\ to\ Peak} = \frac{86.82m}{10m} = 8.682$$

The voltage gain has hardly increased, yet the Beta of the transistor has increased by a factor of 3.

Therefore, we have now achieved what we set out to achieve, an amplifier that does not suffer from thermal runaway and an amplifier whose gain is set purely by the ratio of $R_L$ divided by $R_E$ and is independent from the Beta of the transistor.

# Summary

In this chapter, we have studied the use of the load line and the output characteristics. We have learned about the drawbacks of the fixed bias amplifier and how the stabilized amplifier overcomes those drawbacks.

In the next chapter, we will study the ac analysis of a BJT amplifier. We will look at what bandwidth is and what affects it.

I hope you have found this chapter exciting and informative.

CHAPTER 6

# Further Analysis of the BJT Amplifier

In Chapter 5, we learned about using the load line to design a class A stabilized amplifier. We also studied the DC quiescent conditions of the amplifier. In this chapter, we are going to further our understanding of how the BJT amplifier works as an amplifier of ac signals. We will look at the ac model of an amplifier and compare its use as a method for designing amplifiers with the other methods we will use.

We will study how the amplifier responds to a range of frequencies and so define what is the bandwidth of the amplifier. We will look at the relationship between the voltage gain and the bandwidth of the amplifier.

Finally, we will look more closely into the class B amplifier and then finish off with a look at the push-pull amplifier.

## An Alternative Approach to Designing a Class A Amplifier

In the last chapter, we looked at using the load line to design amplifiers. In this chapter, we will restrict our designs to that of the stabilized amplifier, as we have shown in the previous chapter that this overcomes a lot of the drawbacks of the fixed bias amplifier. The circuit we are going to analyze is shown in Figure 6-1.


H. H. Ward, *Mastering Analog Electronics*, Maker Innovations Series,
https://doi.org/10.1007/979-8-8688-0245-4_6

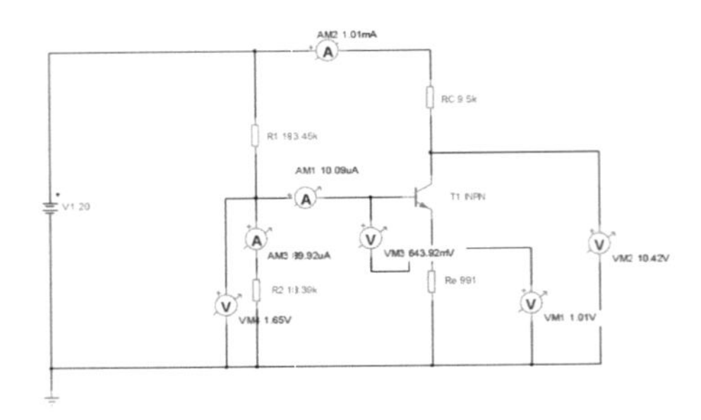


***Figure 6-1.*** *The DC Quiescent Setting for the Class A Amplifier*

We will use this example to describe an alternative approach to designing the amplifier. As before, there has to be a specification, and, in this case, it is as follows:

The DC quiescent collector current is 1mA.

The supply voltage is 20V.

We will set the DC quiescent emitter voltage, $V_E$, to 1V and assume the Beta value of the transistor is 100.

We can now start off by calculating the base and emitter currents flowing through the transistor as follows.

$$I_B = \frac{I_C}{\beta} = \frac{1E^{-3}}{100} = 10uA$$

$$I_E = I_C + I_B = 1E^{-3} + 10E^{-6} = 1.01mA$$

We can now calculate the resistor values around the circuit. First is the collector resistor $R_C$. We can calculate this using

$$R_L = \frac{V_{CC} - V_C}{I_C}$$

This is simply the volt drop across the collector resistor divided by the current flowing through it. Before we can calculate the value for $R_C$, we need to calculate the voltage $V_C$. To ensure the output can swing as far positive as it swings negative at the output, we need the DC quiescent voltage to be half the available voltage, and this can be calculated using

$$V_C = \frac{V_{CC} - V_E}{2} + V_E = \frac{20 - 1}{2} + 1 = 10.5V$$

We need to appreciate that the lowest voltage the collector can go down to, when the transistor turns fully on, is the $V_E$ voltage, as the emitter is not connected to ground. The resistor $R_E$ is between the emitter and ground; see Figure 6-1. This means that as long as current is flowing through the $R_E$ resistor, the emitter must be at a higher voltage than ground. The specification has stated that the DC quiescent voltage $V_E$ should be 1V.

The volt drop across $R_C$ is the $V_{CC}$ voltage minus the voltage at the collector, $V_C$. Using this concept, the value of $R_C$ can be calculated as follows:

$$R_C = \frac{V_{CC} - V_C}{I_C} = \frac{20 - 10.5}{1E^{-3}} = 9.5k$$

To calculate the emitter resistor $R_E$ value, we can use the volt drop across $R_E$, $V_E$, divided by the current flowing through it, $i_E$. Therefore, we have

$$R_E = \frac{V_E}{I_E} = \frac{1}{1.01E^{-3}} = 991\Omega$$

The next resistor to consider is $R_2$. Note $R_1$ and R2 create a voltage divider network that sets the voltage at the base; see Figure 6-1. To calculate the value for $R_2$, we can use the volt drop across $R_2$ divided by the current flowing through it. Therefore, we have

$$R_2 = \frac{V_B}{I_{R2}}$$

The voltage $V_B$ is the voltage at the base of the transistor. The term $V_{BE}$ represents the voltage across the base emitter diode, and it is approximately 0.7V. However, empirical analysis with the TINA software shows that TINA sets this more closely to 0.655V. As the diode is conducting, then this volt drop would be in series with the emitter voltage as shown in Figure 6-1. This means the voltage at the base, $V_B$, can be calculated as follows:

$$V_B = V_{BE} + V_E = 0.655 + 1 = 1.655V$$

The current that will flow through $R_2$ will be the current that flows through $R_1$ minus the base current, $I_B$, that flows into the transistor. The current that flows through $R_1$ is termed the "bleed current." It is normal to set the base current to be approximately 10% of this bleed current. Therefore, the current through the $R_2$ resistor is the remaining 90% of this bleed current. Therefore, we have the following relationships:

$$I_{R2} = 9 \times I_B = 9 \times 10E^{-6} = 90uA$$

$$I_{R1} = 10 \times I_B = 10 \times 10E^{-6} = 100uA$$

Using this value for $I_{R2}$, we can calculate the value for $R_2$ as follows:

$$R_2 = \frac{V_B}{I_{R2}} = \frac{1.655}{90E^{-6}} = 18.39k$$

Similarly, we can determine the value for $R_1$ knowing it is the volt drop across $R_1$ divided by the current flowing through it. This can be done as follows:

$$R_1 = \frac{V_{CC} - V_B}{I_{Bleed}} = \frac{20 - 1.655}{100E^{-6}} = 183.45k$$

We have used the DC quiescent values to calculate the resistor values around the circuit. We have set the resistor values in the circuit shown in Figure 6-1 to those calculated here. The measurements of the DC quiescent voltages and currents around the circuit in Figure 6-1 do agree very closely with the calculations here.

I hope you can see that this method can be used to design a class A amplifier.

# Exercise 6.1

As an exercise, try using this method to design a class A amplifier to the following specification:

- The VCC should be 15V.
- The DC quiescent collector current should be 1.5mA.

- The DC quiescent emitter voltage should be 1V.
- The typical beta value of 100 should be assumed.
- My circuit design will be shown in the appendix.

# The Mid-Frequency ac Model of an Amplifier

We will take a look at the ac model of the amplifier circuit shown in Figure 6-1. We will look at how we can move from the actual circuit, shown in Figure 6-1, to the actual model as shown in Figure 6-2.

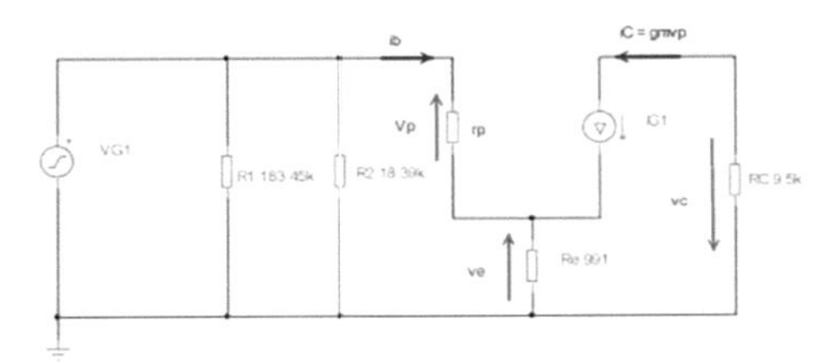


***Figure 6-2.*** *The Mid-Frequency ac Model of the Class A Amplifier*

The first step is to replace all DC sources with a short circuit. This is because they will have a large smoothing capacitor across the output, and so, to ac, this would be a short circuit. Then replace all capacitors in the circuit with a short circuit, as to ac the capacitance has such a large value that their impedance would be zero. Doing this would give us the interim circuit shown on Figure 6-3.

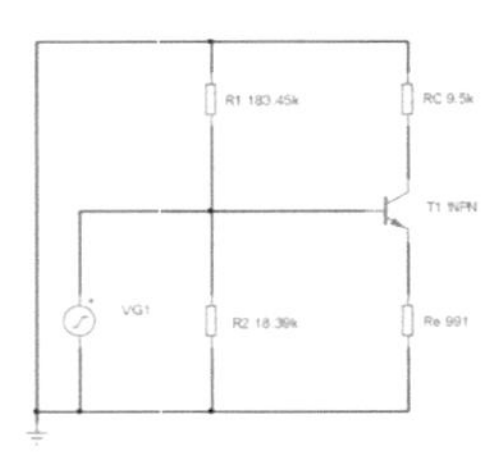


***Figure 6-3.*** *The First Interim Circuit*

From this circuit, we can see that the top rail that was at VCC is connected to the ground rail. This means that the biasing resistor, $R_1$, is in parallel with the other biasing resistor, $R_2$. Also, the collector resistor $R_C$ would be in parallel with the series combination of the transistor, in series with the emitter resistor, $R_E$. This would produce the second interim circuit shown in Figure 6-4.

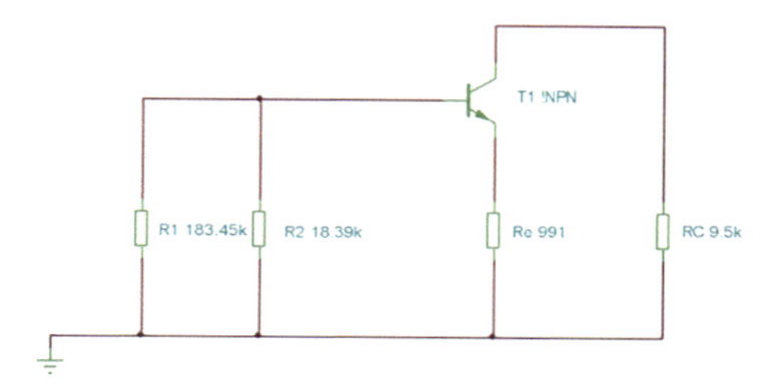


***Figure 6-4.*** *The Second Interim Circuit*

Now all that is left to do is replace the transistor with the ac model of a transistor, which is shown in Figure 6-5.

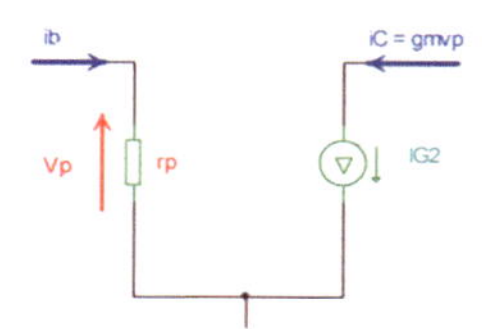


***Figure 6-5.*** *The ac Model of a Transistor*

Putting all this together does produce the circuit as shown in Figure 6-2.

We will use this model, as shown in Figure 6-2, to complete an ac analysis of the amplifier circuit shown in Figure 6-1. We would be making a generic model if we didn't use the resistor values that we have taken from the amplifier circuit shown in Figure 6-1. Before we go too far into looking at the model, we will decide what the main parameters are of the amplifier that are of interest. These are listed here:

1. The input resistance or impedance of the transistor
2. The input resistance or impedance of the amplifier circuit
3. The output resistance or impedance of the amplifier circuit
4. The current gain of the circuit
5. The voltage gain of the circuit

# The Input Resistance of the Transistor

We will look at item 1 now. One method to determine the input resistance is to use the measurements from the circuit shown in Figure 6-1. This is because we are modeling that particular amplifier circuit. The easiest way to determine this input resistance is to measure both the voltage at the input and the current flowing into the input. The input of the amplifier is the voltage at the base of the transistor, that is, $V_B$. From Figure 6-1, this voltage was 1.68V; therefore, $V_B = 1.68V$. We are using the DC quiescent values with the ac input set to zero. This is not an issue as the parameters of the transistor are not reactive, they are resistive, and so the DC resistance would be the same as the ac resistance or impedance.

Again, reading from Figure 6-1, we can see that the input current is simply the base current $I_B$ which is 10.09μA. Using the following relationship

$$R = \frac{V}{I}$$

We can say

$$Transistor\ Input\ R = \frac{Input\ Voltage}{Input\ Current} = \frac{V_B}{I_B} = \frac{1.65}{10.09E^{-6}} = 163.528k\Omega$$

# Determine the Input Resistance of the Transistor $R_b$ Using the ac Model

We can use Thevenin's theory to determine the input impedance. To do this, we remove the VG1 voltage source, as we are looking back into the circuit at those two terminals. We must also replace all sources with their ideal internal impedance. For a voltage source, this would be a short circuit, and for a current source, this would be an open circuit. There is only one source in the ac model, and that is the current source associated with the transistor model. We must replace that current source with an open circuit. This would give us the circuit as shown in Figure 6-6.

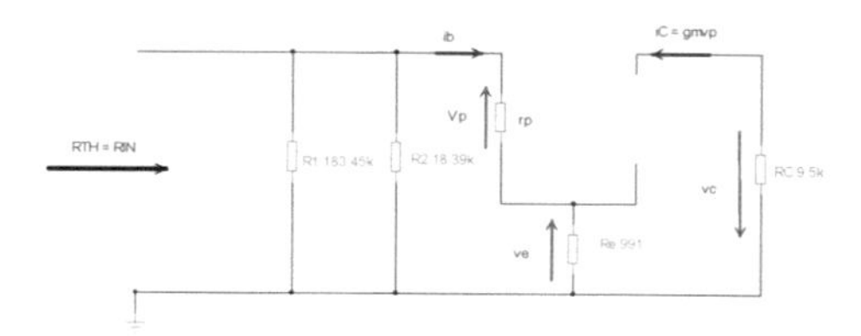


***Figure 6-6.*** *The Input Section of the Class A Amplifier*

Using the circuit shown in Figure 6-6, it should be clear that, due to the open circuit where the current source has been, we don't need to include $R_C$ in this calculation. Indeed, we can see that the input resistance, $R_{IN}$, is simply the parallel combination of $R_1$ and $R_2$ that are also in parallel with the series combination of $r_\pi$ and $r_e$. Note, in the schematics, I am using "$r_p$" instead of "$r_\pi$" as TINA has limited characters we can easily use.

We need to determine the series combination of $r_\pi$ and $r_e$, which we will call $R_b$, but this is not as simple as just adding them together. This is because the emitter resistor, $r_e$, appears in the output side of the amplifier. Indeed, we can see that the current that flows through $r_\pi$ is just the base current $i_b$, whereas the current flowing through the emitter resistor is $i_e$. We can use KVL to help us with this, as the two volt drops, $v_\pi$ and $v_e$, are in the closed loop in which the voltage applied to that loop is the base voltage $v_b$. Using KVL, we can say

$$v_b = v_\pi + v_e$$

We can express the two voltage drops as

$$v_\pi = i_b r_\pi$$

$$v_e = i_e R_e$$

We can express the emitter current as

$$i_e = i_c + i_b$$

We know

$$i_c = \beta i_b$$

This means

$$i_e = \beta i_b + i_b = i_b(\beta + 1)$$

This means we can say

$$v_b = i_b r_\pi + i_b(\beta + 1)R_e$$

If we take the term $i_b$ out as a common factor, we get

$$v_b = i_b\left[r_\pi + (\beta + 1)R_e\right]$$

From this, we can say

$$\frac{v_b}{i_b} = \left[r_\pi + (\beta + 1)R_e\right]$$

This means that the input resistance at the base, which we will call $R_b$, is

$$R_b = \left[r_\pi + (\beta + 1)R_e\right]$$

This would be in parallel with the parallel combination of $R_1$ and $R_2$. This means that the total expression for $R_{IN}$ is $R_1//R_2//R_b$.

We can't evaluate this expression yet as we don't have a value for $r_\pi$. To help us with this, we can look at the expression for the collector current $i_c$. We have two expressions for this current, which are

$$i_c = g_m v_\pi$$

$$i_c = \beta i_b$$

This means

$$\beta i_b = g_m v_\pi$$

We know

$$i_b = \frac{v_\pi}{r_\pi}$$

This means

$$\frac{\beta v_\pi}{r_\pi} = g_m v_\pi$$

From this, we can say

$$\frac{\beta v_\pi}{g_m v_\pi} = r_\pi$$

Therefore, we have

$$r_\pi = \frac{\beta}{g_m}$$

If we could get a value for $g_m$, we could then get a value for $r_\pi$. The term $g_m$ is the transconductance for the amplifier, and it can be calculated using the following expression:

$$g_m = \frac{I_{CQ}}{V_T}$$

where $I_{CQ}$ is the DC quiescent collector current, and $V_T$ is the thermal voltage associated with the conductance of a diode. It has been found that at an ambient temperature of 25°C, $V_T$ has a value of 26mS. Using the values of the DC circuit shown in Figure 6-1, we can see that $I_{CQ}$ = 1.01mA. Using this value for $I_{CQ}$, we can calculate the value for $g_m$ as follows:

$$g_m = \frac{1.01E^{-3}}{26E^{-3}} = 0.0388$$

Knowing β = 100, we can calculate the value for $r_\pi$ as follows:

$$r_\pi = \frac{\beta}{g_m} = \frac{100}{0.0388} = 2.577k$$

We can now calculate the value for $R_b$ as follows:

$$R_b = \left[r_\pi + (\beta + 1)R_e\right] = \left[2.577E^3 + (100 + 1)991\right] = 102.668k$$

We can now calculate the total input resistance using

$$R_{IN} = \frac{1}{\frac{1}{R_1} + \frac{1}{R_2} + \frac{1}{R_b}} = \frac{1}{\frac{1}{183.45E^3} + \frac{1}{18.39E^3} + \frac{1}{102.668E^3}} = 14.374k$$

## The Output Resistance

We can again use Thevenin's theory to determine the output resistance, $R_{OUT}$, of the ac model. The circuit to determine ROUT is shown in Figure 6-7.

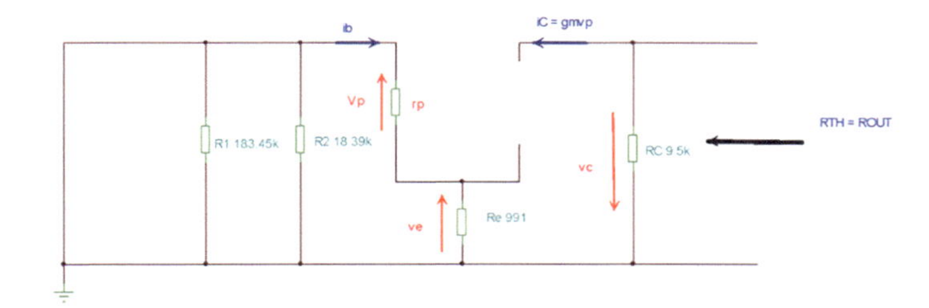


***Figure 6-7.*** *The Circuit to Determine $R_{OUT}$ Using Thevenin's Theory*

We have replaced the voltage source with a short circuit and the current source by an open circuit, their ideal internal impedance. The open circuit between the top of $R_C$ and the emitter resistor $R_e$ means that the only resistor that we would measure at the output would be the collector resistor $R_C$. This means that

$$R_{OUT} = R_C$$

In this case, with the circuit as shown in Figure 6-1, we have

$$R_{OUT} = 9.45k\Omega$$

# The Voltage Gain $A_V$

We should be able to determine an expression for the voltage gain $A_V$ from the ac model. We start with the expression

$$A_V = \frac{V_{out}}{V_{in}}$$

We should derive the expression for the $V_{out}$ and $V_{in}$.
With respect to Vout, we can say

$$v_{out} = -i_c R_c$$

The current "$i_c$" is shown as being negative as, when we look at the ac model, the current "$i_c$" is shown as flowing down through the current source and the emitter resistor. This is correct, as we know the conventional current will flow through the NPN transistor in this direction. We can also see that the arrowhead of the volt drop across the collector resistor is shown as pointing to the ground line. This is also correct as we know the two volt drops "$v_e$" and "$v_c$" will add together. The fact that the positive end of the volt drop is at the ground rail, that is, at 0V, means that the output voltage, "$v_{out}$", which is taken across $R_C$ with respect to the ground rail, must be negative to make the ground rail more positive than the top of $R_C$. This means the following expression must be true:

$$v_{out} = -i_c R_c$$

This means the collector current must be negative. I appreciate this explanation is a bit awkward to read, but if you read it carefully, I hope you will be able to follow it.

We know from before that

$$i_c = g_m v_\pi$$

From before, we know

$$i_b = \frac{v_\pi}{R_\pi}$$

This means

$$v_{out} = -g_m v_\pi R_c$$

If we now look at $V_{in}$, we can say

$$v_{in} = v_\pi + v_e$$

From before, we can say

$$v_e = (i_b + i_c) R_e \; therefore; v_e = \left( \frac{v_\pi}{R_\pi} + g_m v_\pi \right) R_e$$

This means that

$$v_{in} = v_\pi + \left( \frac{v_\pi}{R_\pi} + g_m v_\pi \right) R_e$$

Therefore, taking $v_\pi$ out as a common factor, we have

$$v_{in} = v_\pi \left[ 1 + \left( \frac{1}{R_\pi} + g_m \right) R_e \right]$$

We can use this to create an expression for the voltage gain $A_V$ as follows:

$$A_v = \frac{v_{out}}{v_{in}} = \frac{-g_m v_\pi R_c}{v_\pi \left[ 1 + \left( \frac{1}{R_\pi} + g_m \right) R_e \right]} = \frac{-g_m R_c}{1 + \left( \frac{1}{R_\pi} + g_m \right) R_e}$$

We should appreciate that the term $1/R\pi$ is very small; it can be ignored, for example, in this case:

$$\frac{1}{R_\pi} = \frac{1}{2.577E^3} = 3.88E^{-4}$$

This means the voltage gain can be expressed as

$$A_v = \frac{-g_m R_c}{1 + g_m R_e}$$

Using the values from the circuit, we have

$$A_v = \frac{-0.0388 \times 9.45E^3}{1 + 0.0388 \times 991} = 9.29$$

If we consider the denominator which is

$$1 + 0.0388 \times 991 = 1 + 38.4508 = 39.4508$$

We could ignore the 1 in the denominator. This would mean that the expression for the voltage gain AV is

$$A_v = \frac{-g_m R_c}{g_m R_e} = \frac{-R_c}{R_e}$$

If we use this simplified expression for the voltage gain, we get

$$A_v = \frac{-9.45E^3}{991} = 9.536$$

## Is the ac Model Valid

The ac model shown in Figure 6-3 is an accepted model, but does the calculations match the simulation of the circuit shown in Figure 6-1?

We can check the expression for the gain by simulating the circuit shown in Figure 6-8.

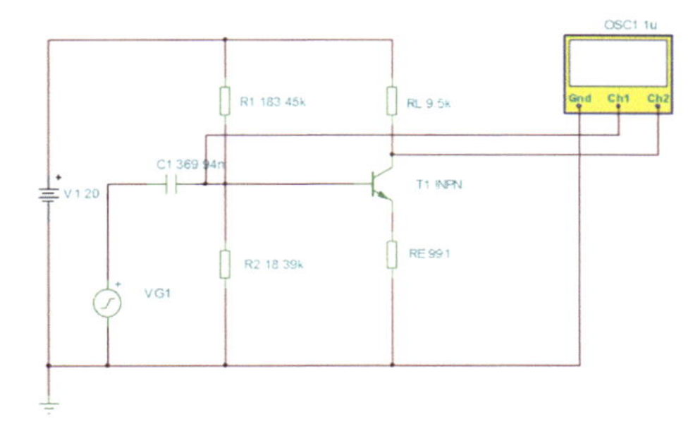


***Figure 6-8.*** *The Test Circuit for the Amplifier Voltage Gain*

We can use the circuit shown in Figure 6-8 to measure the input and output voltage waveforms so that we can determine the voltage gain "$A_v$". The display of the two waveforms is shown in Figure 6-9.

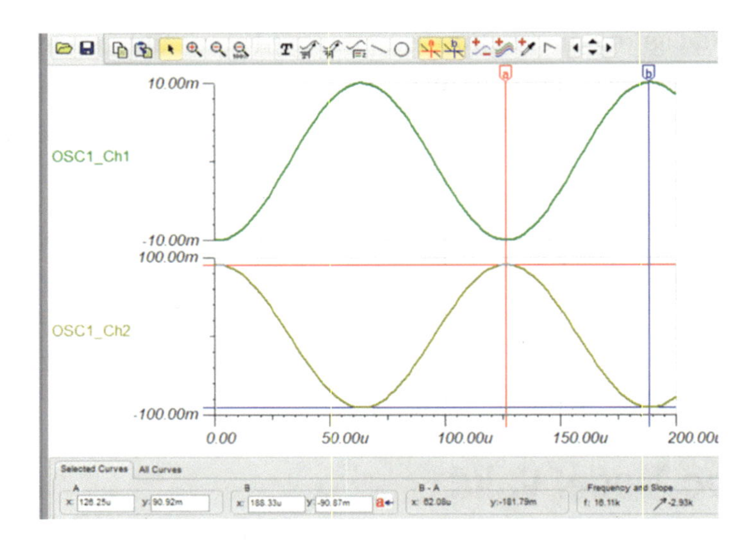


***Figure 6-9.*** *The Input and Output Voltage Traces*

Using Figure 6-9, we can see that the peak-to-peak output voltage is 181.79mV, and the peak-to-peak input voltage is 20mV. This means that the voltage gain is

$$A_v = \frac{Peak - to - Peak\ Vout}{Peak - to - Peak\ Vin} = \frac{181.79mV}{20mV} = 9.0895$$

This is close to the first calculation from the model which gave $A_V$ = 9.29 and also the value of $A_V$ = 9.536. These are 2.2% and 4.91% errors. We could live with them.

If we compare the input and output resistance, we can see that the value for the output resistance is the same when using the ac model or the actual circuit. However, the input resistance values are not the same. Using the values from the circuit, the input resistance of the transistor Rb was

$$R_b = 163.528k$$

However, using the ac model, the value of Rb was

$$R_b = 102.688k$$

If we look at how the two values were obtained, we might see where the problem was. The first value was taken from the readings from the simulation shown in Figure 6-1 as follows:

$$R_b = \frac{v_b}{i_b} = \frac{1.65}{10.09E^{-6}} = 163.528k\Omega$$

Using the ac model, the value for $R_b$ was

$$R_b = \left[r_\pi + (\beta + 1)R_e\right] = \left[2.577E^3 + (100+1)991\right] = 102.668k$$

The term that is not the same as the simulated circuit is the value for $r_\pi$. This was

$$r_\pi = \frac{\beta}{g_m} = \frac{100}{0.0388} = 2.577k$$

To calculate $r_\pi$, we used the value for β and $g_m$. The value for β is one that we set when we simulate the circuit, so that is not the problem. The problem may lie with the value for $g_m$, the transconductance of the transistor. The value for $g_m$ from the ac model was calculated using

$$g_m = \frac{I_{CQ}}{V_T} = \frac{1.01E^{-3}}{26E^{-3}} = 0.0388$$

We could use values from the simulated circuit to calculate $g_m$ as follows:

$$g_m V_\pi = \beta i_b$$

From this, we can say

$$g_m = \frac{\beta i_b}{v_\pi}$$

In the ac model, the voltage $v_\pi$ is shown as being the voltage across the base emitter junction. If we assume that this means

$$v_\pi = v_{be} = 0.64392$$

Then we can say

$$g_m = \frac{100 \times 10.09E^{-6}}{0.64392} = 1.567mS$$

If we use this value to calculate the resistance $r_\pi$, we get

$$r_\pi = \frac{100}{1.567^{-3}} = 63.817k$$

Also, assuming $v_\pi = v_{be}$, we can calculate the value of $r_\pi$ as follows:

$$r_\pi = \frac{v_\pi}{i_b} = \frac{0.64392}{10.09E^{-6}} = 63.817k$$

Using this value for $r_\pi$, we can calculate the input resistance at the transistor, $R_b$, as follows:

$$R_b = \left[r_\pi + (\beta + 1)R_e\right] = \left[63.817E^3 + (100 + 1)991\right] = 163.908k$$

This agrees more closely, indeed very closely, to the value calculated using the base voltage divided by the base current using the readings from the simulation.

When you look at different approaches to determine the same parameters, it is difficult to see which approach you prefer. The use of the ac model has enabled us to derive an expression for the voltage gain that works well. It is really when calculating the value of $r_\pi$ and $g_m$ that the ac modeling has let us down. Really, I think it is the calculation for $g_m$ such as

$$g_m = \frac{I_{CQ}}{V_T}$$

I think the problem is with choosing the value for $I_{CQ}$, the DC quiescent collector current. I think the method I have used to calculate the transconductance $g_m$ is more reliable, that is

$$g_m = \frac{\beta i_b}{v_{be}}$$

This relies on the voltage $v_\pi$ equaling $v_{be}$. It is really up to you which method you use, or perhaps you can explain where the ac model theory is going astray. I am happy to use the simulation results to calculate the quantities along with the various relationships I have shown in this section of the text. The rest of the chapter will use that approach.

## The (b+1) Parameter

Before we leave our look at the ac model, I think it is worthwhile stating that the analysis has shown how we got this multiplier of (β+1) when referring the emitter resistor, "$r_e$", to the base, that is, $(\beta+1)R_e$. This is used in calculating the gain and the input resistance. The ac model analysis has shown us where this has come from.

Also, the analysis has shown us that the dependence on the value of "β" has been removed from the expression for the voltage gain $A_V$ as

$$A_v = \frac{-g_m R_c}{1 + g_m R_e}$$

or

$$A_v = \frac{-g_m R_c}{g_m R_e} = \frac{-R_c}{R_e}$$

The term "β" has been removed from the expressions, especially the last expression.

## The Bandwidth of the BJT Amplifier Circuit

The circuit, shown in Figure 6-1, is using the NPN transistor option from TINA. This is a standard transistor that can be used for general amplification. The Beta, which can vary quite a bit, has been set to 100. We are going to use this circuit to look at the bandwidth of the amplifier. The bandwidth of the amplifier is the band, or range, of frequencies over which the gain of the amplifier is deemed high enough. This is the range of frequencies over which the gain is more than 3dbs below the maximum gain. We stated in Chapter 5 the -3db level is the half power point level, and it is the benchmark that engineers use to measure the performance of electronic circuits against. The Bode plot shown in Figure 6-12 is the plot for the amplifier circuit shown in Figure 6-11. Before we could run the analysis to display the Bode plot, we must first determine the value for the capacitor $C_1$. This capacitor is used to create a high pass filter at the input to the transistor. To determine the value of the capacitor, we need to consider the high pass filter created at the input, as shown in Figure 6-10.

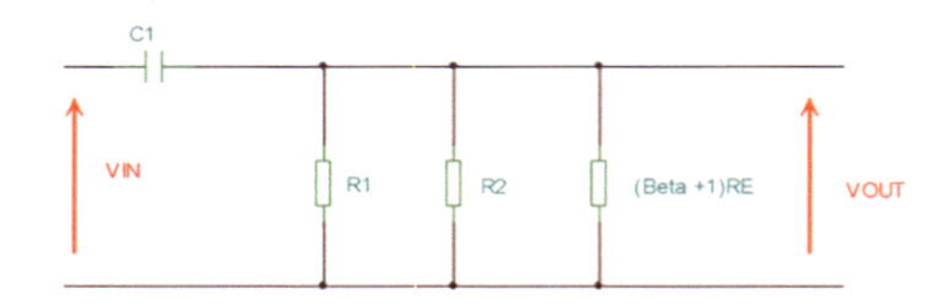


***Figure 6-10.*** *The High Pass Filter Circuit at the Input of the Transistor*

The theory behind creating this CR high pass filter has been described in Chapter 5.

We can say R = $R_1$, $R_2$, and (Beta+1)$R_E$ in parallel with each other, as shown in Figure 6-2. Putting in the value from the circuit, we calculate it as follows:

$$\frac{1}{R} = \frac{1}{R1} + \frac{1}{R2} + \frac{1}{(\beta+1)RE} = \frac{1}{183.45E^3} + \frac{1}{18.39E^3} + \frac{1}{101 \times 991}$$

$$\frac{1}{R} = 5.451E^{-6} + 5.438E^{-5} + 9.991E^{-6} = 6.982E^{-5}$$

Therefore

$$R = \frac{1}{6.982E^{-5}} = 14.32k$$

I feel I should point out that the high pass filter circuit, shown in Figure 6-10, does miss out the $r_\pi$ resistor that is in the ac model. However, most models are using some rules of thumb, and the following simulation does agree closely with our calculations. Again, it is up to you to decide which method you want to use.

We can now calculate the value of the capacitance $C_1$ as follows:

$$C1 = \frac{1}{2\pi \times 30 \times 14.32E^3} = 370.41nF$$

Inserting this capacitor value for $C_1$ into the amplifier circuit and adding an ac signal, we have the test circuit as shown in Figure 6-11.

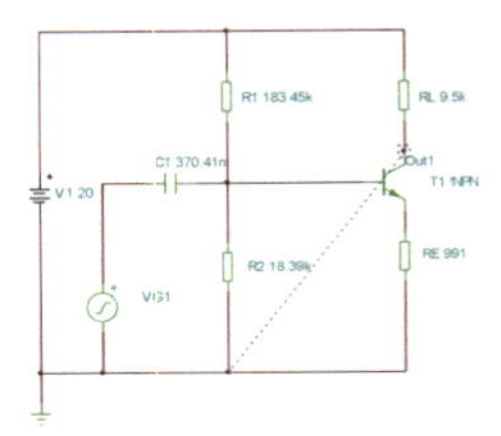


***Figure 6-11.*** *The Test Circuit for the ac Bandwidth*

Then, running the AC Transfer Characteristics analysis allows us to create the Bode plot shown in Figure 6-12.

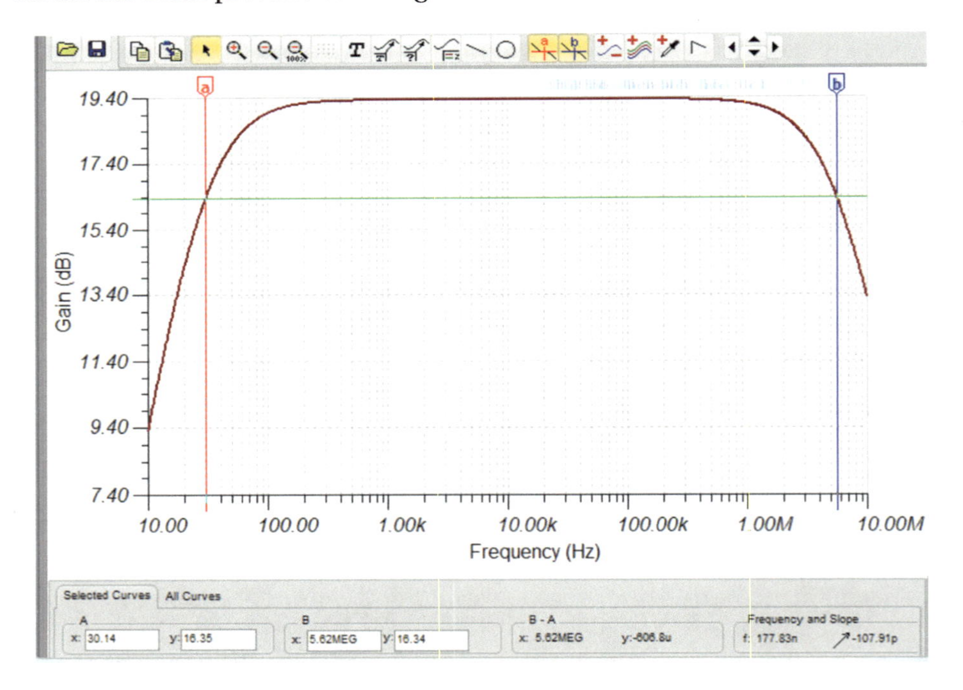


***Figure 6-12.*** *The Bode Plot for the Amplifier Circuit Shown in Figure 6-3*

With this analysis, I have restricted the display to just the display that just shows the amplitude and increased the number of points to 1000. The plot shows how the gain of the amplifier, expressed in dBs, varies as the frequency, displayed on a logarithmic scale, varies. I have extended the end frequency to end at 10Mhz. This is so that we can see how the gain falls off at the high frequencies. Both cursor "A" and cursor "B" have been placed at the point where the gain has fallen down to 16.35dBs. This is 3dBs down from the maximum gain, as the maximum gain was measured at 19.35dBs. This maximum gain is shown as being fairly constant from about 100Hz to around 1MHz.

As the two cursors are at the points where the gain is 3dBs down from this maximum, then the bandwidth is from the frequency at cursor A, that is, 29.99 Hz, to the frequency at cursor B, that is, 5.62MHz. The bandwidth of the amplifier is defined as the range of frequency over which the gain is greater than 3dBs down from the maximum value of the gain. This is simply $F_2 - F_1$ where $F_2$ is the high frequency at which the gain has fallen down by 3dBs and $F_1$ is the lower frequency. In this case, the Bode plot shows that $F_2$ is 5.62 MHz and $F_1$ is 29.99Hz. Therefore, in this example, the bandwidth, BW, is

$$BW = 5.62E^6 - 29.99 = 5.62E^6 = 5.62MHz$$

This is a very wide bandwidth, however, as we are trying to design an audio amplifier then the bandwidth is more than enough for our purpose.

We need to appreciate what effects the roll-off of the bandwidth. We should realize that the early roll-off has the cutoff frequency of 30.14HZ, that is the frequency at which the gain drops down by 3dBs from the maximum. This is due to the CR high pass filter we have created at the input of the transistor. This is what we expected and confirms that our analysis and calculations are correct. So, what allows the roll-off at the upper frequency? Well, we should appreciate that, just like all systems, including humans, electronic circuits will find the easiest route to return the current, that carries the energy around the circuit, to the ground path back to the supply. We should also appreciate that, of the two reactive components, the capacitor and the inductor, it is the capacitor whose impedance reduces as the frequency increases. This is shown in the expression for XC, the impedance of the capacitor, as

$$XC = \frac{1}{2\pi fC}$$

We know there are two diodes, made from the two PN junctions, internal to the transistor. We have shown in Chapter 2 that there is some capacitance across the depletion layer of the diode. This means there are two main stray capacitances within the transistor, and they are

- CBCO: The capacitance between the base and collector with the base open circuited
- CBEO: The capacitance between the base and emitter with the base open circuited

It is the CBEO which is trying to provide a path for the input signal to bypass the transistor and use this path to take the signal to ground via the $R_E$ resistor. In this way, as the frequency rises, this stray capacitor provides an easier path for the signal to use, and so the signal does not go through the transistor, which means it does not get amplified. TINA gives this CBEO capacitor a value of 14.8pF. If we change this value to 14.8nF, then the bandwidth might change. The TINA ECAD software allows us to change this value of capacitance and many other parameters of the transistor. If we change this capacitance value, we can create another Bode plot to confirm if the bandwidth does change. We can use this to confirm the concepts as to why we have the upper roll-off as described here. The new Bode plot is shown in Figure 6-13.

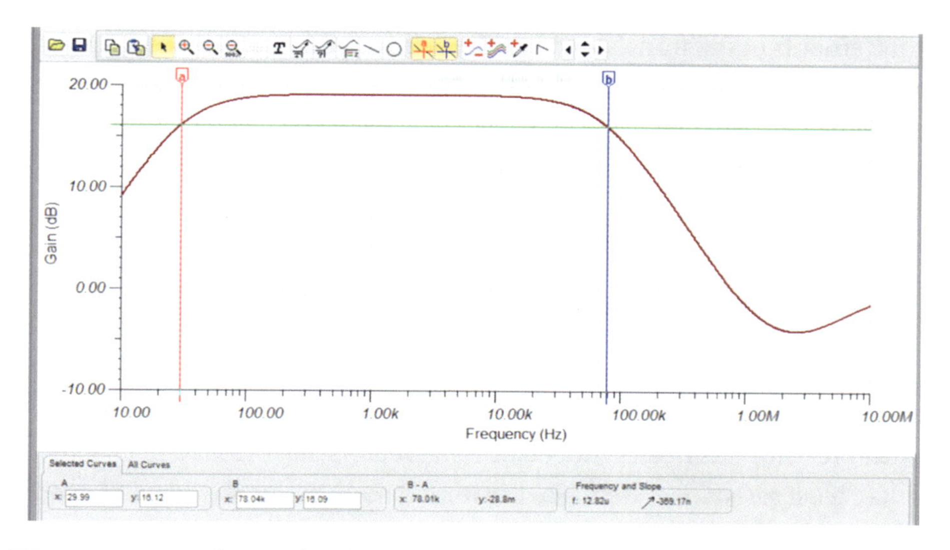


***Figure 6-13.*** *The Bode Plot When the CBEO Value Is Changed to 14.8nF*

Figure 6-13 shows clearly how increasing the value of CBEO reduces the bandwidth to around 78kHz.

We will now move on to see what else we can do with this transistor. Don't forget to return the CBEO back to 14.8pF.

## The Emitter Bypass Capacitor

We have shown that the gain of the amplifier is simply the ratio of $R_C$ divided by $R_E$. That is OK, but unless we have a very large value for $R_C$ and a low value for $R_E$, then the voltage gain can be quite small. We will add a further component to the circuit that might change this. The component is a capacitor that is put in parallel with the emitter resistor, hence the name "The Emitter Bypass Capacitor." To decide what value of capacitance we need to use, we have to understand the reason why we want to place the capacitor in parallel with $R_E$. We have chosen the value of $R_E$, indeed the value of all the resistors in the circuit, based on the DC quiescent requirements of the transistor. If we don't do anything about it, it will be these quiescent conditions that control the ac gain. Really, we need to modify the gain of the circuit to respond to an ac signal without affecting the DC quiescent conditions. It is the ratio of $R_C$ and $R_E$ that controls the DC gain. Hence, if we could reduce the value of $R_E$, we could increase the gain. But how do we do that without changing the DC quiescent operating conditions? The answer is to give the ac signal a path that ignores the emitter resistor $R_E$. That is what the emitter bypass capacitor does; it provides the ac signal with a path that is easier for it to flow through than the $R_E$ resistor. In this way, we could make the ac signal think that the emitter resistor has been reduced or removed. As all current will take the path of least resistance, we make the impedance, XC, of the capacitive path equal to one tenth that of the resistance of $R_E$, that is

$$XC = \frac{R_E}{10}$$

In our example, $R_E$ is 991Ω (see Figure 6-1), and so XC should be set to 100Ω. We need this to happen at the lowest frequency of use, that is, in this case, 30Hz. Therefore, putting these values onto the expression for C, we have

$$C = \frac{1}{2\pi fXC} = \frac{1}{2\pi \times 30 \times 99.1} = 53.53uF$$

If we add this to the circuit, we can simulate it and see what effect it has. The circuit is shown in Figure 6-14.

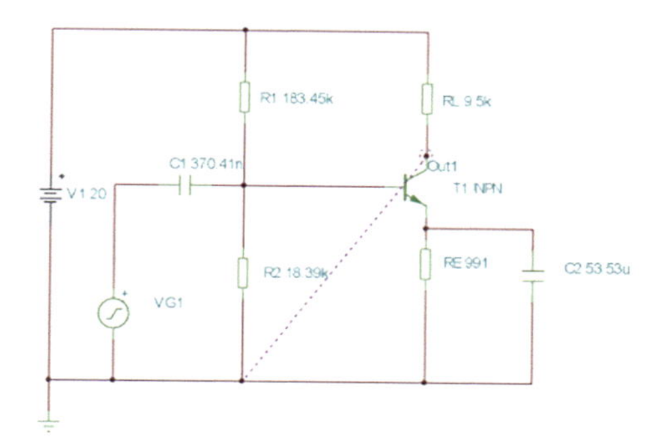


***Figure 6-14.** The Amplifier Circuit with the Emitter Bypass Capacitor Added*

The capacitor $C_2$ is the emitter bypass capacitor, and it will allow the ac signal to bypass the emitter resistor at frequencies of 30Hz and above. This should then mean that the ac voltage gain is more than the 19.14dbs as measured before. If we simulate the circuit, we can examine the Bode plot as shown in Figure 6-15.

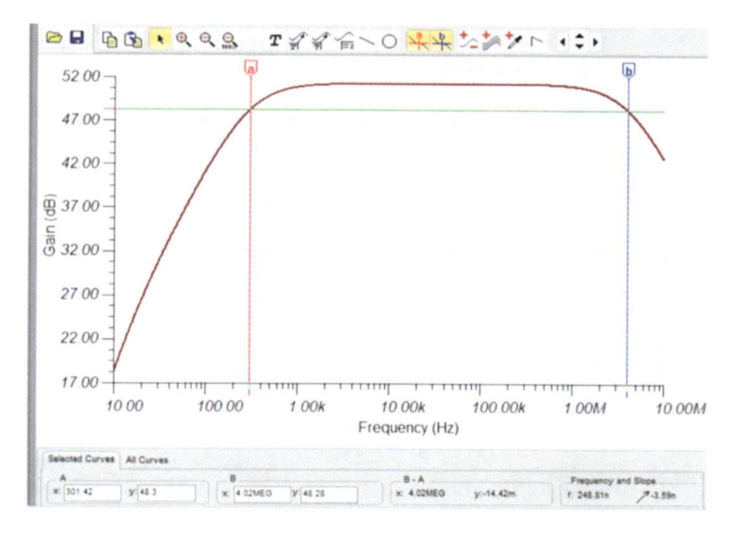


***Figure 6-15.** The Bode Plot After Adding the Emitter Bypass Capacitor*

We can see that the gain has increased as expected; it is 51.29dBs. However, there is a more important change in that the lower roll-off frequency has increased to around 301Hz. This is higher than the minimum audio frequency of around 50Hz, and so we might miss out on some of the sounds we want to hear, especially some of the low bass sounds. Also, we can see that the overall bandwidth has reduced. It is now around 4.02MHz as opposed to the 5.62MHz from before; see Figure 6-12.

We can try to change the bypass circuit by splitting the emitter resistor into two resistors and placing the bypass capacitor across the resistor closest to the ground. This arrangement is shown in the circuit in Figure 6-16.

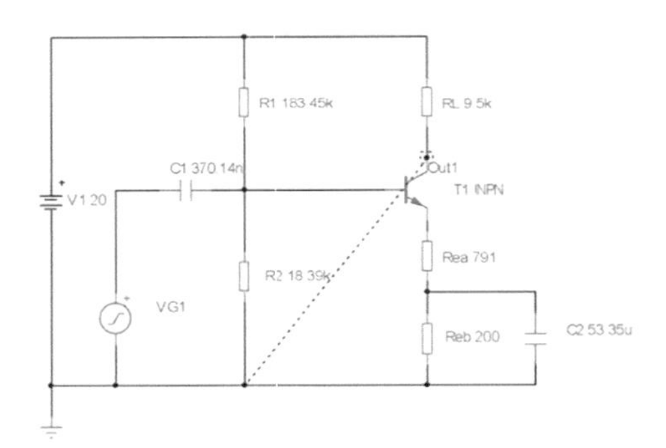


***Figure 6-16.*** *The Split Emitter Bypass Circuit*

The Bode plot for the frequency analysis of this circuit is shown in Figure 6-7.

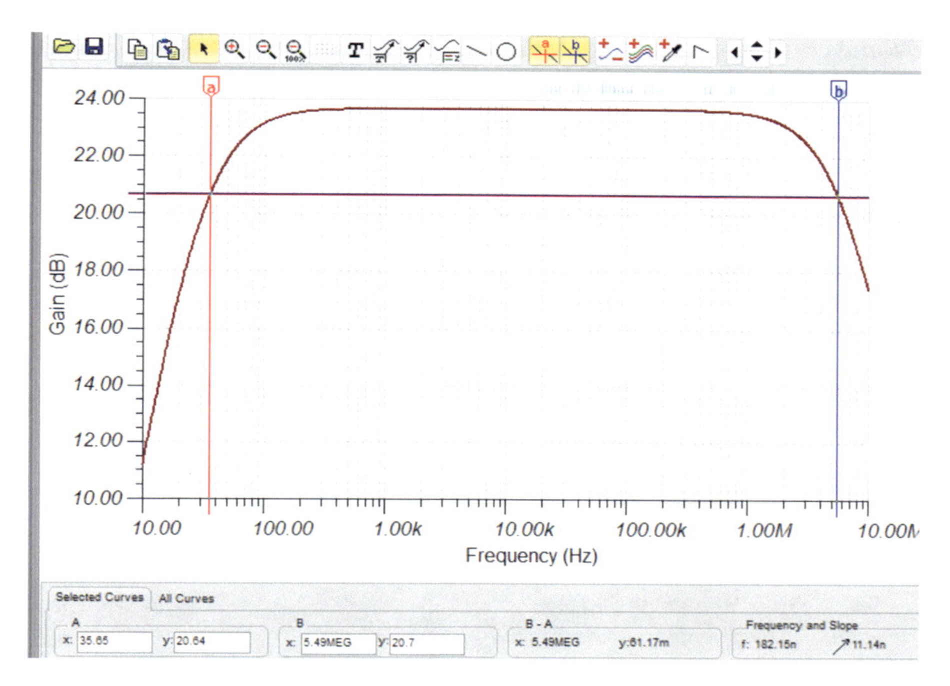


***Figure 6-17.*** *The Bode Plot for the Split Emitter Resistor Circuit*

Using the cursors, the maximum gain was measured at 23.67dBs. This is a reduction in gain, but we can see that the low-frequency roll-off is now at 35.65Hz, which is closer to the 30Hz we designed the circuit for. Note, the BW was measured at 5.49MHz. We can try and calculate the voltage gain using the ratio of the $R_L$ and $R_E$ as before. Therefore

$$V_{Gain} = \frac{R_L}{R_{E1}} = \frac{9.5E^3}{0.591E^3} = 16.07$$

Expressing this in dBs, we have

$$V_{Gain} = 20\,Log\left(16.07\right) = 24.12Bs$$

This value is very close to the 23.67dBs measured in the Bode plot. This could suggest that splitting the emitter resistor as we have done has given us some control of the voltage gain. However, to DC the circuit will see the two resistors in series as just the same 991Ω resistor as before. Therefore, the DC quiescent operating conditions will not have changed.

To test this concept further, we will change the $R_{ea}$ to 791Ω and the $R_{eb}$ to 200Ω. This should change the voltage gain as follows:

$$V_{Gain} = 20\,Log\left(\frac{9.5E^3}{0.791E^3}\right) = 21.591dBs$$

After making those changes and simulating the circuit, the Bode plot gave the following results:

- The maximum gain was 21.2dBs.
- The low end 3dB point was at a frequency of 33.98Hz.
- The BW was 5.62MHz.

Now changing $R_{ea}$ to 200Ω and $R_{eb}$ to 791Ω would give the gain at

$$V_{Gain} = 20\,Log\left(\frac{9.5E^3}{0.2E^3}\right) = 33.53dBs$$

After making those changes and simulating the circuit, the Bode plot gave the following results:

- The maximum gain was 32.4 dBs.
- The low end 3dB point was at a frequency of 53.6Hz.
- The BW was 5.23MHz.

These results do suggest that using the split emitter approach, with the bypass capacitor, does give us a good range of control over the gain of the amplifier while maintaining the bandwidth, with good control over the low-frequency roll-off.

One suggestion we could look at is using the setting of $R_{ea}$ at 791Ω and $R_{eb}$ at 200Ω, as this gave a good gain while still keeping the low-frequency roll-off very close to our desired 30Hz. Then, to get more voltage gain, we could design a two-stage amplifier circuit using the same parameters. The circuit for this two-stage amplifier is shown in Figure 6-18.

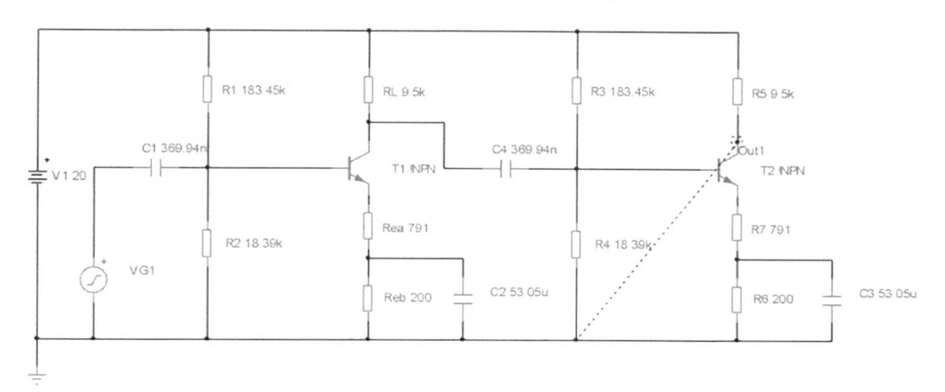


***Figure 6-18.*** *The Two-Stage Amplifier*

We can use the AC Transfer Characteristics analysis to look at the Bode plot to check the overall gain of the circuit. This is shown in Figure 6-22. Using that Bode plot, the voltage gain was measured at 37.93dBs. The lower roll-off of frequency was 44.24Hz, which is just acceptable. The BW was 589.94kHz. We could reduce the frequency by increasing the size of the filter capacitor $C_1$. With a bit of trial and error, we find that changing $C_1$ to 700nF produces a lower roll-off frequency of 29.99Hz.

# The Effects of Interstage Loading

By adding the second amplifier, we are introducing the possibility of interstage loading. This posibbly caused by the input impedance of the second stage is loading, i.e., changing the output impedance of the first stage. If this happens, it would change the gain of the first stage. Using the Bode plot shown in Figure 6-7, the gain of the first stage when it was the only stage in the amplifier was measured at 24.96dBs. However, when we used a Bode plot to display the gain of the first stage, with the second stage added, we found the gain of the first amplifier reduced to 16.54dBs. This is quite a reduction, and we need to learn why it has come about.

To make matters worse, we will add a final stage to the circuit which is just a load. This is shown in Figure 6-19.

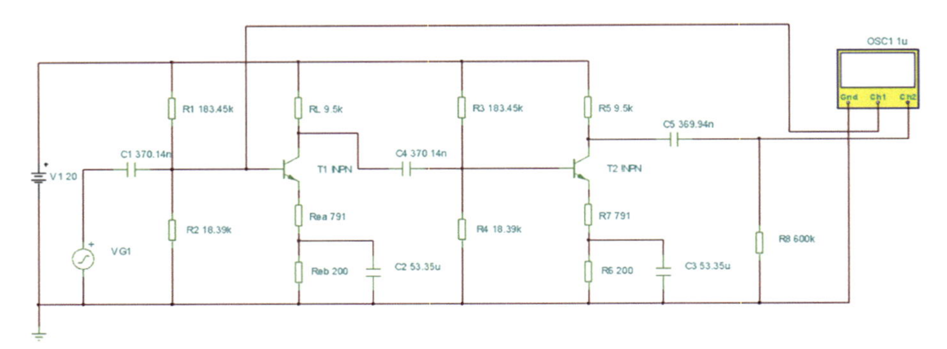


***Figure 6-19.*** *The Two-Stage Amplifier with a 600kΩ Load Added*

We need to determine the impedances starting the furthest end away from the input signal. Also, we will be looking at the ac impedances, as we want to study the effect on the ac gain of the circuit. We know that to ac, the VCC rail and the ground rail are connected together. Also, all the capacitors in the circuit can be considered as a short circuit. However, to show this we could determine the impedance of the lowest value capacitor, the 370.14nF capacitor, at a mid-frequency of say 10kHz, assuming we are designing an audio frequency amplifier circuit. Using these values, the impedance XC can be calculated as follows:

$$XC = \frac{1}{2\pi fC} = \frac{1}{2\pi \times 10E^3 \times 370.14E^{-9}} = 42.99\Omega$$

If we compare this impedance value to the 600kΩ resistor, it is small enough to be ignored, especially as it would be a phasor sum, not scalar, when adding them in series.

The first impedance we will calculate is the output impedance of the second amplifier. This would simply be the parallel combination of $R_5$ and $R_8$. Remember the current source of the amplifier model can be replaced by an open circuit; see Figure 6-7. As there are only two resistors involved, this can be calculated as follows:

$$Output\ Impedance\, T2 = \frac{R_5 R_8}{R_5 + R_8} = \frac{9.5E^3 \times 600E^3}{9.5E^3 + 600E^3} = 9.352k$$

This shows that a 600kΩ load does not load the resistor at the collector of T2 by a huge amount. However, if the load was a small speaker, then that could change the load value from 600k to 8Ω. If we use this 8Ω value, the output impedance of T2 would change to

$$Output\ Impedance\, T2 = \frac{9.5E^3 \times 8}{9.5E^3 + 8} = 7.993\Omega$$

These two calculations conform to the rule that when you add resistors in parallel, the result will always be less than the smallest value. Also, if the value of the impedance loading the first impedance is very large, then the loading can be negligible. That is why we normally want the output impedance of a stage to be low and the input impedance of the next stage to be high. When we look at Opamps in Chapter 9, we will look more closely at this.

The loading effect of the 8Ω speaker is an issue we must overcome, and we will see what can be done later in the chapter.

Now that we have calculated the output impedance of the T2 transistor, we can move on and determine the input impedance of the T2 amplifier. This is because this input impedance would be in parallel with the collector resistor in the T1 amplifier.

However, before we do that, we will calculate the ac gain of the T2 amplifier. We have shown already that the DC and ac gain are related to the ratio of the collector resistor and the emitter resistor of the amplifier circuit. However, with the ac gain, we need to take into account the effect of the interstage loading. This means we must use the loaded value of 9.352k instead of the 9.5k resistance. When considering the emitter resistance, we can ignore the $R_6$ resistance as the bypass capacitor $C_3$ directs all ac current away from flowing through $R_6$. So, we can use the value of $R_7$ at 800 $||R_5$ Ω. However, we have mentioned the inherent resistance "$r_e$" that is inside the diode between the base and emitter in the NPN transistor. Using the Ebers-Moll equation, don't worry, we are not going too deep into this type of analysis; we can determine a value for this internal resistance "$r_e$".

$$I_E = I_{ES}\left(e^{\frac{VBE}{VT}} - 1\right)$$

IE is the emitter current.

IES is the reverse saturation current of the base emitter diode.

VBE is the base voltage across this base emitter diode.

VT is the thermal voltage.

We are going to use the thermal voltage term, $V_T$, as it has been shown, by better engineers or scientist, than me, that if we relate this term to the charge "Q" flowing through a diode, we can come up with this approximation to this internal resistance "$r_e$" of a diode. The expression is

$$r_e = \frac{V_T}{I_{EQ}}$$

The thermal voltage has been found to be 26mV at an ambient temperature of 25°C. Therefore, we can say

$$r_e = \frac{26E^{-3}}{I_E}$$

We can approximate this to the collector current knowing that $I_C$ very closely equals $I_E$. Therefore, we can say

$$r_e = \frac{26E^{-3}}{I_C}$$

However, we should realize that "$r_e$" will change with temperature, but the change would be fairly small, so we can use this expression.

This means that when the collector current is equal to the quiescent current, which in this case is 1mA, the value for this $r_e$ resistance would be

$$r_e = \frac{26E^{-3}}{1E^{-3}} = 26\Omega$$

This would be in series with $R_7$ in the circuit, and so we need to use the 826Ω value to calculate the gain of T2 as follows:

$$T2\ gain = \frac{9352}{826} = 11.322$$

If we use the simulation to measure the input and output of the T2 transistor, we can use the display of the oscilloscope, shown in Figure 6-20, to determine the gain by comparing the peak to peak of both traces.

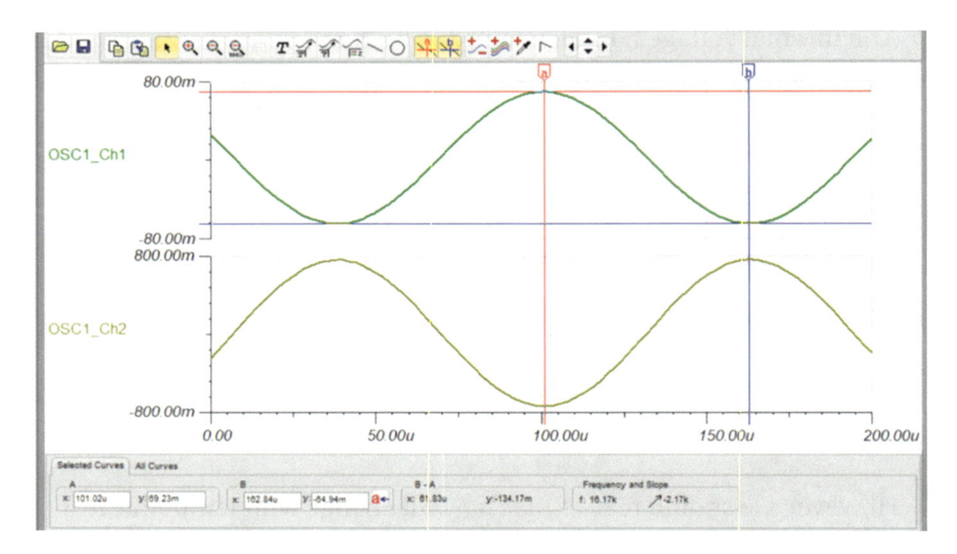


***Figure 6-20.*** *The Input and Output Voltages of the Final Transistor T2*

The output voltage, that is, the Ch2 trace, was measured with a peak to peak 0.149v. The input, that is, Ch1, was measured at 0.13417v. Using two readings, the voltage again of the T2 transistor was

$$V_{Gain} = \frac{Output\ pk\ to\ pk}{input\ pk\ to\ pk} = \frac{1.49}{0.13417} = 11.11$$

This agrees quite closely to our previous calculations.

Now we need to move back through the circuit and determine the output and input impedances of the T1 transistor. However, as T2 is added to the circuit, the output impedance of T1 will see the input impedance of T2 in parallel with it. This means we need to determine the input impedance seen at the base of T2. This will be the combination of $R_3$, $R_4$, and $R_{Base}$ in parallel. The $R_{Base}$ resistance will be the series combination of $r_e$ and $R_7$, but due to the transistor action, they will be multiplied by the beta of the transistor. We are using the value of 100 as this would produce the smallest value for $R_{Base}$ which would load the value of $R_3$ and $R_4$ in parallel the most. This means we can calculate the input impedance of T2 as follows:

$$\frac{1}{R_{in}} = \frac{1}{R_3} + \frac{1}{R_4} + \frac{1}{\beta(r_e + R_7)} = \frac{1}{183.45E^3} + \frac{1}{18.39E^3} + \frac{1}{100(26+800)} = 7.193E^{-5}$$

Therefore

$$R_{in} = \frac{1}{7.193E^{-5}} = 13.901k$$

We can now determine the output impedance of the T1 transistor. This would be the RL in parallel with the Rin of the T2 transistor. This would give a value of

$$Output\ Impedance\ T1 = \frac{R_L R_{in}}{R_L + R_{in}} = \frac{9.5E^3 \times 13.901E^3}{9.5E^3 + 13.901E^3} = 5.643k$$

We can use this value for the output impedance of T1 to calculate the now loaded gain of T1 as follows:

$$V_{GainT1} = \frac{Output_R}{Emitter_R} = \frac{5.643E^3}{26+800} = 6.83$$

This is a very much reduction of the gain from the same circuit as shown in Figure 6-1 when the amplifier was used as a single amplifier.

If we simulate the circuit but now use the oscilloscope to measure the input and output voltage of the first transistor T1, we will get the display as shown in Figure 6-21.

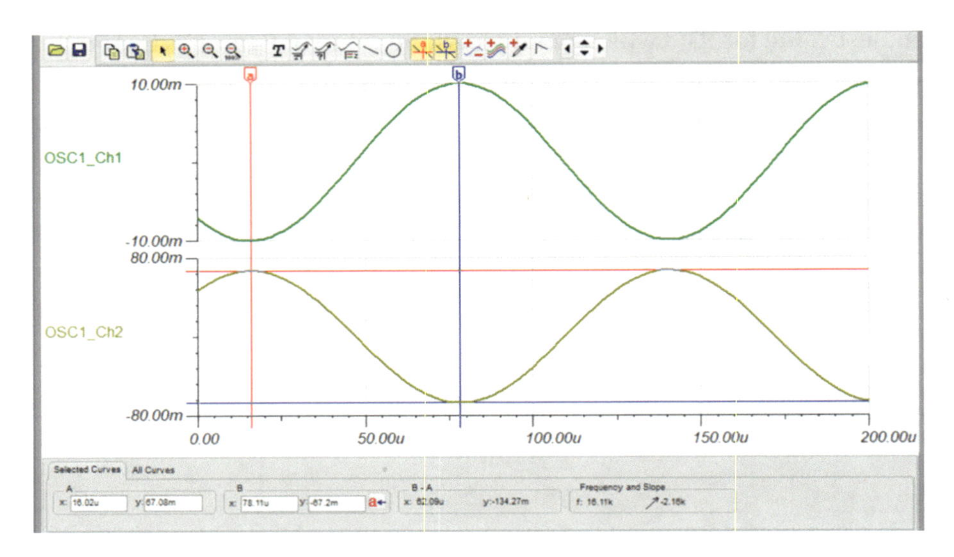


***Figure 6-21.*** *The Input and Output Voltages of the Final Transistor T1*

Using Figure 6-21, the voltage gain of T1 can be calculated as follows:

$$V_{Gain\,T1} = \frac{134.27m}{20m} = 6.71$$

This is again close to what we have calculated.

Using the Figures 6-20 and 6-21, we can see that the output of T1 is, as expected, the input to T2.

We should now be able to calculate the voltage gain of the complete amplifier circuit by simply multiplying the two gains together. This would give

$$Amplifier_{Gain} = V_{GainT1} \times V_{GainT2} = 6.71 \times 11.336 = 76.065$$

We have measured the peak-to-peak voltage at the output as 1.49v, and we know we have set the peak-to-peak 20mv. We can use this to calculate the gain of the amplifier as follows:

$$Amplifier_{Vgain} = \frac{1.49}{20E^{-3}} = 74.5$$

The Bode plot for the two-stage amplifier circuit is shown in Figure 6-22. The maximum gain was measured using this Bode plot, and the value was 37.5dBs.

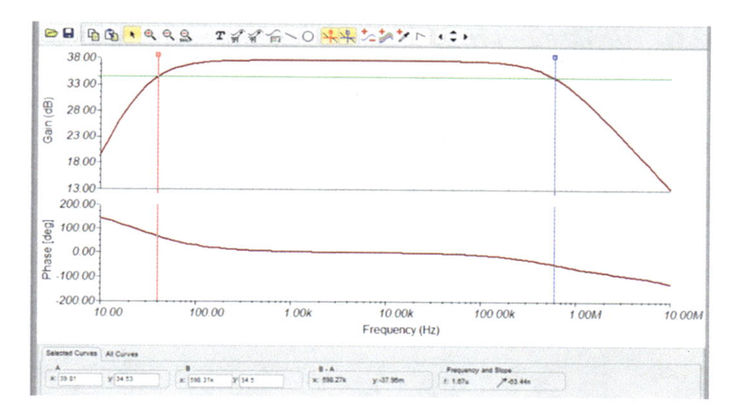


**Figure 6-22.** *The Bode Plot of the Two-Stage Amplifier Circuit*

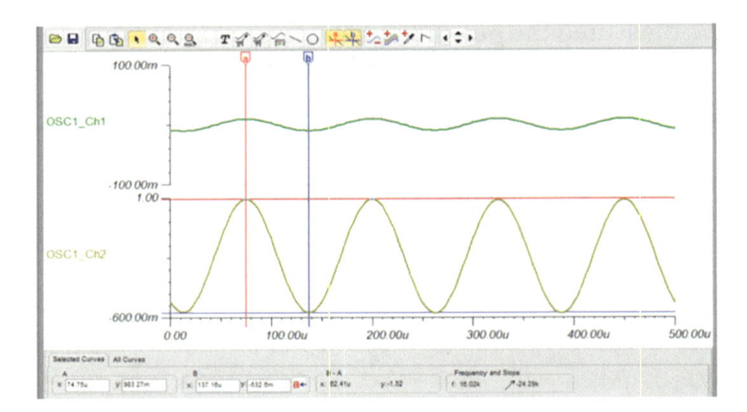


***Figure 6-23.*** *The Oscilloscope Traces of the Input and Output of the Two-Stage Amplifier*

I am also including the display of an oscilloscope used to measure the gain of the circuit using the input and output waveforms. This is shown in Figure 6-23. Using this method to measure the gain of the circuit, we get

$$V_{Gain} = \frac{Output\ Peak\ to\ Peak}{Input\ Peak\ to\ Peak} = \frac{1.52}{20E^{-3}} = 76 = 20\,Log\left(76\right) = 37.62dBs$$

The comparisons are close enough to suggest that the theories and calculations we have used are valid.

# Exercise 6.2

With respect to the class A amplifier circuit shown in Figure 6-24, calculate the following knowing that the DC quiescent $V_E$ voltage is to be around 2V and the beta is 100:

1. The emitter current
2. The base current
3. The base voltage

4. The output impedance
5. The input impedance
6. The voltage gain

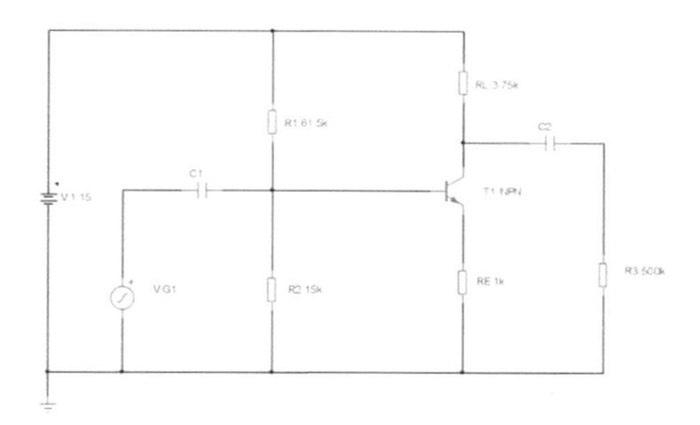


***Figure 6-24.*** *The Exercise 6.1 Circuit*

The answers to this and all other exercises will be in the appendix.

# Exercise 6.3

With respect to the circuit shown in Figure 6-25, determine the input and output impedances of both transistors as well as the voltage gain of each transistor and the overall circuit.

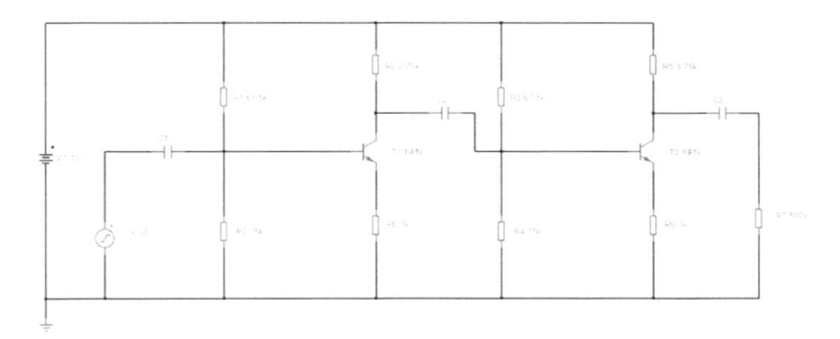

***Figure 6-25.*** *The Exercise 6.2 Circuit*

It has to be said that we are using nonstandard component values and an ECAD package to simulate the circuit. If we were to use standard values, there would be some small changes in the circuit operation. However, in the aspect of comparing the simulation to the real-life circuit, we should note that modern ECAD software uses very good mathematical models for the components. This means that the responses of the simulations are very close to what you would get if you built the circuit using practical components. The software is so good that many engineers use an ECAD package to test out their circuit designs first, as it is much quicker than building the test circuits practically. That being the case, I firmly believe that all engineers should learn how to use an ECAD software package to test their design first. We will be testing out all the theories that we go through in this book, with the TINA ECAD software, and I hope you will find that the test circuits do confirm the theories we go through in this book.

To sum up what we have done so far, in these last two chapters, we have studied how to design a BJT amplifier. We have finally arrived at an approach that produces an amplifier that can cope with thermal runaway and whose gain is set by two resistors external to the transistor and isn't affected by changes in Beta. However, this is not the perfect circuit it might appear. One issue is that the efficiency of the class A amplifier is quite low, around 30%. This is mainly due to the fact that even when there is no ac signal present, the transistor is still conducting, and so really it is wasting power. This would be the DC quiescent current, and in the circuit shown in Figure 6-1, we set this to 1mA. Therefore, we need to look at an improved type of amplifier.

## Moving the Biasing Point

If we consider the circuit shown in Figure 6-1, the DC quiescent voltage at the base is 1.65V. This is when we biased the amplifier at the midpoint on the load line. We should be aware that the base emitter diode, internal to the transistor, requires around 0.7V to forward bias it and turn it on. When we apply an ac signal to the base, we will momentarily raise and lower the voltage at the base. In this way, we can increase or decrease the current flowing through the transistor by basically turning on more or less this diode. If the ac input voltage should reduce this voltage, at the base, to below the 0.7V required to turn on the base emitter diode, then the output voltage would be clipped. This would limit the peak of the input voltage to around 1V, taking the maximum voltage at the base to around 2.65V and the minimum to around 0.65V. Also, as we have biased the amplifier in Figure 6-1 to sit in the middle of its range, then if the voltage at the base raised above 2.65V, that is, with a peak input greater than 1V, then we would see clipping of the output at both peaks. We can test this concept by simulating the circuit as shown in Figure 6-1 with, firstly, a peak input of 1V and then with a peak input of 1.2V at the base voltage. The displays from the oscilloscope are shown in Figures 6-26 and 6-27.

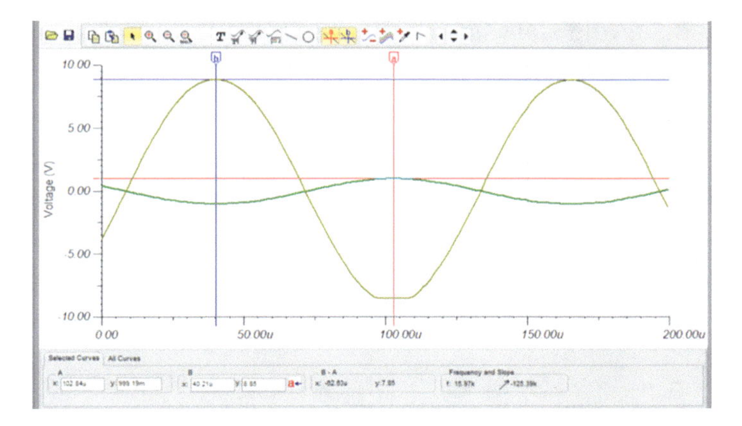


***Figure 6-26.** The Input and Output Traces with a 1v Peak Input*

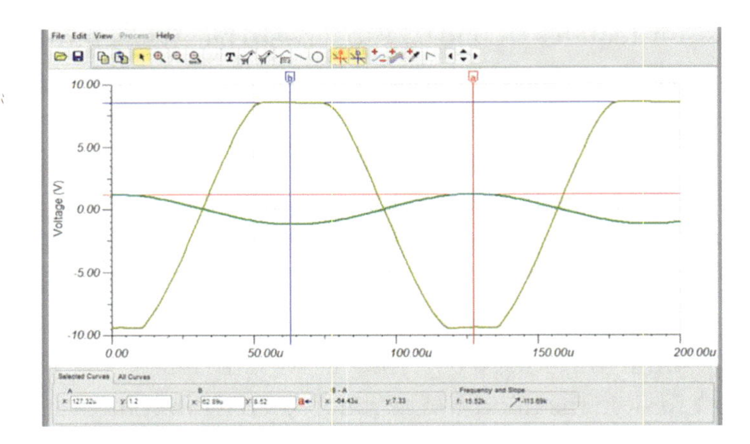


***Figure 6-27.*** *The Input and Output Traces with a 1.2v Peak Input*

If we look at the output trace shown in Figure 6-26, we can see that the positive peak of the output is correct, at 8.85v, with no clipping. However, the negative peak is just starting to show signs of clipping as it peaks at –8.53v and stays flat for a short time. We should remember that the output goes positive when the input goes negative, that is, as we reduce the current flowing through the transistor moving closer to the cutoff point, and the output goes negative when the input goes positive, as we move closer to the saturation point where the current through the transistor reaches a maximum. This would suggest that the biasing point is just a bit higher than the midpoint on the load line. This is shown as Q on the load line in Figure 6-28.

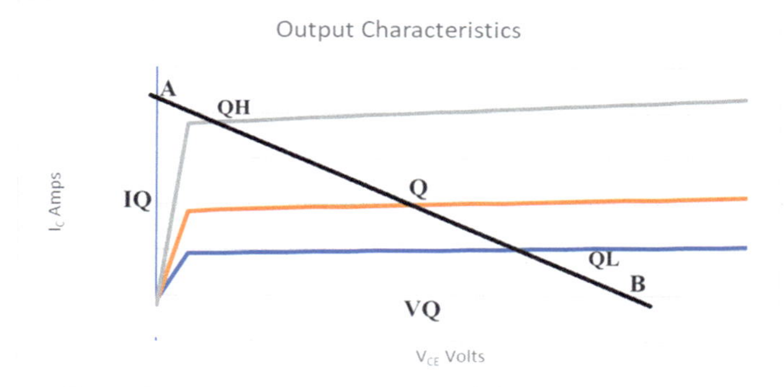


***Figure 6-28.*** *The Load Line on the Output Characteristics*

However, when we increase the input voltage to 1.2v peak, we can see clipping on both the positive and negative peaks of the output. This is shown in Figure 6-27.

This seems to agree with the concept that to achieve a good representation of the input signal at the output, then we should bias the amplifier so that it sits in the middle of the load line. This results in a high degree of what is termed fidelity of amplification. However, if we increase this biasing voltage, we would take the bias point higher up the load line, up toward the vertical axis (see point "QH" in Figure 6-28), and we would start to see clipping of the negative peak. To experiment with this concept, the value of the $R_2$ resistor, in the voltage divider network, was increased to 25k. This raised the DC quiescent voltage to 2.11V, moving the biasing point up the load line. We can see from Figure 6-29 that the negative peak of the output voltage has been clipped. This is because the transistor has been forced into its saturation current early, before the input voltage had reached the positive peak.

Similarly, if we reduce the value of $R_2$ to 11k, we would reduce the DC quiescent voltage to 1.1V (see "QL" in Figure 6-28), and this would clip the positive peak of the output voltage. This is because we would have turned the transistor off before the input signal reached its negative peak. This is confirmed in Figure 6-30.

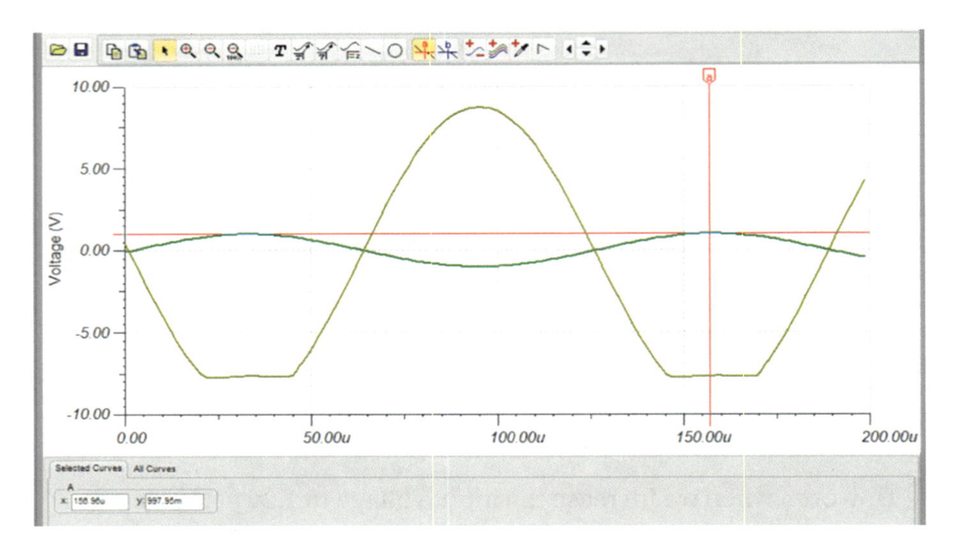


***Figure 6-29.*** *R2 Increased to 25k; Base Voltage Increased to 2.11V*

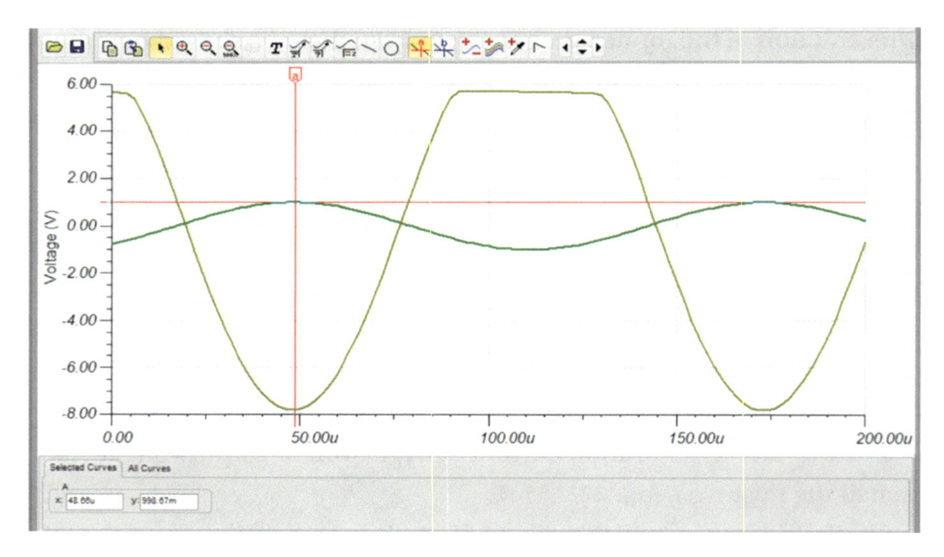


***Figure 6-30.*** *R2 Decreased to 11k; Base Voltage Decreased to 1.1V*

We can see that by varying the base voltage, we can alter how the amplifier works and so change the class of the amplifier. What these simulations have also shown is that moving the biasing point up and down the load line does not alter the maximum peak of the output as it was around the 8.8V peaks in all four displays. As the peak of the input signal was 1V, then this would suggest that the gain of the amplifier was consistent at around 8.8, which is around the value set by the ratio of $R_L$ divided by $R_E$.

## Creating a Class B Amplifier

The class B amplifier is one that amplifies just one half, be it the positive or negative half, of the input signal. The circuit shown in Figure 6-31 shows an NPN transistor biased to the point that is near the upper saturation point on the load line.

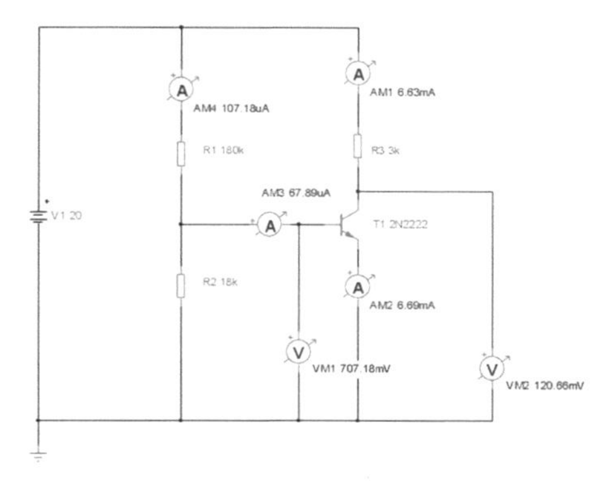


***Figure 6-31.*** *An NPN Class B Amplifier Circuit*

The class B amplifier shown in Figure 6-31 is biased such that the transistor is in the saturation point on the load line. This means that if the base voltage rose in an attempt to pass more current through the amplifier, the transistor could not respond as it is conducting at its maximum already. This means that in this quiescent state, the voltage at the output would be as low as it could be. Then when the input signal went negative, in an attempt to reduce the current through the amplifier, the transistor could respond and reduce the current flowing through it. This would result in the output voltage rising. This response is shown in Figure 6-32 where Ch2 shows the output of the class B amplifier. We can see that the output goes positive when the input goes negative. Also, the peak of the output is very much higher than the input. This means there is some voltage gain with this class B amplifier.

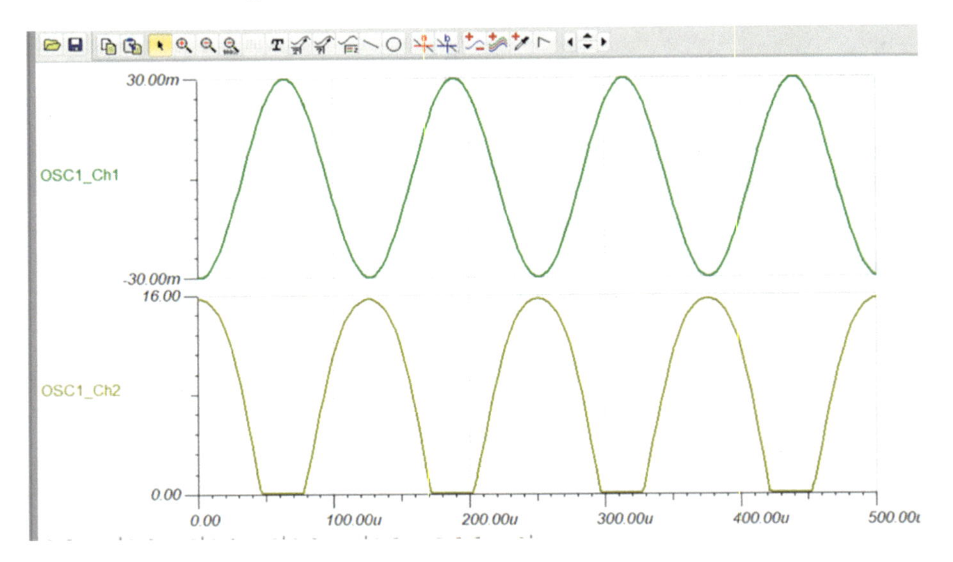


***Figure 6-32.** The Output Voltage of the NPN Class B Amplifier*

Of course, we need to use the other half of the input signal. A circuit that uses a PNP transistor biased up near the saturation point of the load line is shown in Figure 6-33.

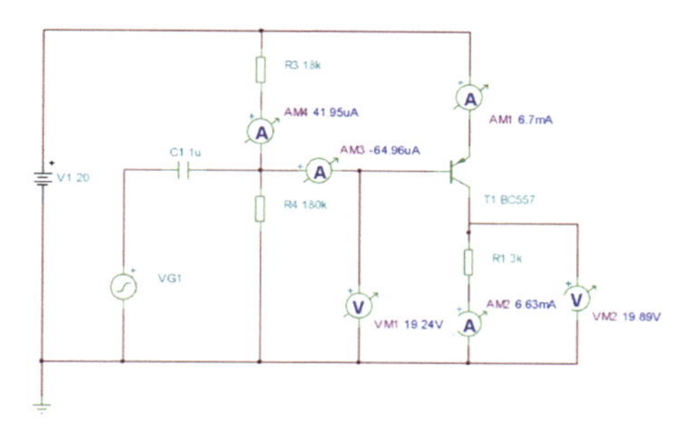


***Figure 6-33.*** *A PNP Class B Amplifier*

This works in basically the same way as the NPN transistor, but because we need to make the emitter more positive than the base to forward bias the emitter to base diode, when the input signal starts to rise the transistor starts to conduct less, and so the output voltage starts to rise. This is shown in the traces of the output and input voltage shown in Figure 6-34.

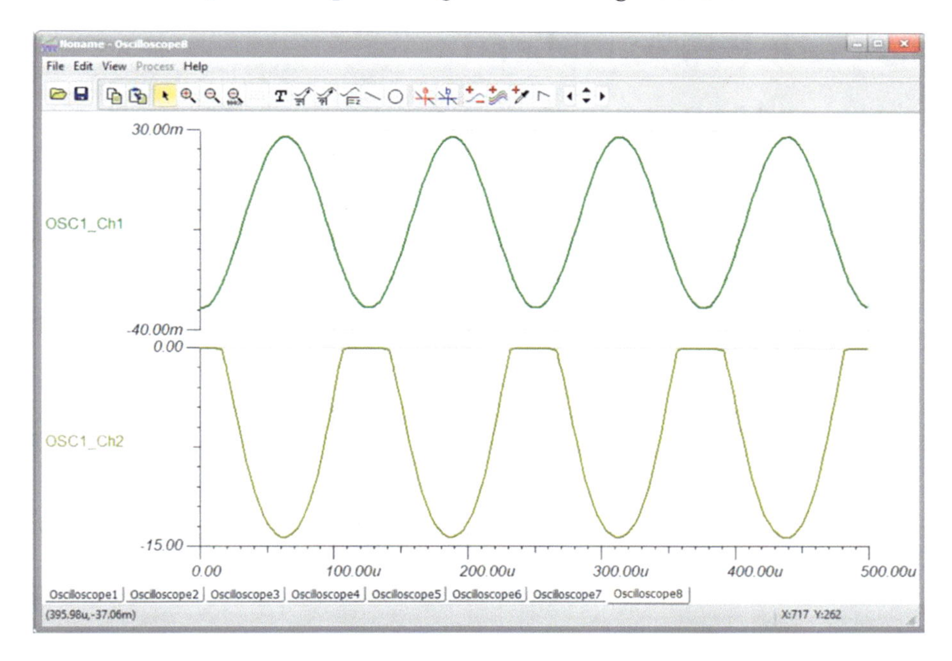


***Figure 6-34.*** *The Input and Output Voltages of the PNP Class B Amplifier*

This work shows that we can configure the basic amplifier to make use of the class B type amplifier. This would allow us to amplify the input signal and possibly get an even larger output going to both positive $V_{CC}$ and negative $V_{CC}$. However, there are a few issues with this approach, and the main ones are

1. We need to bring the two halves together.
2. We would need a dual rail supply.
3. We would waste too much power as both transistors could be working at maximum current even when no signal had been applied.

The waste of power is the main reason we don't use this approach. A more efficient approach is looked at next.

## The Push-Pull Amplifier

This is an amplifier that will push the output down toward the ground voltage when the input voltage rises. Then it will pull the output up as far as it can toward the $V_{CC}$ rail when the input signal reduces, hence the name for the amplifier. Our work with moving the bias point up and down the load line would suggest we could make a class B amplifier, which is an amplifier that could push the output to the ground or pull it up toward $V_{CC}$. Indeed, we could use two class B amplifiers, one to amplify the positive half only and the other to amplify the negative half only, but as we have seen, they would waste too much power. Another issue is that of crossover distortion which comes about when we would need to join the two outputs together to reproduce the amplified version of the signal at the output. A typical push-pull amplifier is shown in Figure 6-35.

Although we call this circuit an amplifier, it does not increase the amplitude of the input signal. This is because it is basically two BJTs configured in the common collector mode. The T1 transistor is an NPN transistor (we are using the basic 2N2222), and the T2 transistor is a PNP transistor (we are using a basic BC557).

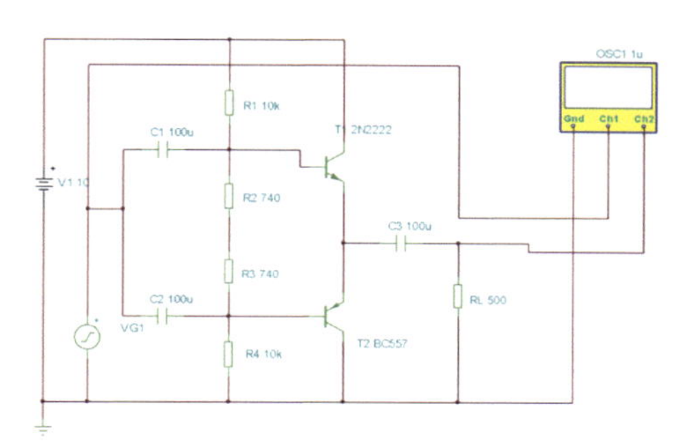


***Figure 6-35.*** *A Typical Push-Pull Amplifier*

Figure 6-36 shows the voltage divider network of the resistors: $R_1$, $R_2$, $R_3$, and $R_4$. We are using this network to provide the base voltage to the two transistors. Using the voltage divider rule, we can calculate the voltage across $R_2$ and $R_3$ as follows:

$$V_{R2} = \frac{V \times R_2}{R_1 + R_2 + R_3 + R_4} = \frac{20 \times 740}{10E^3 + 740 + 740 + 10E^3} = 689.01mV$$

The voltage across the $R_3$ resistor is calculated the same way, so it too will equal 689.01mV. This is very close to what is measured in Figure 6-36, which is just the DC network.

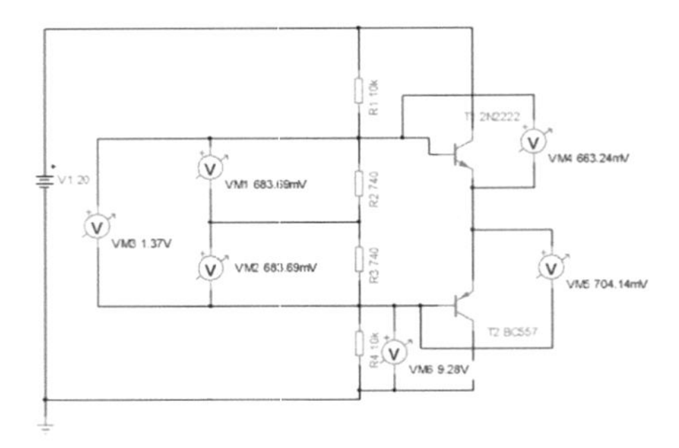


***Figure 6-36.*** *The DC Quiescent Circuit*

Figure 6-36 shows that due to the diode action inside transistors, the actual base emitter voltage at each transistor is virtually the same as the volt drop across $R_2$ and $R_3$. This can be confirmed if we change the value of these two resistors. I will leave that for you to do, but when we do change the resistor values, we see the base voltage changes and the diode voltage changes in line with the base voltage changes.

Figure 6-36 uses the NPN transistor to pass the positive half of the input onto the load resistor, while the PNP transistor passes the negative half. We need to appreciate that, due to the internal diode between the base and emitter of a transistor, any transistor requires the base voltage to reach about 0.7v to start turning on. From Figure 6-36, we can see that we have nearly got the bases of both transistors to that turn on voltage. Therefore, with the NPN transistor, when the input signal rises above this 0.663mv (see VM4 in Figure 6-36), it turns the T1 transistor on and allows current to flow through $R_L$ from the emitter end of $R_L$ to ground.

When the input signal goes below this 0.663mv, the NPN transistor turns off and so stops conducting. Then, when the input voltage goes more negative than the 0.704mv (see VM5 in Figure 6-36), the PNP transistor turns on and allows the current to flow through the load resistor $R_L$, in the direction from ground end of $R_L$ through the emitter of the PNP transistor. This means that the current flows through the load resistor as shown in Figure 6-37.

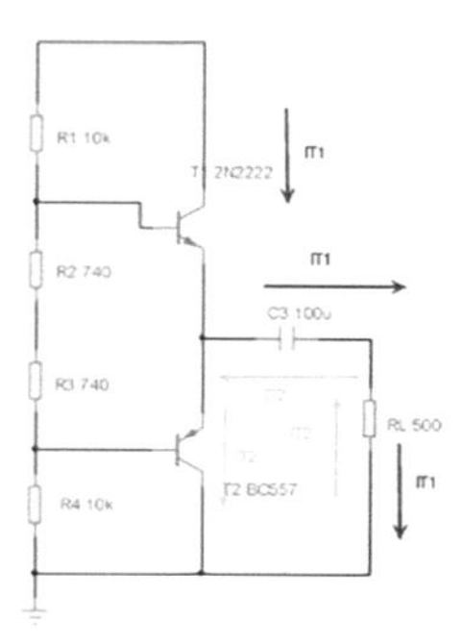

***Figure 6-37.*** *The Flow of Current Through the Load Resistor*

The dark blue arrows show the direction of current through $R_L$ when the NPN transistor T1 is conducting. The light blue arrows show the direction of current through $R_L$ when the PNP transistor T2 is conducting.

Figure 6-37 shows that the current flowing through the load resistor does change direction which means the voltage across it must be positive when the NPN transistor conducts and then negative when the PNP transistor conducts.

When we look at Figure 6-36, we can see that both the transistors are configured as common collectors. Indeed, the output is from the emitters of the transistors showing that they are set up as emitter followers. This means that they will not amplify the input signal. This is confirmed when we look at the oscilloscope traces of the input and output as shown in Figure 6-38.

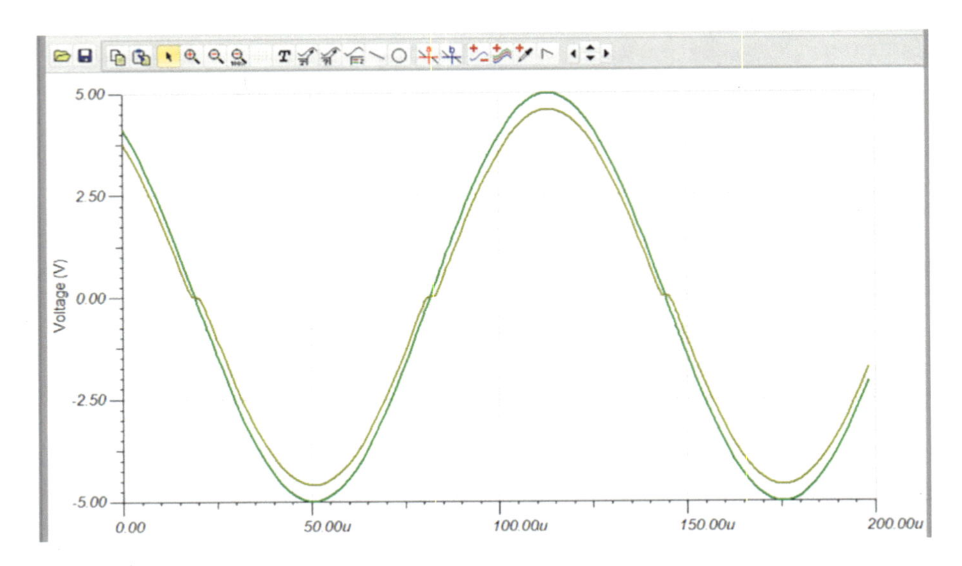


***Figure 6-38.*** *The Input and Output Displays of the Push-Pull Amplifier*

Figure 6-38 shows that the output voltage peaks at a slightly lower voltage than the input. This is due to the diode drop between the base and the emitter of the transistors. Also, if we look at the zero crossing levels, we see there is some distortion. This is due to the fact that the NPN transistor turns off when the input falls below around the 0.633mv, yet the PNP transistor is not ready to turn on yet as the input voltage has not fallen below around the 0.704mv needed to turn that transistor on. We should appreciate that to turn an NPN transistor on, we can use the acronym FEN, Forward Emitter Negative. This means the emitter needs to be around 0.7v more negative than the base. However, for the PNP, we use FEP, Forward Emitter Positive. This means for the PNP the emitter needs to be around 0.7v more positive than the base. I hope we can see how this can be applied to the circuit shown in Figure 6-36.

With the resistor network arrangement as shown in Figure 6-36, we have already set the base voltage at a level close to the approximate 0.7V required to turn the transistors on, that is, at 0.684V, and so this crossover distortion is not too bad. However, if we could set this a bit higher, by increasing the values of $R_1$ and $R_2$, we could get a better output with less crossover distortion. Setting the resistors to 1.4k would give a base voltage of

$$V_{Base} = \frac{20 \times 1.4E^3}{10E^3 + 1.4E^3 + 1.4E^3 + 10E^3} = 1.23V$$

This might mean we will be turning the transistors on a little early, and so we would amplify some of the other half cycle as well, but it does reduce the crossover distortion. This is confirmed in Figure 6-40. This type of biasing is termed class AB, when all of the positive or negative half is amplified but with a small amount of the other half being amplified as well. When this type of biasing is applied to the circuit, the output we get is as shown in Figure 6-39.

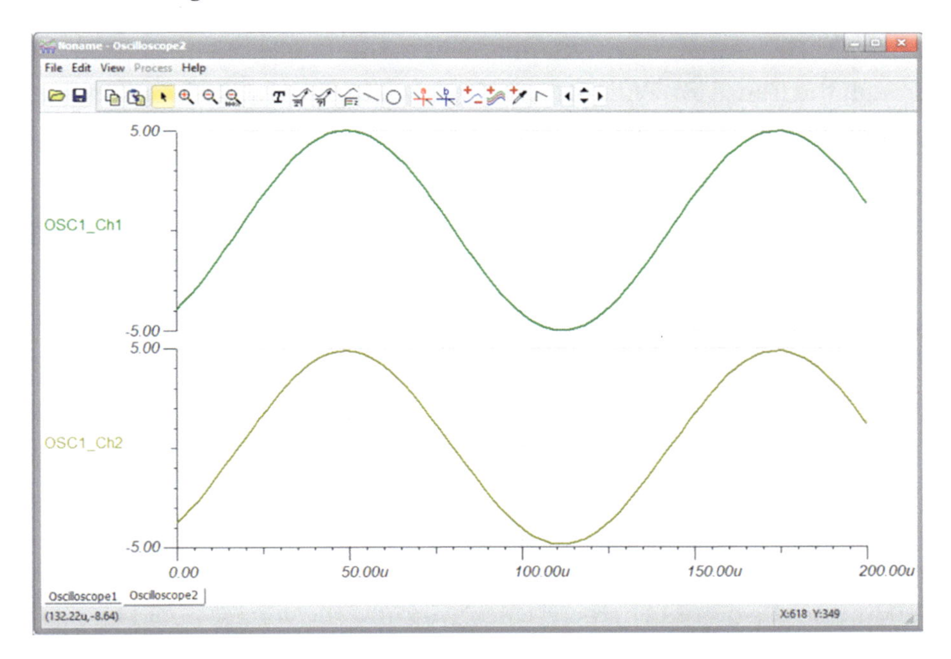


***Figure 6-39.** The Input and Output Voltage When R2 and R3 Were 1.4k*

The Ch2 trace does show a voltage waveform with virtually no crossover distortion. This shows that using a class AB type biasing with the push-pull amplifier, we can get a good reproduction of the input signal. This should produce an output with a high level of fidelity.

# The Power Amplifier

We have seen that the typical push-pull amplifier uses emitter follower type configuration. This means there is no real voltage gain; the output is slightly less than the input. So why do we use these push-pull type amplifiers? The reason is that they are more efficient than the stabilized class A amplifier. We stated earlier that the efficiency of the class A amplifier is around 30%. This is pretty low, and it is basically because the class A amp is turned on, with current flowing through it all the time, even when there is no signal being present at the input, that is, the base of the transistor. This is due to the DC quiescent current that is always flowing through the transistor. With the circuit shown in Figure 6-1, this was set at 1.01mA. Knowing that, we can calculate the DC power using

$$Power = VI\ watts$$

The DC power is

$$P_{DC} = V_{CC} \times I_{CQ}$$

$$P_{DC} = 20 \times 1.01E^{-3} = 20.2mW$$

To calculate the power out, we can multiply the maximum peak-to-peak voltage by the maximum peak-to-peak current. If we consider the load line on the output characteristics, we know the voltage could ideally swing from 0V to $V_{CC}$, and the current could swing from 0 to $\frac{V_{CC}}{(R_L + R_E)}$.

As this is an ac voltage or current, we need to use the rms of these values, as the rms value relates the ac voltage and current to the equivalent DC voltage and current to achieve the same power. This means the output power of the amplifier can be calculated using

$$Pout = rmsV_{CE(pk-pk)} \times rmsI_{C(pk-pk)}$$

Assuming these are sinusoidal waveforms which is OK, we can calculate the rms of a sine wave using

$$rms = \frac{Peak}{\sqrt{2}} = 0.707 \times Peak$$

Therefore, knowing the Peak is simply the pk-pk divided by 2, we can calculate the power out using

$$Pout = 0.3535V_{CE(pk-pk)} \times 0.3535I_{C(pk-pk)}$$

Using the value from our circuit, we get

$$Pout = 0.3535 \times 20 \times 0.353 \times 2.08^{-3} = 5.199mw$$

We can now calculate the efficiency $'\eta'$ of our class A amplifier as follows:

$$\eta = \frac{Pout}{P_{DC}} = \frac{5.199E^{-3}}{20.8E^{-3}} = 0.25\ or\ 25\%$$

This shows that the class A amplifier is not very efficient.

Well, we now need to see if the push-pull amplifier is more efficient than the standard class A amplifier. To calculate the efficiency $'\eta'$, we again use

$$\eta = \frac{Pout}{P_{DC}}$$

We relate the power to DC power, which is simply the voltage multiplied by the current. In the input, the voltage is simply VCC. The current is the current flowing through each of the transistors which is only when the transistor is conducting. As each transistor simply conducts for one half cycle of the input, then the DC input current for a half sine wave can be calculated using

$$I_{DC} = \frac{I_{Cpk}}{\pi}$$

As there are two transistors, then the DC power can be calculated using

$$P_{DC} = \frac{2 \times I_{Cpk} \times V_{CC}}{\pi}$$

To calculate the power out, we simply use the rms of both the current and voltage. As they are both sinusoidal and there are full sine waves, then the power out can be calculated using

$$P_{Out} = \frac{I_{Cpk}}{\sqrt{2}} \times \frac{V_{CC}}{\sqrt{2}}$$

As the efficiency is simply the ratio of power out divided by power in, we have

$$\eta = \frac{I_{CPk} \times V_{CC}}{\sqrt{2} \times \sqrt{2}} \div \frac{2 \times I_{Cpk} \times V_{CC}}{\pi}$$

To divide by a fraction we invert and multiply, this gives

$$\eta = \frac{I_{Cpk} \times V_{CC}}{\sqrt{2} \times \sqrt{2}} \times \frac{\pi}{2 \times I_{Cpk} \times V_{CC}} = \frac{\pi}{\sqrt{2} \times \sqrt{2} \times 2} = \frac{\pi}{4} = 0.785 = 78.5\%$$

This shows that the push-pull amplifier has a much greater efficiency than the class A amplifier; that is why, sometimes, it is called a power amplifier.

## A Simple Audio Amplifier with a Small Speaker Output

In this last circuit in this chapter, we are going to look at how we can get our simple class A amplifier, with its high output impedance, to drive a small speaker that has an impedance of 8Ω. The circuit for this system is shown in Figure 6-40.

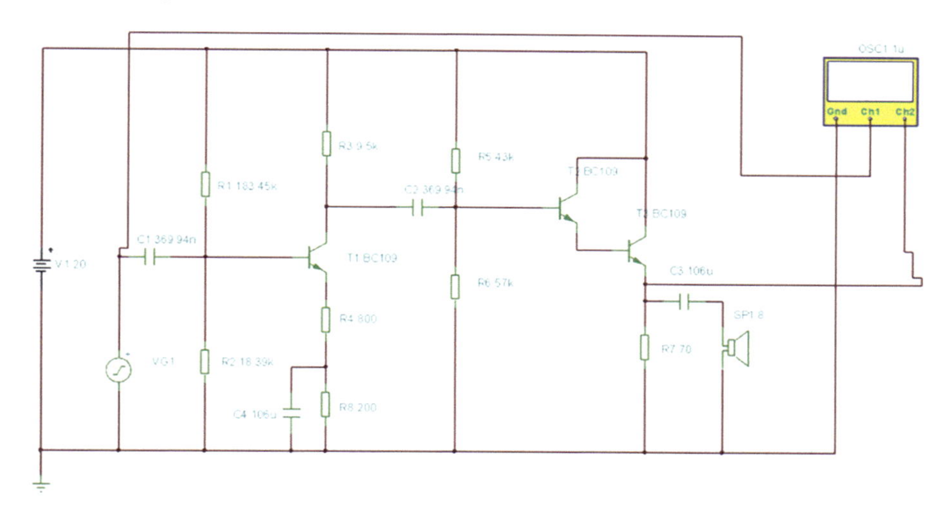


***Figure 6-40.*** *A Simple Audio Amplifier Driving an 8Ω Speaker*

I am not suggesting that this is a great hi-fi circuit that will astound you. I am no expert at designing audio systems; there is an awful lot more work and experience you would need to design the best hi-fi systems. I am just trying to sum up some of the work we have gone through in this chapter. I hope when you do go through the chapter and the book, it will give you a good foundation in designing analog circuits. We will use this last circuit to see if we can overcome the issue of the 8Ω speaker loading the amplifier circuit as shown earlier in the chapter.

The circuit is using two BC109 NPN transistors arranged as a basic Darlington transistor. This is because I can't find a Darlington transistor, that is, the typical TIP 122 to TIP 127, in Tina. However, a typical Darlington transistor is simply two transistors connected as shown in Figure 6-40. I have set the beta for each of the transistors to 100. This would give the Darlington arrangement a gain of around 10,000, which is pretty typical for a Darlington Pair. We are using this Darlington arrangement to interface the 8Ω speaker into the circuit without loading the amplifier. To appreciate how this happens, we will determine the ac impedances in the circuit.

Starting at the load, that is, the speaker, we can see that, to ac, the 8Ω speaker is in parallel with the 70Ω resistor. The output ac impedance of the Darlington would be

$$Z_{Outdar} = \frac{8 \times 70}{8 + 70} = 7.179\Omega$$

This impedance is multiplied by the beta of the Darlington. We could assume this would be 10,000. However, we can use the current measurements of the base and collector currents of the Darlington to get a better idea of the current gain. Using the values shown in Figure 6-42, this would give

$$Beta_{dar} = \frac{I_C}{I_B} = \frac{A_{M1}}{A_{M2}} = \frac{138.3E^{-3}}{10.27E^{-6}} = 13466.41$$

This shows that the beta of the Darlington is 13466.41, a really high value. However, it is this high value of beta that allows us to use the Darlington to reduce the loading effect of the 8Ω speaker. That is because the input impedance of the Darlington is made up of $R_5$, $R_6$, and $(\beta + 1)Z_{Outdar}$. This is ignoring the effect of the two internal diodes inside the two NPN transistors of the actual Darlington. It is OK to ignore these two "$r_e$" values, as the collector current is very high at 138mA. This would make the "$r_e$" values low. That being the case, we calculate the input impedance of the Darlington as follows:

$$\frac{1}{Z_{indar}} = \frac{1}{43E^3} + \frac{1}{57E^3} + \frac{1}{13466.41 \times 7.179} = 5.1144E^{-5}$$

Therefore

$$Z_{indar} = \frac{1}{5.1144E^{-5}} = 19.552k$$

We can now calculate the output impedance of the class A amplifier knowing $Z_{out1}$ is $R_3$ in parallel with this $Z_{indar}$, therefore, using

$$Z_{out1} = \frac{R_3 \times Z_{indar}}{R_3 + Z_{indar}} = \frac{9.5E^3 \times 19.552E^3}{9.5E^3 + 19.552E^3} = 6.394k$$

We can use this to calculate the voltage gain of the class A amplifier as follows:

$$A_{V1} = \frac{Z_{out1}}{(r_e + R_4)}$$

The resistor "$r_e$" is the inherent resistance of the base emitter diode inside the class A NPN transistor. The value of "$r_e$" can be calculated as follows:

$$r_e = \frac{26^{-3}}{I_C} = \frac{26E^{-3}}{1.04E^{-3}} = 25$$

Therefore, we have

$$A_{V'} = \frac{6.394E^3}{(25+800)} = 7.75$$

We can also calculate the input impedance of the class A amplifier. This would be the resistors $R_1$, $R_2$, and (beta + 1 x$R_4$) all in parallel with each other.

$$Z_{in1} = \frac{1}{\frac{1}{183.45E^3} + \frac{1}{18.39E^3} + \frac{1}{100 \times 825}} = 13.896k$$

We can check these calculations from the simulation.

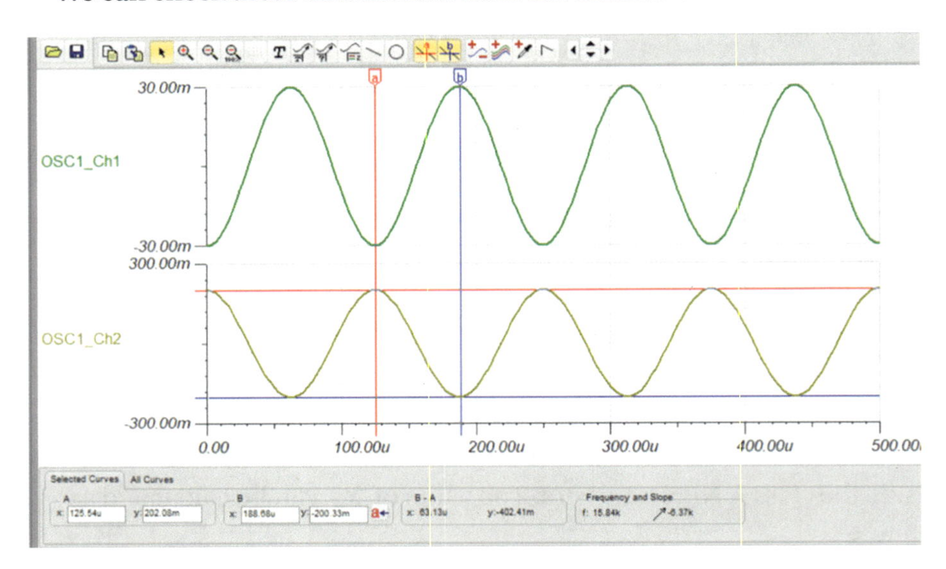


***Figure 6-41.*** *The Input and Output Voltages*

Using the peak-to-peak voltage of the traces shown in Figure 6-41, the voltage gain can be calculated as follows:

$$A_V = \frac{402.41E^{-3}}{60E^{-3}} = 6.707$$

However, this is the gain of the overall circuit as Ch2 of the oscilloscope display was measuring the voltage at the emitter of T3, the output of the Darlington. The Darlington is arranged as an emitter follower which has a voltage gain of just less than 1. If we measured the voltage at the input of the Darlington, at the base of T2, then the gain measured by the oscilloscope would be

$$A_{V'} = \frac{453.32E^{-3}}{60E^{-3}} = 7.556$$

This is closer to the calculated gain of 7.75.

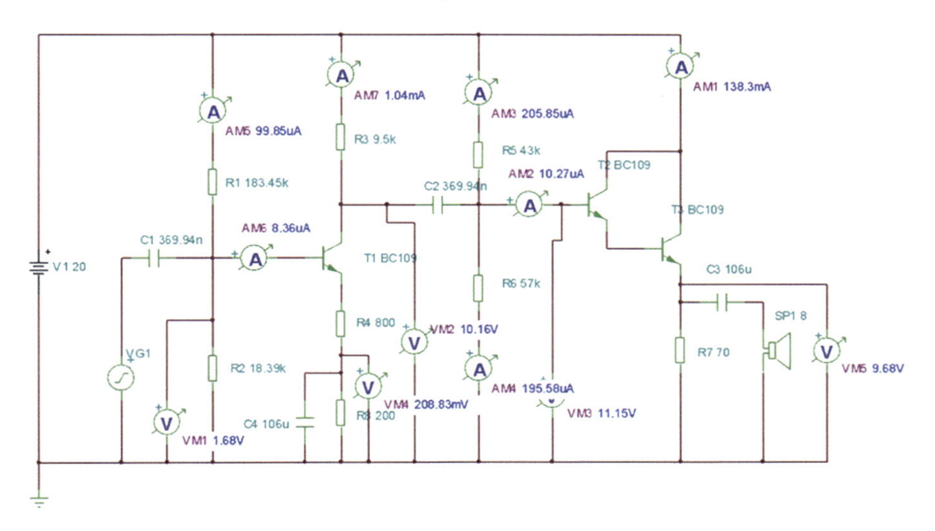


***Figure 6-42.*** *The Simulated Results of the DC Quiescent Values Around the Circuit*

We can calculate some of the DC quiescent values around the circuit and compare them with the simulated results shown in Figure 6-42. As we are designing a class A amplifier, then we would want the emitter of the Darlington to be sat at around half $V_{CC}$. This would put it at around 10V in this case. This would mean that the current flowing through the emitter would be

$$I_{EDAR} = \frac{V_{EDAR}}{R_7} = \frac{10}{70} = 142.86mA$$

If we assume there is one diode between the base and emitter of the two transistors that make up the Darlington, then the voltage at the base of the Darlington would be

$$V_{BDAR} = 0.655 + 0.655 + 10 = 11.31V$$

The current flowing through $R_6$ can be calculated as follows:

$$I_{R6} = \frac{V_{BDAR}}{R_6} = \frac{11.31}{57E^3} = 198.42uA$$

We can calculate the current flowing through $R_5$ as follows:

$$I_{R5} = \frac{V_{CC} - V_{BDAR}}{R_5} = \frac{20 - 11.31}{43E^3} = 209.7uA$$

This means we can calculate the current flowing into the base of the Darlington as follows:

$$I_{BDAR} = I_{R5} - I_{R6} = 209.7E^{-6} - 198.42E^{-6} = 11.28uA$$

If we now consider the class A amplifier at the beginning of the circuit, we would want the collector voltage $V_C$ to be at half $V_{CC}$, that is, 10V. This means we can calculate the collector current $I_{C1}$ as follows:

$$I_{C1} = \frac{V_{CC} - V_C}{R_3} = \frac{20 - 10}{9.5E^3} = 1.05mA$$

This means we can calculate the voltage at the emitter as follows:

$$V_{E1} = I_{C1} \times \left(R_4 + R_8\right) = 1.05E^{-3} \times \left(800 + 200\right) = 1.05V$$

Using this value for $V_{E1}$, we can calculate the voltage at the base of T1 as follows:

$$V_{B1} = 0.655 + 1.05 = 1.705V$$

Now we can calculate the currents flowing through the voltage divider network of $R_1$ and $R_2$ and into the base of the class A amplifier, T1, as follows:

$$I_{R2} = \frac{V_{B1}}{R_2} = \frac{1.705}{18.39E^3} = 92.7134uA$$

$$I_{R1} = \frac{V_{CC} - V_{B1}}{R_1} = \frac{20 - 1.705}{183.45E^3} = 99.727uA$$

$$I_{B1} = I_{R1} - I_{R2} = 99.727E^{-6} - 92.7134E^{-6} = 7.593\text{uA}$$

These calculations are very close to the simulated result shown in Figure 6-42. It is good practice to carry out these calculations as the more you do them, the easier it is to understand the theories involved.

If the rms input current was 1.5µA and the rms input voltage was 21.21mV, then the rms input power would be 31.815nW. Knowing the rms output voltage across the speaker was 142.25mV, then the rms current would be 17.78mA, and the output power would be 2.53mW. This would give an overall power gain of 79503.9, a very good power gain which is another reason for using the Darlington transistor as a driver for the amplifier.

The rms output voltage was calculated as follows:

$$rms = \frac{V_{pk-pk}}{2} \times 0.707 = \frac{402.41E^{-3}}{2} \times 0.707 = 142.25mV$$

The rms current in the speaker was calculated using

$$rmsI_{out} = \frac{rms_{Vout}}{8} = \frac{142.25E^{-3}}{8} = 17.78mA$$

One word of caution is that you must make sure that the power supply you use to supply the $V_{CC}$ has enough current capacity to supply that output power. You should appreciate that the gain in power must come from somewhere, and in all amplifiers, it must come from the power supply.

## Exercise 6.4

To try and confirm that the Darlington transistor does allow the speaker to be added to the amplifier without loading the amplifier, the circuit shown in Figure 6-43 was simulated. The voltage gain was measured using the oscilloscope, and the peak-to-peak output voltage was measured at 574.6µV. This means the circuit actually attenuated the input signal. Use the theories you have studied to determine the ac impedances of the circuit and so confirm that this attenuation would happen.

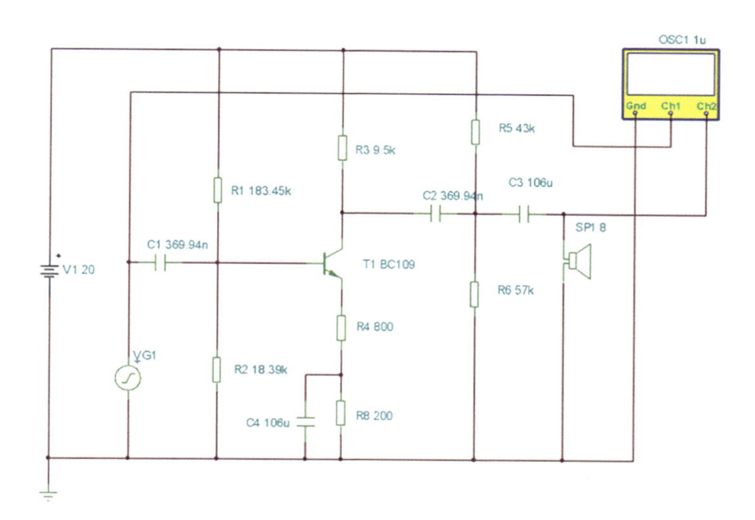


***Figure 6-43.*** *The Circuit Diagram for Exercise 6.3*

# Summary

In this chapter, we have studied an alternative method for designing a class A amplifier. We have studied the use of the ac amplifier model and made some comparisons with the class A amplifier.

We have improved our understanding of the load line and used it to look at the class B amplifier. We then went on to study the push-pull amplifier and compared the efficiency of the push-pull amp with the standard class A amp.

We have learned about the bandwidth of an amplifier and what affects it. Finally, we looked at a simple audio circuit driving a small 8Ω speaker.

In the next chapter, we will look at the FET, the field effect transistor.

I hope you have found this chapter informative and useful. I do try to cover all aspects of circuit analysis, but there is always more we can do. I hope you can use this chapter as a good foundation for your own investigations into this exciting and challenging topic.

CHAPTER 7

# The Field Effect Transistor

In this chapter, we will look at the FET, or field effect transistor. This is an alternative semiconductor device to the BJT. We will study the makeup of the various FETS to understand how they work. Then we will carry out some experiments to determine the main operating characteristics of the different FETs and so learn how to use them. It is important to appreciate how these devices work at the semiconductor level, as it is only then that the engineer can appreciate why these devices work in the way they do.

## Introduction

The basic FET is the JFET, that is, the Junction or Jug FET. This is a three-terminal device similar to the BJT. They are manufactured in two main types, which are n-channel and p-channel. The symbols for these basic JFETS are shown in Figure 7-1.


H. H. Ward, *Mastering Analog Electronics*, Maker Innovations Series,
https://doi.org/10.1007/979-8-8688-0245-4_7

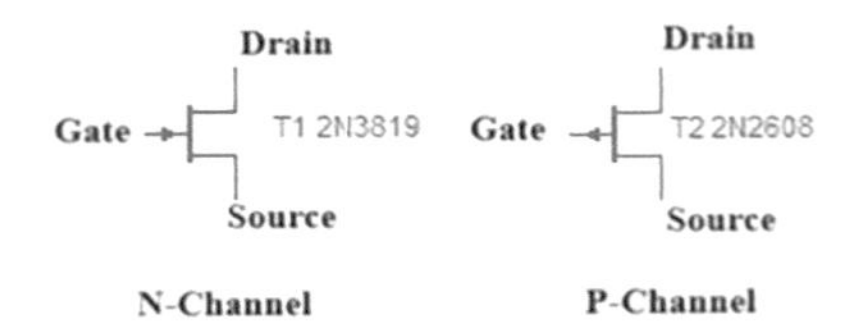


***Figure 7-1.*** *The Symbols for the Two Types of JFETS*

The three terminals of the JFET relate to the three terminals of the BJT, bipolar junction transistor, as stated in Table 7-1.

***Table 7-1.*** *The Comparison of the Terminals on the JFET and BJT*

| JFET | BJT |
|---|---|
| Drain | Collector |
| Gate | Base |
| Source | Emitter |

This relationship is true for all the other types of FET as well.

# The Makeup of the JFET

We need to examine how the JFET is constructed. The term junction relates to the fact that there is a PN junction, with respect to the n-channel, and an NP junction, with respect to the p-channel, inside the JFET. In the simplest JFET, there is just one junction, and this type of FET is referred to as the UJFET, the Unijunction FET. However, the UJFET is not part of this analysis, as it is not in common use now; this analysis will concentrate on the JFET and MOSFET devices.

In the JFET and MOSFET, there are two junctions; however, as the gate terminal is connected to the two junctions, these FETS are still three-terminal devices. The construction of the two types of JFETS, or FETs as they are more commonly referred to, is shown in Figure 7-2.

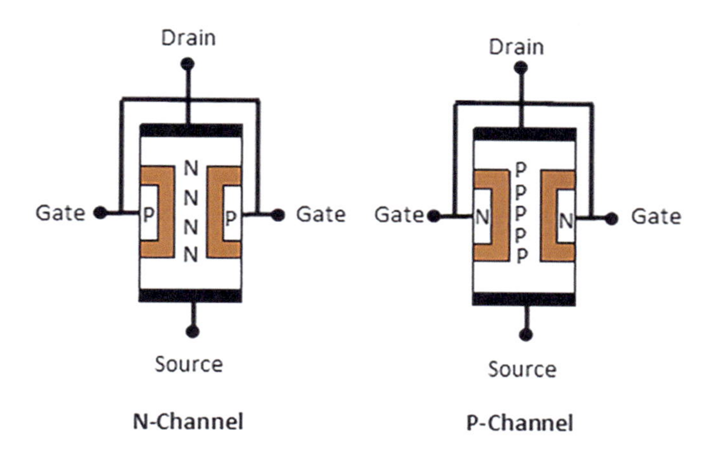


***Figure 7-2.*** *The Construction of the N-Channel and P-Channel JFET*

The main part of the JFET is the region in the central area, which makes a channel of extrinsic, that is, doped, semiconductor material. This is sandwiched between two areas of semiconductor material that are doped in the polarity that is opposite to that of the channel; see Figure 7-2. The channel region is where the JFETs get their respective names: N-channel, where this central channel is made up of N-type material, and P-channel, where the area is made up of P-type material. Just to recap, N-type material is a semiconductor material that has an excess of electrons, that is, it has been doped, or infused, with a pentavalent material such as Antimony. This means it has a negative charge. On the other hand, a P-type semiconductor material has a shortage of electrons, that is, it has been doped with a trivalent material such as Boron. This means it has a positive charge on it as it has a shortage of electrons.

To appreciate how the FETs work, this text will discuss the N-channel as this is the most common type of FET.

In the n-channel, the material has an excess of electrons, which means it has a negative charge. On either side of this channel, there is an area of p-type material, which has a positive charge. Where the two different types of material meet, electrons from the N-type material drift across into the p-type material. This then sets up an area around the junction, where the two materials meet, that has no charge within it. This is called a depletion area as it is depleted of charge. This area then forms a barrier between the two materials, and the drifting of electrons stops. Note this action happens with no potential applied to the JFET. Therefore, in its stable, disconnected, state, the JFET now has two depletion areas around the two junctions. This is shown in Figure 7-3 as the dark thick line around the two P-type areas at the gate terminals.

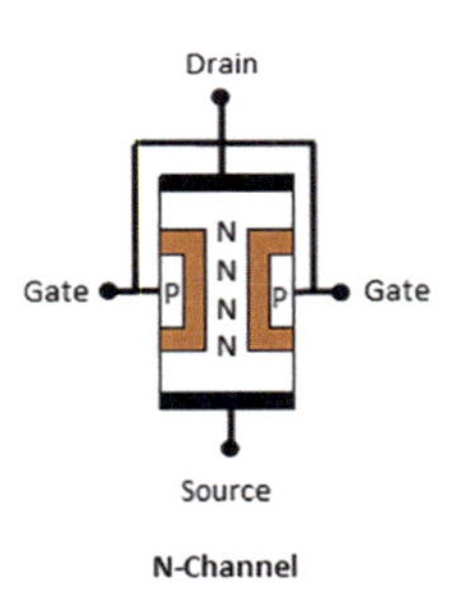


***Figure 7-3.*** *The Depletion Areas in the Channel of the JFET*

## Testing the JFET

The first test is to apply a supply to the JFET. When we carry out any kind of experiment to enable us to determine a relationship for any device to any stimulus, it is best to only change one item at a time and note the response of the test item. Therefore, the first test is to connect the gate and the source to 0V and then increase the drain voltage in small steps. This test circuit is shown in Figure 7-4.

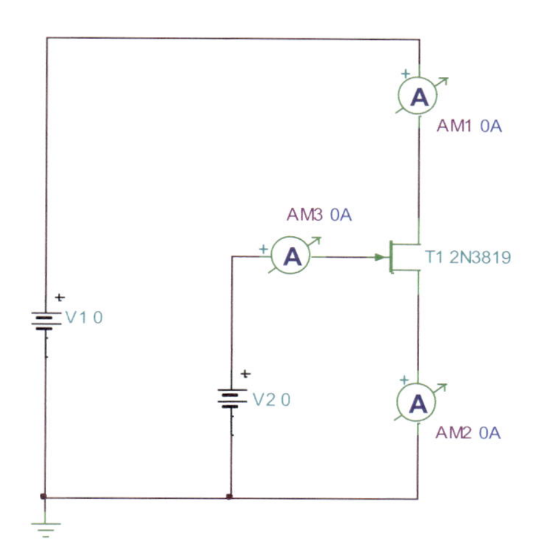


***Figure 7-4.*** *The Test Circuit to Determine the Output Characteristics of the N-Channel JFET*

The battery, V1, is the voltage across the drain to source connection. This is termed $V_{DS}$, voltage drain to source.

The ammeter will measure the current flowing from the drain to source, and this is termed the drain current $I_D$.

The procedure to carry out the experiment is to vary the $V_{DS}$ from 0V to 5V, in steps of 0.25V, and record the current flowing through the FET. Record the results, then plot a graph of the results.

The results are listed Table 7-2.

***Table 7-2.*** *Results of the Test Circuit Shon in Figure 7-4*

| VDS | IDS | | | |
|---|---|---|---|---|
| | VGS = 0 | VGS = –1 | VGS = –1.5 | VGS = –2 |
| 0 | 0 | 0 | 0 | 0 |
| 0.25 | 1.85mA | 1.21mA | 890uA | 567.52uA |
| 0.5 | 3.54mA | 2.26mA | 1.62mA | 973.78uA |
| 0.75 | 5.07mA | 3.16mA | 2.19mA | 1.22mA |
| 1 | 6.54mA | 3.89mA | 2.6mA | 1.3mA |
| 1.25 | 7.76mA | 4.46mA | 2.84mA | 1.3mA |
| 1.5 | 8.73mA | 4.87mA | 2.93mA | 1.3mA |
| 1.75 | 9.63mA | 5.12mA | 2.93mA | 1.3mA |
| 2 | 10.37mA | 5.21mA | 2.93mA | 1.3mA |
| 2.25 | 10.95mA | 5.21mA | 2.93mA | 1.3mA |
| 2.5 | 11.36mA | 5.21mA | 2.93mA | 1.3mA |
| 2.75 | 11.62mA | 5.12mA | 2.93mA | 1.3mA |
| 3 | 11.72mA | 5.21mA | 2.93mA | 1.3mA |
| 3.25 | 11.72mA | 5.21mA | 2.93mA | 1.3mA |
| 3.5 | 11.73mA | 5.21mA | 2.93mA | 1.3mA |
| 3.75 | 11.74mA | 5.12mA | 2.93mA | 1.3mA |
| 4 | 11.74mA | 5.21mA | 2.93mA | 1.3mA |
| 4.25 | 11.75mA | 5.21mA | 2.93mA | 1.3mA |
| 4.5 | 11.76mA | 5.21mA | 2.93mA | 1.3mA |
| 4.75 | 11.76mA | 5.21mA | 2.93mA | 1.3mA |
| 5 | 11.77mA | 5.21mA | 2.93mA | 1.3mA |

This is the table of results of the preceding experiment. The graph of the results is shown in Figure 7-5.

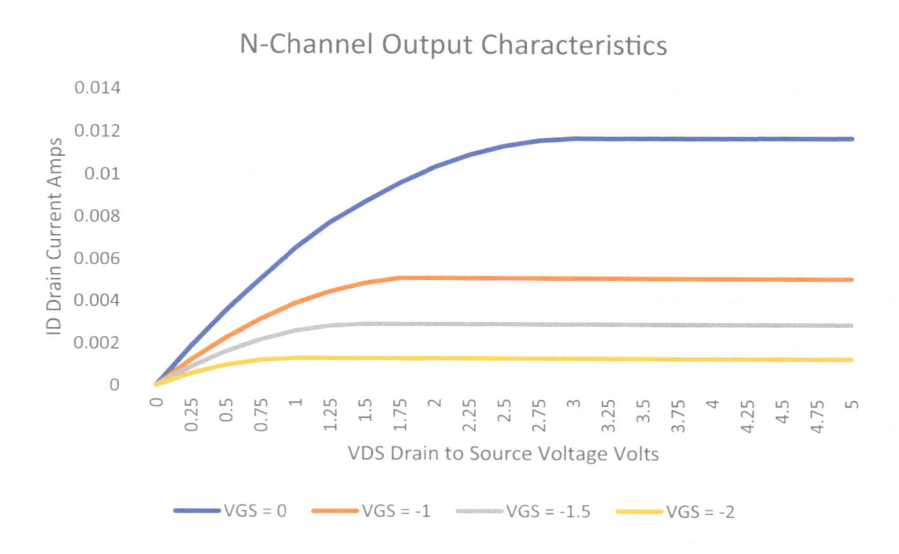


***Figure 7-5.*** *The Output Characteristics of the N-Channel JFET*

From an inspection of the graph, it can be seen that from 0v to 1.25V the graph is fairly linear. This is termed the "Ohmic region." After that, the graph rolls off, and when $V_{DS}$ gets to around 2.75V, the graph flattens out, and further change in $V_{DS}$ does not result in any increase of the current $I_D$. This voltage of around 2.75V is termed the pinch-off voltage.

It should be appreciated that the gradient of the graph is expressed as

$$gradient = \frac{\text{change in } I_D}{\text{change in VDS}} = \frac{I}{V}$$

This is actually the inverse of resistance, as resistance, according to Ohm's Law, can be expressed as voltage divided by current.

Therefore, we can say

$$gradient = \frac{I}{R}$$

$$\therefore R = \frac{\text{change in VDS}}{\text{change in } I_D}$$

Therefore, taking some values within the linear region of the graph, we get

$$R = \frac{0.9}{8E^{-3}} = 112.5\Omega$$

However, in the later part of the graph, the gradient is zero, and for this to happen, the resistance must rise to infinity. This is with the gate voltage VGS set at 0V.

When the gate voltage is made negative, as with the –1v, –1.5v, and –2v series, the Ohmic regions are smaller, and the pinch-off voltage is reduced.

What is really happening?

## The Semiconductor Analysis of the JFET

In Chapter 1, we looked at the PN junction and how it can be biased. Figure 7-3 shows us that there are two such PN junctions inside the N-channel JFET. If we look at this JFET as a semiconductor material, then we might be able to appreciate how the JFET works. To help us, we will consider the circuit as shown in Figure 7-6.

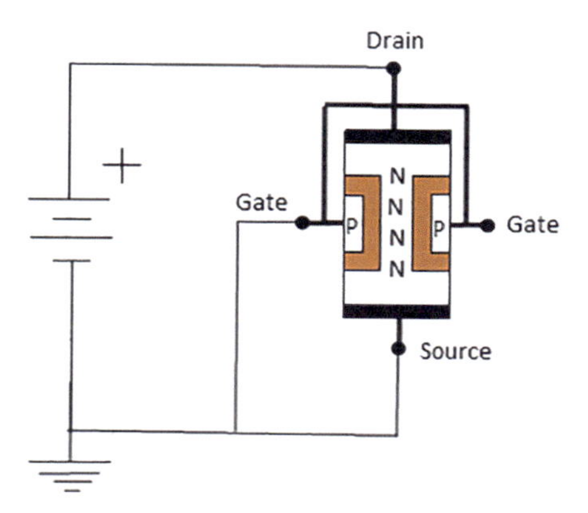


***Figure 7-6.*** *The JFET As Two PN Junctions with Biasing*

The drain is connected to the positive terminal, and the gate and source are both connected to the ground. This means that the N-type material is connected to the positive terminal, and the P-type material is connected to the negative terminal. This is the condition for the reverse biasing of these two PN junctions, which in turn means that the depletion area around the p-type materials will expand.

However, it should be noted that there is a voltage gradient across the N-channel as the top of the N-channel can be connected to a voltage greater than 0V, say 2V, for example, while the other end of the n-channel is connected to ground. This will mean that the force creating this depletion layer is higher at the positive end of the n-channel compared to the other end which is at 0V, and so at the source end of the channel, there will be no force creating the depletion layer.

This then means that the increasing of the depletion layer is greater at the top of the JFET compared to the bottom of the JFET. This will result in a depletion layer formed as shown in Figure 7-7.

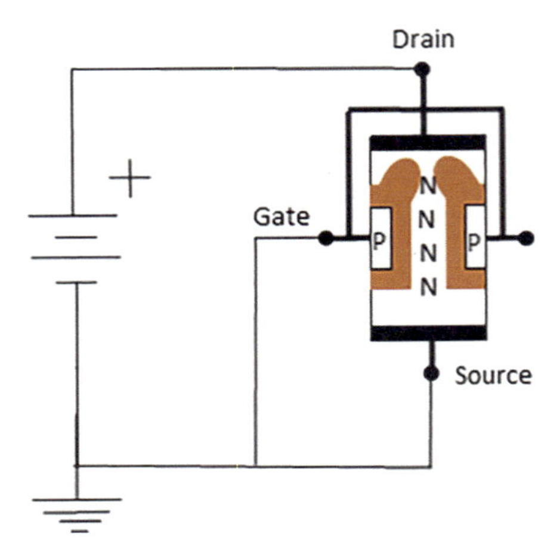


***Figure 7-7.** The Depletion Layer Growing More Larger at the Top of the JFET*

Note some current will flow through the n-channel, simply because the electrons in the n-channel are attracted by the positive potential connected to the drain, and this will increase linearly, as shown in the graph in Figure 7-5. However, as the current is increasing, because the applied voltage $V_{DS}$ is increasing, then the depletion area around the p-type material will also increase.

## The Pinch-Off Voltage

As there are two depletion areas that, when we increase the voltage $V_{DS}$, are increasing, then there will be a point when the two depletion layers meet. This is termed "pinch off," and when this happens, the flow of current stops increasing; no matter how much more the $V_{DS}$ increases, the current no longer increases. This is because there now is a depletion layer going right across the n-channel, as shown in Figure 7-8, and the normal electron flow inside the channel now has to overcome this depletion layer. However, the more the supply tries to overcome this depletion layer, the bigger it gets, so current flow can no longer increase. The voltage at which this meeting of the two depletion layers occurs is termed "the pinch-off voltage." With respect to the graph shown in Figure 7-5, this is around the point when the $V_{DS}$ voltage is at 2.75V.

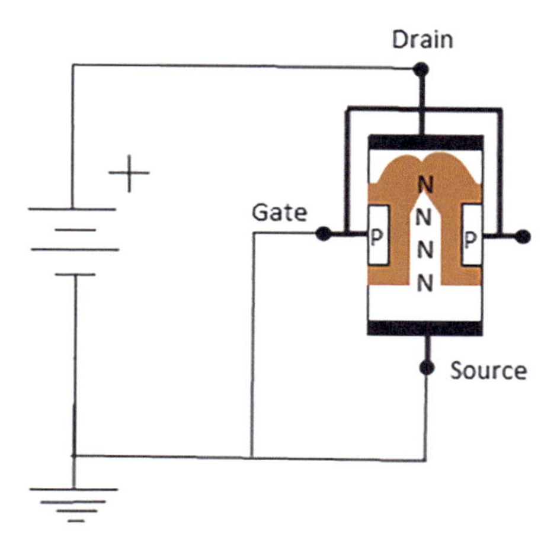


***Figure 7-8.*** *The Two Depletion Layers Meet, Pinch Off*

# Turning On and Off the JFET

One of the main uses of FETs is in switching applications. To that end, we should investigate how we can switch these FETs on and off. We can investigate the turning on and off of the n-channel JFET by simulating the circuit shown in Figure 7-9.

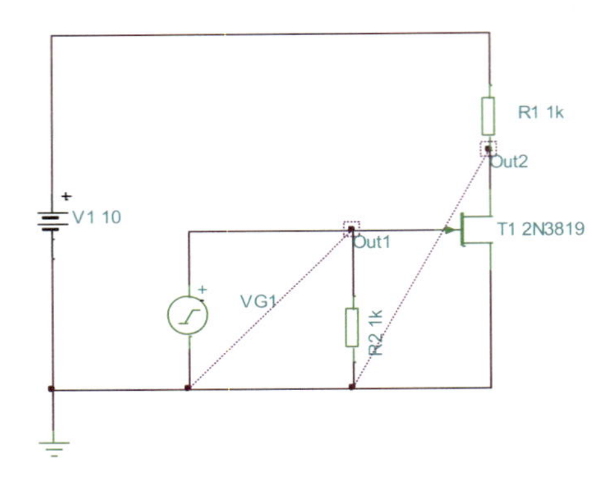


***Figure 7-9.*** *The Test Circuit for Turning On and Off the JFET*

The voltage traces to show how the FET turns on and off are shown in Figure 7-10.

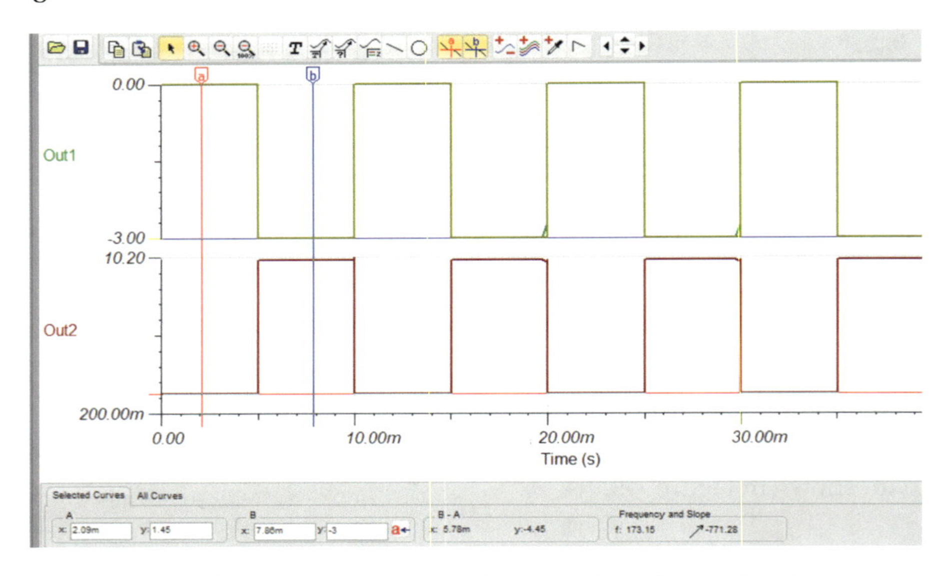


***Figure 7-10.*** *The Gate and Drain Voltage Waveforms for the N-Channel JFET*

The negative going voltage is the gate voltage, which is being used to control when we turn on and off the JFET. This is OUT1 on the circuit shown in Figure 7-9 and the top trace on the display shown in Figure 7-10. The bottom trace, which is OUT2 on the circuit, shows the voltage at the drain of the JFET. We can see that initially the drain voltage is low at around 1.5V; this is cursor "A" in Figure 7-10. This shows that the JFET is initially turned on. Then, when the gate voltage drops to -3V, cursor "B" in Figure 7-10, we can see that the drain voltage goes high to 10V. This is the same as the supply voltage and shows that the JFET has turned off.

## The Threshold Voltage $V_T$

All FETs have a parameter called "the threshold voltage $V_T$." This is the voltage at which the FET must change state, in that if it was turned on, then when the gate terminal reaches this threshold voltage it would turn off. If the FET was turned off, then it would turn on when the gate voltage reached this threshold voltage. Different FETs will have different threshold voltage levels.

If we look at the traces in Figure 7-10, we can see that the FET does turn off when the gate voltage, OUT1, reaches -3V. This suggests that the "threshold voltage $V_T$" for this FET is -3V. If we look at the model parameters that the software Tina uses for the 2n3819, we can see that it sets the threshold voltage for this JFET at -2.9985V. See Figure 7-11.

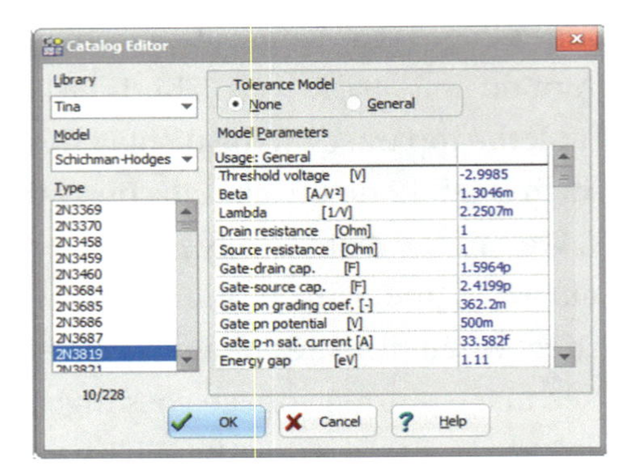


***Figure 7-11.*** *The Model Parameters for the 2N3819 JFET*

# Turning On and Off a P-Channel JFET

We can carry out a similar experiment with a p-channel FET. We will start with the circuit shown in Figure 7-12.

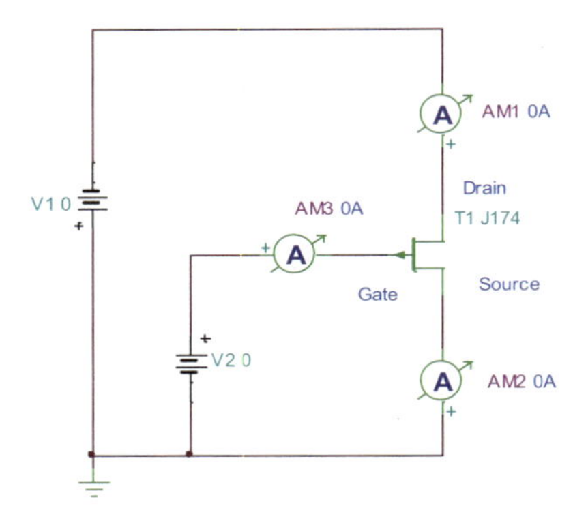


***Figure 7-12.*** *The Test Circuit for the P-Channel JFET*

When we simulate the circuit, we can complete the table of results as shown in Table 7-3.

***Table 7-3.*** *Table of Results for the Test Circuit Shown in Figure 7-11*

| **VDS** | **IDS** | | | |
|---|---|---|---|---|
| | **VGS = 0** | **VGS = 1** | **VGS = 4** | **VGS = 6.54** |
| 0 | 0 | 1.46pA | 4.46p | 7pA |
| 0.25 | 2.58mA | 2.19mA | 986.58uA | 75.61nA |
| 0.5 | 5.11mA | 4.31mA | 1.89mA | 76.36nA |
| 0.75 | 7.59mA | 6.38mA | 2.71mA | 77.11nA |
| 1 | 10.01mA | 8.38mA | 3.44mA | 77.86nA |
| 1.25 | 12.37mA | 10.32mA | 4.08mA | 78.61nA |
| 1.5 | 14.67mA | 12.18mA | 4.62mA | 79.35nA |
| 1.75 | 16.91mA | 13.98mA | 5.06mA | 80.1nA |
| 2 | 19.07mA | 15.69mA | 5.41mA | 80.85nA |
| 2.25 | 21.17mA | 17.33mA | 5.64mA | 81.6nA |
| 2.5 | 23.2mA | 18.89mA | 5.78mA | 82.35nA |
| 2.75 | 25.14mA | 20.36mA | 5.83mA | 83.1nA |
| 3 | 27.01mA | 21.75mA | 5.88mA | 83.85nA |
| 3.25 | 28.8mA | 23.04mA | 5.94mA | 84.6nA |
| 3.5 | 30.5mA | 24.25mA | 5.99mA | 85.35nA |
| 3.75 | 32.12mA | 25.35mA | 6.04mA | 86.1nA |
| 4 | 33.64mA | 26.35mA | 6.09mA | 86.95nA |
| 4.25 | 35.07mA | 27.26mA | 6.15mA | 87.6nA |
| 4.5 | 36.4mA | 28.05mA | 6.2mA | 88.34nA |
| 4.75 | 37.63mA | 28.74mA | 6.25mA | 89.09nA |
| 5 | 38.76mA | 29.32mA | 6.3mA | 89.84nA |

From the table of results, we can see that as the gate voltage increases, the drain current, ID, is lower for the same value of VDS. At the specific gate voltage of 6.54V, the drain current is virtually zero for all values of the VDS voltage. This is the specified threshold voltage that TINA uses for this JFET. At this voltage, the depletion layer would have covered the channel, and this would make it difficult for any current to flow through the JFET.

Using the results from Table 7-3, we can create the graph of the output characteristics for the p-channel JFET as shown in Figure 7-13.

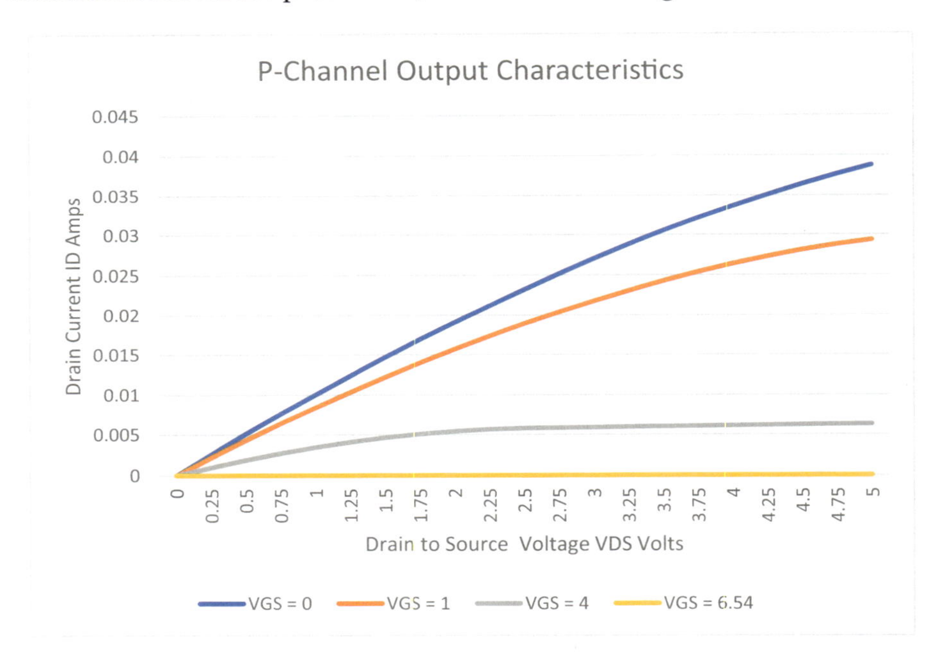


***Figure 7-13.*** *The Output Characteristics for the P-Channel JFET*

The graph shown in Figure 7-13 does confirm that we can control the amount of current flowing through the JFET with the gate voltage. If we consider the trace when VGS was set to 4V, we can see that the graph has flattened off. This suggests that the JFET, with the gate voltage at 4V, has reached its saturation, and the $I_D$ current has reached a maximum of around 6.5mA. If we wanted to determine the saturation current when the gate voltage was set to 0V and 1V, we would need to extend the results. However, that is something I will leave for you to do if you so wished to do so.

The graph shown in Figure 7-13 does show the drain current stops rising as steeply as at the start, when the VDS voltage reaches around 3 to 4V. What is happening inside the JFET is that the depletion layers are expanding closer to the point where we get pinch off when the depletion layers meet.

Using these results, we can try and explain the difference between "pinch-off voltage" and "threshold voltage." The "pinch-off voltage" is the drain to source voltage, $V_{DS}$, that causes the depletion layers inside the FET to meet across the channel; see Figure 7-8. The "threshold voltage" is the voltage at the gate, the $V_{GS}$ voltage, that causes the FET to change from either conducting to not conducting or from not conducting to conducting.

Using Table 7-3, we can suggest that the threshold voltage for the J174 JFET is 6.54V. We can test this concept further using the test circuit shown in Figure 7-14.

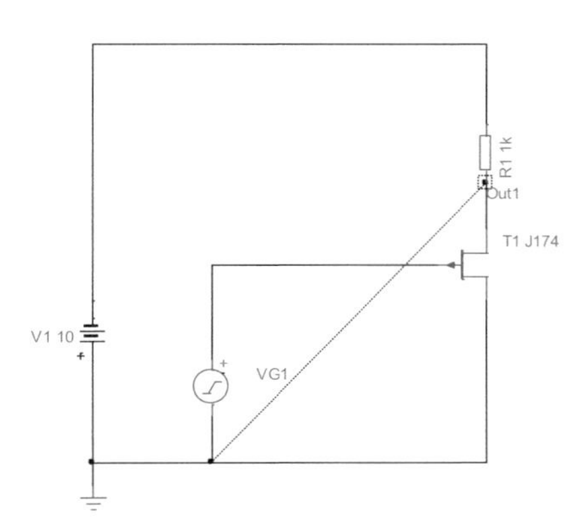


***Figure 7-14.*** *The Test Circuit for the P-Channel JFET*

The p-channel JFET works in the same way as the n-channel except that the polarities are reversed. This means that the p-channel will also have a pinch-off voltage, but it will be a positive voltage.

We are again using a transient analysis to examine the voltage at the drain over time. The waveforms from the analysis are shown in Figure 7-15.

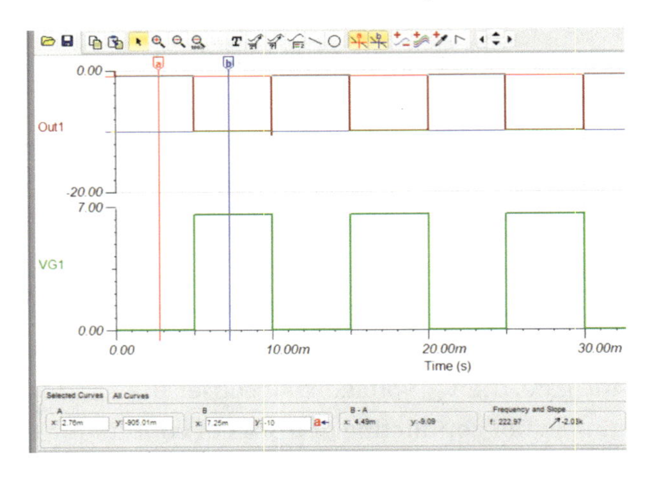


a

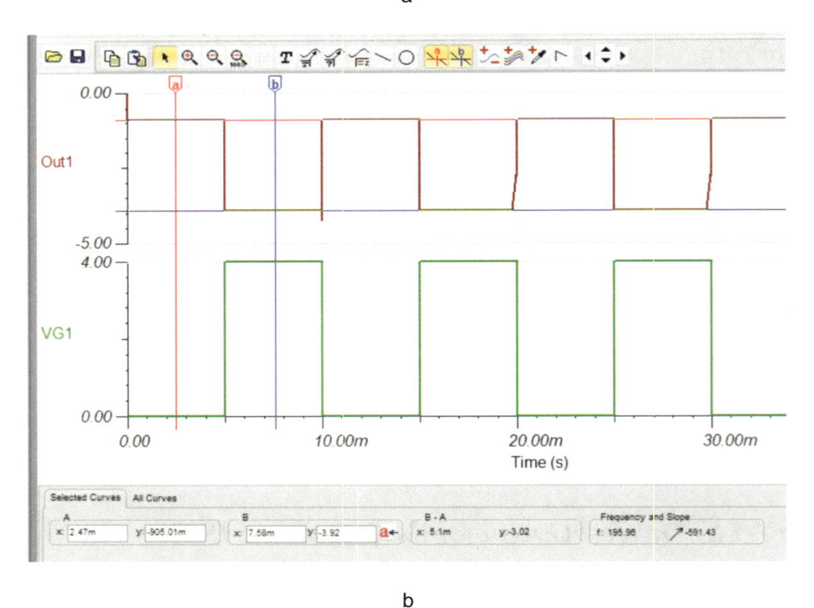


b

***Figure 7-15.*** *The Voltage Waveforms from the Test Circuit Shown in Figure 7-14*

The waveforms shown in Figure 7-15a are when the gate voltage was set to 6.54V. This is the expected threshold voltage as shown in the parameter setting for the J174 JFET in TINA. The VG1 trace is the voltage applied to the gate of the JFET, while the OUT1 is the voltage at the drain terminal of the JFET. We can see that when the gate voltage is 0V, the JFET is turned on as the drain voltage has risen toward 0V at -905.01mV. This shows that the JFET is turned on and current is flowing through R1. Then, when the gate voltage rises to 6.54V, the drain voltage drops to -10V. This shows that there is no current flowing through R1 and the JFET has turned off.

The waveforms shown in Figure 7-15b show that the JFET does turn on when the gate voltage is at 0V. However, when the gate voltage rises to only 4V, as that is the peak voltage of the gate voltage, the JFET has not fully turned off. This can be seen because the drain voltage has only dropped to -3.92V. This means there must still be current flowing through R1 as the other 6.08V had been dropped across R1.

This simple test does confirm that the threshold voltage for this JFET, the J174, according to TINA, is 6.54V as the JFET has turned off. I should point out that there is a wide range of different JFETS all with their own characteristics such as threshold voltage and typical drain currents. You must examine the datasheets to decide which device you want to use for your application.

We can see that both the n-channel and p-channel JFETs are normally turned on when the gate voltage is at 0V. Then we have to drive the voltage at the gate terminal either negative, as with the n-channel, or positive, as with the p-channel, to turn the JFETs off. This is because they are both depletion-type JFETs, in that the channel is already doped and there is a depletion layer that can be increased until the JFETs turn off. In this way, because the width of the depletion layer is controlled by the voltage at the gate terminal, these JFETs can be viewed as a voltage-controlled resistor.

These JFETS can be used as a switch to turn a load on and off. However, one problem, in the JFETS we have looked at, is that even when the JFET is switched on, there is still some voltage dropped across the JFET. In Figure 7-15, we can see this was 905mV with the J174. The issue here is that the JFET could dissipate too much power and so overheat and destroy itself. However, as power, in general, is the product of voltage and current, then if the current is kept low, at around 60mA, then the power dissipated would be

$$Power = vi = 905^{-3} \times 60^{-3} = 54.3mW$$

This would not be too much of a problem, but if we were using higher current loads, then we might need a more efficient device. The MOSFET that we will look at next is one such device.

## The MOSFET

The makeup of the MOSFET is shown in Figure 7-16.

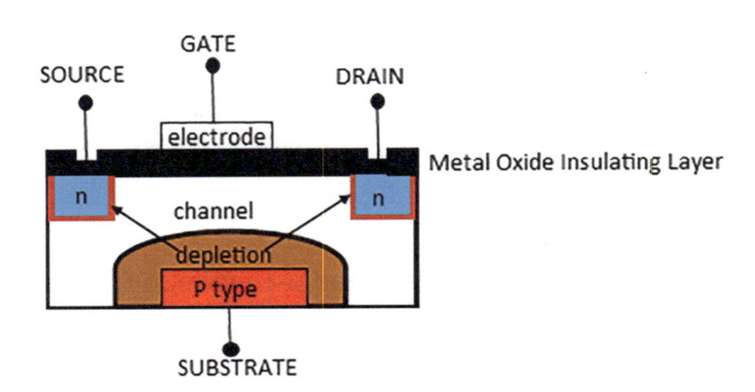


***Figure 7-16.*** *The Basic Construction of a MOSFET*

MOSFET stands for metal oxide semiconductor field effect transistor. The metal oxide is actually a physical layer inserted between the gate terminal and the channel. As the gate is the normal input to the FET, then no current can flow through this insulating layer, and so the input impedance of the MOSFET is very high in the order of 10000s of Meg Ohms. This explains one of the major differences between the BJT and the FET. With the BJT, we force current to flow into the transistor, and that is why the BJT can be viewed as a current-controlled device. However, with the MOSFET no current flows into the transistor, and it is the voltage applied at the gate that creates the field that controls the conduction of these FETs, hence the name field effect transistor.

Due to this insulating layer, which is basically between two conductive materials, there is some capacitance at the input of the MOSFET. This can cause a problem with static, and you may have to take some precautions when using MOSFETs.

It is normal to have the substrate terminal connected internally to the source terminal. This means that just as with the JFET and the BJT, the MOSFET is a three-terminal device. The construction of the main types of MOSFET and their associated circuit symbols are shown in Figure 7-17. The symbols for the MOSFET show that the substrate is connected to the source.

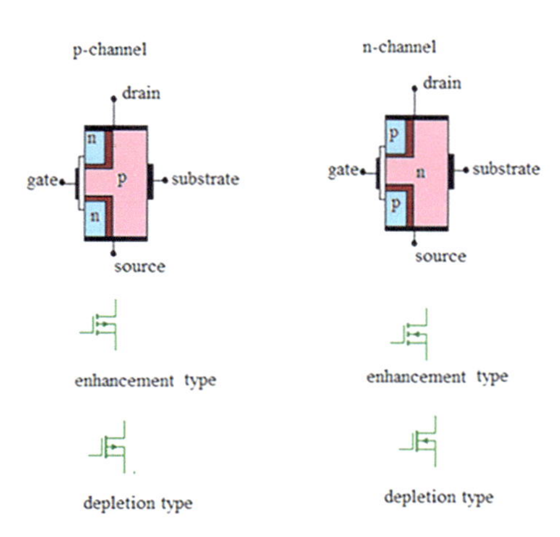


***Figure 7-17.*** *The Main Types of MOSFETS*

The main difference between the types of MOSFETs, which are enhancement mode and depletion mode, is that with the enhancement mode the channel is very lightly doped or even has no doping in it. This means that the channel is in a stable state and is not conducting with the gate voltage at zero. The gate voltage must be raised either positively with the n-channel MOSFET or negatively with the p-channel MOSFET to turn the MOSFET on, whereas with the depletion type the channel is fully doped, and so the channel is already conducting. With the depletion type, we must apply a voltage to turn the MOSFET off.

The symbols for the MOSFETs are different as well, as these differences, in the symbols, are used to try and show differences in their characteristics. The first is the insulating gate barrier at the input of the MOSFETs. With the enhancement mode MOSFETs, this line is dashed. This is to indicate that the MOSFET is only lightly doped and is not conducting, and we need to do something to enhance it and make it conduct. When the line for the insulating gate barrier is a solid bar, as with the depletion mode MOSFETs, it is indicating the MOSFET is already conducting, and we need to increase the depletion layer to turn the MOSFET off.

You should note that the bar at the gate for both of the JFET symbols is a solid bar because both types of JFET are depletion type.

# Turning On the Enhancement-Type MOSFET

The circuit used for this simulated experiment is shown in Figure 7-18.

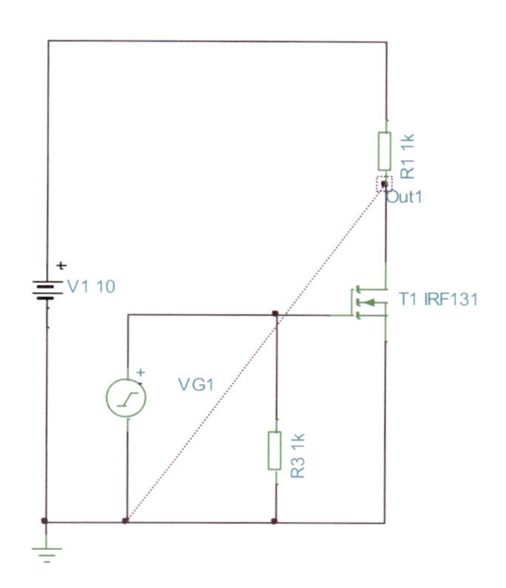


***Figure 7-18.*** *The Test Circuit for the N-Channel MOSFET Enhancement Mode*

The MOSFET used in this experiment is the IRF131, which is an enhancement-type n-channel MOSFET. From the datasheet, we can see that the gate threshold voltage is between 2V and 4V. TINA sets this parameter at 3.406V. We will apply a square wave at the gate terminal of the circuit shown in Figure 7-18 and set the peak voltage to 2.5V and then 3.5V. We will again use the transient analysis with the OUT1 probe measuring the voltage at the drain terminal of the MOSFET. The two waveforms are shown in Figure 7-19.

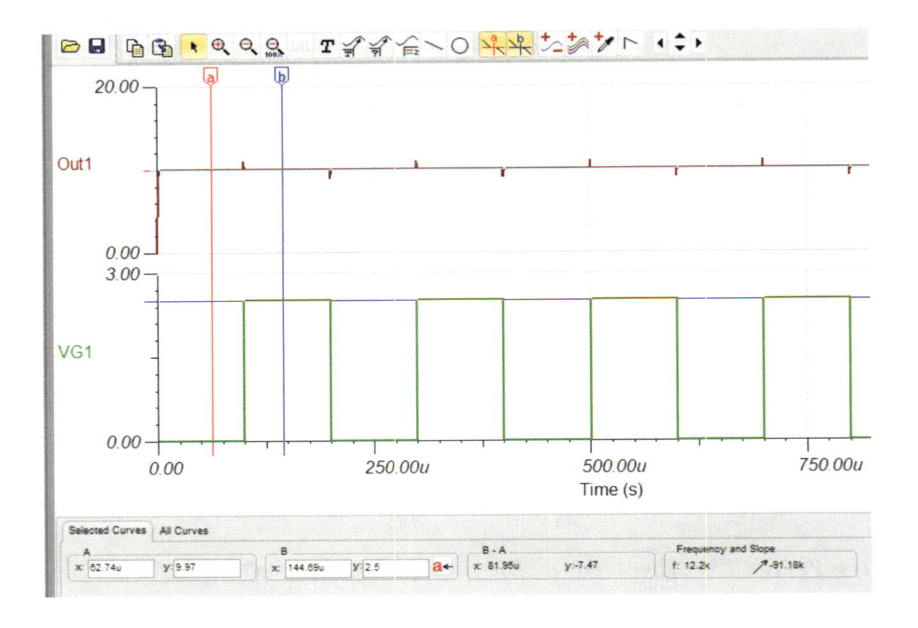


a

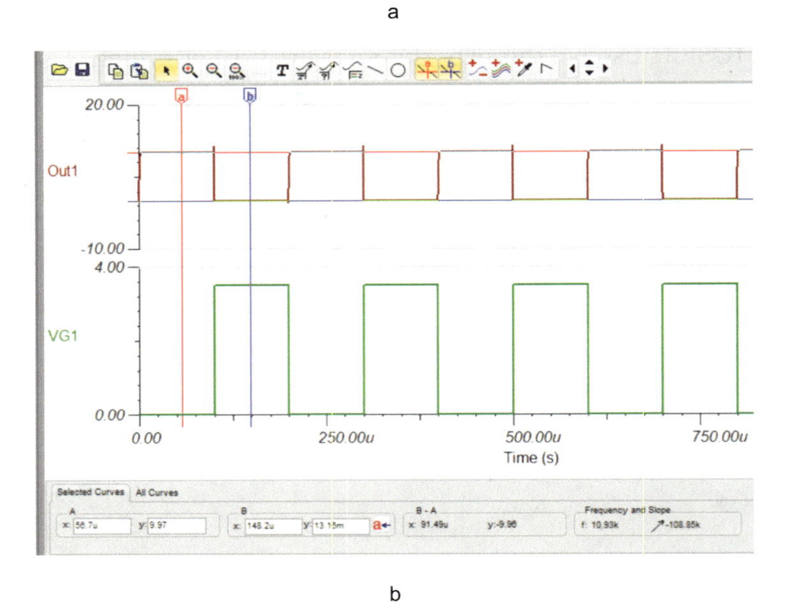


b

***Figure 7-19.*** *The Voltage Waveforms for the Test Circuit Shown in Figure 7-18*

Figure 7-19a shows the waveforms when the peak of the gate voltage was set at 2.5V. We can see from the OUT1 trace, which displays the voltage at the drain terminal of the MOSFET, that the MOSFET does not turn on. This is because the gate voltage is too low.

Figure 7-19b shows the waveforms when the peak of the gate voltage is set to 3.5V. From the OUT1 trace, we can see that the MOSFET now does turn on. This is because the drain voltage does fall to around 13.15mV when the gate voltage goes to 3.5V.

This does confirm that with the enhancement mode, the MOSFET does not conduct when there is no gate voltage applied or indeed if the gate voltage is below its threshold. Also, this particular MOSFET has a threshold voltage at around 3.5V on the gate terminal.

Now let's think about what is happening inside the MOSFET. For this, we need to consider the state where no voltage is applied to the MOSFET, that is, $V_{DS} = 0$ and $Vg = 0$.

In this state, as with any PN junction, a depletion layer has formed around the two PN junctions. Now as we apply the positive voltage between the drain and source terminals, $V_{DS}$, it should be clear that the top and bottom PN junctions are reverse biased (see the n-channel MOSFET shown in Figure 7-17) as the source terminal is at 0V. However, we should also appreciate that there is a voltage gradient throughout the n-channel, just as there was with the JFET.

Now if a positive voltage is applied to the gate, it should be clear that the lower PN junction is now becoming forward bias, but the upper junction is still reverse biased. Now as the gate voltage is increased, the lower PN junction becomes more forward biased to the point that the depletion layer across that PN junction is overcome. This is when the gate voltage has reached the threshold voltage. Electrons can now enter into the lower region of the channel heading toward the gate terminal. However, instead of flowing out of the gate terminal, these electrons become more attracted by the $V_{DS}$ voltage, and an avalanche effect occurs as they rush through the n-channel toward the $V_{DS}$ at the drain terminal, and so we have current flow.

# The Depletion-Type MOSFET

The same type of experiment can be used to confirm the operation of the depletion-type n-channel MOSFET. The test circuit for this experiment is shown in Figure 7-20.

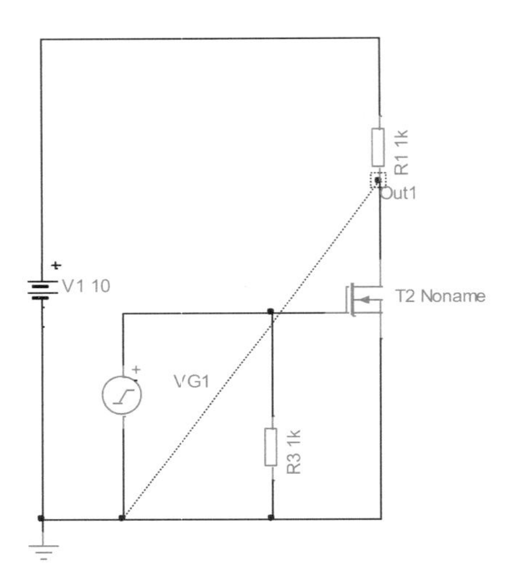


***Figure 7-20.*** *The Test Circuit for the Depletion Mode N-Channel MOSFET*

Unfortunately, TINA only has a generic model for the depletion mode N-channel MOSFET. However, as we are really only studying how we can turn the MOSFET on and off, it will serve our purpose. We will again use a transient analysis with OUT1 displaying the drain voltage waveform. Figure 7-21 shows the two sets of waveforms from the experiment. Figure 7-21a shows the drain voltage when the gate voltage has a peak of –1V. The Out1 trace shows that the MOSFET is conducting when the gate voltage is 0V. However, it is still conducting when the gate voltage goes to –1V.

Figure 7-21b shows what happens when the gate voltage is set to peak at –2V. We can see that the MOSFET is turned on when the gate voltage is 0V, but it does now turn off when the gate voltage goes to –2V.

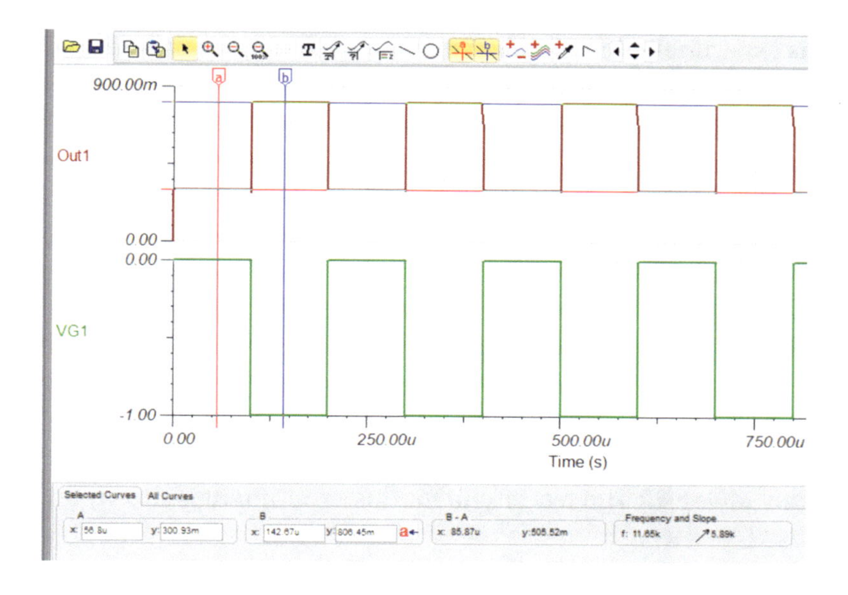


a

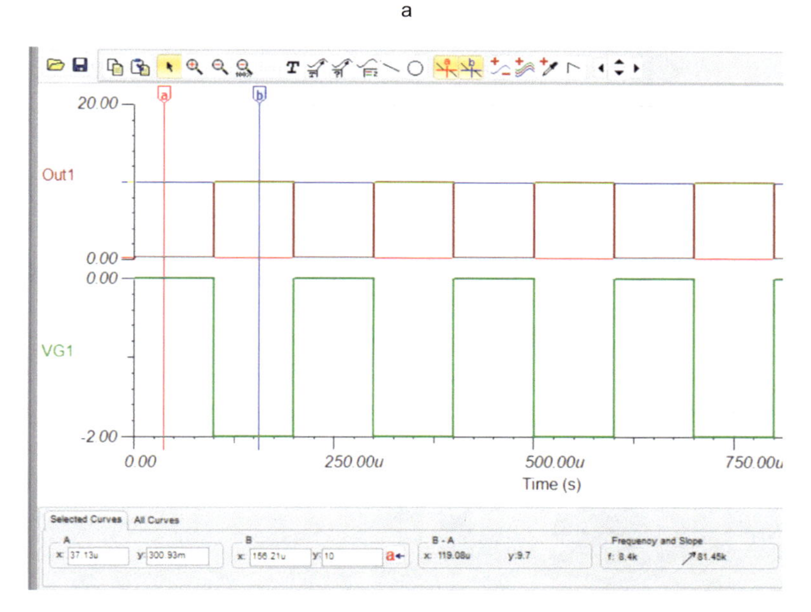


b

***Figure 7-21.*** *The Two Displays of the Drain Voltage for the Test Circuit Shown in Figure 7-20*

This experiment does confirm that the depletion mode MOSFET is conducting when the gate voltage is 0V. It also confirms that we have to drive the gate voltage negative, which reverse biases the PN junction between the gate and source terminals (see Figure 7-17), to turn the MOSFET off. The value of -2V is the threshold voltage set by TINA for this depletion mode n-channel MOSFET.

# The P-Channel MOSFET

Now we will carry out similar experiments to see how the P-channel MOSFET turns on and off. Firstly, we will look at the enhancement mode P-channel MOSFET. The test circuit for this experiment is shown in Figure 7-22.

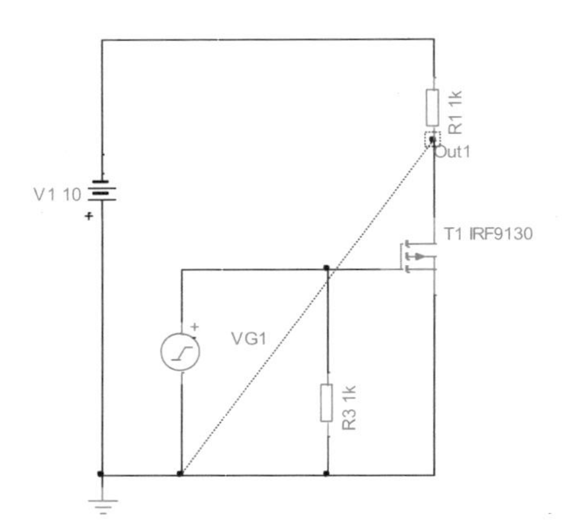


***Figure 7-22.*** *The Test Circuit for the Enhancement Mode P-Channel MOSFET*

The first thing we can see is that the $V_{DS}$ is negative as the source has to be more positive than the drain. If we consider the figure of the P-channel MOSFET (see Figure 7-17), we can see that this would reverse bias the upper PN junction. Also, as the source is at 0V, the lower PN junction is not forward biased enough to make that junction conduct. That is why we have to drive the gate negative enough to forward bias this lower PN junction. When that happens, current can attempt to flow between the gate and the source, but as the electrons move from the base to the source, the electrons that have been trying to enter the channel via the upper PN junction can now flow through the path now opened by the gate to source PN junction being overcome, and so the avalanche effect happens and current can flow through the channel from source to drain. This can be confirmed if we simulate the test circuit shown in Figure 7-23.

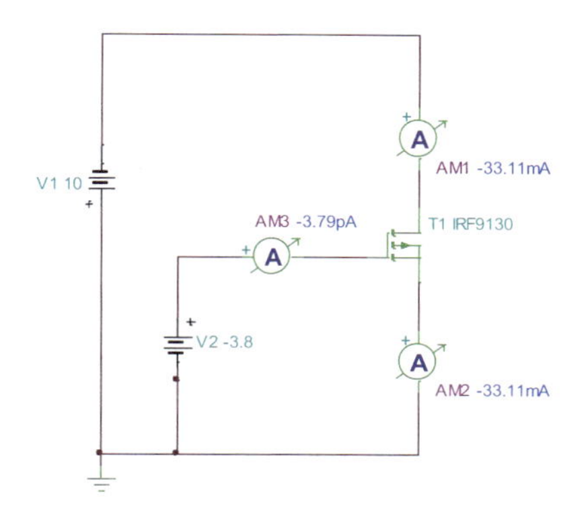


***Figure 7-23.*** *The Test Circuit to Show the Current Flow Through the Channel*

Figure 7-23 shows that when the gate voltage is at –3.8, a substantial current is flowing through the channel. However, if the gate voltage is kept below the threshold voltage, which TINA has set at –3.702V for the IRF9130, the current is too low at around 22.5μA. As an example, even when the gate voltage was set at –3.5, the MOSFET did not allow much current to flow through it.

However, we are interested in how we turn the MOSFET on and off. When we simulate the test circuit shown in Figure 7-23, we can display the voltage at the drain terminal, and the waveforms are shown in Figure 7-24.

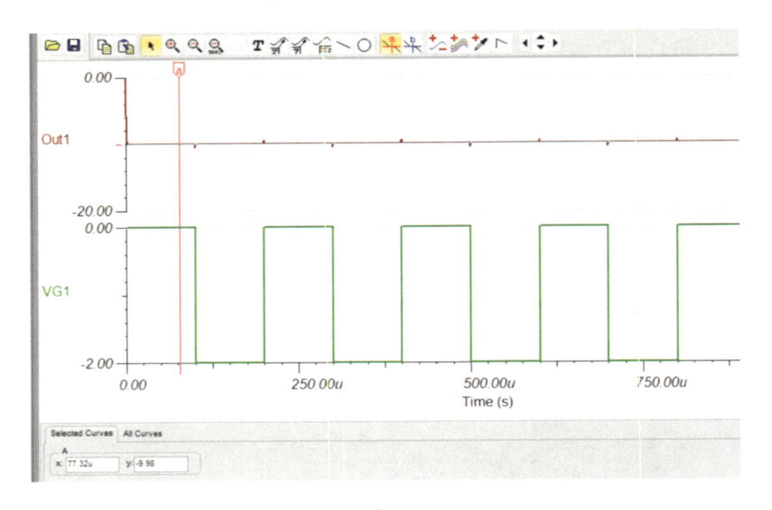


a

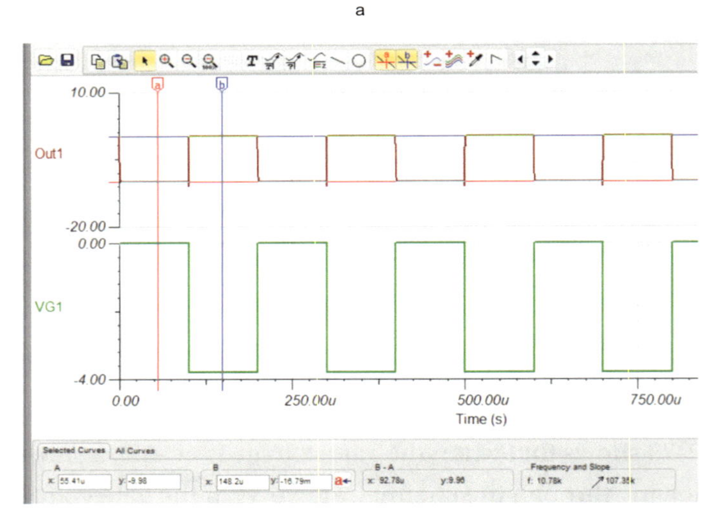


b

***Figure 7-24.*** *The Waveforms for the Test Circuit Shown in Figure 7-22*

Figure 7-24a shows the drain voltage when the gate voltage peaks at –2V. This is below the threshold voltage of –3.702V, and so the MOSFET does not turn on. Figure 7-24b shows the drain voltage when the gate voltage peaks at –3.8V. This is more than the threshold voltage, and so the MOSFET does turn on. This does confirm that the enhancement mode P-channel MOSFET does not conduct when the gate voltage is at 0V. It also confirms that we must drive the gate voltage negative to turn the MOSFET on and that for this particular MOSFET the threshold voltage is –3.702. The datasheet for the IRF9130 states that the threshold voltage should be between –2 and –4 volts.

Now we can test the depletion mode MOSFET. As this is a depletion mode MOSFET, it should be conducting when the gate voltage is 0V, and we should drive the gate positive by the specified threshold voltage to turn it off. The test circuit for this experiment is shown in Figure 7-25.

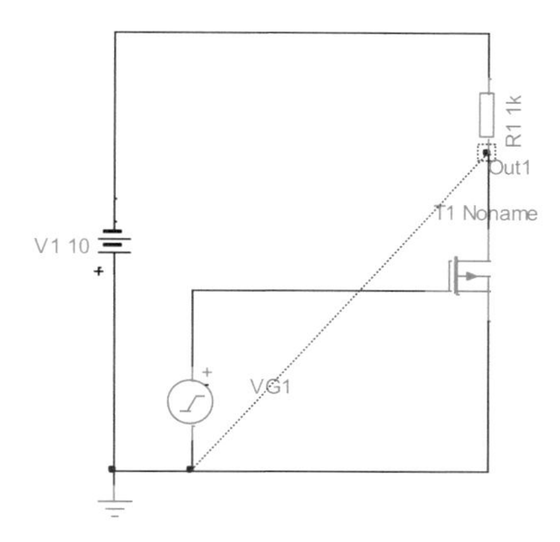


***Figure 7-25.*** *The Test Circuit for the Depletion Mode P-Channel MOSFET*

We will test the circuit shown in Figure 7-25 using the transient analysis as before. The waveforms for the two settings of the gate voltage are shown in Figure 7-26.

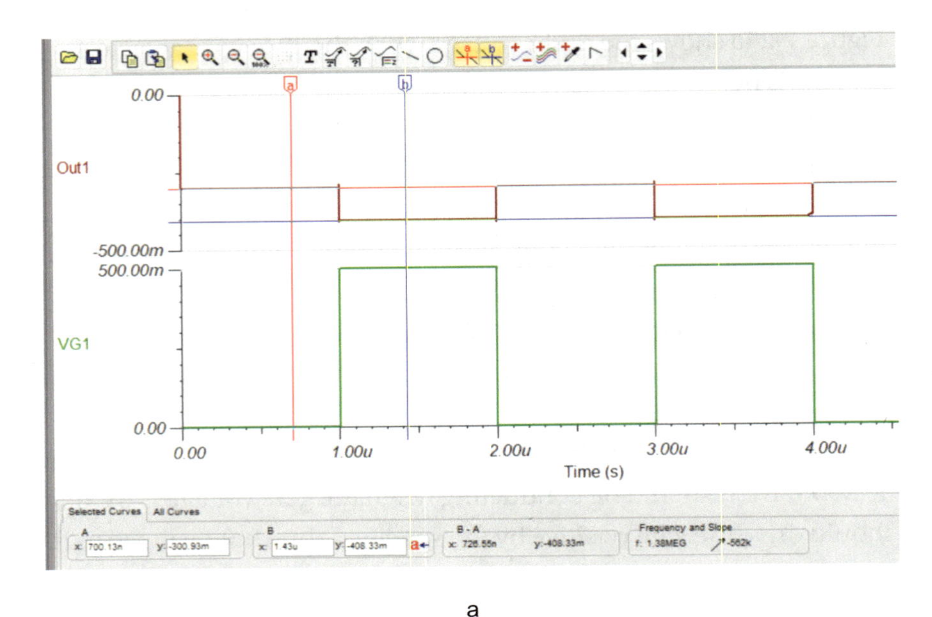


a

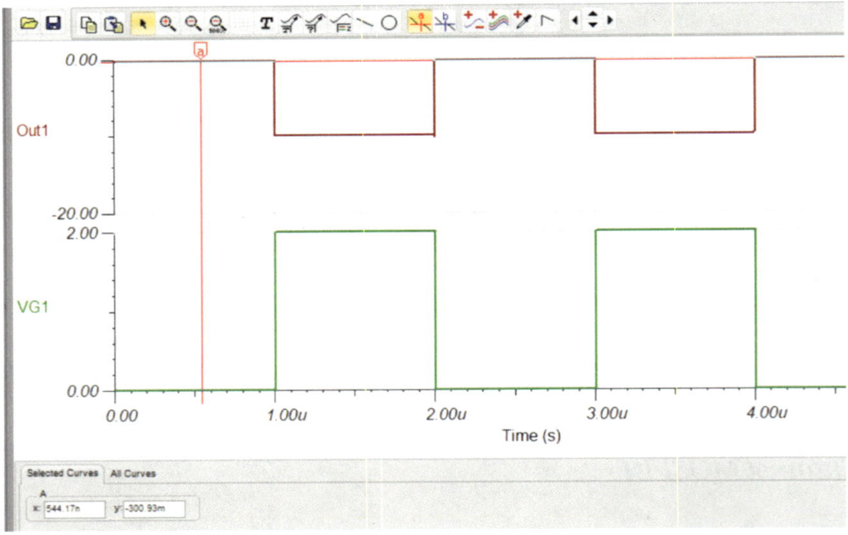

b

***Figure 7-26.*** *The Waveforms for the Test Circuit Shown in Figure 7-25*

Figure 7-26a shows the drain voltage when the peak voltage at the gate is 0.5V. This is lower than the threshold voltage of 2V set by TINA for this MOSFET. We can see when the gate voltage is at 0V, the MOSFET is conducting as the drain voltage is at –0.3V. When the gate voltage goes to 0.5V, the drain voltage only rises to –0.4V, which means the MOSFET is still turned on.

Figure 7-26b shows the drain voltage when the peak of the gate voltage goes to 2V. This shows that the MOSFET does turn off when the gate voltage rises to 2V.

The experiments we have carried out show us how we can turn on and off the various FETS and MOSFETs. We can see that they can all be used in a switching application; however, the MOSFETs may be better than the JFETs as a switch because we can see that when the MOSFETs turn on, there is only a small voltage dropped across the MOSFETs, typically 300mV or less (see Figures 7-18, 7-21, 7-23, and 7-26), whereas the JFETs have a higher voltage across them.

Before we leave this investigation into turning on and off a MOSFET, we should consider the stray capacitance around the MOSFET. We won't take a detailed look at this, as it is really a book on its own. However, the stray capacitances affect the turn on and off of the MOSFET as the gate source has to charge the input capacitance up to the required gate voltage before the MOSFET can turn on. This would take some time. Also, and perhaps more important, we must provide a path to discharge this input capacitance to a point below the required gate voltage, and this will take some time. The capacitor would hopefully discharge through the internal impedance of the gate source, but this could be too high and so take too long. In that case, you might need to provide a discharge path for this capacitance to discharge it. To fully understand the effect of these stray capacitances would involve too much for what I want to do with this book. The datasheet for the MOSFET will detail these capacitances and also the turn on and turn off times for the MOSFET under certain test conditions.

Now we will consider if we can use the FETs as an amplifier.

# The MOSFET Amplifier

We will concentrate on the enhancement mode MOSFETs as these are more commonly used. The MOSFET amplifier we will use is the 2n6577 enhancement mode MOSFET. This means that the amplifier will be turned off when the gate voltage is at 0V. We will need to raise the gate voltage to a suitable voltage that is higher than the threshold voltage for the MOSFET. We will carry out some different experiments to determine some important parameters of the MOSFET. The circuit we will use to test the first parameter, which will show how the drain current $I_D$ is related to the gate voltage $V_{GS}$, is shown in Figure 7-27.

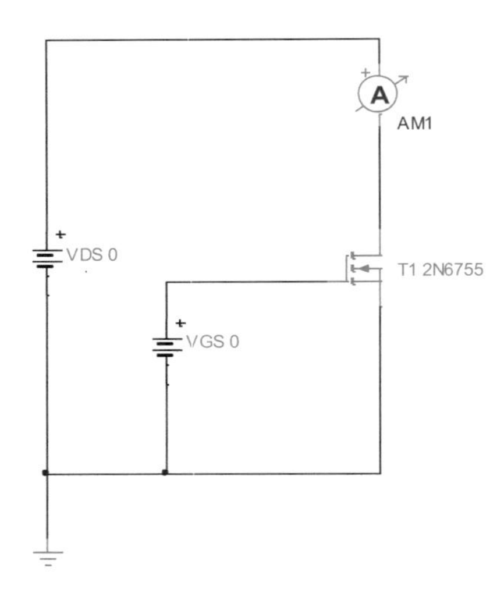


***Figure 7-27.** The Test Circuit to Determine the Input Characteristics for the 2N6755 Enhancement MOSFET*

With this test circuit, we will set the drain to source voltage to 10V. We will then vary the gate to source voltage from 0 to 10V in steps of 0.5V. The results were recorded in Table 7-4.

***Table 7-4.*** *The Results for the Test Circuit Shown in Figure 7-27*

| VGS | ID VDS =10V |
|---|---|
| 0 | 16.67E-6 |
| 0.5 | 16.67E-6 |
| 1 | 16.67E-6 |
| 1.5 | 16.67E-6 |
| 2 | 16.67E-6 |
| 2.5 | 16.67E-6 |
| 3 | 16.67E-6 |
| 3.5 | 0.545 |
| 4 | 2.17 |
| 4.5 | 4.27 |
| 5 | 6.65 |
| 5.5 | 9.22 |
| 6 | 11.93 |
| 6.5 | 14.47 |
| 7 | 17.65 |
| 7.5 | 20.26 |
| 8 | 23.65 |
| 8.5 | 26.74 |
| 9 | 29.87 |
| 9.5 | 33.04 |
| 10 | 36.25 |

We can use these results to create a graph that will help us determine the transconductance "gm" for this MOSFET. Note, it does not really matter what $V_{DS}$ we use as long as it is a reasonable positive value. You could carry out the same experiment with a VDS of 9, 12, and 15, and you would get very similar results especially once you have reached the threshold voltage of the MOSFET. The graph of the input characteristics for this MOSFET is shown in Figure 7-28.

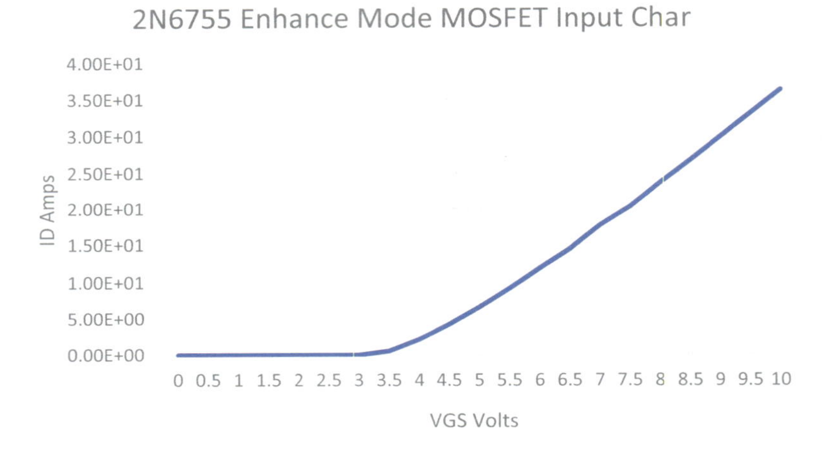


***Figure 7-28.*** *The Input Characteristics for the 2N6755 Enhance Mode MOSFET*

The gradient of this graph is the transconductance "gm" of the MOSFET. It gives a value, in siemens, that shows how the drain current, ID, is related to the field created across the MOSFET by the voltage at the gate. Using the graph shown in Figure 7-28, this can be calculated as

$$gm = \frac{change\ in\ ID}{change\ in\ VGS} = \frac{36.25 - 0.545}{10 - 3.128} = 5.196\ S$$

The accurate values for the drain current were really taken from the table of results, and the $V_{GS}$ voltage of 3.128V was the threshold voltage for the 2n6755 according to the parameters set by TINA.

We can now use this value for "gm," that is, 5.196S, to determine the minimum voltage for $V_{GS}$ that will enable the MOSFET to conduct:

$$V_{GS} = V_{TH} + \sqrt{\frac{I_D}{G_m}} = 3.128 + \sqrt{\frac{20E^{-3}}{5.196}} = 3.128 + \sqrt{0.003849} = 3.19V$$

We can test this idea of a minimum $V_{GS}$ voltage with the test circuit shown in Figure 7-29.

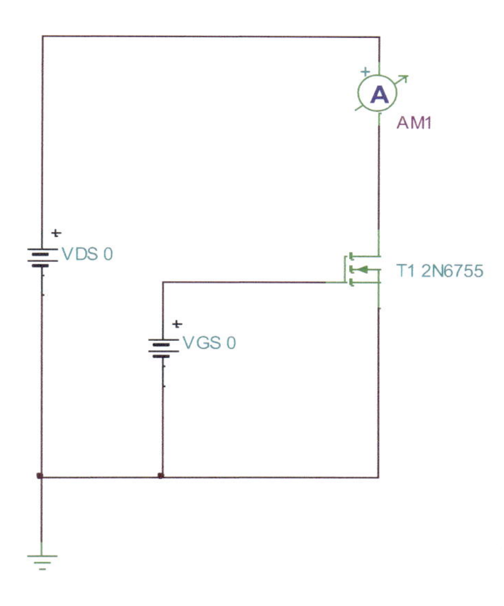


***Figure 7-29.*** *The Test Circuit for the Output Characteristics of the 2N6755*

This is the same circuit as that shown in Figure 7-28 except that the simulation was to vary the $V_{DS}$ voltage from 0 to 10 in steps of 0.5V for different settings of the $V_{GS}$ voltage. The measured results of the drain current $I_D$ are shown in Table 7-5.

With this circuit, we will simply vary the drain to source voltage $V_{DS}$ from 0 to 10V for different settings of the gate to source voltage, VGS, of 0V, 3V, 4V, 4.5V, and 5V. We will measure the drain current $I_D$ using the AM1 ammeter. The results were recorded in Table 7-5.

***Table 7-5.*** *The Table of Results for the Test Circuit Shown in Figure 7-29*

| VDS | IDS mA | | | | |
|---|---|---|---|---|---|
| | VGS = 0 | VGS = 3 | VGS = 4 | VGS = 4.5 | VGS = 5 |
| 0 | 0 | 0 | 0 | 0 | 0 |
| 0.5 | 833.33E-9 | 833.33E-9 | 1.5 | 1.88 | 2.08 |
| 1 | 1.67E-6 | 1.67E-6 | 2.17 | 3.38 | 3.94 |
| 1.5 | 2.5E-6 | 2.5E-6 | 2.17 | 4.27 | 5.47 |
| 2 | 3.33E-6 | 3.33E-6 | 2.17 | 4.27 | 6.46 |
| 2.5 | 4.17E-6 | 4.17E-6 | 2.17 | 4.27 | 6.65 |
| 3 | 5E-6 | 5E-6 | 2.17 | 4.27 | 6.65 |
| 3.5 | 5.83E-6 | 5.83E-6 | 2.17 | 4.27 | 6.65 |
| 4 | 6.67E-6 | 6.67E-6 | 2.17 | 4.27 | 6.65 |
| 4.5 | 7.5E-6 | 7.5E-6 | 2.17 | 4.27 | 6.65 |
| 5 | 8.3E-6 | 8.3E-6 | 2.17 | 4.27 | 6.65 |
| 5.5 | 9.17E-6 | 9.17E-6 | 2.17 | 4.27 | 6.65 |
| 6.0 | 10E-6 | 10E-6 | 2.17 | 4.27 | 6.65 |
| 6.5 | 10.83E-6 | 10.83E-6 | 2.17 | 4.27 | 6.65 |
| 7 | 11.67E-6 | 11.67E-6 | 2.17 | 4.27 | 6.65 |
| 7.5 | 12.5E-6 | 12.5E-6 | 2.17 | 4.27 | 6.65 |
| 8 | 13.33E-6 | 13.33E-6 | 2.17 | 4.27 | 6.65 |
| 8.5 | 14.17E-6 | 14.17E-6 | 2.17 | 4.27 | 6.65 |
| 9 | 15E-6 | 15E-6 | 2.17 | 4.27 | 6.65 |
| 9.5 | 15.83E-6 | 15.83E-6 | 2.17 | 4.27 | 6.65 |
| 10 | 16.67E-6 | 16.67E-6 | 2.17 | 4.27 | 6.65 |

We can use the results from this table to create a graph of the output characteristics which is shown in Figure 7-30.

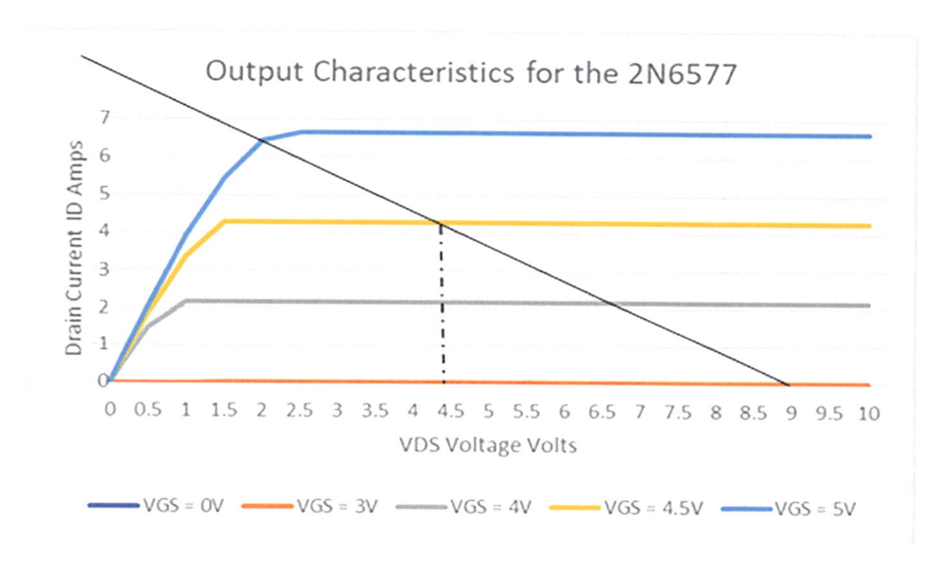


***Figure 7-30.*** *The Output Characteristics for the 2N6577*

We have added the load line to the graphs just to show we can set a quiescent point just like we did for the BJT. We have chosen a $V_{DS}$ voltage of 4.5 Volts as this is the ideal voltage if the source voltage was set to 0V. However, we are going to use the amplifier circuit shown in Figure 7-31, in which the source is not connected to the ground.

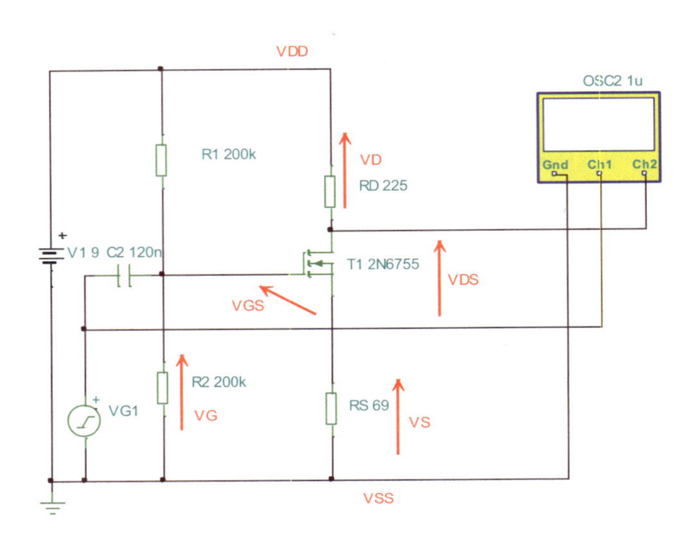


***Figure 7-31.** The Test Circuit for the Class A MOSFET Amplifier*

The main idea is to set the voltage at the drain to approximately 4.5V, which is half of $V_{DD}$, so that we could swing more positive and negative by the same amount, if the source was connected to ground. This means that the volt drop across $R_D$ would be 4.5V as well.

We know that our minimum voltage for $V_{GS}$, to start the MOSFET conducting, must be 3.19 volts. We should choose an appropriate value for $V_G$ the gate to ground voltage. This is not the same as $V_{GS}$; $V_{GS}$ is the voltage between the gate and the source as shown in Figure 7-31. The $V_G$ voltage can be set to any reasonable positive voltage, as long as it is high enough to forward bias the gate to source junction. The easiest choice would be to set this to 4.5V. That is because the voltage divider network of $R_1$ and $R_2$ will be used to set this value. This would simply mean we need to make the two resistors the same value, as this would halve the $V_{DD}$ voltage. To try and ensure that we have a reasonably high input impedance, we should set these two resistors to 200KΩ each. That would set the $V_G$ voltage to 4.5V and the input impedance to 100kΩ; we need to remember that to ac $R_1$ and $R_2$ are in parallel with each other.

Now we must choose an appropriate current to flow through the MOSFET. This would depend upon your requirement, but we will set this, in this example, to 20mA. We know that because of the insulating gate material across the input to the MOSFET, the source current, $I_{S,}$ must be the same as the drain current $I_D$, as no current flows into the MOSFET via the gate terminal. We know we want the volt drop across the $R_D$ resistor, that is VD in Figure 7-31, must be 4.5V to ensure the voltage at the drain terminal is 4.5V. Therefore, we can calculate the value of the drain resistor $R_D$ as follows:

$$R_D = \frac{V_D}{I_D} = \frac{4.5}{20E^{-3}} = 225\Omega$$

To calculate the value of the source resistor, $R_S$, we need to determine the value of the volt drop across this resistor, that is, $V_S$. Applying Kirchhoff's Voltage Law to the closed loop at the gate of the MOSFET, we can see that

$$V_G = V_{GS} + V_S$$

This means that we can calculate the source voltage as follows:

$$V_S = V_G - V_{GS} = 4.5 - 3.19 = 1.31$$

Knowing that the current flowing through $R_S$ must be the same 20mA as the current $I_D$, then we can calculate the value of $R_S$ as follows:

$$R_S = \frac{V_S}{I_S} = \frac{1.31}{20E^{-3}} = 65.5\Omega$$

We have now calculated all the resistor values for the circuit. The final component we need to determine the value of is the decoupling capacitor that is part of the high pass filter designed to remove any DC voltage from the input signal that could affect the quiescent DC biasing we have with $R_1$ and $R_2$. This is calculated in the same fashion as we did with the BJT amplifier in Chapter 6. We will set the low frequency cutoff value of 20Hz. Knowing that, the high pass filter circuit would be made up as shown in Figure 7-32.

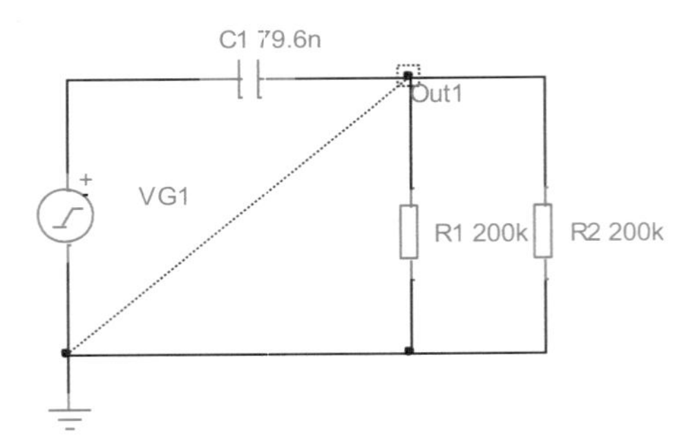


***Figure 7-32.*** *The High Pass Filter at the Input of the Amplifier*

The value of the capacitor depends upon the chosen cutoff frequency. In this example, we will choose a cutoff frequency of 20Hz. This means that any signal with a frequency of less than 20Hz will not be amplified enough to be of any use.

# The –3db Point Benchmark

It will be at this cutoff frequency that the gain will be 3dbs down from the maximum gain of the circuit. The circuit shown in Figure 7-32 is a simple filter, and the maximum gain will actually have a value of 1 as there will be no gain. In dBs, a gain of one is zero dbs. Therefore, a value of 3dbs down from this would be the –3db point. This is the standard benchmark for any electronic system. At this –3db point, the system will not produce an output that would be of any use. We say –3dBs, but really it is 3dbs down from the maximum gain in dbs.

We need to be able to calculate the value of this capacitor. This will be when the impedance of the capacitor equals the parallel combination of R1 and R2 in the case of the filter shown in Figure 7-32. This will mean that

$$\frac{1}{2\pi FC} = \frac{R_1 R_2}{R_1 + R_2} = \frac{200E^3 \times 200E^3}{200E^3 + 200E^3} = 100E^3$$

Knowing the cutoff frequency is 20Hz, then the value of the capacitance can be calculated as follows:

$$C = \frac{1}{2\pi \times 20 \times 100E^3} = 79.6nF$$

If we carry out an AC Analysis in TINA for the circuit shown in Figure 7-32, we will get the Bode plot as shown in Figure 7-33.

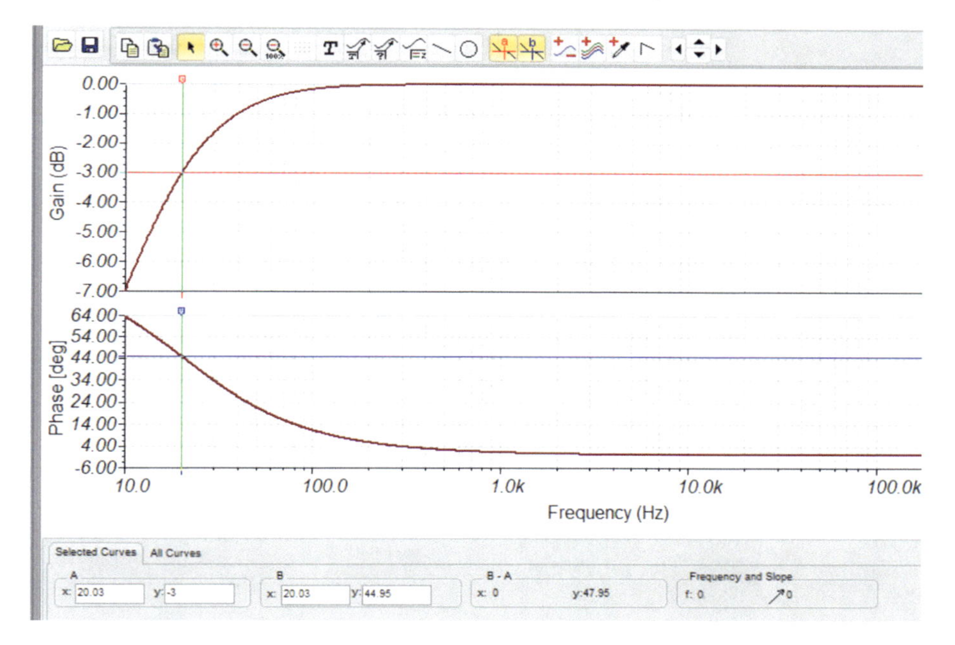


***Figure 7-33.*** *The Bode Plot for the High Pass Filter in Figure 7-32*

Using the Bode plot shown in Figure 7-33, we can see that the maximum value is 0dbs. This is equivalent to a value of 1 for the actual gain of the circuit. Then, using cursor "A," we can see that the gain drops to -3dBs when the frequency is 20Hz. This is the response we have designed the high pass filter to give. One other aspect of the filter is that the phase shift produced by the circuit is 45° at the cutoff frequency of 20Hz.

We now have all the components we need to complete the single-stage amplifier, and the circuit is shown in Figure 7-34.

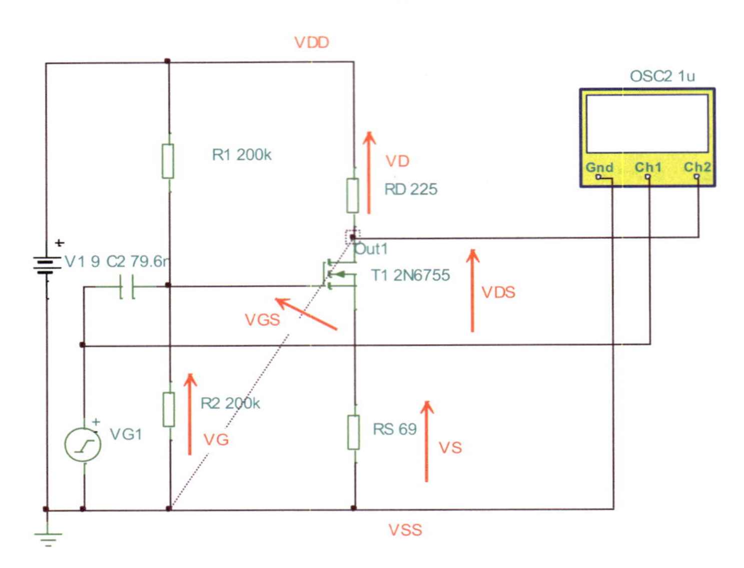


***Figure 7-34.*** *The Single-Stage MOSFET Amplifier*

We can use the oscilloscope to display the input and output waveforms of the amplifier. These waveforms are shown in Figure 7-35.

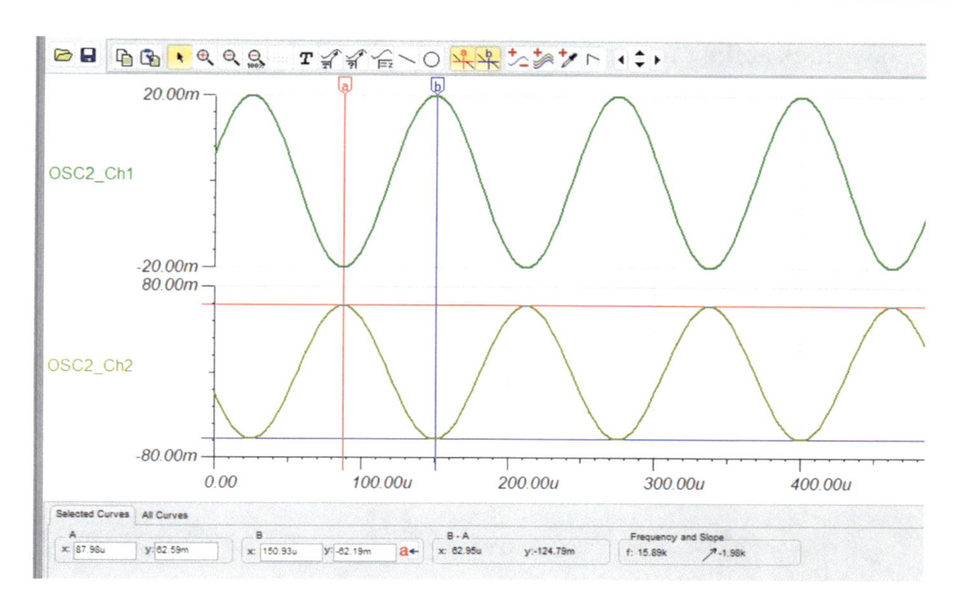


***Figure 7-35.*** *The Input and Output Waveforms of the MOSFET Amplifier*

The output waveform is shown on the channel 2 trace of the oscilloscope. The first thing we can see is that, just like the BJT amplifier, the output is 180° out of phase with the input. Using the two cursors, we can measure the peak-to-peak voltage of the output wave, and this is shown to be 124.79mV. We can use the channel 1 trace to measure the peak-to-peak voltage of the input, and this is shown to be 40mV. Using these two values, we can determine the voltage gain of this MOSFET amplifier as follows:

$$Vgain = \frac{Output}{Input} = \frac{Output\ pk-pk}{Input\ pk-pk} = \frac{124.79E^{-3}}{40E^{-3}} = 3.12$$

If we look at the ratio of the drain resistor, $R_D$, and the source resistor, $R_S$, we can see that similar to the BJT amplifier the gain of this MOSFET can be expressed as

$$Vgain = \frac{R_D}{R_S} = \frac{225}{69} = 3.26$$

The slight difference could be accounted if we consider the ON resistance, $R_{DS(ON)}$, which is the resistance of the channel when the MOSFET is conducting. A typical value for this $R_{DS(ON)}$ is 3Ω, and this would be in series with the source resistance RS. Adding this value to the 69Ω would give a voltage gain of

$$Vgain = \frac{225}{69 + 3} = 3.125$$

This value of voltage gain is still quite low, but, as with the BJT, we could put two stages together. The circuit for the two-stage MOSFET amplifier is shown in Figure 7-36.

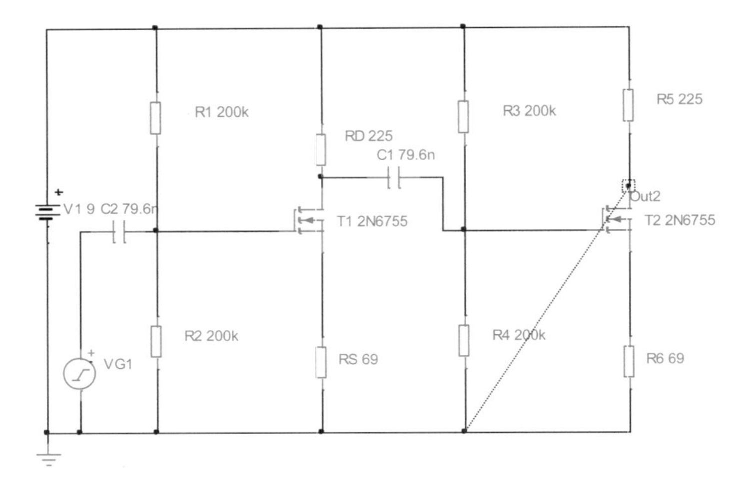


***Figure 7-36.*** *The Two-Stage MOSFET Amplifier*

Using this approach, the voltage gain was increased to 9.73. If we added a third stage, the gain would be 30.37. We need to add the decoupling capacitor between the drain of the first amplifier and the gate of the second amplifier to ensure no DC is superimposed on the gate of the second amplifier.

# The Bypass Capacitor

We used a bypass capacitor to increase the voltage gain of the BJT amplifier. The MOSFET amplifier works in a similar way, and so we should be able to add a bypass capacitor to the MOSFET amplifier. The circuit for this arrangement is shown in Figure 7-37.

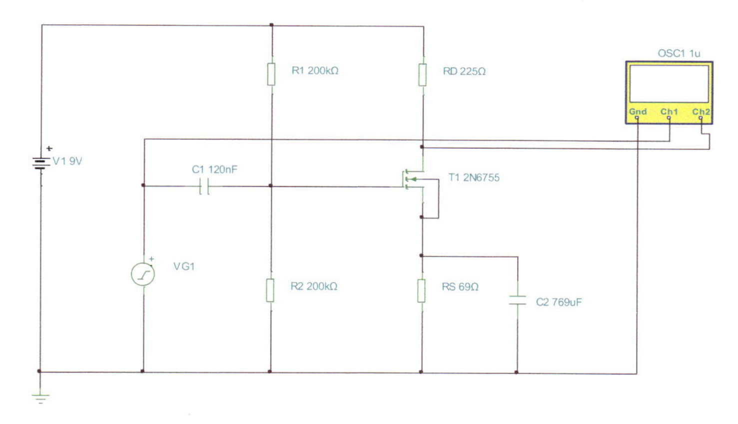


***Figure 7-37.*** *The MOSFET Class A Amplifier with the Bypass Capacitor Added*

We calculate the value of the bypass capacitor in the same way as we did with the BJT. Therefore, the impedance of the capacitor is set at a tenth of the source resistance RS, and using the low frequency of 30Hz, the value of C2, the bypass capacitor, is

$$C_2 = \frac{1}{2\pi \times 30 \times 6.9} = 769\mu F$$

Using this bypass capacitor to ac, the source resistance can be ignored when calculating the voltage gain. However, we cannot ignore the On resistance, $R_{DSON}$, of the MOSFET as this is internal to the MOSFET and so not bypassed by the capacitor. Using the value of 3Ω for $R_{DSON}$, the gain was calculated as

$$V_{gain} = \frac{R_D}{R_{ON}} = \frac{225}{3} = 75$$

When we simulated the circuit, the measured voltage gain was 98.6, so we can see that the gain can be increased with the use of a bypass capacitor. However, the difference between the two gain values shows us that the $R_{DSON}$ resistance is not as straightforward as the single value of 3Ω suggests, but that's for a more detailed book on FETs.

# Exercise 7.1

Design a MOSFET amplifier with the following specification:

VDD = 15V

ID = 5mA

The minimum gate voltage is 3.19 as we are using the 2N6755 MOSFET.

Calculate the voltage gain using a value of 3Ω for RDSON.

Hint: Set the gate voltage, $V_{GS}$, to 5V.

Then add a bypass capacitor if the lower frequency of interest was 25Hz.

# Summary

In this chapter, we have looked at the various types of field effect transistors (FETs) that are available to us. We have studied how they work on a subatomic level and studied how to use them in their simply switching applications. Finally, we have studied how we can design a basic class A amplifier using the MOSFET.

In the next chapter, we will go back to looking at the BJT, as we look at the differential amplifier, or long-tailed pair. This is in readiness for moving on to the Opamp or Operational Amplifier.

I hope you have found this study of the FET interesting and useful although it has been limited to simulations within the ECAD software Tina. I will be looking at some practical circuits, as with the BJTs, in the appendix as it is always useful to see the devices working practically.

CHAPTER 8

# The Beginnings of the Operational Amplifier

In this chapter, we will look at the long-tailed pair, sometimes referred to as the differential amplifier. We will look at the two main uses of the differential amplifier and learn about the term CMRR, common mode rejection ratio. We will study a simple constant current source and the current mirror circuit.

## The Differential Amplifier

There will be many occasions when we need to amplify the difference between two voltages, such as when we are measuring the output of some transducer, such as a simple RTD, resistance temperature detector, as shown in Figure 8-1.


H. H. Ward, *Mastering Analog Electronics*, Maker Innovations Series,
https://doi.org/10.1007/979-8-8688-0245-4_8

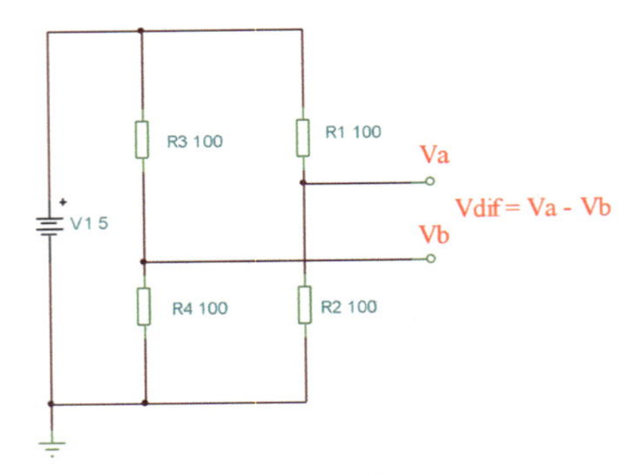


***Figure 8-1.*** *The Wheatstone Bridge Circuit*

Figure 8-1 is a typical Wheatstone bridge arrangement where $R_1$ would be the transducer, such as a PT100 for temperature transducer or a strain gauge for measuring weight or mass. When the temperature or mass is zero, then Va would equal Vb, and there would be no difference voltage. The difference voltage will typically be very small, and so we would need an amplifier to increase the output voltage.

Another situation where we would need a differential amplifier might be with a simple comparator circuit that will compare the output of a transducer against some reference.

The basic circuit of a differential amplifier is shown in Figure 8-2.

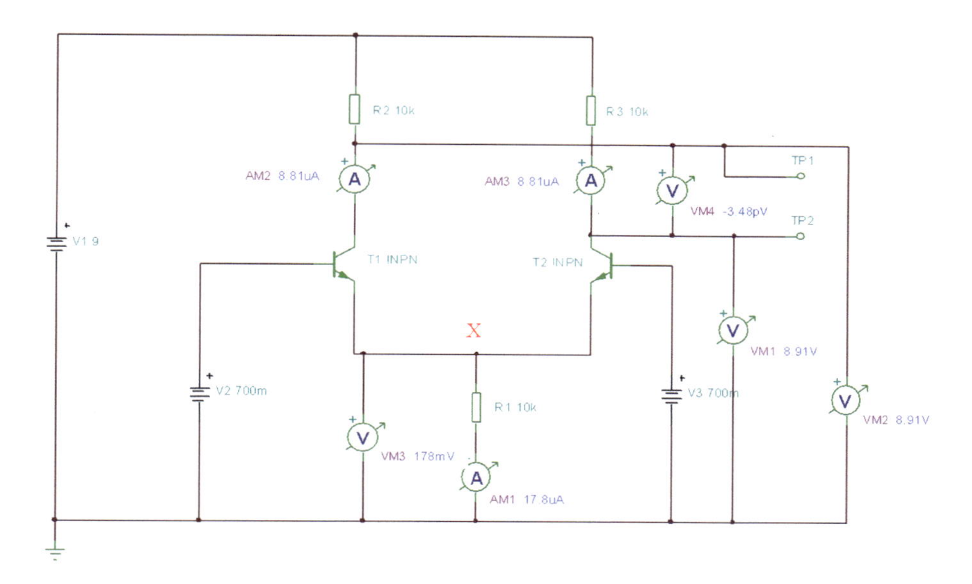


***Figure 8-2.*** *A Basic Differential Amplifier*

This circuit could be viewed as two standard common emitter amplifiers except that share the one emitter resistor. They have to share the same emitter resistor, $R_1$ in the circuit shown in Figure 8-2, because they will use the current flowing through, $R_1$, according to how each is biased due to the two input voltages. The situation in Figure 8-2 is that both transistors are biased equally, as both the voltages at the bases are 700mv. This means that they share the current flowing through $R_1$ equally. We should appreciate that the current flowing through $R_1$ is set by the voltage at point "X" in the circuit. From our work with the common emitter amplifier, we should appreciate that the voltage at point "X" would be equal to the voltage at the base minus $V_{BE}$, which is why the voltage at point "X" is 178mV as measured by VM3. This then sets the current through $R_1$.

In Figure 8-2, the two base voltages are the same, so the 17.8uA flowing through $R_1$ is divided equally between the two transistors, that is, the 8.81uA, as shown in Figure 8-2.

If one of the base voltages was greater than the other, then that transistor would take more of the 17.8uA flowing through $R_1$ than the other. This is shown in Figure 8-3 where the base voltage for $T_1$ is 750mv and the base voltage for $T_2$ remained at 700mv.

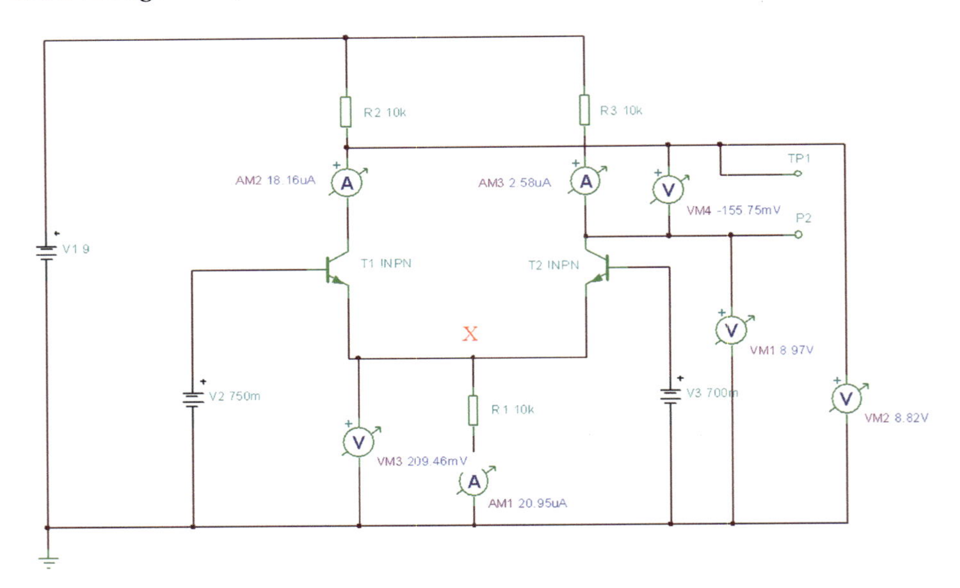


***Figure 8-3.*** *The Differential Amplifier with a Difference Voltage*

We can see that the current through $R_1$ has increased to 20.95uA. This is because the voltage at point "X" has increased due to the increase in the base voltage at $T_1$. However, we can also see that the transistor $T_1$ is now taking a greater share of the current as it is taking 18.16uA, while transistor $T_2$ is taking only 2.58uA. Note the sum of the two transistor currents still equates very closely to the current flowing through $R_1$. The fact that $T_1$ is drawing more current than $T_2$ means that the voltage at the collector of $T_1$, which is test point TP1 on the circuit, is now lower than the voltage at the collector of $T_2$. This difference which is VM2 – VM1 is now the output of the amplifier. We can see that the difference is –155.75mV at the output. This then is an amplification of around 3, as the difference between the two base voltages, V2 – V3, is only 50mV.

So, this would suggest that the circuit shown in Figure 8-2 can be used as a differential amplifier. However, there are some constraints on this circuit in that the voltage at the base of at least one of the transistors must be high enough to turn one of the transistors on before any current will flow through $R_1$. Also, the gain is not linear.

## The Long-Tailed Pair

One possible improvement was to pull the resistor $R_1$ down to a negative supply. This arrangement is shown in Figure 8-4, and it was given the name the long-tailed pair – the long tail because the resistor $R_1$ looks like a tail. However, an essential aspect of the differential amplifier is that the two transistors are perfectly matched. This is to ensure their value for Beta and their VBE are the same and their temperature dependence is the same. This is rather difficult to guarantee if we use individual transistors, but they can be bought on a single IC such as a transistor array. The THAT 300 is one such IC you could use. It has four matched NPN transistors on it, and it comes in dual in-line 14-pin package.

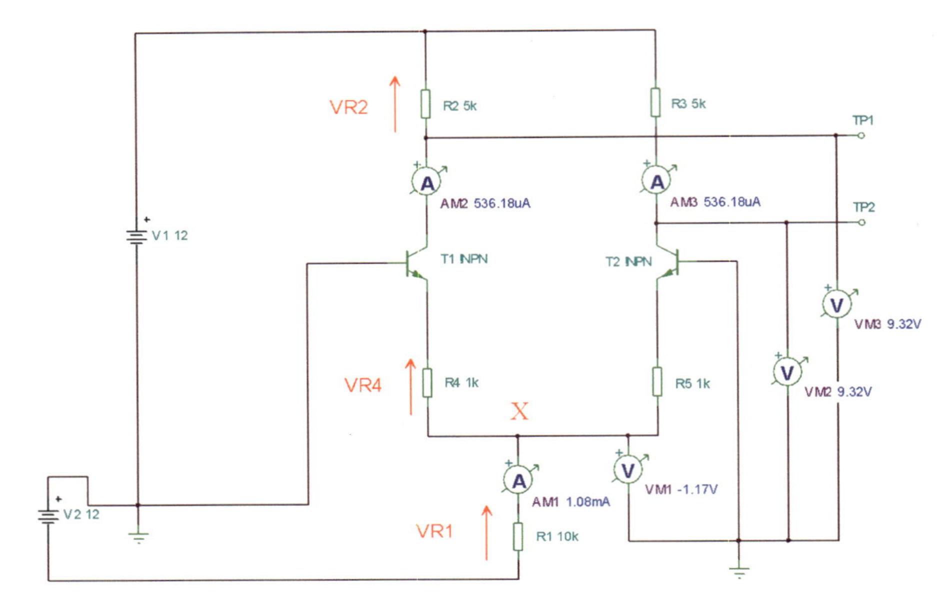


***Figure 8-4.*** *The Test Circuit for the Long-Tailed Pair Diff Amplifier Quiescent Conditions*

There is a lot to look at in this test circuit, but we are trying to understand the basic operation of this differential amplifier. In this analysis, we have added individual emitter resistors so that we can look at the differential amplifier as two common emitter amplifier circuits that have one extra resistor which is termed the "tail." This resistor is shared between the two amplifiers. Also, this extra resistor is taken down to its own negative supply.

Just as we did with the common emitter amplifier, we will first look at the quiescent operating conditions for the circuit. When there is no signal inputted to the circuit, indeed, the bases of both transistors are grounded, and this will set up the quiescent operating conditions. With the base of each resistor connected to ground, that is, 0V, and the emitters connected via their own emitter resistors and the shared tail resistor $R_1$, down to -12V, then the diodes in the base emitter junction of the transistors are both forward biased. This means there will be an emitter current flowing through the transistors. This current will be set by the volt drop across $R_1$, that is, VR1 in the circuit. The volt drop VR1 will be the voltage at "X" -12V. The voltage at each of the emitters must be around -0.6 to -0.7V due to the volt drop across their base emitter diodes. This means there must be at least 1mA of current flowing through $R_1$ of which half will flow through $R_4$ and $R_5$. This means there must be at least 0.5V dropped across $R_4$ and $R_5$. This then means that the voltage at "X" must be around -1.2V. This then means the current flowing through $R_1$ will be

$$I_{Tail} = \frac{-1.2 - -12}{10E^3} = 1.08mA$$

We have called it "$I_{Tail}$" as it flows through the tail resistor. This is confirmed with the simulation, where we can see AM1 measures 1.08mA.

Assuming we can match our transistors well enough, we can say that this 1.08mA will be shared equally between the two transistors. Therefore, under these quiescent conditions, we can say

$$I_Q = \frac{1.08mA}{2} = 0.54mA$$

We need to express the quiescent current in mA as we will use those units later when we calculate the intrinsic resistance of the transistors $r_e$.

Now we will move on and apply some signals to the amplifier. The circuit we will use to test the response of the amplifier is shown in Figure 8-5.

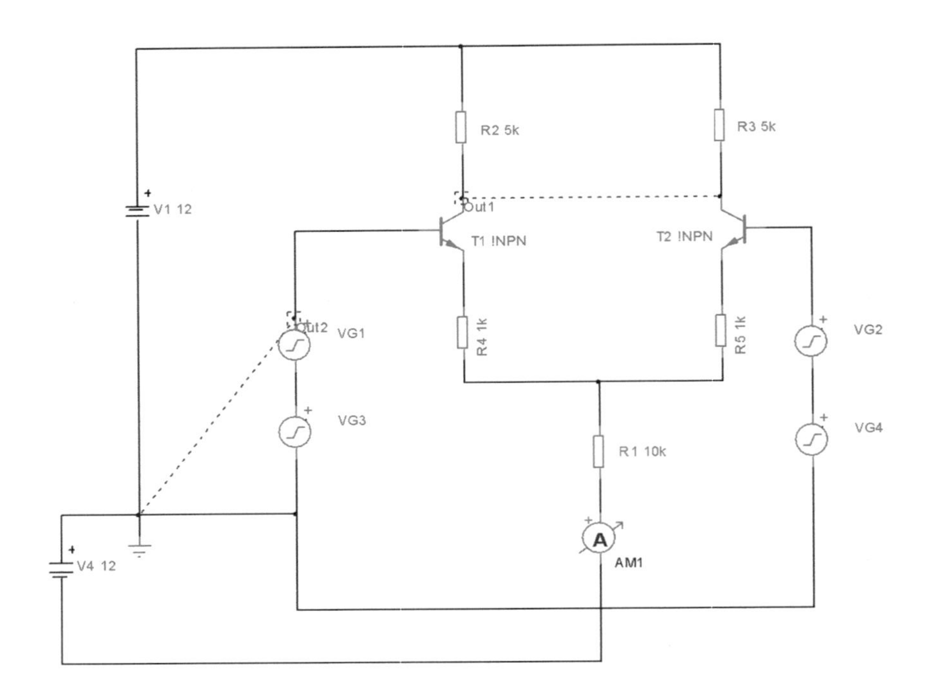


***Figure 8-5.*** *The Test Circuit for the ac Response of the Differential Amplifier*

When we simulated the circuit, the waveforms recorded were as shown in Figure 8-6.

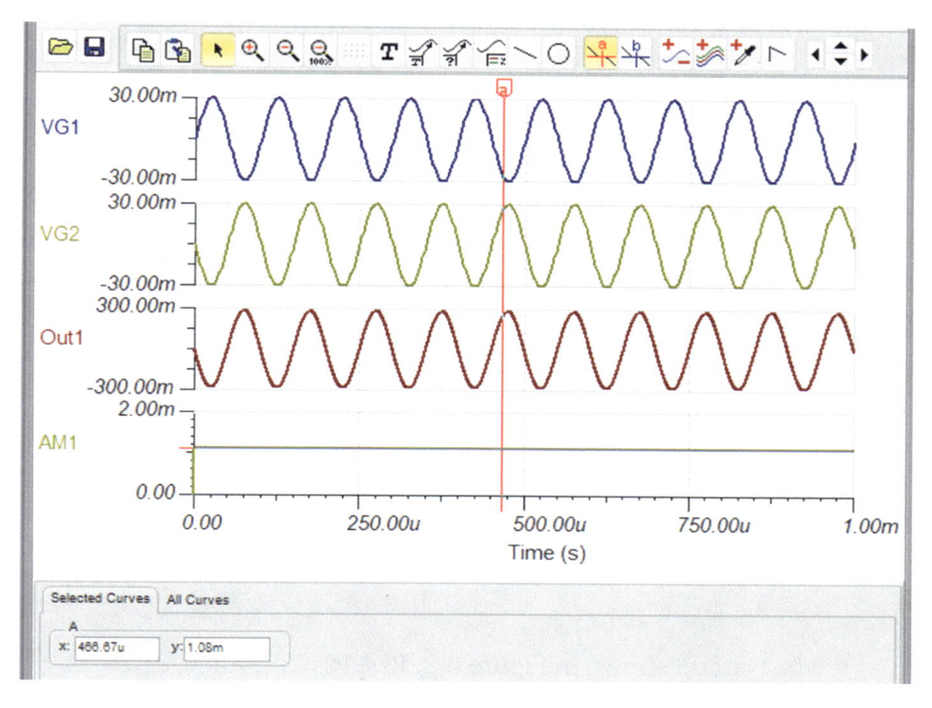


***Figure 8-6.** The Waveforms for the Test Circuit Shown in Figure 8-5*

The ammeter AM1 is measuring the tail current flowing through $R_1$. As we can see, it is constant at 1.08mA. This is what we expected. There are four voltage sources as we want to supply a signal to each transistor - these are VG1 and VG2, respectively - but we will also be introducing some noise to the signals. These are with VG3 and VG4. Initially, we will set the noise to 0V so that we can see how the differential amplifier deals with a clean difference voltage. If we look at the traces for VG1 and VG2, we see they are completely in antiphase with each other. If we describe the waveforms using the peak-to-peak value, then we can say each signal has a 60mv peak-to-peak value. This means that the difference, which is what this amplifier will amplify, is a 120mv peak-to-peak signal.

When VG1 is +30mV, VG2 is -30mV; therefore, the difference is +30 – -30 = 60mV. When Vg1 is –30mV, Vg2 is +30mV; therefore, the difference voltage is –30 – 30 = –60mV. This confirms the difference voltage, which is what the amplifier actually amplifies, goes from +60mV to -60mV; therefore, it has a 120mV peak-to-peak voltage.

When we look at the $O_{UT1}$ trace, which is the output of the amplifier, we can see this has a value of approximately 600mv peak to peak. Using these two values, we can calculate the difference voltage gain as

$$V_{DiffGain} = \frac{v_{out\ Peak-Peak}}{v_{in\ Peak-Peak}} = \frac{600E^{-3}}{120E^{-3}} = 5$$

If we apply the same analysis for the gain of the differential amplifier as we did with the basic common emitter amplifier, then we can say

$$v_{DiffGain} = \frac{R_C}{R_e}$$

In the test circuit shown in Figure 8-3, $R_C = R_2 = 5k$ and $R_e = R_4 = 1k$. Therefore, we can calculate the gain as

$$v_{DiffGain} = \frac{R_C}{R_e} = \frac{5E^3}{1E^3} = 5$$

The tail resistor plays no part in the gain calculations because the voltage at “X” does not change. It can be used as our reference. We know it is around –1.2V below our usual ground, but that would be the same for both input and output, and so the –1.2V can be ignored. The two values for the gain agree with each other. With that in mind, we can try to improve the gain by changing the values of the resistors to those shown in Figure 8-7.

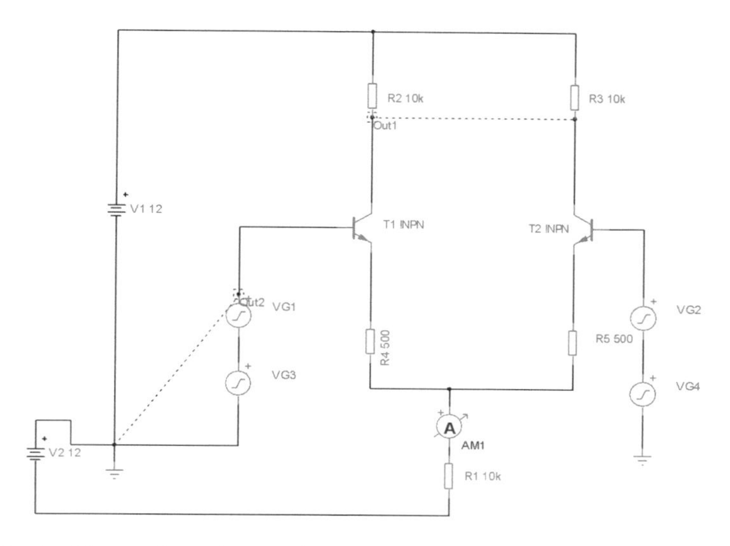


***Figure 8-7.*** *Changing the Resistors to Increase the Gain*

This simulation showed that the tail current had increased very slightly to 1.11mA, but the output voltage had changed to 2.17v peak to peak. This means the gain had increased to

$$V_{DiffGain} = \frac{v_{out\ Peak-Peak}}{v_{in\ Peak-Peak}} = \frac{2.17}{120E^{-3}} = 18.083$$

However, using

$$v_{DiffGain} = \frac{R_C}{R_E} = \frac{10E^3}{0.5E^3} = 20$$

This means we need to take something else into account. The something else is the intrinsic resistance $r_e$. We can calculate the value of this intrinsic resistance as follows:

$$r_e = \frac{25}{I_Q}$$

where $I_Q$ is the quiescent current flowing through the transistor but stated in the units of mA. Earlier, we recorded this current at 0.549mA. We have looked at this concept of calculating the value of $r_e$ in Chapter 6. Therefore, using this value, we can calculate the value of $r_e$ as

$$r_e = \frac{25}{0.549} = 45.54$$

We can use this in a more complete expression for the voltage gain as

$$v_{gain} = \frac{R_c}{\left(R_E + r_e\right)}$$

Therefore, we have

$$v_{DiffGain} = \frac{R_c}{\left(R_E + r_e\right)} = \frac{10^3}{\left(500 + 45.54\right)} = 18.33$$

This is much closer to the other calculation using the output and input waveforms.

# Adding Noise to the Signal

Most signals have some noise aspect added to them. Those of you who can remember the old stereo radiograms when they first came out. The cheaper ones had that low main's hum on the sound which came from the 100Hz rectifying signal on the power supply. Well, we soon filtered that out. However, there are many sources of noise on electrical signals, and the really good hi-fi systems should be able to remove them from their output. One way which this could be achieved would be to amplify the pure signal and attenuate the noise content. That is what the differential amplifier will do. The differential amplifier will amplify any differences between the two inputs, but anything that is common to both inputs will be attenuated. Any noise that appears on the signal will normally appear in the same format on both of the inputs to the differential amplifier.

That means the noise would be common to both inputs, and so the differential amplifier will attenuate it. The ability of the differential amplifier to attenuate this common noise content is referred to as CMRR (common mode rejection ratio).

We will use our test circuit to see how the differential amplifier will reject this common noise content. To add noise to the signal, we will use the VG3 and VG4 voltage sources. We will set them both to input a 10mv peak noise signal at 100kHz. This is because a noise signal is usually small and at a high frequency. If we simulate the test circuit with the noise added, we get the waveforms as shown in Figure 8-8.

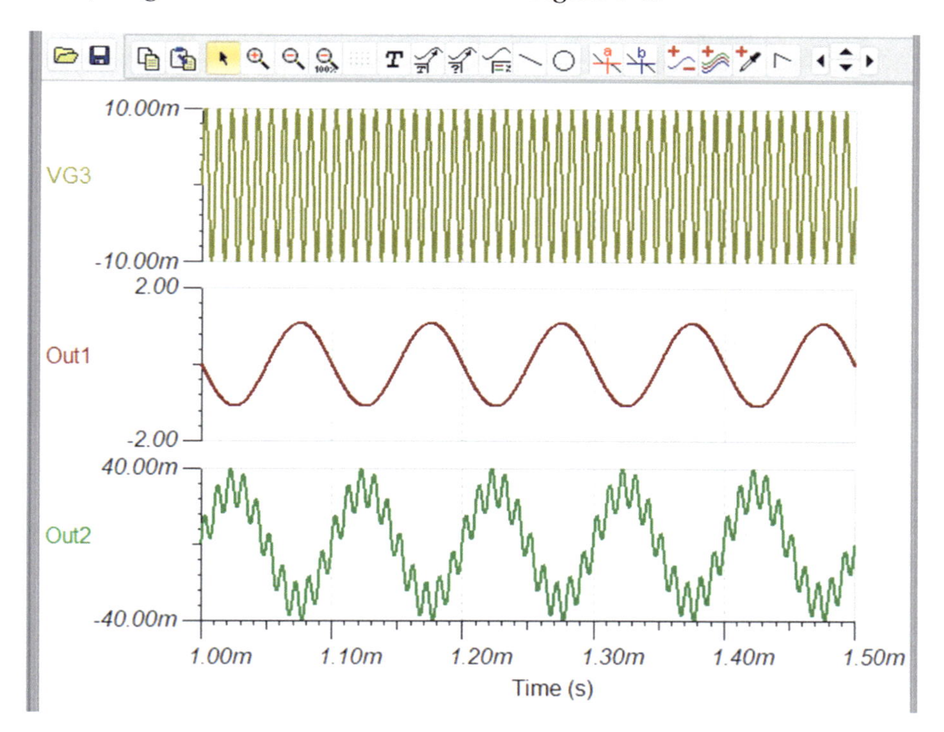


***Figure 8-8.** The Waveforms for the Test Circuit with Noise Added*

The trace VG3 is the actual noise signal that is added to the wanted signal. The trace Out2 shows how the noise corrupts the wanted signal. However, we can see with trace Out1 that the output of the differential amplifier is unchanged. There appears to be little or no noise present on the output. That's exactly what we want. The gain of the wanted signal, the difference between the two inputs, is still the same at around 18. So, why has the noise been attenuated? To understand why, we need to look at the circuit slightly differently. This is shown in Figure 8-9.

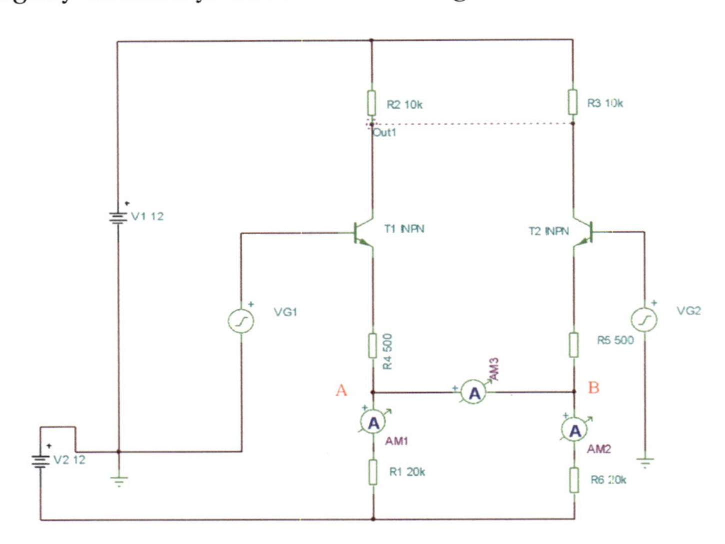


***Figure 8-9.** An Alternative Circuit for the Long-Tailed Pair Diff Amplifier*

The thinking behind this representation of the tail resistance is that we can split a resistor into two resistors in parallel with each other. In this way, the single tail resistor with a value of 10kΩ can be replaced by two 20k resistors in parallel, as shown by $R_1$ and $R_6$ in Figure 8-9. The circuit will work in the same way, so it is a valid change to the differential amplifier. However, we now have only two sources, VG1 and VG2. These are both set to the noise signal of 10mv at 100kHz. As the two sources are the same, then the voltages at points "A" and "B" will be the same, that is, they will move up and down by the same amount at the same time. This means that no current will flow through the ammeter AM3. If we look at the waveform shown in Figure 8-10, we should be able to confirm what is happening.

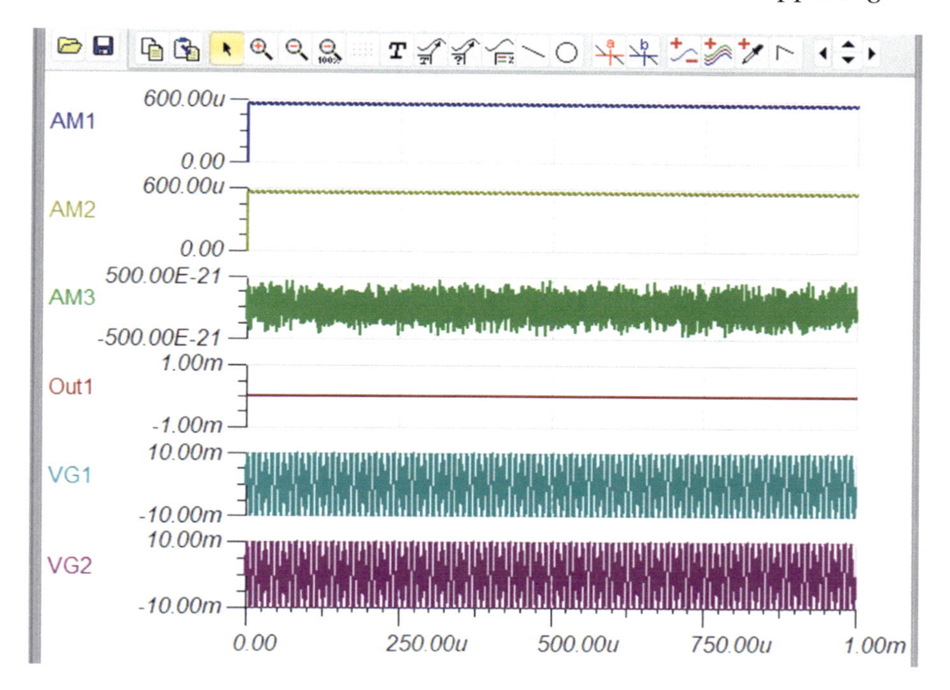


***Figure 8-10.*** *The Waveforms of the Circuit Shown in Figure 8-9*

This confirms that there is no current flowing through AM3 and the output, shown as trace Out1, is virtually zero, showing that this circuit works the same. It also confirms that the current flowing through both emitters is still around the 550mA as it was in the original differential amplifier circuit.

The fact that no current flows through AM3 suggests that we can remove that connection and treat the two amplifiers as two separate amps with the same input voltage. We can calculate the gain for the two amplifiers as

$$v_{CommGain} = \frac{R_c}{(R_E + r_e + R_1)} = \frac{10E^3}{(500 + 45.54 + 20E^3)} = 0.487$$

This means that this noise signal would be reduced to about half. This does seem to suggest that the differential amplifier does amplify the difference voltage and attenuates any noise content. This is what we want from the differential amplifier.

# Improving the Gain

The next thing we will look at could increase the gain of the circuit. We have shown that the differential gain is set by

$$v_{gain} = \frac{R_c}{(R_E + r_e)}$$

Why not see what happens if we remove the emitter resistor altogether? First, we should look at the quiescent conditions. This is done with the test circuit as shown in Figure 8-11.

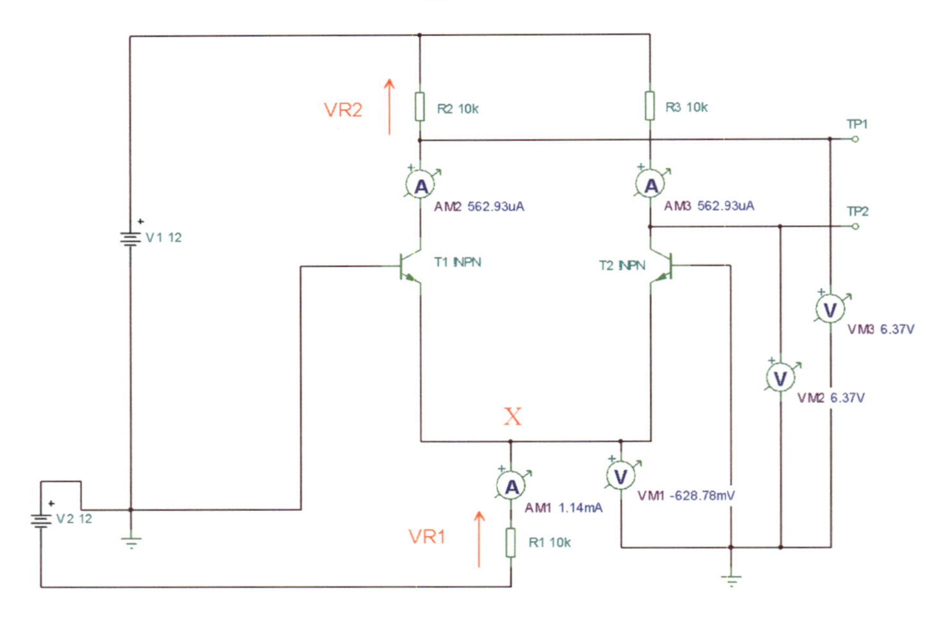


***Figure 8-11.*** *The Test Circuit for the Quiescent Operating Conditions*

We can use this new quiescent current, $I_Q$ value of 562.93uA, to calculate the intrinsic resistance when we have removed the emitter resistor. Note, we use a value of 0.563 to relate the current to mA:

$$r_e = \frac{25}{I_Q} = \frac{25}{0.563} = 44.405$$

This means the voltage gain becomes

$$v_{gain} = \frac{R_c}{r_e} = \frac{10E^3}{44.405} = 225.2$$

This is a very high gain, and if we kept the difference voltage at 120mv peak to peak, the output would be clipped as it would try to go to 13.5v peak. The output of any amplifier is restricted by the supply rails. At the moment, the supply rails are set at 12v. To accommodate this, we changed the input voltage to a peak of 20mV. This meant that the peak-to-peak input was 80mV. When we changed this and simulated the circuit, the waveforms created were as shown in Figure 8-12.

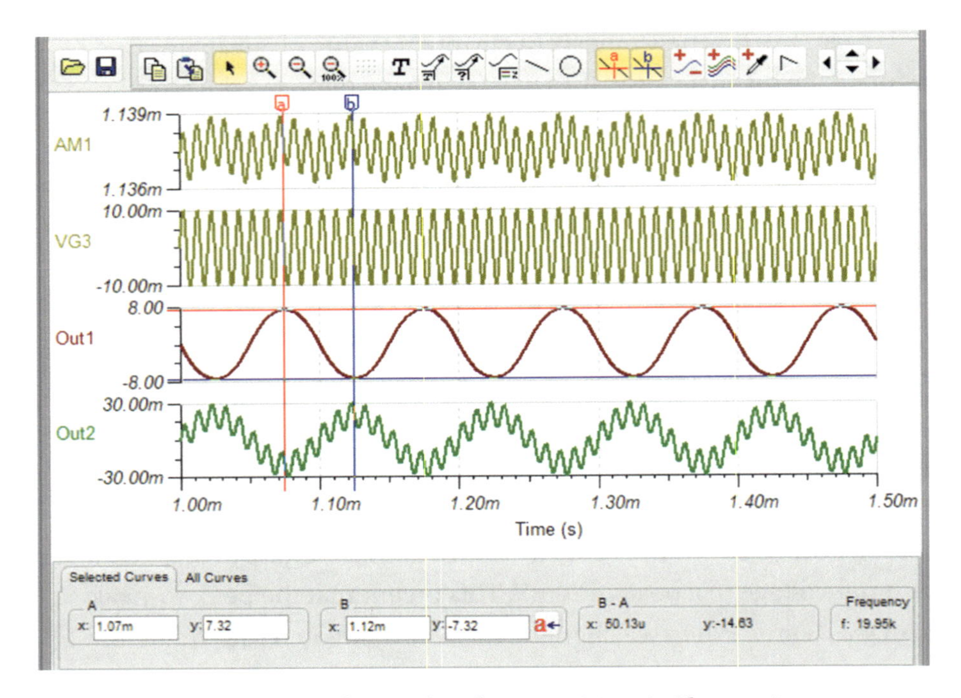


***Figure 8-12.*** *The Waveforms for the Test Circuit Shown in Figure 8-11*

This shows that the gain had increased but only to around 183, not quite the same as calculated, but we are using the standard value of 25 from the term VT as detailed in Chapter 6, so there will be some leeway in our calculations. Everything else was good; there was no noise on the output even though there was noise on the input.

What we have now shown is that the gain is a factor of the intrinsic resistance. The intrinsic resistance is controlled, to some extent, by the DC quiescent current. However, this DC quiescent current is half that of the tail current. This seems to suggest that the gain of the differential amplifier can be controlled by the value of the tail current. So, if we can control this tail current, then perhaps we can improve the gain of the differential amplifier. For example, if we could set the tail current at 2mA, then the quiescent current through the transistors would be 1mA, that is, half the tail current. This would mean the intrinsic resistance would be

$$r_{e'} = \frac{25}{I_Q} = \frac{25}{1} = 25$$

Therefore, the projected gain would be

$$v_{gain} = \frac{R_c}{r_e} = \frac{10E^3}{25} = 400$$

When we look at the actual Opamp, as in Chapter 9, we will see that the open-loop gain is very high around 100,000. This means that our investigation into the differential amplifier is only a small part of what goes into the Opamp circuit. However, it was only meant as an introduction into the birth of the Opamp, and I hope it has given you some insight into the differential amplifier.

## The Constant Current Source

Another useful circuit in analog electronics is that of the constant current source. We will look at how a simple constant current source circuit can be created, and one such circuit is shown in Figure 8-13.

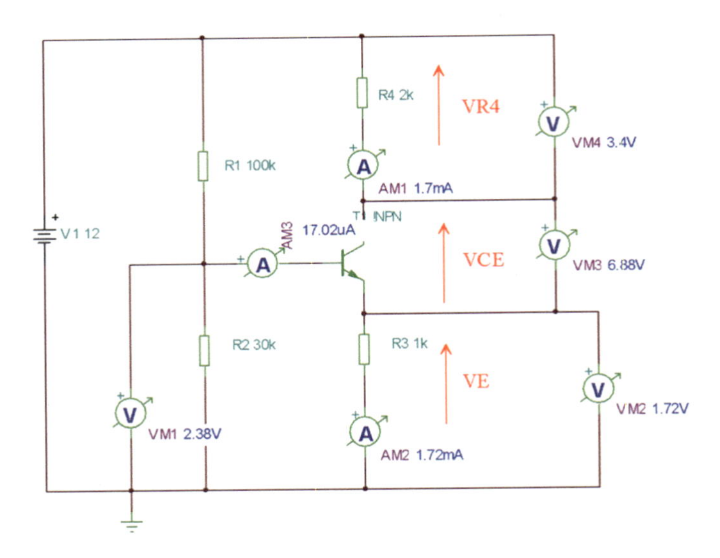


***Figure 8-13.** A Simple Constant Current Source*

I hope you can see that the circuit shown in Figure 8-13 is the basic common emitter amplifier. We should be able to appreciate that the three currents have the following relationship:

$$I_E = I_C + I_B$$

where

- $I_E$ is the emitter current.
- $I_C$ is the collector current.
- $I_B$ is the base current.

The emitter current is set by the voltage $V_E$, measured by VM2 in Figure 8-13, divided by the value of $R_E$. However, from our work in Chapters 5 and 6, we know that the voltage $V_E$ is

$$V_E = V_B - V_{BE}$$

The voltage $V_B$ is set on the whole by the voltage divider network of $R_1$ and $R_2$. This means that the voltage $V_B$ cannot change, and so the voltage $V_E$ cannot change. This means that unless we change the value of the emitter resistor, then the emitter current is fixed. This in turn means the collector current is fixed. This is true, but there are some constraints. The main one is that the voltage left after the voltage dropped across the collector voltage is high enough for the emitter voltage to be at this set voltage of $V_B - V_{BE}$. For example, the circuit shown in Figure 8-13 shows that the voltage at the emitter, VM2, is at 1.72V. The voltage between the collector and the emitter, VM3, is 6.88V. The volt drop across $R_4$ is 3.4V. The sum of these three volt drops must add up to the supply voltage, that is, $V_{CC}$ = 12V. We can check this using

$$V_{CC} = V_{R4} + V_{CE} + V_E = 3.4 + 6.88 + 1.72 = 12V$$

This is OK; however, if we change the value of $R_4$ to say 7kΩ, then assuming the collector current was still at 1.7mA, this would mean that the volt drop across $R_4$ would be

$$V_{R4} = R_4 \times I_C = 7E^3 \times 1.7E^{-3} = 11.9V$$

This would not leave enough voltage from the emitter voltage to remain at 1.72V. This means the emitter and collector currents must reduce. We could change the value of $R_4$ to 7k and simulate the circuit. When we did, the following results were obtained:

- IE = 1.51mA
- IC = 1.49mA
- VE = 1.51V
- VCE = 88.45mV
- VR4 = 10.4

The sum of VR4 + VCE + VE is 10.4 + 0.08845 + 1.51 = 11.99845 close enough to VCC. This shows that as long as the value of the collector resistor, $R_4$ in this circuit, is not too high, then the circuit can work as a constant current source. However, there is another issue that you need to consider, the smaller we make the value of the collector resistor, then the smaller the volt drop across that resistor. As the $V_E$ voltage cannot change, then this means that the $V_{CE}$ voltage must increase. The $V_{CE}$ voltage is the voltage dropped across the transistor which is allowing current to pass through it. This means the transistor would be dissipating power across it, and it would get hot. We need to ensure the power being dissipated is within the SOAF, Safe Operating Area in Forward mode. This will be stated within the specification. It may mean that we would have to employ a heat sink. However, that would be for more specialized circuit designs, and I will not go into that in this book.

We could look at designing a more efficient constant current source such as one that uses an Opamp. However, I am only introducing the concept of a constant current source, and we will look at using the Opamp in Chapter 12.

# The Current Mirror

An extension of the constant current source is the current mirror. This is where one transistor is used to source or set the current that is mirrored in a second transistor or indeed multiple transistors. A simple current mirror circuit is shown in Figure 8-14.

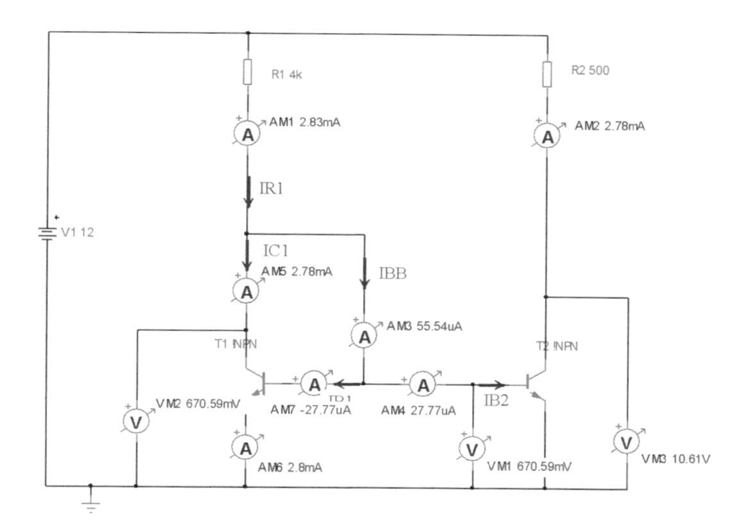


***Figure 8-14.*** *A Simple Current Mirror Circuit*

The two transistors, or multiple transistors, must be matched so that their beta and other parameters are as identical as possible. The first transistor, $T_1$, is the controlling transistor in that it sets the current that is mirrored in the other transistor, $T_2$, or other transistors if there are more than two. Figure 8-14 has a lot going on, but I hope we can use it to show how the mirror circuit works.

If we look at the currents, we can see that by applying KCL we can say

$$I_{R1} = I_{C1} + I_{BB}$$

Also, we can see that

$$I_{BB} = I_{B1} + I_{B2}$$

As the two transistors are matched, we can say

$$I_{B1} = I_{B2}$$

This means we can say

$$I_{BB} = 2I_{B1}$$

Knowing that any collector current, $I_C$, is simply the base current $I_B$, multiplied by the beta of the transistor, we can say

$$I_{C1} = \beta I_{B1}$$

From this, we can say

$$I_{B1} = \frac{I_{C1}}{\beta}$$

This then means that

$$I_{BB} = 2\frac{I_{C1}}{\beta} = \frac{2I_{C1}}{\beta}$$

Substituting this into the expression for $I_{R1}$, we have

$$I_{R1} = I_{C1} + \frac{2I_{C1}}{\beta}$$

Therefore, we can say

$$I_{R1} = I_{C1}\left(1 + \frac{2}{\beta}\right)$$

which we can transpose for $I_{C1}$ as

$$I_{C1} = \frac{I_{R1}}{1 + \frac{2}{\beta}}$$

It is the current $I_{C1}$ that is mirrored into the collector current or currents of the other transistor or transistors. This is shown in Figure 8-14 as the ammeters AM5 and AM2 have the same reading of 2.78mA. The current $I_{R1}$ is set by the volt drop across $R_1$ divided by the value of $R_1$. The volt drop across $R_1$ is set by $V_{CC}$ minus the base emitter voltage of the transistor $T_1$. This is because the collector of $T_1$ is connected to the base of $T_1$. This means that assuming the base emitter voltage $V_{BE}$ is approximately 0.65 and the $V_{CC}$ is 12V, then the current $I_{R1}$ is

$$I_{R1} = \frac{V_{CC} - V_{BE}}{R_1} = \frac{12 - 0.65}{4E^3} = 2.84mA$$

Assuming the beta of the transistors is 100, then we can calculate the collector current $I_{C1}$ as

$$I_{C1} = \frac{I_{R1}}{1 + \frac{2}{\beta}} = \frac{2.84E^{-3}}{1 + \frac{2}{100}} = \frac{2.84E^{-3}}{1.02} = 2.78mA$$

The readings from the simulated circuit, shown in Figure 8-14, do match very closely these calculated values. We can vary the load resistor, that is, the collector resistor $R_2$ in Figure 8-14, of the mirror transistor, and the current will not change as long as the volt drop across the load resistor does not exceed the VCC - VBE, in this case, 11.3V. This means that with a current of 2.78mA, the maximum value of the load resistor in the mirror circuit would be

$$R_{LMAX} = \frac{V_{CC} - V_{BE}}{I_{C1}} = \frac{12 - 0.65}{2.78E^{-3}} = 4.082k$$

One thing the mirror circuit allows is that the $V_{CC}$ of the mirrored circuit does not need to be the same as the control circuit. However, both circuits must share the same ground. For example, we could set the VCC of the mirror circuit to be 9V. This would give a maximum load resistance of

$$R_{LMAX} = \frac{V_{CC} - V_{BE}}{I_{C1}} = \frac{9 - 0.65}{2.78E^{-3}} = 3k$$

This calculation was confirmed with a simulation.

If you wanted to mirror the current in a number of other transistor circuits, then we would calculate the collector current $I_{C1}$ as

$$I_{C1} = \frac{I_{R1}}{1 + \frac{N}{\beta}}$$

where "N" is the total number of transistors, including the controlling transistor. For example, if you wanted to create two mirrors of the controlling current, then N would equal 3.

## Exercise 8.1

Design a current mirror circuit to source a 3mA current with a VCC of 15V for the control circuit and a VCC of 9V for the mirror circuit. Determine the maximum value of the load resistor for the mirror circuit. Assume the beta of the transistors was 100.

## Summary

In this chapter, we have started an investigation into the birth of the Opamp. We looked at a basic differential amplifier and extended it to the long-tailed pair. We considered the concept of CMMR and how the differential amplifier removed noise from a signal. We then went on to look at two new circuits, the constant current source and the current mirror. I hope you have found the analysis both useful and interesting.

In the next chapter, we will look at the Operational Amplifier, the Opamp for short.

CHAPTER 9

# The Operational Amplifier

In Chapter 8, we looked at the long-tailed pair and the differential amplifier. This was an introduction to the Operational Amplifier or Opamp. In this chapter, we will study the Opamp. We will learn how it works and how it can be used in a range of analog applications.

## The Opamp

The Opamp, or Operational Amplifier, is one of the most important, if not the most important, breakthroughs in analog electronics. It led to the creation of the analog computer as it could perform a range of mathematical operations, hence its name. It has since gone on to be used in many analog applications. We will look at a range of these applications in this chapter.

The Opamp is an integrated circuit, or IC, and the most common Opamp IC is the LM741. If you examine the datasheet for the LM741, you will see that the circuitry of the Opamp uses around 20 different transistors and 12 resistors, which means it is a rather complex IC. Therefore, we will use the standard symbol that represents the Opamp. The symbol is shown in Figure 9-1.


H. H. Ward, *Mastering Analog Electronics*, Maker Innovations Series,
https://doi.org/10.1007/979-8-8688-0245-4_9

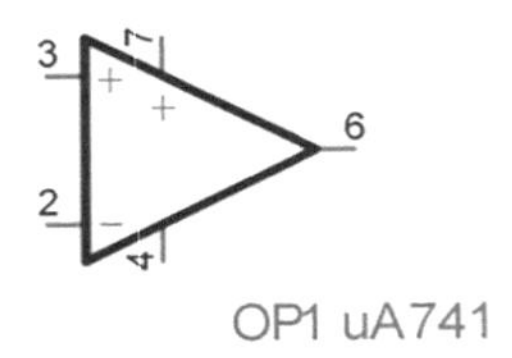


***Figure 9-1.*** *The Standard Symbol for the Opamp*

It comes in a variety of packages, and the basic 741 Opamp can come in an eight-pin dual in-line package as shown in Figure 9-2.

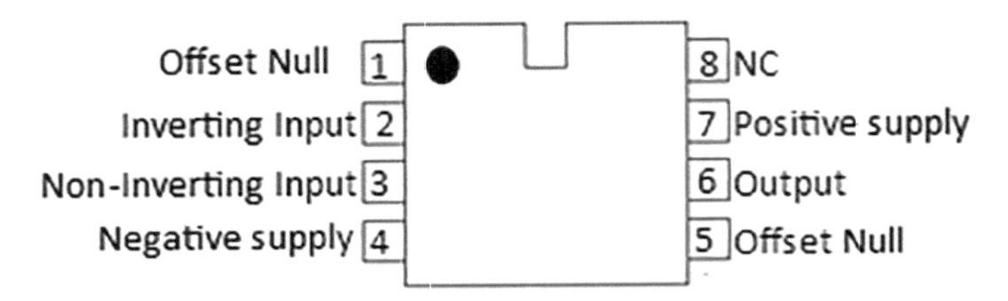


***Figure 9-2.*** *A Standard Eight-Pin Dual In-Line Package*

To use the Opamp, it is normal to have a dual supply with both a positive and a negative voltage. A typical voltage level would be +15V and -15V. However, the Opamp can be used with lower voltages or indeed a little higher. The 741 can be used with a maximum supply of 22V. There are two inputs, one called the inverting and the other called the non-inverting. However, there is only one output. There are also two offset null inputs which we will explain the use of later in this chapter. The Opamp can be connected in an open-loop or closed-loop configuration, and we will look at the open-loop configuration first. This is shown in Figure 9-3.

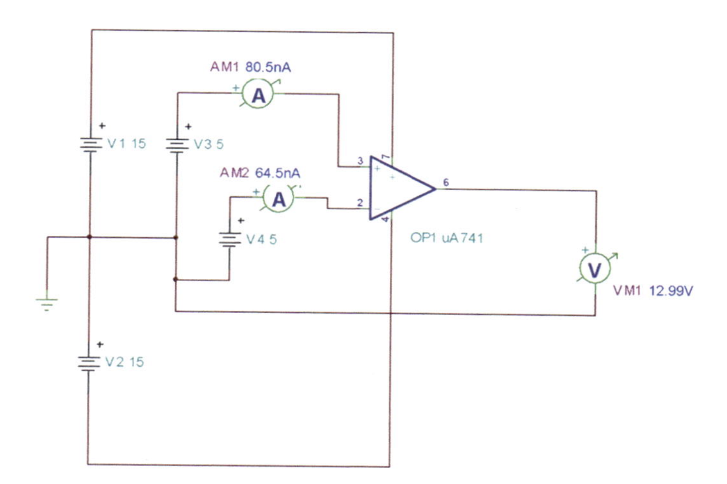


***Figure 9-3.** The Basic Open-Loop Configuration*

The concept of this Opamp, being in the open-loop configuration, means that none of the output is fed back to the input, which is what does happen in the closed-loop configuration. The main principle of operation is that the Opamp will simply amplify the difference voltage between its two inputs.

In all the following circuit diagrams, the negative supply must be connected to pin 4 of the Opamp and the positive supply connected to pin 7. In Figure 9-3, $V_3$ and $V_4$ are the two inputs, and at present they are both set to 5V. This would mean the output should be zero. However, we can see that it is actually showing 12.99V. Another interesting fact that we can see from Figure 9-3 is that both inputs of the Opamp are taking very little current into the Opamp. Both ammeters read only a few nano amps. Not much current at all.

If we change both $V_3$ and $V_4$ to 1mv, which means they are still both equal and the difference voltage is still zero, we see that the output voltage does not change at all. Let's try and find out what is going on by introducing some difference between the two input voltages. We will simulate the circuit after each change and record the results in Table 9-1.

***Table 9-1.*** *Table of Results for Open-Loop Opamp*

| Row | V3 Volts | V4 Volts | AM1 nAmps | AM2 nAmps | VM1 Volts |
|---|---|---|---|---|---|
| 1 | 1m | 1.01m | 80.49 | 59.41 | 12.98 |
| 2 | 1m | 1.1m | 80.45 | 59.55 | 12.98 |
| 3 | 1m | 1.2m | 80.4 | 59.6 | 12.98 |
| 4 | 1m | 1.5m | 80.25 | 59.75 | 12.97 |
| 5 | 1m | 2m | 80 | 60 | 0 |
| 6 | 1m | 3m | 79.5 | 60.5 | –12.99 |
| 7 | 1m | 5m | 78.5 | 61.51 | –13.01 |
| 8 | 1m | 5V | –2420 | 2560 | –13.1 |
| 9 | 1.01m | 1m | 80.51 | 59.5 | 12.99 |
| 10 | 1.1m | 1m | 80.55 | 59.45 | 12.99 |
| 11 | 1.2m | 1m | 80.6 | 59.4 | 12.99 |
| 12 | 1.5m | 1m | 80.75 | 59.25 | 13 |
| 13 | 2m | 1m | 81 | 59 | 13 |
| 14 | 3m | 1m | 81.5 | 58.5 | 13.01 |
| 15 | 5m | 1m | 82.5 | 57.5 | 13.03 |
| 16 | 5V | 1m | 2580 | –2440 | 13.1 |

Using Table 9-1, we can make some interesting observations. Firstly, knowing the Opamp amplifies the difference between the two input voltages, and the fact that the maximum voltage the output can swing to is the voltage rails, well actually just short of the voltage rails, then we can see, from row 1 of Table 9-1, the Opamp does not need much difference between the two inputs before the output swings to a maximum. In row 1, $V_3$ is 1mv and $V_4$ is 1.01mv, which is only a 10μV difference. The output has gone to 12.98V, which is very close to the maximum as the output can only go to the voltage rails minus the saturation voltage of the transistors at the output of the Opamp. This saturation voltage is approximately 1.8 to 2V.

If we treat the input, which we will call "$V_{DIFF}$", as the difference between the two inputs, then knowing that $V_{OUT}$, the output voltage, is the input multiplied by the open-loop gain "$A_V$" of the Opamp, then we can say

$$V_{OUT} = A_V V_{DIFF}$$

From this, we can say the voltage gain can be expressed as

$$A_V = \frac{V_{OUT}}{V_{DIFF}}$$

Then using the two values from row 1, we can say

$$A_V = \frac{12.98}{10E^{-6}} = 1.298E^6$$

Similar values can be seen in row 9 when $V_3 = 1.01$mV and $V_4 =$ 1mV. This means the $V_{DIFF}$ was 10μV when the $V_{OUT}$ was 12.99. This would make the voltage gain a value of $1.299E^6$.

These values would suggest the voltage gain of the Opamp in open-loop configuration is in the order of $10E^5$, that is, 1,000,000, $1E^6$ or more; the results suggest the gain is 1,298,000. The theory suggests the open-loop gain is in the order of $10E^5$, that is, 1,000,000, or $1E^6$.

If we use the value for the open-loop gain of 1,000,000 and the ideal maximum output voltage is the power rails, then using a power rail of 15V we can determine what the maximum voltage difference that the Opamp can respond to at its inputs as follows:

$$Maximum\, V_{OUT} = Maximum\, V_{DIFF} \times A_V$$

From this, we can say

$$Maximum\ V_{DIFF} = \frac{Maximum\ V_{OUT}}{A_V} = \frac{15}{10E^5} = 15\mu V$$

If we use the voltage gain calculated from the results listed in Table 9-1, we get

$$Maximum\ V_{DIFF} = \frac{15}{1.298E^6} = 11.57\mu V$$

Both of these are very small, virtually 0V. Consider the 15μV; this is 0.000015, which is virtually 0, and 11.57μV is 0.00001157v, even closer to 0V. What this actually means is that we can consider the two voltages at the input as being virtually the same, which we interpret as there being a "virtual short circuit across the two inputs." Rubbish I hear you say. Surely, we can see in rows 8 and 16 there is a massive difference between the two inputs. In both cases, one input is at 1mv, while the other is at 5V, a massive difference. That is true, but if we look at the output voltage, we can see that in both rows the output is 13.1V, while, when the difference is only 10μV, as in rows 1 and 9, the output voltage is virtually the same at 12.98 and 12.99V. This shows that the Opamp treats the two different voltages, that is, the 10μV and the 5V, as if they were the same. This is because they are very close or greater than the maximum $V_{DIFF}$ the Opamp can cope with. That is why we say there is "Always a Virtual Short Circuit Across the Input Terminals."

If we now look at the current flowing into the two inputs, as measured with AM1 and AM2, we see that they are very small currents around the 80 and 60 nano amps. What we can say, because of these small currents, is that virtually no current, even when, as with rows 8 and 16, the currents are around 2.5μA, flows into the Opamp. We can then interpret this as there being a very high, ideally infinite, internal impedance at the input terminals of the Opamp.

Before we sum up the results of this simulated experiment, I feel I should say that the fact the Opamp cannot tell the difference between a voltage difference of 10μV and 5V, the output being the same in both cases makes the Opamp pretty useless. Well useless in this open-loop configuration, but, as you will see, we normally use the Opamp in what is termed a closed-loop configuration. There is one circuit where we might use the Opamp in the open-loop configuration, but we will look at that later in this chapter.

To sum up what this experiment has shown us, we can make the following statements:

1. The open-loop voltage gain, $A_V$, is very high. We will use the figure of $10E^5$.

2. The input impedance is also very high. We will use the value of infinity. The output impedance is very low, in the order of 75Ω, although we have not studied this in this experiment.

3. There is a virtual short circuit across the input terminals.

We will use these statements when we come to analyze how the following Opamp circuits work.

# The Offset Null Inputs

When the two inputs to the Opamp are zero, you would expect the output of the Opamp to be zero also. However, if you tried this, you would find that the output would be some voltage other than zero. To try and ensure the output can be set to zero, when the two inputs are zero, Opamps usually come with a voltage offset facility. This is normally an extra input into which you can supply a variable voltage, which can be either positive or negative, to reduce this unwanted output to zero volts. If we look at Figure 9-9, we can expect the voltage at point "X" to be 0V. This is due to the virtual short circuit across the input and the other input being connected to ground, that is, 0V. However, if we look at the voltmeter VM3, we see that the voltage at "X" is 1.02mV. This is called the offset voltage, and it is due to the internal circuitry in the Opamp. With the basic inverting and non-inverting Opamps, this offset voltage is not a real problem, and we can ignore it. However, with the integrating Opamp, this would cause a problem, and you would have to use some offset voltage at the extra input to counteract this small voltage.

Under the same conditions, you would expect the input current to be zero. Also, as we have stated that the input impedance of the Opamp is ideally infinity, then under any conditions there should be no current going into the Opamp. Again, if we look at Figure 9-9, we can see that the ammeter AM2 measures a current of 60nA flowing into the Opamp, but the theory says there should be no current flowing into the Opamp. Therefore, there is an offset null current input for you to use to ensure no current flows into the Opamp if you thought you would need it. When using the basic Opamps, this is not an issue, but where it might become an issue, there is an option with the other extra input to the Opamp to offset this small current.

Before we leave the results of the open-loop configuration, we should try and decide which way the Opamp creates the voltage difference. Is it V4 – V3 or V3 – V4? If we assume it is V4 – V3, then when V4 is greater than V3 we would expect a positive voltage at the output. From Table 9-1, we see that we do get a positive output for the first four rows. However, at row 5 the output is actually zero, and then for rows 6, 7, and 8 the output voltage is actually negative. Also, in rows 9 to 16, the output voltage is positive, but V4 is at 1mV and V3 is higher, going from 1.01mV in row 9 to 5V in row 16. These last seven results would suggest that the difference is V3 – V4.

However, if that was the case, then why in rows 1 to 4 was the output voltage positive when V4 was greater than the 1mV at V3? The answer is that the voltage at the "+" or so called non-inverting terminal was made up of the 1mV at V3 plus this internal error voltage that is created by the internal circuitry of the Opamp. The results suggest this internal error voltage is around 1mV, making the actual input voltage at the "+" terminal 2mV. That is why when, as in row 5, V4 is 2mV, the output voltage goes to zero. Then from then on up to row 9, the output voltage was negative as V4 was now greater than V3.

This is just an example of the fact that there is this internal error voltage at the input of the Opamp that we might have to offset.

# The Unity Gain Buffer

This is the first use of the Opamp we will look at. The circuit for the unity gain buffer is shown in Figure 9-4.

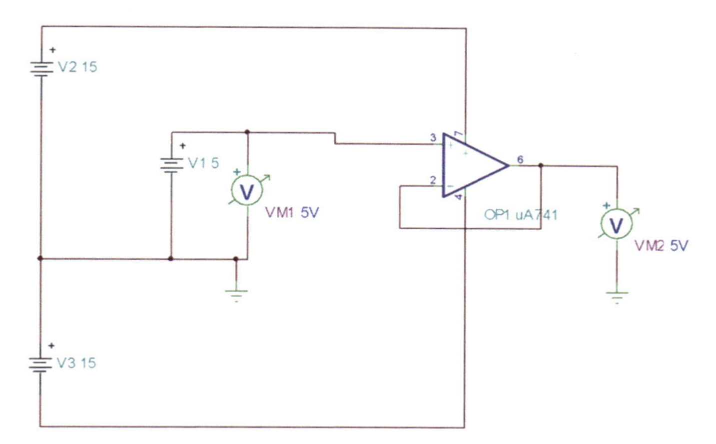


***Figure 9-4.*** *The Unity Gain Buffer*

The input voltage is 5V, and we can see that the output voltage is also 5V. This shows that the voltage gain $V_{GAIN}$ is

$$V_{GAIN} = \frac{V_{OUT}}{V_{IN}} = \frac{5}{5} = 1$$

Hence the name unity gain, but why a buffer? Well, to appreciate that, we will look at the circuit shown in Figure 9-5.

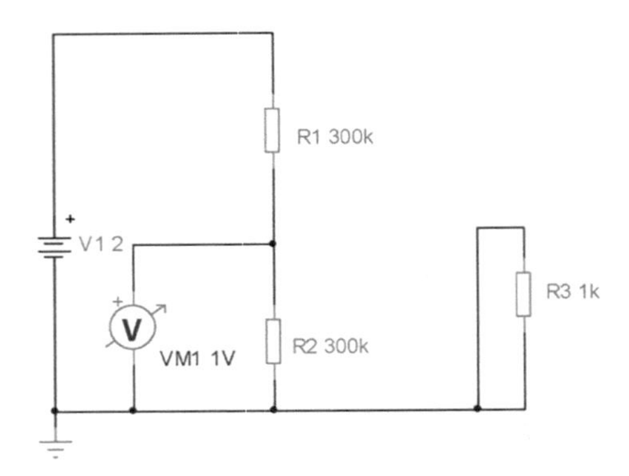


***Figure 9-5.*** *The Test Circuit for Interstage Loading*

The two resistors, $R_1$ and $R_2$, make up the first stage of a system. The resistor $R_3$ makes up the second stage. The system is pretty useless, but I am only using it to explain what interstage loading is and how the Opamp shown in Figure 9-4 can be called a buffer.

The purpose of the first stage is to provide an input of 1V to the second stage. We can see that with the second stage disconnected, the first stage is doing what is asked of it. Let's now use the circuit as shown in Figure 9-6 to see what happens when we connect the second stage.

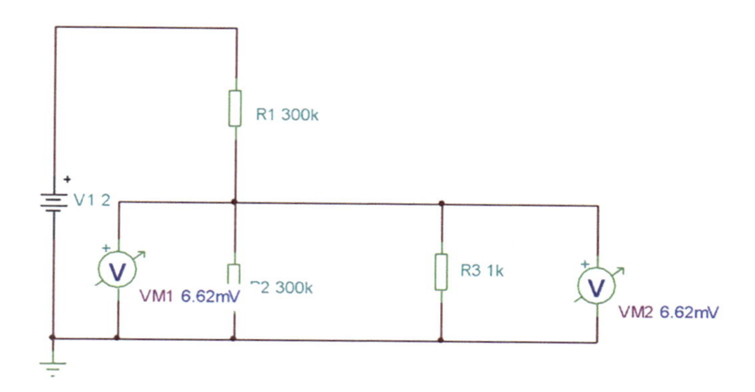


***Figure 9-6.*** *The Two Stages Connected Together*

We can see that when we connect the second stage, the 1v has been reduced to 6.2mV. In this way, we can say that the second stage has loaded the first stage, that is, this is an example of interstage loading.

The reason why this has happened is that we have put a low resistance in parallel with a high value resistance, and the combination has effectively reduced how the first stage sees the 300k resistor. If we combine the two resistors $R_2$ and $R_3$ in parallel, we will see what this effective value is as follows:

$$R_{COMB} = \frac{R_2 \times R_3}{R_2 + R_3} = \frac{300E^3 \times 1E^3}{300E^3 + 1E^3} = 996.68\Omega$$

If we now use the voltage divider rule to calculate the voltage across $R_2$ using this new effective value, we get

$$V_{R2EFF} = \frac{V_1 \times R_{2EFF}}{R_1 + R_{2EFF}} = \frac{2 \times 996.68}{300E^3 + 996.68} = 6.62mV$$

I hope this explains why the second stage has loaded the first stage. This actually can be used to represent a real experience I have shown to my students when we used an old AVO 7 to measure the voltage across two 300k resistors as shown. The problem with the AVO 7 was that with such a low voltage to be displayed, the input impedance of the AVO 7 was about 1kΩ. As we had to put the AVO across the resistor R2 to measure the voltage, the AVO itself loaded the 300k resistor, and the measured voltage was practically 0V.

The solution, and this is the solution I did practically, was to insert the unity gain buffer between $R_2$ and $R_3$. In practice, placing the unity gain buffer circuit between the resistor $R_2$ and the AVO 7 voltmeter used to measure the voltage across $R_2$. The result is shown in Figure 9-7.

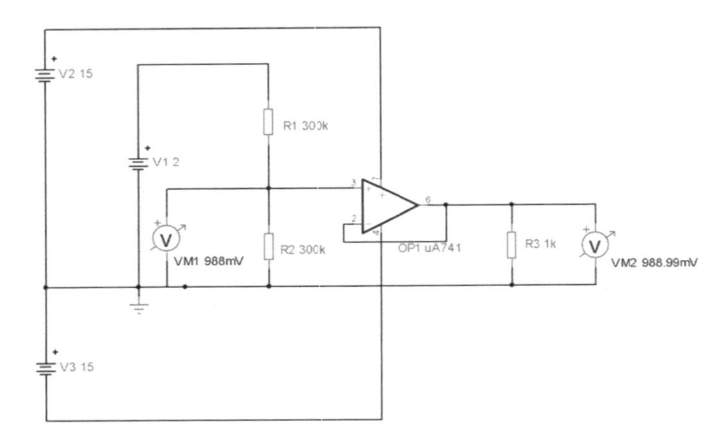


***Figure 9-7.** The Loading Cured by Inserting the Unity Gain Buffer*

We can see that the voltage across $R_3$ is now 988.99mV or 1V, and the volt drop across $R_2$ is 988mV or 1V. This is as it should be, and it shows how the Opamp has acted as a buffer between the two stages. The reason why it works as a buffer is the infinite input impedance of the Opamp does not alter the value of the 300k resistor. The first stage still sees $R_2$ as the 300k it should be. Also, the low output impedance of the Opamp, around 75Ω, is not loaded by the 1k value of $R_3$. So, I hope you can see why the circuit shown in Figure 9-4 is called the "unity gain buffer."

# The Inverting Opamp

The next Opamp circuit we will look at is shown in Figure 9-8.

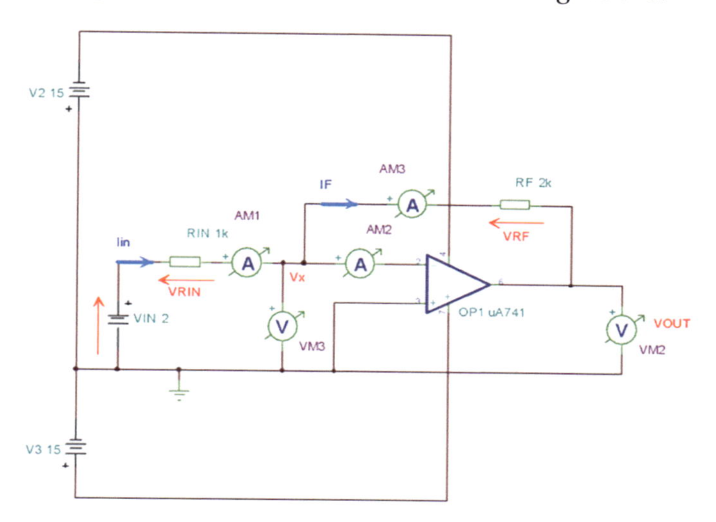


***Figure 9-8.*** *The Inverting Opamp*

Before we simulate the circuit, we will try to analyze how the circuit works using some of the statements we derived using the test circuit shown in Figure 9-3. The first thing we can see is that there is a path from the output, via the resistor $R_F$, that feeds some of the output back to the input. This is why this configuration is termed "closed loop" as there is a loop that feeds back some of the output to the input. This is called "feedback," and we have labeled the resistor $R_F$ as it is in the feedback path.

The next thing we will look at is determining what the voltage is at the point "Vx" shown on the circuit. Knowing there is a virtual short circuit across the two input terminals and seeing that the input on pin 3 of the Opamp is connected to ground, that is, 0V, then we can say that the voltage at "Vx" must also be at 0V.

If we consider the voltage arrows drawn on the circuit, using the convention that the arrowhead indicates the more positive end of the volt drop, it is, I hope, fairly obvious why the arrowhead on $V_{RIN}$ is pointing toward $V_{IN}$, as that is the more positive end of the volt drop. However, it is not as obvious as to why the arrowhead on the volt drop, $V_{RF}$, across $R_F$ is pointing toward the junction of $R_{IN}$ and pin 2 of the Opamp. We need to appreciate the direction of the current flow in the circuit to understand why the volt drop is labeled in that direction. In electrical circuit analysis, it is usual to use "conventional current flow" that flows from positive to negative and always flows into the positive end of a voltage drop.

We will consider the current "$I_{IN}$" first. We can use Ohm's Law which states that

$$I = \frac{V}{R}$$

where the "V" is the volt drop across the resistance in question. The volt drop is simply the voltage at one end minus the voltage at the other end. This means that in the case of $I_{IN}$ the volt drop is simply

$$Voltdrop = V_{IN} - V_X$$

However, using the virtual short circuit across the input terminals, we have stated that $V_X = 0V$. This means that the volt drop, across $R_{IN}$, that is, $V_{RIN}$, is simply $V_{IN}$. This means that the current $I_{IN}$ can be calculated as

$$I_{IN} = \frac{V_{IN}}{R_{IN}} = \frac{2}{1E^3} = 2mA$$

Now we have stated that no current flows into the Opamp because of the infinite input impedance across the input terminals of the Opamp. This means that the only place the current can flow is up into the resistor in the feedback path, $R_F$. Therefore, we can say

$$I_{IN} = I_F$$

Also, as conventional current must flow into the positive end of the volt drop, then the junction of $R_F$ and pin 2 of the Opamp must be the more positive end of the volt drop, and so the arrowhead for $V_{RF}$ is pointing correctly in the direction shown.

As to the voltage "$V_{OUT}$", this is measured across the other end of $R_F$ and ground. It is termed $V_{OUT}$ as it is measured across the output of the Opamp and ground. We know that one end, the more positive end of the $R_F$, is connected to the point "$V_X$". We also know that $V_X$ is at 0V. This means that for that end to be more positive than the output end, it must mean that the output must be negative, which means $V_{OUT}$ is $-V_{OUT}$. We should be able to see that the voltage $-V_{OUT}$ is equal to the volt drop $V_{RF}$. Using Ohm's Law, we can say that the volt drop across a resistor is equal to the current flowing through it times the value of the resistor. This means we can say

$$V_{RF} = I_F R_F$$

Therefore, we can say

$$-V_{OUT} = I_F R_F$$

From this, we can say

$$V_{OUT} = -I_F R_F$$

Now we can consider the voltage gain. As always, the voltage gain, "$V_{GAIN}$", can be expressed as

$$V_{GAIN} = \frac{V_{OUT}}{V_{IN}}$$

We have expressed $V_{OUT}$ as $-I_F R_F$, and similarly we can express $V_{IN}$ as $I_N R_{IN}$. Therefore, we can say

$$V_{OUT} = \frac{-I_F R_F}{I_{IN} R_{IN}}$$

However, we have shown that $I_F = I_{IN}$, and so we can say

$$V_{Gain} = \frac{-I_{IN} R_F}{I_{IN} R_{IN}}$$

The two $I_{INs}$ can cancel out, and so the expression for the voltage gain of this Opamp circuit becomes

$$V_{Gain} = \frac{-R_F}{R_{IN}}$$

Using the values from the circuit, we can calculate the voltage gain as

$$V_{GAIN} = \frac{-2E^3}{1E^3} = -2$$

The negative sign means the output is an inversion of the input.

If we simulate the circuit, we will get the results as shown in Figure 9-9.

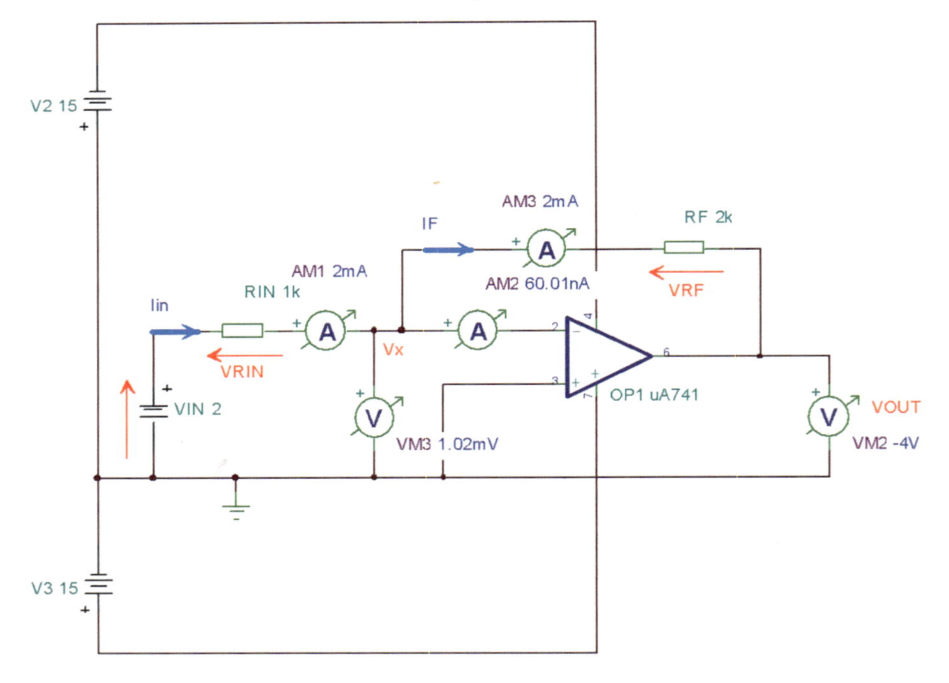


***Figure 9-9.** The Test Circuit for the Inverting Opamp*

We can create a table of results from which we can make our comparisons. The table is shown in Table 9-2.

***Table 9-2.*** *The Results for the Simulation of the Test Circuit Shown in Figure 9-9*

| Parameter | Calculated Value | Simulated Result |
|---|---|---|
| $V_X$ | 0V | 1.02mV |
| $I_N$ | 2mA | 2mA |
| $I_F$ | 2mA | 2mA |
| Current into Opamp | 0 | 60.01nA |
| $V_{OUT}$ | –4 | –4 |
| $V_{GAIN}$ | –2 | –2 |

The simulated result for the voltage gain was determined using

$$V_{GAIN} = \frac{-V_{OUT}}{V_{IN}} = \frac{-4}{2} = -2$$

The comparisons, using Table 9-1, are very close to each other and so show that the calculations using what is basically Ohm's Law work very well. We will use the same approach when analyzing the remaining Opamp circuits.

# An Alternative Analysis

Before we do move on to the next circuit, we will go through an alternative analysis of the inverting amplifier. This uses the superposition rule to determine the voltage "$V_X$" due to $V_{IN}$ and $V_{OUT}$. The principle of the superposition rule is that we determine the voltage in question with all but one source replaced with their ideal internal impedance. We do this for all sources and then add the results together. With the Opamp circuit shown in Figure 9-9, there are only two sources that affect the voltage "$V_X$", and they are $V_{IN}$ and $V_{OUT}$.

## $V_X$ Due to $V_{OUT}$

To determine the voltage $V_X$ due to $V_{OUT}$, we must replace $V_{IN}$ with a short circuit, as this is the ideal internal impedance of a voltage source. This would produce the circuit shown in Figure 9-10.

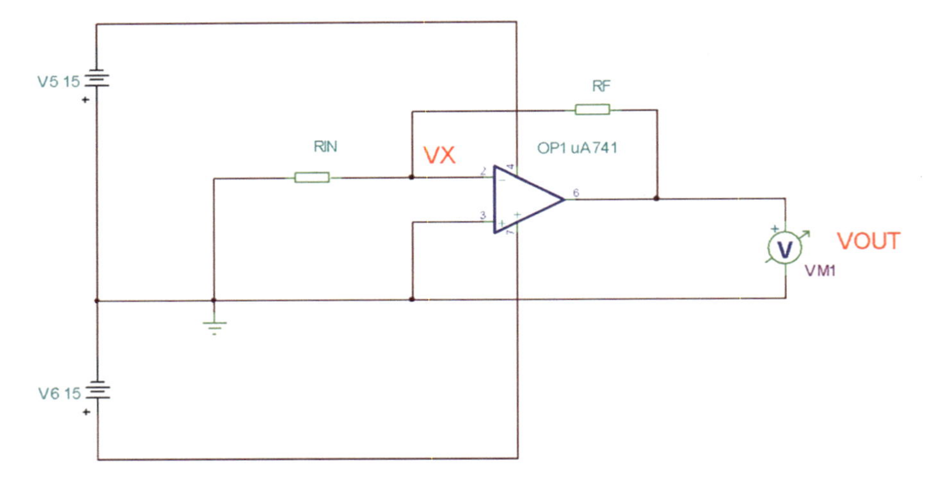


***Figure 9-10.*** *VX Due to VOUT*

This circuit shows us that the voltage $V_{OUT}$ is the output of the amplifier. The input to the amplifier is "$V_X$", and using the open-loop gain of the amplifier, we can say

$$V_{OUT} = A_V V_X$$

From the circuit shown in Figure 9-10, we can see that the two resistors $R_{IN}$ and $R_F$ form a voltage divider network with $V_{OUT}$ as the source. This concept can be shown in the circuit in Figure 9-11.

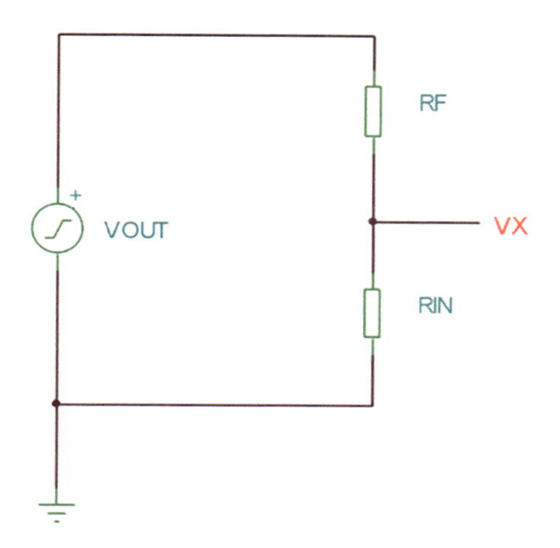


***Figure 9-11.*** *The Resistor Divider Network*

Using the voltage divider rule, we can say

$$V_{XDUE\ TO\ VOUT} = \frac{V_{OUT} \times R_{IN}}{R_F + R_{IN}}$$

## $V_X$ Due to $V_{IN}$

To calculate "$V_X$" due to $V_{IN}$, we replace the $V_{OUT}$ with a short circuit, and this is shown in Figure 9-12.

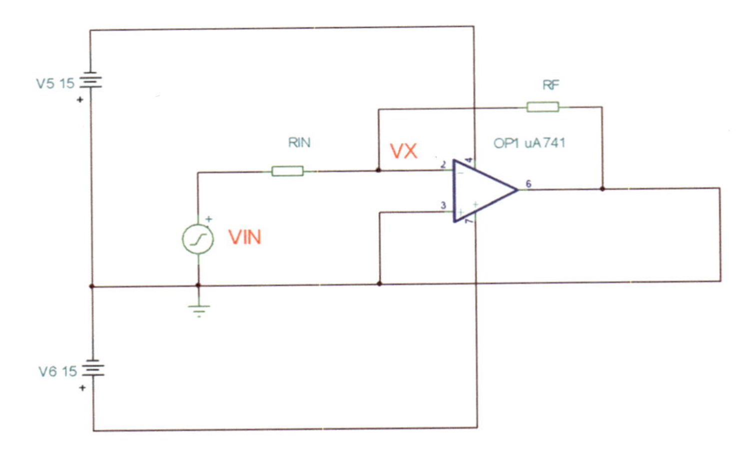


***Figure 9-12.*** *Vout Replaced with a Short Circuit*

The two resistors would now create the voltage divider network as shown in Figure 9-13.

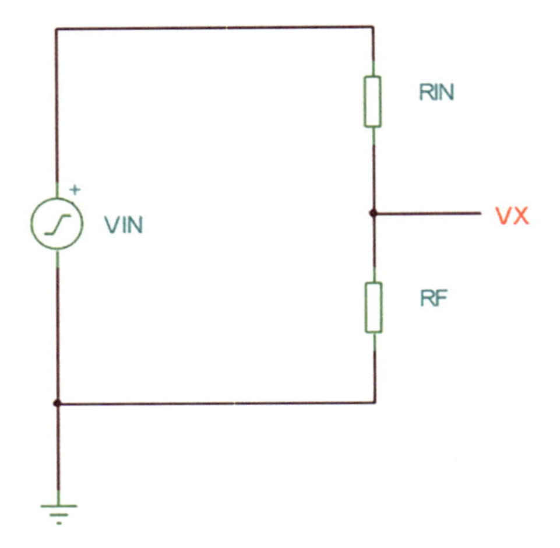


***Figure 9-13.*** *The Voltage Divider for $V_{IN}$*

Again, using the voltage divider rule, we can say

$$V_{X\ DUE\ TO\ VIN} = \frac{V_{IN} R_F}{R_{IN} + R_F}$$

# The Complete Expression for $V_X$

Now we need to add the two terms together to create an expression for $V_X$ as follows:

$$V_X = \frac{V_{OUT} R_{IN}}{R_{IN} + R_F} + \frac{V_{IN} R_F}{R_{IN} + R_F}$$

# The Expression for the Voltage Gain

We can now use the expression for $V_X$ to determine the expression for the voltage gain.

Knowing

$$V_{OUT} = A_V V_X$$

We can say

$$V_X = \frac{V_{OUT}}{A_V}$$

This then means we can say

$$\frac{V_{OUT}}{A_V} = \frac{V_{OUT} R_{IN}}{R_{IN} + R_F} + \frac{V_{IN} R_F}{R_{IN} + R_F}$$

Subtracting the first term $\frac{V_{OUT} \times R_{IN}}{R_{IN} + R_F}$ from both sides, we get

$$\frac{V_{OUT}}{A_V} - \frac{V_{OUT} R_{IN}}{R_{IN} + R_F} = \frac{V_{IN} R_F}{R_{IN} + R_F}$$

Multiplying through by $A_V$, we get

$$V_{OUT} - \frac{A_V V_{OUT} R_{IN}}{R_{IN} + R_F} = \frac{A_V V_{IN} R_F}{R_{IN} + R_F}$$

We can take $V_{OUT}$ as a common factor on the LHS gives

$$V_{OUT}\left[1-\frac{A_V R_{IN}}{R_{IN}+R_F}\right]=\frac{A_V V_{IN} R_F}{R_{IN}+R_F}$$

Combining the terms in the bracket gives

$$V_{OUT}\left[\frac{R_{IN}+R_F-A_V R_{IN}}{R_{IN}+R_F}\right]=\frac{A_V V_{IN} R_F}{R_{IN}+R_F}$$

Now divide both sides by the term $\left[\frac{R_{IN}+R_F-A_V R_{IN}}{R_{IN} R_F}\right]$ to transpose for $V_{OUT}$, and knowing that when dividing with a fraction we invert and multiply gives

$$V_{OUT}=\frac{A_V V_{IN} R_F}{R_{IN}+R_F}\times\frac{R_{IN}+R_F}{R_{IN}+R_F-A_V R_{IN}}$$

This cancels down to

$$V_{OUT}=\frac{A_V V_{IN} R_F}{R_{IN}+R_F-A_V R_{IN}}$$

If we divide both sides by $V_{IN}$, we get

$$\frac{V_{OUT}}{V_{IN}}=\frac{A_V R_F}{R_{IN}+R_F-A_V R_{IN}}$$

As the open-loop gain $A_V$ is around $10E^5$ and as long as the sum of $R_{IN}+R_F$ is a lot smaller than $A_V$, we can ignore the $R_{IN}+R_F$ in the denominator, and this then gives

$$\frac{V_{OUT}}{V_{IN}}=\frac{A_V R_F}{-A_V R_{IN}}=-\frac{A_V R_F}{A_V R_{IN}}$$

The $A_V$ terms cancel out, and we end up with

$$\frac{V_{OUT}}{V_{IN}} = -\frac{R_F}{R_{IN}}$$

As the term $\frac{V_{OUT}}{V_{IN}}$ is the voltage gain of the Opamp, we can say

$$V_{GAIN} = -\frac{R_F}{R_{IN}}$$

This is the same expression that we arrived at using the initial Ohm's Law approach. This derivation has been a long process, but when you get two approaches producing the same result, it gives you confidence in using the result you have arrived at. Both approaches rely on the condition that the open-loop gain for the Opamp is very high; some say it's $10E^5$; our simulation of the circuit in Figure 9-3 produced a value of $1.298E^6$.

We can carry out some simulations on the circuit shown in Figure 9-14 to further confirm the expression for the voltage gain of the inverting Opamp.

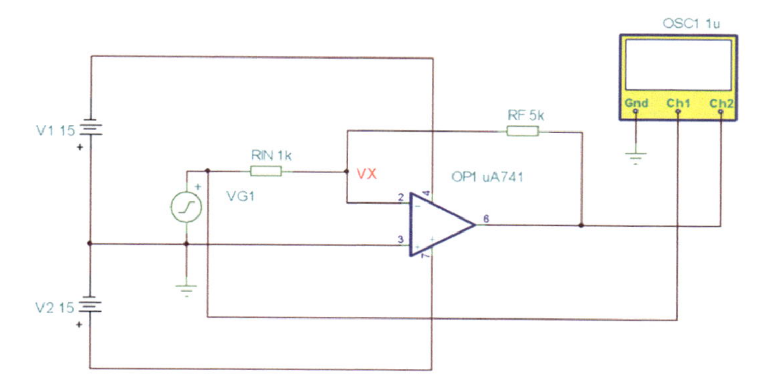


***Figure 9-14.*** *The Test Circuit for the Inverting Amplifier*

We have replaced Vin with an ac source set to 20mV peak to peak at 8kHz, and we are using the oscilloscope to measure the input and output waveforms. This is to show that the expression for the voltage gain works for ac signals as well as DC voltages. The display for the ac test, with $R_{IN}$ set to 1k and $R_F$ set to 2k, is shown in Figure 9-15.

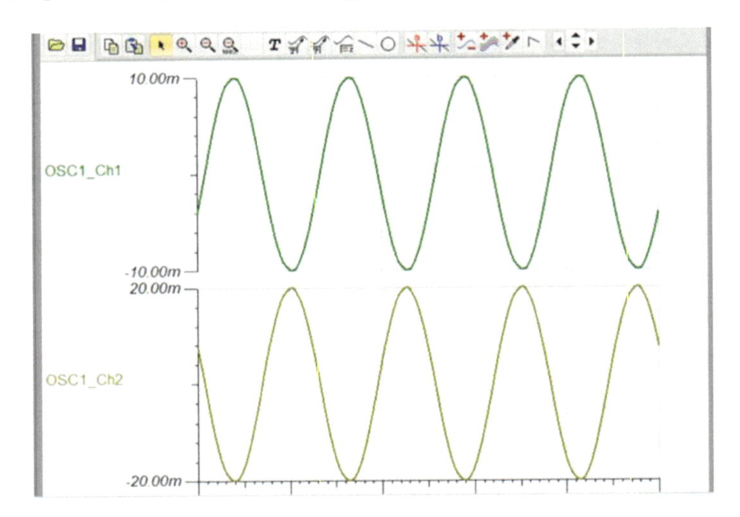


***Figure 9-15.** The Input and Output Waveforms for the Test Circuit Shown in Figure 9-14*

Channel 1 displays the input voltage, and channel 2 displays the output voltage. We can see that the two waveforms are 180° out of phase which agrees with the negative sign on the expression for the voltage gain. Also, using the peak-to-peak voltage, the gain is

$$V_{Gain} = -\frac{V_{OUT} Peak\ to\ Peak}{V_{IN} Peak\ to\ Peak} = -\frac{40E^{-3}}{20E^{-3}} = -2$$

This is what we expect. We will now change the resistor values and record the simulated voltage gain so that we can compare it with the calculated voltage gain as shown in Table 9-3.

***Table 9-3.*** *The Table of Results for the Test Circuit Shown in Figure 9-14*

| Row | $R_{IN}$ | $R_F$ | Calculated $V_{GAIN}$ | Simulated $V_{GAIN}$ |
|---|---|---|---|---|
| 1 | 1k | 4k | –4 | –4 |
| 2 | 1k | 5k | –5 | –5 |
| 3 | 3k | 5k | –1.67 | –1.67 |
| 4 | 3k | 9k | –3 | –3 |

The results do confirm the theory, and so we can now use all the expressions we have derived.

# The Input Impedance of the Inverting Opamp

You will normally be putting an amplifier onto the output of a first stage that will create the signal we want to amplify. If that was an audio signal, then the first stage could be a simple microphone. However, no matter what the first stage is, we need to consider the possibility of interstage loading as described in Figure 9-6. This means we need to know what the input impedance of the amplifier is and how to control it. We have seen that with the open-loop configuration, the input impedance is very high; typically, we can say it has an infinite input impedance. However, with the inverting Opamp, we include two external resistors, one which completes the feedback path of the Opamp and so closes the loop. The other is at the input of the Opamp; see Figure 9-8. It will be this input resistance that sets the input impedance of the Opamp. The way in which it does that is shown in Figure 9-16.

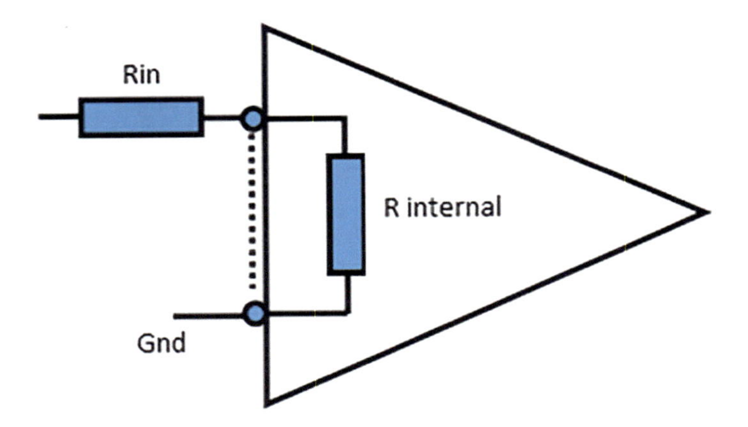


***Figure 9-16.*** *The Input Impedance of the Inverting Opamp*

Figure 9-16 shows the input resistor, Rin, connected to the Opamp. It also shows the internal impedance across the input terminals of the Opamp. As the Opamp does not draw any current into the amplifier, then this internal resistance is said to have an infinite value. There is a dashed line outside the Opamp which represents the fact that there is a virtual short circuit across the two inputs of the Opamp. It is not a real short circuit as this would remove the workings of the Opamp. However, because of the high open-loop gain and the maximum voltage difference that the Opamp can distinguish, as described earlier, there is an electrical short circuit across the outside of the Opamp. This actually means that the end of the Rin resistor at the Opamp is connected to ground. This then means that the internal resistance of the Opamp is electrically taken out of the circuit. This means we can say that the input impedance of the Opamp is set by the value of Rin. This can be confirmed if we look at the circuit shown in Figure 9-8. This shows that when the input voltage was 2V, the input current was 2mA. We should appreciate that, by rearranging Ohm's Law, we can determine the input impedance of a system by measuring the input current for a known input voltage. Using this concept, we can determine the input impedance as

$$R_{IN} = \frac{V_{IN}}{I_{IN}} = \frac{2}{2E^{-3}} = 1E^{3} = 1k$$

This is the value of $R_{IN}$ as shown in Figure 9-8. This confirms that the input impedance of the inverting Opamp is simply the value of $R_{IN}$.

## Designing an Inverting Opamp

Using the concepts we have looked at so far, we can design an inverting Opamp to comply with some simple specifications. For example, design an inverting Opamp that has a gain of –3 with an input resistance of 10kΩ.

The specification gives us the value for $R_{IN}$ as we know it is the value of $R_{IN}$ that sets the input impedance of the Opamp. This means that $R_{IN}$ must be set at 10kΩ. All that remains now is the value of the feedback resistor $R_F$. To set this, we can use the expression for the voltage gain which is

$$V_{GAIN} = \frac{-R_F}{R_{IN}}$$

We know this must be set to –3, and so if we rearrange the expression for the gain to an expression for $R_F$, we get

$$R_F = -R_{IN} V_{GAIN}$$

Putting our values in, we get

$$R_F = -10E^{3}(-3) = 30k$$

We can simulate the test circuit shown in Figure 9-17 to confirm that the circuit works as expected.

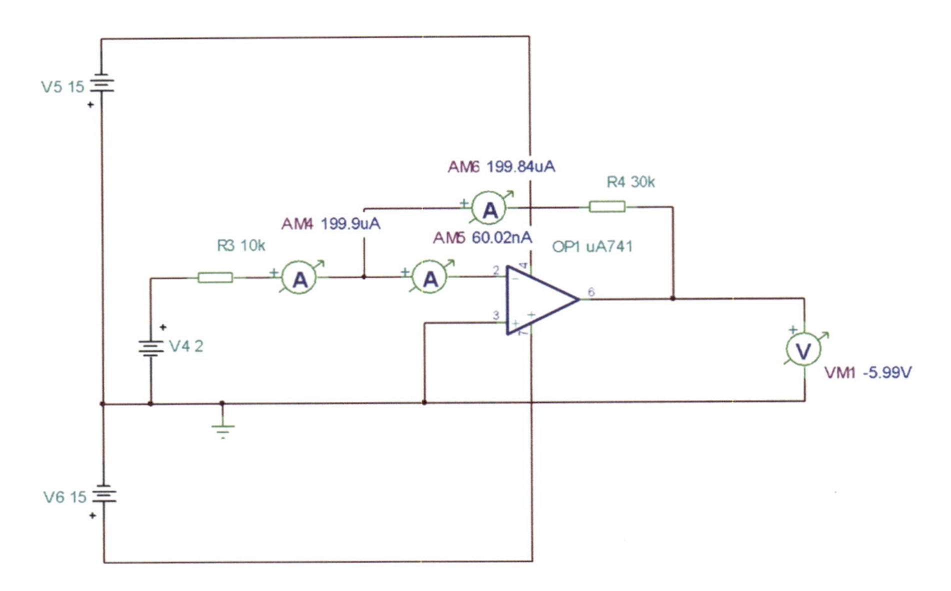


***Figure 9-17.*** *The Test Circuit for Example 1*

The test circuit for example 1 shows that the output voltage was –6V when the input was 2V. This confirms the gain is set to –3. The input current was 200µA when the input voltage was 2V. Using these values, the input resistance can be calculated as

$$R_{IN} = \frac{V_{IN}}{I_{IN}} = \frac{2}{200E^{-6}} = 10E^{3} = 10k$$

# Exercise 9.1

Design an inverting Opamp for the two following specifications:

1. The input impedance is 15kΩ, and the gain is –4.
2. The gain is –5, and the feedback resistor $R_F$ is 15kΩ.

# The Non-inverting Opamp

As there is an inverting Opamp, it is only right there should be a non-inverting Opamp. The circuit we are going to look at is shown in Figure 9-18.

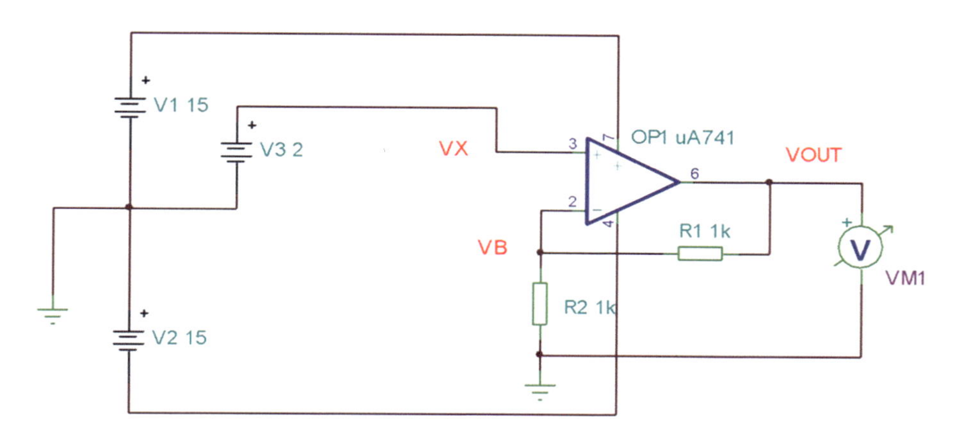


***Figure 9-18.*** *The Non-inverting Opamp*

We are again starting off with a DC input voltage. This is $V_3$ which we will call the input voltage $V_{IN}$. We can see that $V_X$ will be at the same voltage as $V_{IN}$, and so $V_X$ will, in this case, be 2V. Then knowing there is a virtual short circuit across the input terminals, we can say $V_B = V_X = 2V$. The two resistors, $R_1$ and $R_2$, create a voltage divider network that divides $V_{OUT}$ between them to obtain the voltage $V_B$ as

$$V_B = \frac{V_{OUT} R_2}{R_1 + R_2}$$

However, we know $V_B = V_X = V_{IN}$, so we can say

$$V_{IN} = \frac{V_{OUT} R_2}{R_1 + R_2}$$

If we divide both sides by the fraction $\frac{R_2}{R_1 + R_2}$ knowing we must invert and multiply, when dividing by a fraction, we get

$$V_{IN} \times \frac{R_1 + R_2}{R_2} = \frac{V_{OUT} R_2}{R_1 + R_2} \times \frac{R_1 + R_2}{R_2}$$

This cancels down to

$$V_{IN} \times \frac{R_1 + R_2}{R_2} = V_{OUT}$$

If we now divide both sides by $V_{IN}$, we get

$$\frac{R_1 + R_2}{R_2} = \frac{V_{OUT}}{V_{IN}}$$

This means that the expression for the voltage gain $V_{GAIN}$ is

$$V_{GAIN} = \frac{V_{OUT}}{V_{IN}} = \frac{R_1 + R_2}{R_2} = \frac{R_1}{R_2} + \frac{R_2}{R_2} = \frac{R_1}{R_2} + 1 = 1 + \frac{R_1}{R_2}$$

This shows that the voltage gain for the non-inverting Opamp can be expressed as

$$V_{GAIN} = 1 + \frac{R_1}{R_2}$$

If we put the values into this expression, we get the value for the circuit shown in Figure 9-18 is

$$V_{GAIN} = 1 + \frac{1E^3}{1E^3} = 2$$

This means that for the circuit shown in Figure 9-18, the output voltage can be calculated as

$$V_{OUT} = V_{IN} \times V_{Gain} = 2V \times 2 = 4V$$

When we simulate the circuit, we see that the output voltage is 4V. As there is no minus sign associated with the expression for the voltage gain, then this means there is no inversion at the output, and the output voltage would be in phase with the input voltage. We can test this further with the test circuit shown in Figure 9-19.

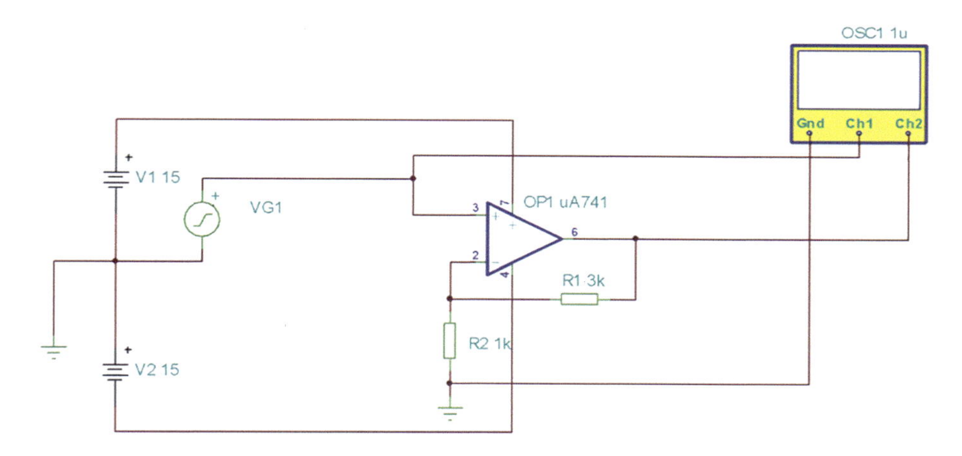


***Figure 9-19.*** *An ac Input Voltage for the Non-inverting Opamp*

We can determine the gain for the circuit using the expression for the gain and putting in the values for the resistors as

$$V_{GAIN} = 1 + \frac{R_1}{R_2} = 1 + \frac{3E^3}{1E^3} = 4$$

The input voltage was set to 20mv peak to peak, and the circuit was simulated. The waveforms displayed were as shown in Figure 9-20.

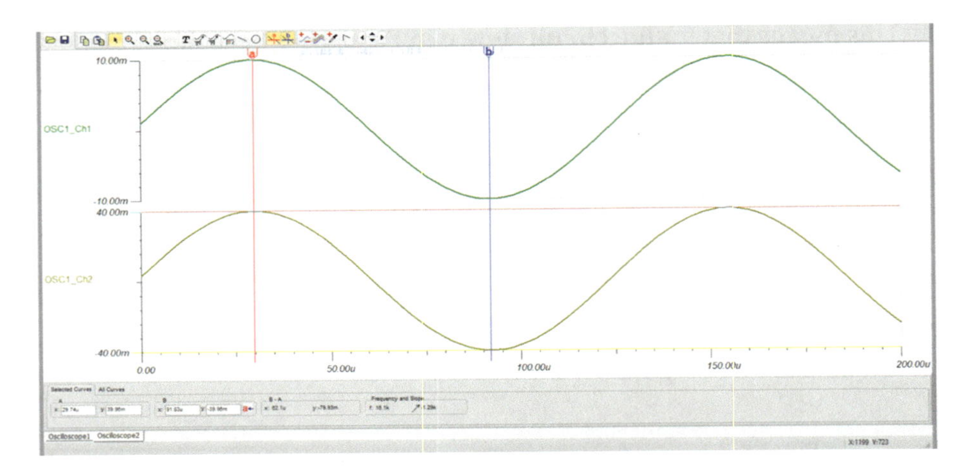


***Figure 9-20.*** *The Waveforms of the Circuit Shown in Figure 9-19*

We can see that channel OSC1_Ch1 shows the input voltage, and channel OSC1_Ch2 shows the output waveform. The peak-to-peak output voltage is 79.93mv, so say 80mv. We can calculate the voltage gain as

$$V_{GAIN} = \frac{V_{Out\ Peak-To-Peak}}{V_{IN\ Peak-To-Peak}} = \frac{80E^{-3}}{20E^{-3}} = 4$$

This confirms what we expect the gain to be.

## The Input Impedance of the Non-inverting Opamp

This is similar to the inverting Opamp except that there is no input resistor to put in parallel with the internal impedance of the Opamp. This means that the input impedance of the non-inverting Opamp is simply set by the internal impedance of the Opamp. We could say that the resistor in the position of $R_2$ (see Figure 9-19) is in series with that internal impedance as it is in the path down to ground for the input. However, a value of infinity plus a small 1k or 100k resistor is still at a value of infinity. This means that the input impedance is ideally at infinity. That is why we can use an Opamp, set up as a non-inverting Opamp, to work as a buffer; see Figure 9-4.

## Exercise 9.2

Design a non-inverting Opamp that has a gain of 5 if the value of the resistor in the position of $R_2$ (see Figure 9-19) was 2kΩ.

Redesign the Opamp if we wanted to use a 15kΩ resistor in the position of $R_1$ (see Figure 9-19), and the gain required was a gain of 4.

## The Operations of the Operational Amplifier

We said earlier that the Opamp got its name because it could carry out a range of mathematical operations. In this section of the chapter, we will look at some of those mathematical operations.

# The Summing Opamp

We will start with an amplifier that can sum a number of inputs. The circuit for this summing Opamp is shown in Figure 9-21.

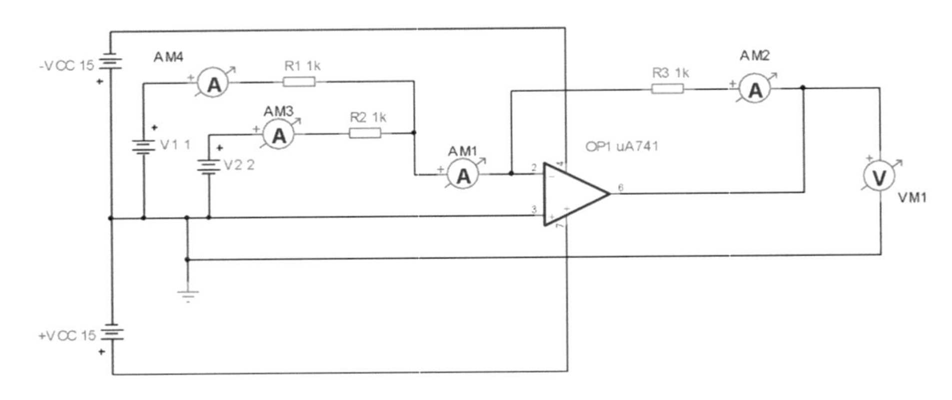


***Figure 9-21.*** *The Basic Summing Opamp*

The purpose of this Opamp circuit is to amplify the sum of the input voltages. We will analyze this circuit to see how it does that.

This is a very similar circuit to the inverting Opamp except that there are now two inputs being applied to pin 2 of the Opamp. We know that whatever current flows toward the Opamp must flow through the resistor in the feedback path, which is $R_3$ in this circuit. We also know that it will flow in the direction from pins 2 toward the output on pin 6. This then means that the output voltage $V_{OUT}$ can be expressed as

$$V_{OUT} = -I_F R_3$$

We also know that $I_F = I_{IN}$. In this circuit, the input current $I_{IN}$ equals $I_{R1} + I_{R2}$. Also, because of the virtual short circuit across the input terminals, the voltage at pin 2 of the Opamp is 0V. This is because the voltage at pin 3 is at ground or 0V. Using Ohm's Law which states

$$I = \frac{V}{R}$$

This really means

$$\text{The current flowing through a component} = \frac{\text{The voltage at one end of the component minus the voltage at the other end}}{\text{Impedance of the component}}$$

Using this interpretation of Ohm's Law, we can say

$$I_{R1} = \frac{V_1 - 0}{R_1} = \frac{V_1}{R_1}$$

If we apply this to the current flowing through $R_2$, we can say

$$I_{R2} = \frac{V_2 - 0}{R_2} = \frac{V_2}{R_2}$$

These two currents both flow into the node at pin 2 of the Opamp, and the feedback current $I_F$ flows out of the node, as no current can flow into the Opamp due to the infinite input impedance of the Opamp. Therefore, using Kirchhoff's Current Law (KCL), we can say

$$I_F = I_{R1} + I_{R2}$$

We know the $V_{OUT} = -I_F R_3$ and so we can say

$$V_{OUT} = -\left(I_{R1} + I_{R2}\right) R_3$$

Substituting the expressions for the two currents, we have

$$V_{OUT} = -\left(\frac{V_1}{R_1} + \frac{V_2}{R_2}\right) R_3$$

If we ensure that $R_1$ has the same value as $R_2$ and we call this $R_{IN}$, we can say

$$V_{OUT} = -(V_1 + V_2)\frac{R_3}{R_{IN}}$$

From this, we can say

$$\frac{V_{OUT}}{V_1 + V_2} = -\frac{R_3}{R_{IN}}$$

As $V_1$ and $V_2$ are the input voltages, then we see that this circuit will be summing the separate input voltages to create the input to the Opamp. Also, we can say the voltage gain of the summing Opamp is

$$V_{Gain} = -\frac{R_3}{R_{IN}}$$

However, this is only true as long as the values of the input resistors are all set to the same value which would be equal to $R_{IN}$.

With respect to the circuit shown in Figure 9-21, we can see that $R_1$ and $R_2$ are both set to 1k, and so we can say $R_{IN}$ = 1k. Then knowing that $R_3$ is also set to 1k, we can say the voltage gain of the Opamp is

$$V_{Gain} = -\frac{R_3}{R_{IN}} = -\frac{1E^3}{1E^3} = -1$$

This means we can calculate the output voltage "$V_{OUT}$" as follows:

$$V_{OUT} = -\left(\frac{V_1}{R_1} + \frac{V_2}{R_2}\right)R_3 = -\left(\frac{1}{1^3} + \frac{2}{1^3}\right)1E^3 = -\frac{3}{1E^3}1E^3 = -3V$$

As $R_1 = R_2 = R_{IN}$, we could have calculated the $V_{OUT}$ as follows:

$$V_{OUT} = -(V_1 + V_2)\frac{R_3}{R_{IN}} = -(1+2)\frac{1E^3}{1E^3} = -3V$$

We can also calculate the value of the currents in the circuit as follows:

$$I_{R1} = \frac{V_1}{R_1} = \frac{1}{1E^3} = 1mA$$

$$I_{R2} = \frac{V_2}{R_2} = \frac{2}{1E^3} = 2mA$$

Knowing

$$I_F = I_{R1} + I_{R2}$$

Therefore

$$I_F = 1mA + 2mA = 3mA$$

When we simulate the circuit shown in Figure 9-21, we can see the results as shown in Figure 9-22.

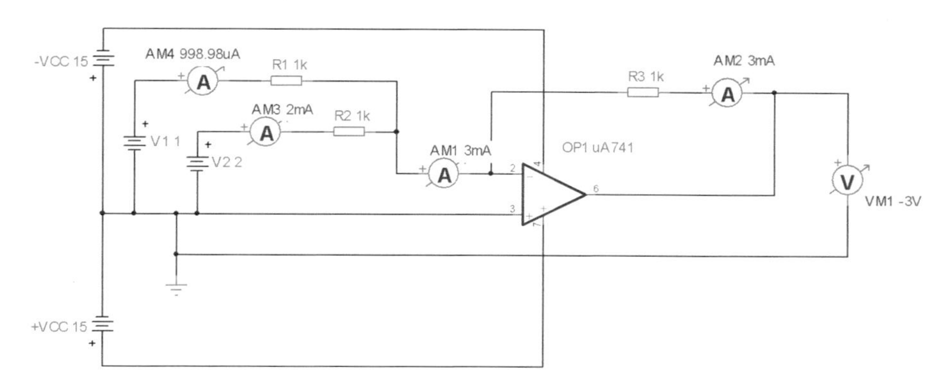


***Figure 9-22.** The Simulated Summing Opamp*

The simulated results are the same as the calculated results, and so they confirm the calculations.

The summing Opamp gives us the ability to vary the gain of the individual inputs. This is when we don't make the input resistors have the same values. This is called a weighted summing Opamp. An example of this is shown in Figure 9-23.

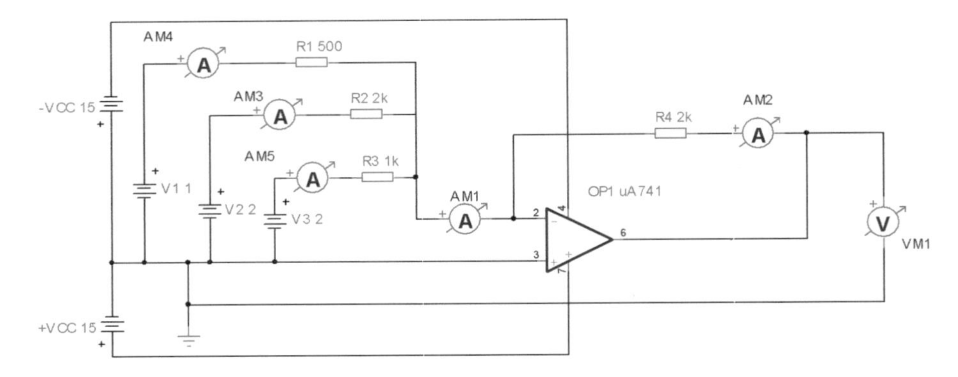


***Figure 9-23.*** *The Weighted Summing Opamp*

The feedback resistor $R_4$ is the multiplier factor of the Opamp. The expression for the output voltage $V_{OUT}$ is

$$V_{OUT} = -\left(\frac{V_1}{R_1} + \frac{V_2}{R_2} + \frac{V_3}{R_3}\right) R_4$$

Putting in the values from the circuit, we have

$$V_{OUT} = -\left(\frac{1}{500} + \frac{2}{2E^3} + \frac{2}{1E^3}\right) 2E^3 = -\left(2E^{-3} + 1E^{-3} + 2E^{-3}\right) 2E^3 = -5E^{-3} 2E^3$$

$$Therefore,\ V_{OUT} = -10V$$

We can calculate the currents flowing around the circuit as follows:

$$I_{R1} = \frac{V_1}{R_1} = \frac{1}{500} = 2mA$$

$$I_{R2} = \frac{V_2}{R_2} = \frac{2}{2E^3} = 1mA$$

$$I_{R3} = \frac{V_3}{R_3} = \frac{2}{1E^3} = 2mA$$

$$I_F = I_{R1} + I_{R2} + I_{R3} = 2mA + 1mA + 2mA = 5mA$$

The simulated results are shown in Figure 9-24.

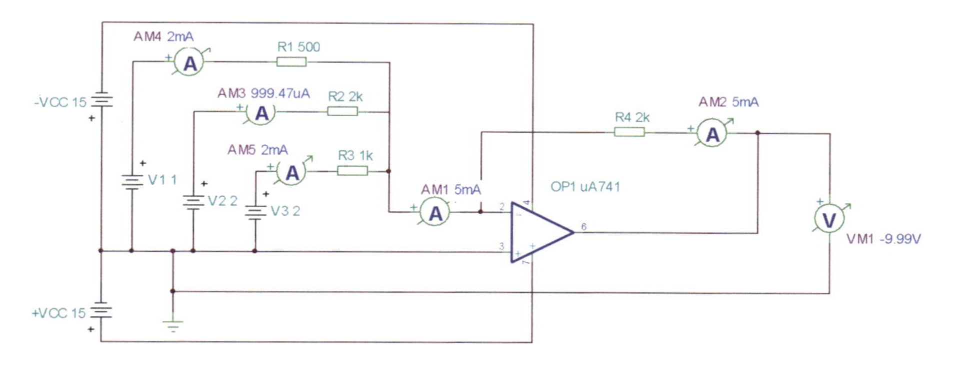


***Figure 9-24.*** *The Simulated Results of the Weighted Summing Opamp*

The simulated results are very close to the calculated results, and so they confirm the expressions are correct.

One application of the summing Opamp could be as the summing junction of an analog control system. The basic block diagram of an analog control system is shown in Figure 9-25.

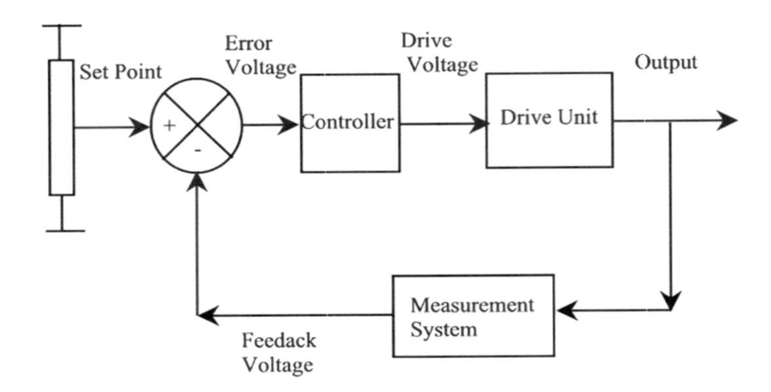


***Figure 9-25.*** *The Block Diagram of a Basic Control System*

The summing junction is the circular block which will sum the set point with the feedback voltage.

# Exercise 9.3

Design a summing Opamp to conform to the following specifications. The supply rail for the Opamps was the normal $\mp 15V$.

1. A summing Opamp has the following three input voltages: V1 = 2V, V2 = 0.5V, and V3 = 1V. The input impedance for each of the inputs should be 5k, and the voltage gain of the Opamp should be -3.

2. A weighted summing Opamp had the same three inputs as those in specification 1. The feedback resistor was chosen to be 9k. The gain for V1 was to be -1, the gain for V2 was to be -3, and the gain for V3 was to be -2.

3. In the design for specification 1, explain what the issue would be if the V1 input voltage was changed to 5V.

# The Differential Amplifier

This is the next operation we will look at that can be implemented with an Opamp. It is probably one of the first applications of the Opamp which followed on from the long-tailed pair circuit as discussed in Chapter 8. The circuit for the differential amplifier is shown in Figure 9-26.

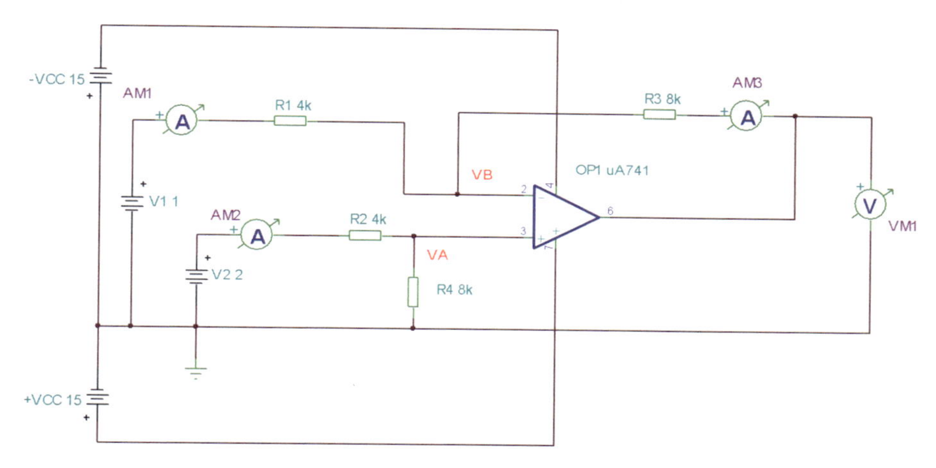


***Figure 9-26.*** *The Differential Amplifier*

The purpose of this Opamp circuit is to amplify the difference between the two input voltages. We will analyze this circuit to see how it does that.

Knowing there is a virtual short circuit across the input terminals means that $V_B$ must equal $V_A$. Using the voltage divider rule, we can calculate the voltage $V_A$ as follows:

$$V_A = \frac{V_2 R_4}{R_2 + R_4}$$

Putting the values from the circuit into the expression, we have

$$V_A = \frac{2 \times 8E^3}{4E^3 + 8E^3} = \frac{16E^3}{12E^3} = 1.333V$$

This means $V_B = 1.333V$.

This means we can now calculate the current flowing through $R_1$, that is, $I_{R1}$, as follows:

$$I_{R1} = \frac{V_1 - V_B}{R_1} = \frac{1 - 1.333}{4E^3} = -83.25\mu A$$

This means that the current $I_{R1}$ is flowing from pin 2 of the Opamp toward the input voltage $V_1$. This is because the voltage $V_B$ is greater than the voltage $V_1$.

We know this current must flow through the feedback resistor $R_3$ as the feedback current $I_F$. This means that

$$I_F = -83.25\mu A$$

This also means that the feedback current must be flowing back from $V_{OUT}$ toward pin 2 of the Opamp. This means that $V_{OUT}$ must be more positive than $V_B$. Using the expression for $I_F$, we can calculate VOUT as follows:

$$I_F = \frac{V_{OUT} - V_B}{R_3}$$

Multiplying both sides by $R_3$, we get

$$I_F R_3 = V_{OUT} - V_B$$

Therefore, we have

$$V_{OUT} = I_F R_3 + V_B$$

Putting the values from the circuit into the expression gives

$$V_{OUT} = 83.25^{-6} 8E^3 + 1.3333 = 1.999V$$

We don't include the minus sign for the current $I_F$ as we are saying the current flows from $V_{OUT}$ toward pin 2 of the Opamp.

The simulated circuit showing the results is shown in Figure 9-27. Note, we have turned the ammeters AM1 and AM3 around to accommodate the direction of current flow.

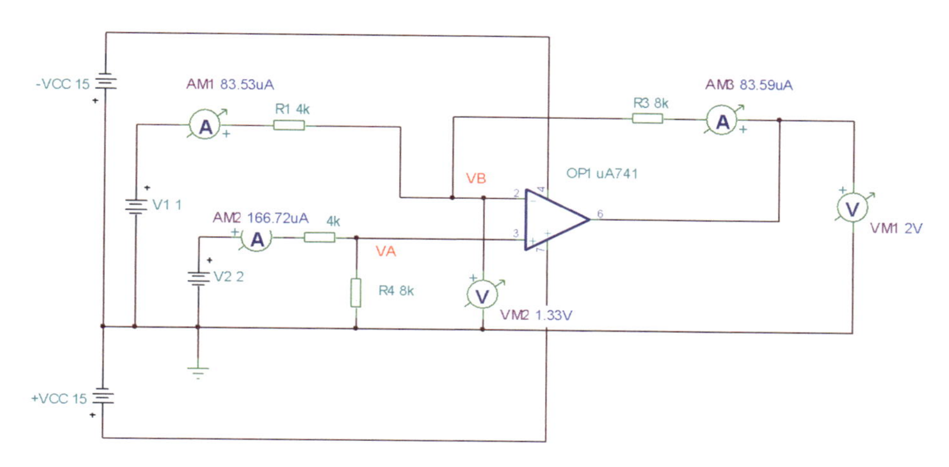


*Figure 9-27. The Simulated Results for the Differential Amplifier*

We need to relate the output voltage to the input voltages $V_1$ and $V_2$. To do this, we use the fact that $I_F = I_{R1}$. The expression for $I_{R1}$ and so $I_F$ is

$$I_{R1} = \frac{V_B - V_1}{R_1} = I_F$$

This assumes that $V_B$ is greater than $V_1$; if it isn't, then the result will be a negative value for $I_F$. This would mean the current would flow in the opposite direction. The direction of the current we are assuming with this expression is that it flows from the direction of $V_{OUT}$ toward $V_B$. Note the orientation of the ammeters AM1 and AM3 in Figure 9-27.

We know $V_B = V_A$ and that the expression for $V_A$ is

$$V_A = \frac{V_2 R_4}{R_2 + R_4} = V_B$$

The feedback current $I_F$ can be expressed as

$$I_F = \frac{V_{OUT} - V_B}{R_3}$$

As $I_F$ is the same as $I_{R1}$, then we can say

$$\frac{V_B - V_1}{R_1} = \frac{V_{OUT} - V_B}{R_3}$$

We can get rid of the two denominators by multiplying both sides by $R_1R_3$. You need to remember you can do any legal mathematical operation so long as you do it to both sides of the equal sign. This gives

$$\frac{(V_B - V_1)R_1R_3}{R_1} = \frac{(V_{OUT} - V_B)R_1R_3}{R_3}$$

This allows the denominators to cancel out as follows:

$$(V_B - V_1)R_3 = (V_{OUT} - V_B)R_1$$

We can now expand the brackets which gives

$$V_BR_3 - V_1R_3 = V_{OUT}R_1 - V_BR_1$$

Adding the $V_BR_1$ term to both sides gives

$$V_BR_3 - V_1R_3 + V_BR_1 = V_{OUT}R_1 - V_BR_1 + V_BR_1$$

This simplifies to

$$V_BR_3 + V_BR_1 - V_1R_3 = V_{OUT}R_1$$

We can take the $V_B$ term out as a common factor on the LHS as follows:

$$V_B(R_3 + R_1) - V_1R_3 = V_{OUT}R_1$$

When we create the circuit, it is normal to let $R_1 = R_2$ and $R_3 = R_4$.

This means we can rewrite this expression as

$$V_B\left(R_4 + R_2\right) - V_1 R_4 = V_{OUT} R_2$$

We have an expression for $V_B$ which is

$$V_B = \frac{V_2 R_4}{R_2 + R_4}$$

Substituting this into the expression, we get

$$\frac{V_2 R_4}{R_2 + R_4}\left(R_4 + R_2\right) - V_1 R_4 = V_{OUT} R_2$$

This now cancels down to

$$V_2 R_4 - V_1 R_4 = V_{OUT} R_2$$

We can take the term $R_4$ out as a common factor which gives

$$R_4\left(V_2 - V_1\right) = V_{OUT} R_2$$

If we divide both sides by $R_2$, we get

$$\frac{R_4}{R_2}\left(V_2 - V_1\right) = V_{OUT}\frac{R_2}{R_2}$$

which cancels down to

$$\frac{R_4}{R_2}\left(V_2 - V_1\right) = V_{OUT}$$

This can be written as

$$V_{OUT} = \left(V_2 - V_1\right)\frac{R_4}{R_2}$$

If we now divide both sides by $(V_2 - V_1)$ and cancel out, we get

$$\frac{V_{OUT}}{V_2 - V_1} = \frac{R_4}{R_2}$$

The term $V_2$-$V_1$ is the differential input and so the expression

$$\frac{V_{OUT}}{V_2 - V_1}$$

This is an expression for the output/input which is the gain of the differential amplifier.

This means the expression for the voltage gain is

$$V_{Gain} = \frac{R_4}{R_2}$$

This expression assumes $R_1$ is the same value of $R_2$, and $R_3$ has the same value as $R_4$.

Using the expression for $V_{OUT}$ and using the values from the circuit shown in Figure 9-26, we can say

$$V_{OUT} = (V_2 - V_1)\frac{R_4}{R_2} = (2-1)\frac{8^3}{4^3} = 2V$$

The voltage gain $V_{GAIN}$ is

$$V_{GAIN} = \frac{R_4}{R_2} = \frac{8^3}{4^3} = 2$$

To reflect this and the fact that $R_1 = R_2$ and $R_3 = R_4$, we can simulate the circuit shown in Figure 9-27.

The simulated results confirm that $V_{OUT}$ = 2V. Also, the $V_A$ voltage, VM3, and $V_B$ voltage, VM2, are the same at 1.33V.

To change the gain of the differential amplifier, we must change the value of $R_4$ and $R_2$, but at the same time, we must ensure that $R_3$ has the same value of $R_4$ and $R_1$ has the same value of $R_2$. This assumes that the numbered resistors are in the positions as shown in Figure 9-27. As an example, to create a gain of 5, we can set $R_4$ and $R_3$ to 10k and keep the value of $R_2$ and $R_1$ at 2k.

## Exercise 9.4

Design a differential amplifier that has a gain of 6 when the two input resistors $R_1$ and $R_2$ were 3k. Calculate what the expected output voltage would be if $V_2$ was 250mv and $V_1$ was 100mV. Calculate the current $I_F$.

# A Useful Application of the Differential Amplifier

In Chapter 8, we looked at using the basic differential amplifier to amplify the output of a temperature transducer placed in a Wheatstone bridge arrangement. The Opamp is essentially an improved differential amplifier. The open-loop gain of the Opamp has been increased to around $10E^5$ to $1E^6$. This next part of this chapter will look at using the Opamp to amplify the output of a transducer in that arrangement. This arrangement is shown in Figure 9-28.

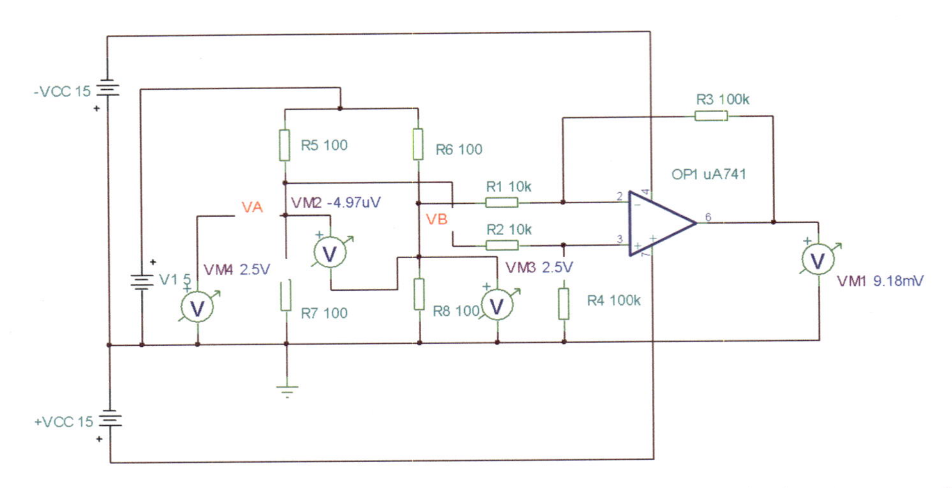


***Figure 9-28.** The Differential Amplifier Used to Amplify the Output of a Temperature Transducer*

The temperature transducer used in this application is a typical PT100, and it is represented by $R_6$ in the circuit. The Wheatstone bridge arrangement is the four resistors: $R_5$, $R_6$, $R_7$, and $R_8$. They are all precision resistors, except R6 which is the PT100, each with a value of 100Ω. This is because, at 0°C, the resistance of the PT100 is 100Ω. This means that, at 0°C, both pairs of resistors in the bridge, R5 with R7 and R6 with R8, form a voltage divider network with the 5V supply, $V_1$. This means that at 0°C the voltage at "$V_A$" and "$V_B$" should be equal at 2.5V. This would mean that the difference voltage would be 0V. However, the voltmeter VM2, which will measure this voltage difference, is showing a value of –4.97μV. This is due to the internal circuitry of the Opamp creating this error voltage. This could mean that we would need to apply an offset voltage to the Opamp to zero this error voltage. It does depend upon how accurate you want the temperature measurement to be.

The PT100 is a resistive temperature detector which has a positive coefficient. This means its resistance increases when the temperature increases. The relationship of the RTD is

$$R = R_0\left(1 + \alpha T\right)$$

where $R_0$ is the resistance at 0°C, which for the PT100 is 100Ω. The term "T" is the temperature in °C, and the term "α" is the temperature coefficient for the RTD. In reality, there are more temperature coefficients in the expression, but the others are so small they can be ignored. A typical value for "α" would be 25mΩ per °C rise in temperature. This would mean that at a temperature of 50°C, the resistance of the RTD, $R_6$, would be

$$R_6 = 100\left(1 + 0.025 \times 50\right)$$

Therefore

$$R_6 = 100\left(1 + 1.25\right) = 225\Omega$$

This would change the voltage at point $V_B$ as the expression for the voltage at $V_B$ is

$$V_B = \frac{V_1 R_8}{R_6 + R_8}$$

Putting the values in, we get

$$V_B = \frac{5 \times 100}{225 + 100} = 1.538V$$

This means that the voltage difference would be

$$V_{DIFF} = V_A - V_B = 2.5 - 1.538 = 961.54mV$$

Note, $V_A$ will stay constant at half of $V_1$, that is, 2.5V.

Knowing that, the gain of the amplifier is set by

$$V_{Diffgain} = \frac{R_4}{R_2} = \frac{100E^3}{10E^3} = 10$$

Therefore, with a $V_{Diff}$ of 961.54mV, the output voltage would be

$$V_{Out} = V_{Diff} \times 10 = 961.54E^{-3} \times 10 = 9.6154V$$

If we change $R_6$ to 225Ω and simulate the circuit, we see that the output voltage is 9.56V, very close to the calculated value.

## Exercise 9.5

Using the same value for the PT100 and the components in the circuit shown in Figure 9-28, determine the output voltage of the amplifier when

1. T = 12°C.
2. T = 75°C.

There are two issues with this use of the differential amplifier. The first is that you must ensure the differential amplifier does not load the Wheatstone bridge. This is because we are putting the differential amplifier in parallel with the Wheatstone bridge. With respect to the circuit in Figure 9-28, we can see that the input connected to $R_1$ would not be loaded by the Opamp as the path to ground would put $R_1$ in series with the infinite internal impedance of the Opamp itself and $R_4$. However, the other input connected to $R_2$ has only the value of $R_4$ in series with it to the path to ground. This means that the $V_A$ input has the voltage divider network as shown in Figure 9-29.

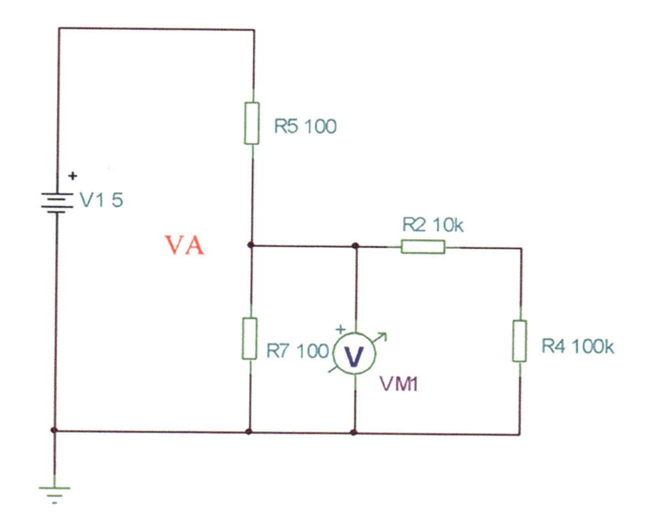


***Figure 9-29.** The Loading Circuit of the Opamp for the Voltage at VA*

In this circuit, we can see that $R_7$ is in parallel with the series combination of $R_2$ and $R_4$. We can calculate the value of $R_7$ that accounts for the loading effect of $R_2$ and $R_4$ using

$$R_7 = \frac{R_7\left(R_2 + R_4\right)}{R_7 + R_2 + R_4}$$

Putting the current values in, we get

$$R_7 = \frac{100\left(10E^3 + 100E^3\right)}{100 + 10E^3 + 100E^3} = 99.91\Omega$$

Not much loading, but $R_4$ is 100k, not a small value of resistance.

# Exercise 9.6

Show that if $R_2$ was 1k and $R_4$ was 10k, which gives the same voltage gain, remember we must make $R_2 = R_1$ and $R_3 = R_4$, the voltage $V_A$ would reduce to 2.489V. Not much difference, but then an error of 11.31mV would produce an error in the temperature reading of 0.363°C.

The second issue with using the circuit as shown in Figure 9-28 is not to do with the Opamp, but it is that this arrangement of the RTD in the Wheatstone bridge would not produce a linear response. This would also introduce some small errors, and so some RTDs have an extra linearizing input to try and compensate for this. But that is really for a different book.

# The Integrating Opamp

The next Opamp circuit we are going to look at shows more clearly why the Opamp is an Operational Amplifier. The output of the circuit would be the integration of the input. The circuit for the integrator is shown in Figure 9-30.

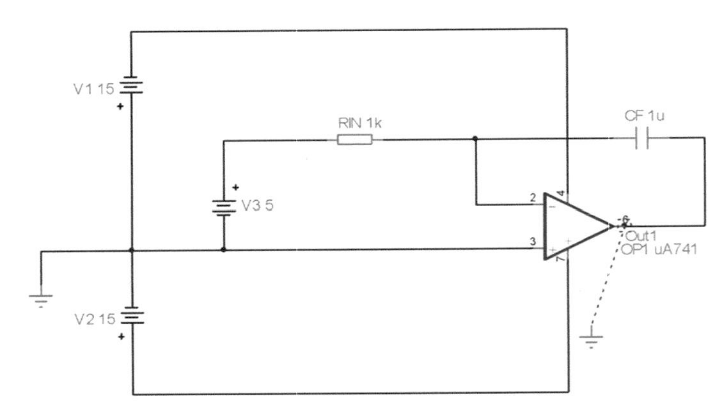


***Figure 9-30.*** *The Integrating Opamp*

It actually carries out an integration of the signal presented at its input. To appreciate how the integrating Opamp works and what it is trying to achieve, it would be useful to recap on what integration is. Integration is not just a mathematical operation dreamed up to make students suffer. It has real meaning, like everything else we engineers do.

Integration is the act of determining the area under a graph or curve for the expression that describes the quantity we are looking at. The best way to understand what integration is would be to look at an electrical quantity. We will consider a DC current that is a constant at 5mA for all time. We can describe the quantity as

$$I = 5mA$$

Using this description, we can draw a graph or curve of the expression for the current "I." The graph would simply be a straight line as shown in Figure 9-31.

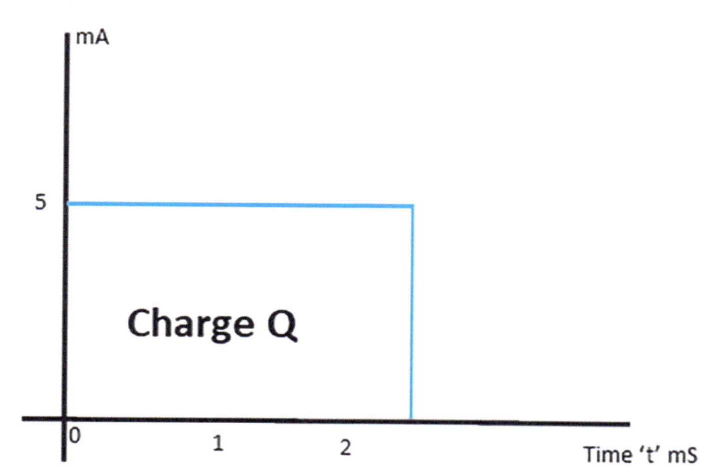


***Figure 9-31.** The Graph Showing a Constant Current of 5mA for 2.5mS*

The graph shows a 5mA current applied for approximately 2.5ms. The area under the graph or curve of the blue line is simply the area of the rectangle, and this can be determined as Area = 5t, that is, 5 times t. If you have done integration before, then you should know that if you integrate 5 with respect to "t," you get the result 5t.

The preceding example should help to make it clear what integration is. However, it does two more things; firstly, it shows that the integral of a constant produces a straight-line graph that starts at zero. This is shown in Figure 9-32.

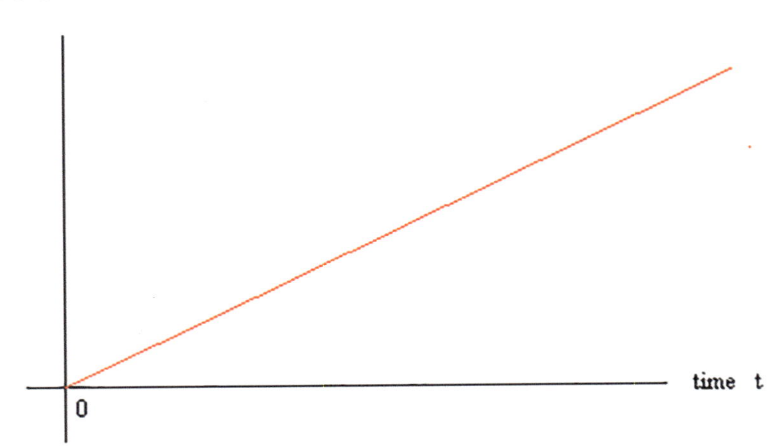


***Figure 9-32.*** *The Graph of the Charge Q = 5t*

The graph is correct because when t = 0, then 5 x t = 5 x 0 = 0, and so the area, A, is 0. Then as time goes on, the result of 5t increases linearly to produce the straight-line graph. This really shows that the area under the curve is growing linearly, from 0, as time moves on.

We know that when we force current into a device, such as a capacitor, or a battery, we make charge, "Q" in Coulombs, flow into it. Even if the current is not charging anything up, the flow of current is the movement of charge. This should help to appreciate that the area under the curve of the current, shown in Figure 9-31, is actually the amount of charge being moved. When t = 0, no charge has been moved, but as time goes on, the charge being moved must increase. Therefore, the integral of the current is the charge "Q."

Knowing this, we can say that

$$Q = \int i \ dt$$

There is also another standard expression that relates charge "Q" to the voltage across a capacitor, and that is

$$Q = CV$$

where C is the capacitance and V is the voltage across the capacitor.

We can use the preceding two expressions to derive an expression for the output voltage of the integrating circuit shown in Figure 9-30. We know from our analysis of the inverting Opamp that the input current that flows through $R_1$ must flow around the feedback path through the capacitor $C_1$, as no current flows into the Opamp. We also know that, as there is a virtual short circuit across the input terminals of the Opamp, and that pin 3 is at ground, that is, 0V, the volt drop across RIN is VIN - 0. This means the expression for the input current is

$$I_{IN} = \frac{V_{IN} - 0}{R_{IN}} = \frac{V_{IN}}{R_{IN}}$$

To derive the expression for the feedback current, "$I_F$", we need to appreciate that the current flowing into a capacitor will push charge into it and so charge the capacitor up. This means we can use the expression for charge "Q" which is

$$Q = \int I_F \, dt$$

It is not easy to measure the charge in a capacitor, but we can measure the voltage across the capacitor as it charges up. The relationship we can use for this is

$$Q = CV_C$$

where $V_C$ is the voltage across the capacitor. The two expressions for charge "Q" must equal each other. Therefore, we can say

$$C_F V_C = \int I_F \, dt$$

If we multiply both sides by $\frac{1}{C_F}$, we will remove the "$C_F$" from the LHS. This is shown as follows:

$$\frac{1}{C_F} C_F V_C = \frac{1}{C_F} \int I_F \, dt$$

The two "$C_{Fs}$" cancel out which gives

$$V_C = \frac{1}{C_F} \int I_F \, dt$$

However, we know

$$I_F = I_{IN} = \frac{V_{IN}}{R_{IN}}$$

Therefore, we can say

$$V_C = \frac{1}{C_F} \int \frac{V_{IN}}{R_{IN}} \, dt$$

We can take the $\frac{1}{R_{IN}}$ out of the integral as it is a constant. This then gives

$$V_C = \frac{1}{C_F R_{IN}} \int V_{IN} \, dt$$

The voltage "$V_C$" is the same as the output voltage "$V_{OUT}$" except that to maintain the direction of the current the output voltage must be negative. Therefore, the expression for $V_{OUT}$ is

$$V_{OUT} = -\frac{1}{C_F R_{IN}} \int V_{IN} \, dt$$

This confirms that the output voltage is the integral of the input voltage. However, the output is tempered by the term $-\frac{1}{C_F R_{IN}}$. This will invert the integral output and slow its rise down. If the input voltage "$V_{IN}$" was a constant, then the output would be a straight line, as shown in Figure 9-32, but the gradient would be negative, and the gradient would be controlled by the $\frac{V_{IN}}{C_F R_{IN}}$ term. If we use a transient analysis to study the output of the circuit shown in Figure 9-30, we would get the display shown in Figure 9-33.

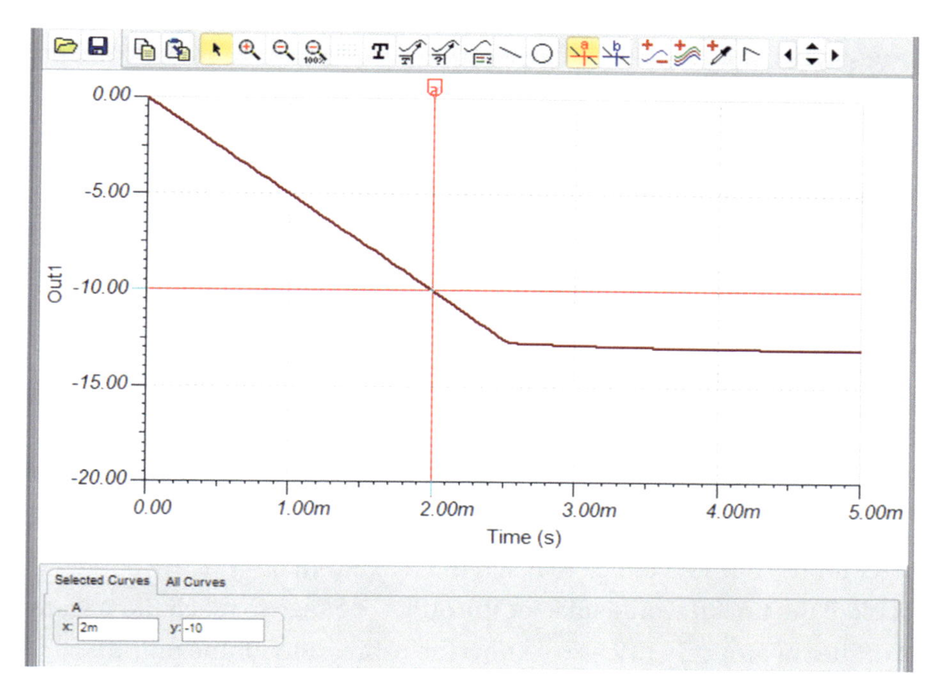


***Figure 9-33.*** *The Output Voltage of the Integrating Opamp*

The transient analysis only looks at the first 5ms after switching the circuit on. This is because, approximately 2.5ms after turning the circuit on, the output voltage has reached its maximum output of around 13V. However, from 0 to 2.5ms, the output is linear with an expression that is

$$V_{OUT} = -\frac{V_{IN}}{C_F R_{IN}} t = -\frac{5}{1E^{-6} \times 1E^{3}} t = -5E^{3} t$$

Using this expression, we can determine the output voltage when t = 1ms and t = 2ms. This would give the following:

When t = 1ms:

$$V_{OUT} = -5E^{3} \times 1E^{-3} = 5V$$

and

when t = 2ms:

$$V_{OUT} = -5E^{3} \times 2E^{-3} = 10V$$

This can be confirmed by inspection of the output voltage in Figure 9-33. When t = 1ms, we can see that the voltage is 5V as expected and 10V when t = 2ms.

To confirm the relationship for the gradient, we will change the resistor to 5kΩ. This would reduce the gradient which means it would take longer for the output voltage to reach the maximum of –13V. We could calculate the time it would take for the output voltage to reach the negative maximum of around –13V as we know the expression for the output voltage would now be

$$V_{OUT} = -\frac{V_{IN}}{C_F R_{IN}} t = -\frac{5}{1E^{-6} \times 5E^{3}} t = -1E^{3} t$$

This expression can be rearranged for time "t" as follows:

$$t = \frac{V_{OUT}}{-1E^{3}} = \frac{-13}{-1E^{3}} = 13ms$$

We can now simulate the circuit extending the transient analysis to

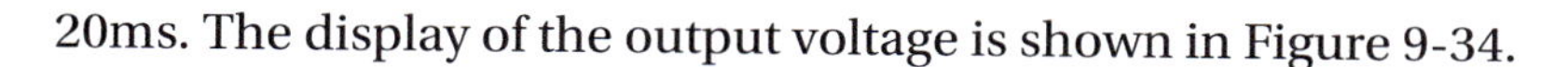

20ms. The display of the output voltage is shown in Figure 9-34.

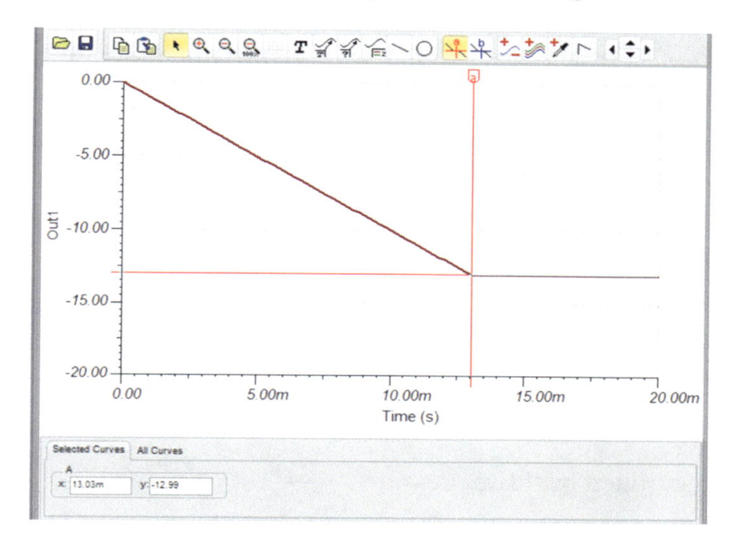


***Figure 9-34.*** *The Output Voltage When RIN = 5k*

We can see from Figure 9-34 that it now takes 13.03ms for the output voltage to reach −12.99V. This agrees with our calculations and so confirms the theory of the integrating Opamp.

# Exercise 9.7

Determine the gradient and the output voltage after 250μs and 300μs if the following changes to the circuit shown in Figure 9-30 were made:

VIN changed to 7.5V

R1 changed to 4.7k

C1 changed to 100nF

# A Sawtooth Generator

When we apply a square wave signal, we can turn the integrating Opamp into a sawtooth generator. This is basically because with the square wave we are applying two DC signals to the integrator. We will call the period that the square wave is positive as $D_{C1}$ and when it is negative $D_{C2}$. Assuming that the square wave is a basic one which has a 50/50 duty cycle and has a peak value of 1V, then we can say $D_{C1} = +1V$ and $D_{C2} = -1V$. We know that when we integrate a constant, the expression we will get is that of a straight line. This means that during the time the square wave is at $D_{C1}$, the output voltage will be $-D_{C1}t$, and during the time the square wave is at $D_{C2}$, then the output would be $-D_{C2}t$. However, the gradient of the output would be controlled by the expression $\frac{1}{C_{IN}R_F}$.

We will test the circuit by applying a 50/50 square wave with a peak voltage of 1V at a frequency of 50Hz. This means the time period for each of the linear outputs would be 10ms. We will use the test circuit as shown in Figure 9-35.

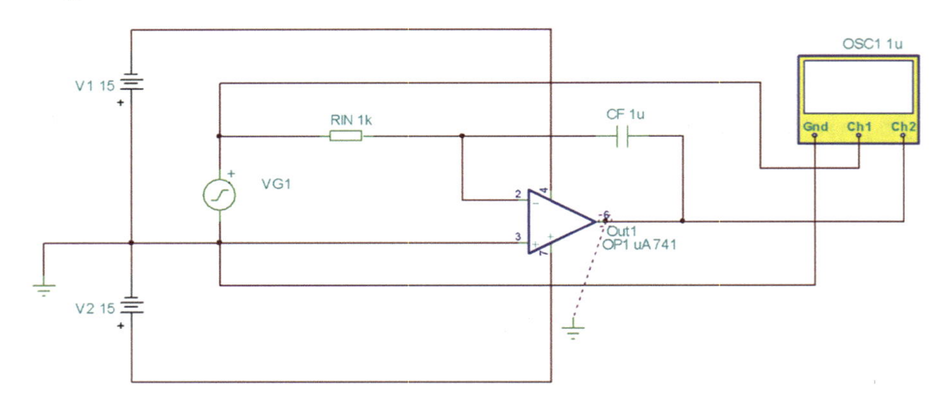


***Figure 9-35.*** *The Test Circuit for the Integrating Opamp with a Square Wave Input*

We have already shown that the expression for the output voltage will be

$$V_{OUT} = -\frac{V_{IN}}{C_F R_{IN}} t$$

Therefore, during the first time period when the square wave was at DC1, the expression for VOUT is

$$V_{OUT} = -\frac{D_{C1}}{C_F R_{IN}} t$$

This will exist for half of the periodic time.

During the second time period, the expression would be

$$V_{OUT} = -\frac{-D_{C2}}{C_F R_{IN}} t$$

Putting in the values from the test circuit, during the first period we get

$$V_{OUT} = -\frac{1}{1E^{-6} \times 1E^{3}} t = -1E^{3} t$$

During the second period, the output voltage would be

$$V_{OUT} = -\frac{-1}{1E^{-6} \times 1E^{3}} t = 1E^{3} t$$

This shows that during each period, the output would swing 10V as

$$V_{OUT} = -1E^{3} t = -1E^{3} \times 10E^{-3} = -10$$

This means the output voltage would start off with a negative gradient going from t = 0 to t = 10ms. Then the output would change to a positive gradient going from t = 0 to t = 10ms; note, at each change, the respective time "t" must start again. Ideally, this sawtooth waveform would settle with a middle voltage of 0v, which would mean the waveform would go from +5 down to –5 and back to +5 at a frequency of 50Hz. However, when we simulate the circuit, the waveforms produced are as shown in Figure 9-36.

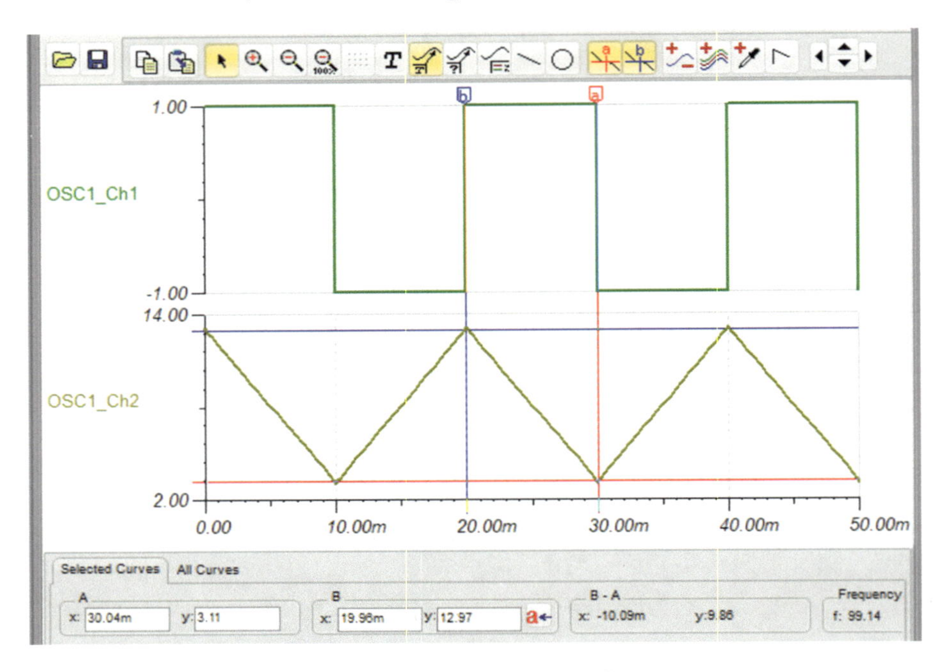


***Figure 9-36.*** *The Waveforms of the Sawtooth Generator*

The output voltage is indeed a sawtooth as required, but it swings from 13V down to 3V, a 10V swing but not around 0V. This is due to the internal error voltage producing a steady DC input that sends the output toward the voltage rails.

Before we look at a possible solution, we need to consider the values of the capacitor $C_F$ and the resistor $R_{IN}$. If we make these values too small, then the gradient of the expression could be so steep that the output voltage would reach the supply rails before the square wave changed its polarity. Indeed, if we want to create a sawtooth that has a 50/50 ratio, then the square wave input must have a 50/50 duty cycle, and the gradient of the output should be such that the output voltage would not reach the supply rails in half of the periodic time of the square wave.

To help explain this, we will use a square wave input of 2V peak at a frequency of 1kHz. The periodic time for a 1kHz waveform is 1ms. Therefore, half the periodic time will be 500μs. Knowing that, the expression for the output voltage is

$$V_{OUT} = -\frac{D_{C1}}{C_F R_{IN}} t$$

You need to decide what pk-to-pk value you want the waveform to move through. Then knowing the output will swing through this pk-to-pk value in half the periodic time of the input frequency, we can derive an expression for the CR product as

$$C_F R_{IN} = \frac{D_{C1}}{(pk - to - pk)2f}$$

We know that $D_{C1}$ is 2V at a frequency of 1kHz, and the pk-to-pk of the output is 10V; therefore, putting the values in, we get

$$C_F R_{IN} = \frac{2}{(10)2 \times 1^3} = 100E^{-6}$$

If we use a 1k resistor for $R_{IN}$, the value of the capacitor would be 100nF. If we change the input and capacitor values for the circuit shown in Figure 9-35 and simulate the circuit, the waveforms obtained are shown in Figure 9-37.

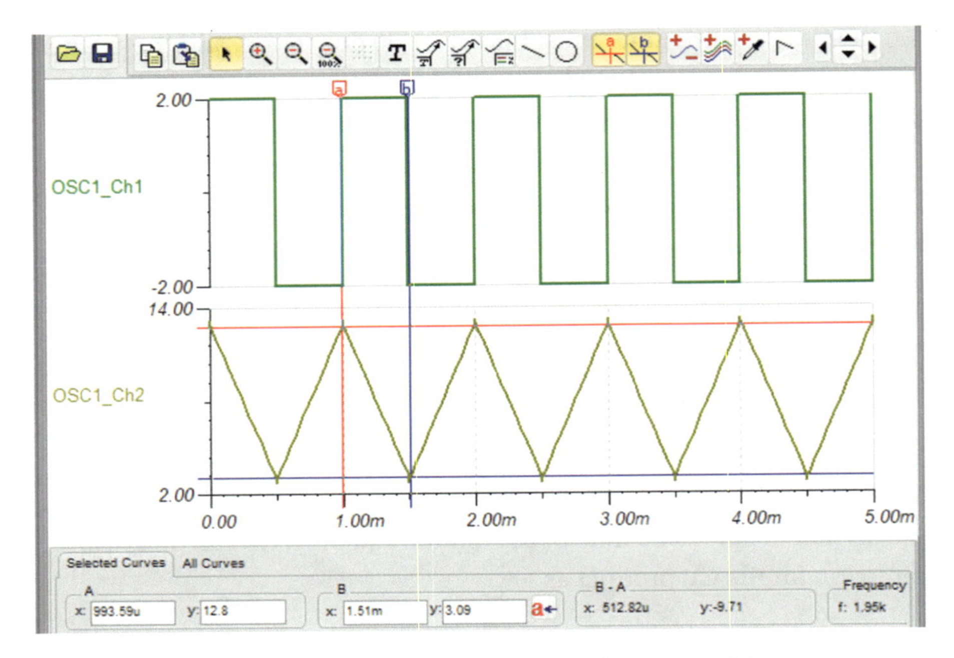


***Figure 9-37.*** *The Waveforms with a 2V Peak Square Wave at a Frequency of 1kHz*

The channel 2 trace shows the output voltage, and it is indeed the sawtooth waveform we expect with a pk-to-pk voltage of approximately 10V as required. However, we can see that it is not centered around the 0V line, but it has been raised up to be centered around the 8V line. This is because the internal error voltage that is created by the Opamp itself represents a DC input that will be integrated by the Opamp, producing a rise that takes the waveform up to this false DC level. We could try using an offset voltage to try and counteract this error voltage, but an alternative approach might be to modify the feedback circuit as shown in Figure 9-38.

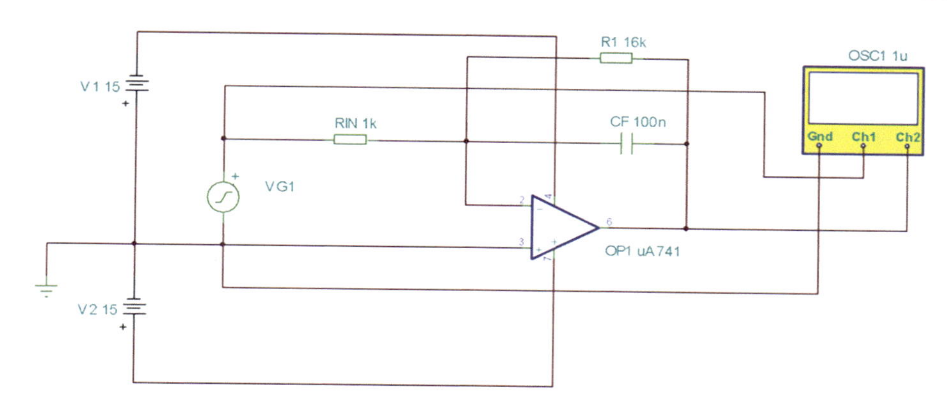


***Figure 9-38.*** *The Feedback Circuit Changed to Include the Resistor R1*

When we simulate the circuit, the waveforms obtained were as shown in Figure 9-39.

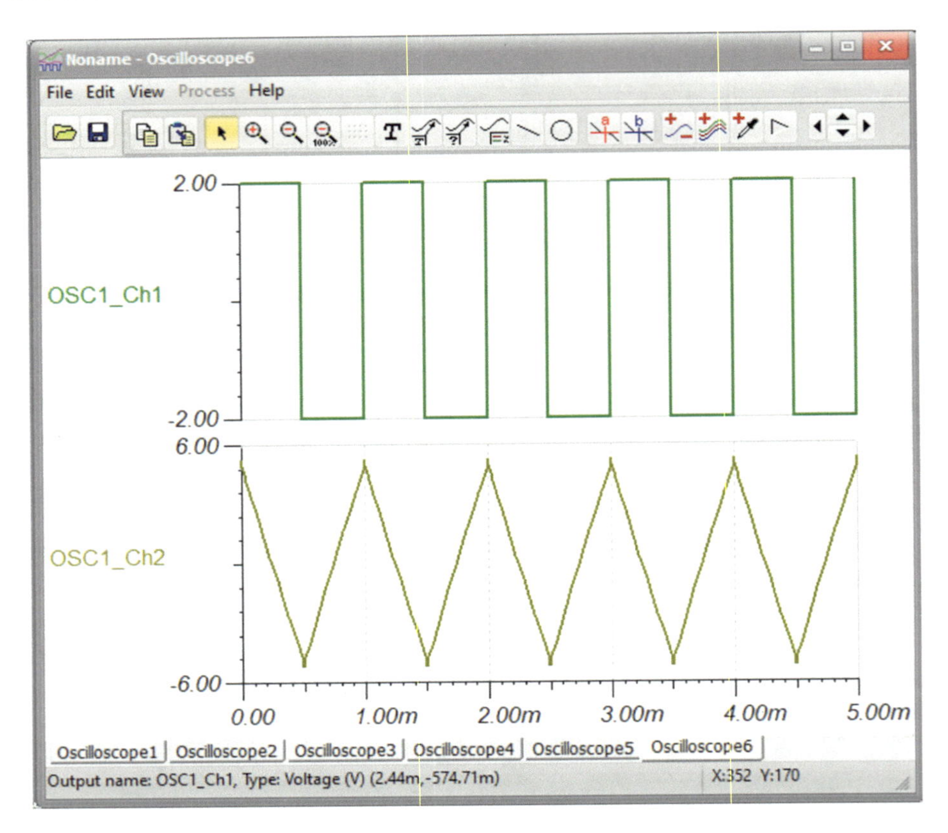


***Figure 9-39.*** *The Waveforms for the Modified Integrating Opamp*

Channel 2 shows us that the pk-to-pk value of the output is still the 10V we wanted, but now it is centered around the 0V line, which is much better. To understand why this has happened and how we have decided upon the value of 16k for $R_1$, we need to look at the feedback path.

There are basically two paths now, one through the capacitor and one through the resistor $R_1$. This is because the Opamp has two signals at its input, one being of an ac signal, that is, the square wave input, and the other being a DC signal, that is, the error voltage from the Opamp itself. As one feedback path includes a capacitor, then, as the frequency of DC is 0Hz, the capacitor will block DC. This means that the DC current must pass through the resistor $R_1$. This means that to DC the Opamp acts like an inverting Opamp, and the gain of the Opamp to any DC signal is

$$DC_{Gain} = -\frac{R_1}{R_{IN}}$$

Using the values in the circuit, this would mean

$$DC_{Gain} = -\frac{30E^3}{1E^3} = -30$$

This means that the DC output would simply be

$$DC_{VOUT} = -30DC_{IN} = -30 \times -1.06E^{-3} = 31.8mV$$

This then would lift the output waveform from swinging around 0V to swinging around 31.8mV. We would not really notice this slight error.

However, we know from before that if the resistor $R_1$ was not there, the DC output would be a linear expression equal to Vt where "V" is the DC voltage. This would mean the DC output would simply rise up toward the supply rails.

The ac voltage applied to the circuit shown in Figure 9-39 would flow through the capacitor as normal, and so the ac output voltage would be the integral of the ac input voltage.

In determining the value of the resistors and the capacitor, we have already shown how to determine the value of $R_{IN}$ and the capacitor, $C_F$. The process in determining the value of the DC feedback resistor $R_1$ is to make the resistance value ten times the value of the impedance of $C_F$ at the frequency of the input waveform. We can calculate the impedance of the capacitor using

$$X_C = \frac{1}{2\pi fC} = \frac{1}{2\pi \times 1E^3 \times 100E^{-9}} = 1.591k$$

This assumes we know the frequency of the input signal. Therefore, the value of $R_2$ will be 10 times 1.59k, so use a 16k resistor as shown in Figure 9-38.

## Exercise 9.8

Design a sawtooth generator that uses a 5kHz square wave with a peak voltage of 5V. The pk-to-pk voltage of the sawtooth should be 6V. The capacitor must be 100nF as the company has hundreds of them spare. The design should employ a resistor in the feedback path to ensure the output centers around 0V.

# The ac Voltage Gain of the Integrator Circuit

We will use the test circuit as shown in Figure 9-38 to determine the voltage gain relationship for the integrator circuit.

We can combine the two impedances in the feedback path into one parallel combination which we will call $Z_F$. Using the product over sum rule, we can combine the two impedances as

$$Z_F = \frac{R_1 \dfrac{1}{j\omega C_F}}{R_1 + \dfrac{1}{j\omega C_F}} = \frac{\dfrac{R_1}{j\omega C_F}}{R_1 + \dfrac{1}{j\omega C_F}}$$

If we divide the top and bottom by the numerator $\dfrac{R_1}{j\omega C_F}$, we get

$$Z_F = \frac{1}{\dfrac{R_1}{\dfrac{R_1}{j\omega C_F}} + \dfrac{\dfrac{1}{j\omega C_F}}{\dfrac{R_1}{j\omega C_F}}} = \frac{1}{\dfrac{R_1}{1}\dfrac{j\omega C_F}{R_1} + \dfrac{1}{j\omega C_F}\dfrac{j\omega C_F}{R_1}} = \frac{1}{\dfrac{j\omega C_F}{1} + \dfrac{1}{R_1}} = \frac{1}{\dfrac{1 + j\omega C_F R_1}{R_1}}$$

Therefore, we have

$$Z_F = \frac{R_1}{1 + j\omega C_F R_1}$$

The output voltage $V_{OUT}$ is simply this impedance multiplied by the feedback current $I_F$. However, we know $I_F = -I_{IN}$ and $I_F = -\frac{V_{IN}}{R_{IN}}$.

This means the expression for $V_{OUT}$ is

$$V_{OUT} = -\frac{V_{IN}}{R_{IN}} \frac{R_1}{1 + j\omega C_F R_1}$$

If we divide both sides by $V_{IN}$, we get

$$\frac{V_{OUT}}{V_{IN}} = -\frac{1}{R_{IN}} \frac{R_1}{1 + j\omega C_F R_1} = -\frac{R_1}{R_{IN}} \frac{1}{1 + j\omega C_F R_1}$$

This means the expression for the ac voltage gain is

$$V_{acGAIN} = -\frac{R_1}{R_{IN}} \frac{1}{1 + j\omega C_F R_1}$$

This is basically the DC voltage gain modified by the ac transfer function. The transfer function is

$$\frac{1}{1 + j\omega C_F R_1}$$

If we convert this to polar format and carry out the division, we get

$$TF = \frac{1}{\sqrt{1^2 + (\omega C_F R_1)^2}} \langle -Tan^{-1}(\omega C_F R_1) \rangle$$

The complete expression for the ac voltage can now be expressed as

$$V_{acGAIN} = -\frac{R_1}{R_{IN}}\left(\frac{1}{\sqrt{1^2 + (\omega C_F R_1)^2}}\langle -Tan^{-1}(\omega C_F R_1)\rangle\right)$$

This can be split into two parts, the magnitude and the phase. The magnitude is

$$V_{acGAIN\ Magnitude} = \frac{R_1}{R_{IN}}\left(\frac{1}{\sqrt{1^2 + (\omega C_F R_1)^2}}\right)$$

There is no need to include the minus sign as magnitude has no sign. The phase expression is

$$phase = -\frac{R_1}{R_{IN}}\left(\langle -Tan^{-1}(\omega C_F R_1)\rangle\right)$$

We can evaluate these two expressions when f = 0 and f = infinity. When f = 0, we know

$$\omega C_F R_1 = 0$$

When f = infinity, we know

$$\omega C_F R_1 = inifinity$$

With these values we can work out the magnitude when f = 0 and when f = infinity.

When f = 0

$$V_{acGAIN\ Magnitude} = \frac{R_1}{R_{IN}}\left(\frac{1}{\sqrt{1^2 + (0)^2}}\right) = \frac{R_1}{R_{IN}} = \frac{16^3}{1^3} = 16$$

When f = infinity

$$V_{acGAIN\ Magnitude} = \frac{R_1}{R_{IN}}\left(\frac{1}{\sqrt{1^2+(\infty)^2}}\right) = \frac{R_1}{R_{IN}}\frac{1}{\infty} = \frac{R_1}{R_{IN}} \times 0 = 0$$

We are really only interested in the magnitude of the gain in this chapter. We will be looking more closely at transfer functions and how we can use complex numbers to derive the transfer function in Chapter 11 when we look at filters.

There is one more frequency we need to consider when we use the transfer function, and this is the frequency at which the gain falls by 3dbs. This is termed the cutoff frequency, and it will occur when the "j" term in the transfer function equals 1. We can use this relationship to determine the frequency of cutoff using

$$\omega C_F R_1 = 1$$

As ω is just a shorthand way of writing 2πf, then we can say

$$2\pi f C_F R_1 = 1$$

From this, we can say

$$f_{CUTOFF} = \frac{1}{2\pi C_F R_1}$$

Putting the values in from the circuit, we get

$$f_{CUTOFF} = \frac{1}{2\pi \times 100E^{-9} \times 16^3} = 99.47Hz$$

We can carry out a frequency analysis on the circuit shown in Figure 9-38 and create a Bode plot. This will show how the gain in dBs varies against the log of the frequency and so allow us to confirm our calculations. We have to express our calculated gains in dBs, which we do using

$$V_{Gain} = 20log(gain)\ dbs$$

Putting our calculated value of 16 in, we get

$$V_{Gain} = 20log(16)\ dbs = 24.08\ dBs$$

We know at the cutoff frequency the gain will have fallen by 3dBs; this will be explained in Chapter 11. This means that when f = 99.47Hz, the gain should be 21.08dBs. When we simulated the circuit, the Bode plot that was produced is shown in Figure 9-40.

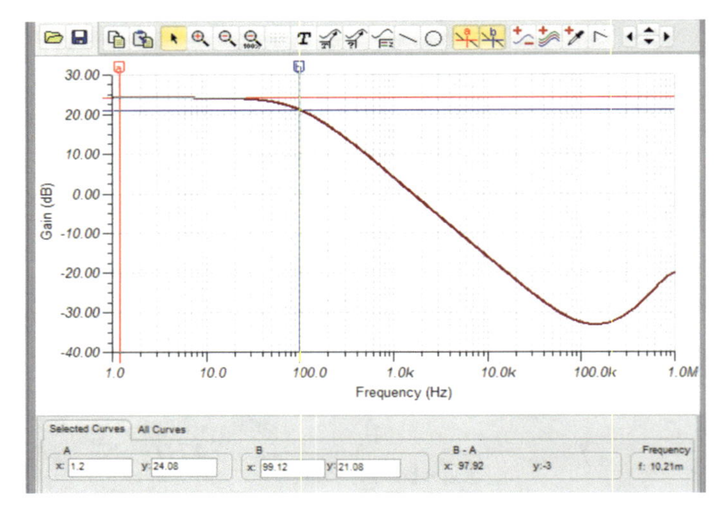


***Figure 9-40.** The Bode Plot for the ac Integrating Opamp*

Using that Bode plot, we can see that cursor “a” shows the gain at the lowest frequency is 24.08dBs. Also, at a frequency of 99.12Hz, the gain has fallen to 21.08 dBs. These two readings are the same as our calculations, and so they should confirm our analysis.

When we applied a sine wave input of 20mV at a frequency of 1kHz, we used the oscilloscope to compare the output voltage with the input. The displays from the oscilloscope are shown in Figure 9-41.

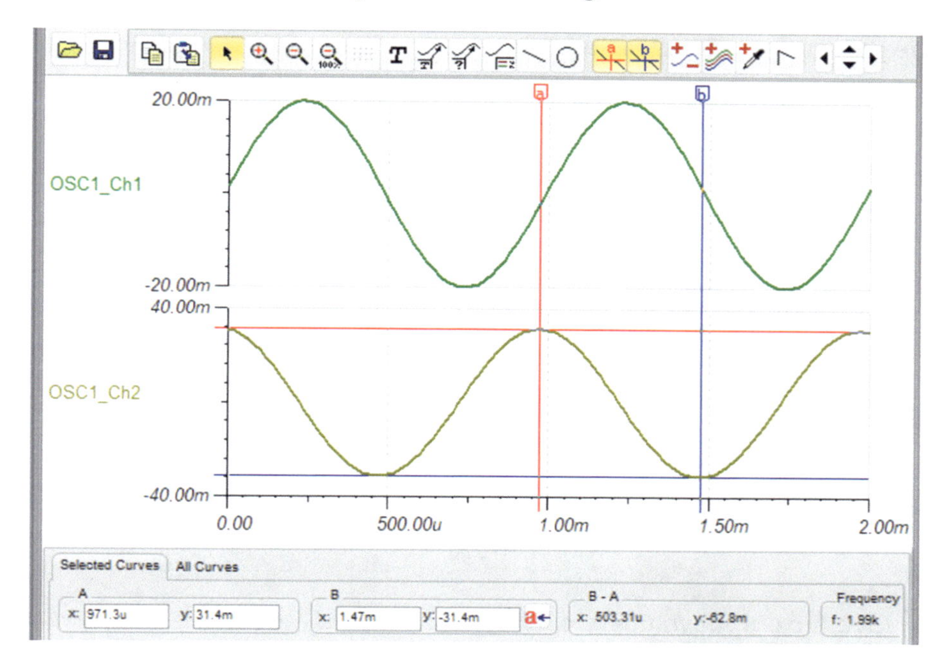


***Figure 9-41.** The Oscilloscope Display of the Integrating Opamp with a Sine Wave Applied*

Channel 1 shows the input which has a pk-to-pk value of 40mV. Channel 2 shows the output voltage which has a pk-to-pk value of 62.8mV. We can also see that the output waveform leads the input waveform by approximately 90°. That is because the input is a sine wave, and when you integrate a sine wave, you get -cosine, which should lag the sine by 90°. However, this is essentially an inverting Opamp, and so the lag becomes a lead of 90°. So, I hope this result does confirm that the output of this Opamp is the integral of the input.

# The Differentiator Opamp

The next Opamp circuit is the differentiating Opamp. Differentiation is the mathematical opposite of integration. When we differentiate a mathematical expression, we will create an expression that describes how the gradient, or rate of change, of that expression changes. For example, if we differentiate the following general expression:

$$y = 5x + 2$$

I hope the resultant expression would just be a constant of "5" as the gradient of a linear expression does not change, and it is the multiplying number "m" in front of the independent variable "x".

This means that if we supplied a differentiating Opamp with a steady ramp wave of

$$v = 5t$$

then the output voltage would be a constant of the value 5.

The differentiating Opamp circuit is shown in Figure 9-42.

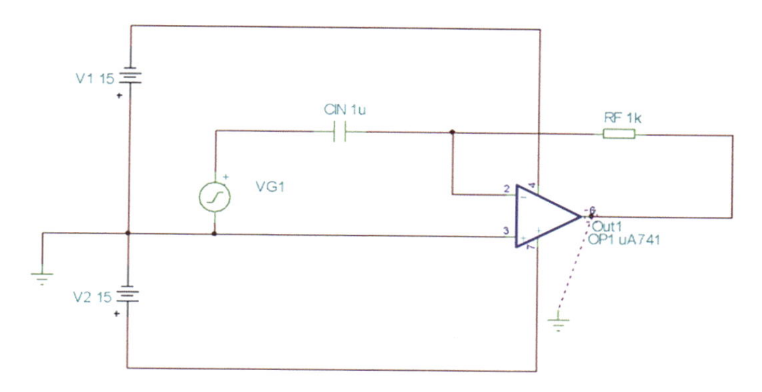


***Figure 9-42.*** *The Differentiator Opamp*

The input voltage is set to rise from 0V to 250V in 50ms. This makes the gradient 5000v per msec which relates to 5V in 1 sec and so a gradient of 5. When we simulate the circuit, we can see that the output voltage is a constant of –5V. This is shown in Figure 9-43.

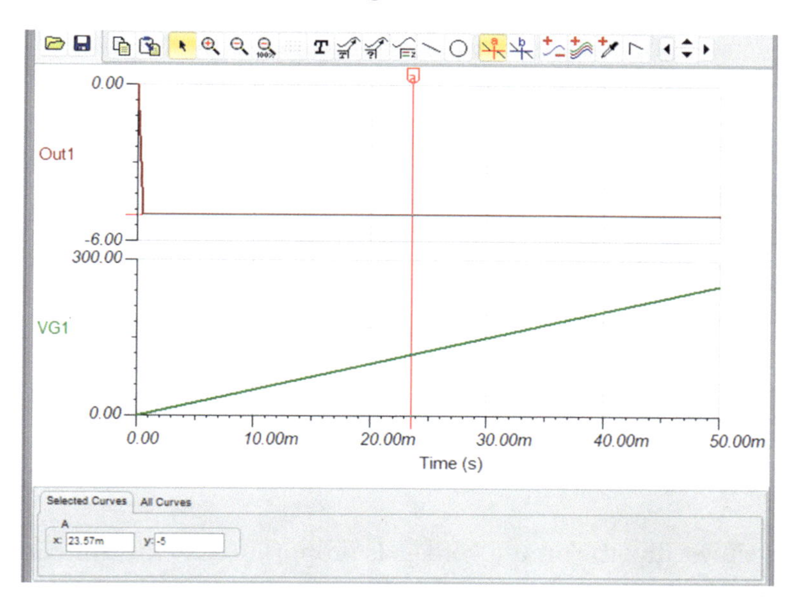


***Figure 9-43.*** *The Input and Output Voltages for the Differentiator Opamp*

We know from our previous analysis that we can express the output voltage as

$$V_{OUT} = -I_F R_F$$

We also know that

$$I_F = I_{IN}$$

From our analysis of the integrator, we know that when we force current to flow into a capacitor, as with $C_{IN}$, we will charge the capacitor up. Using the two basic expressions for the charge in a capacitor, we can say

$$C_{IN} V_{IN} = \int I_{IN}\, dt$$

If we differentiate both sides, we get

$$I_{IN} = C_{IN} \frac{dV_{IN}}{dt}$$

This then means

$$I_F = C_{IN} \frac{dV_{IN}}{dt}$$

We know that the output voltage, VOUT, can be expressed as

$$V_{OUT} = -I_F R_F$$

Therefore, we can say

$$V_{OUT} = -C_{RIN} R_F \frac{dV_{IN}}{dt}$$

This shows that the output voltage is proportional to the differential of the input voltage.

The input voltage of the differentiating Opamp in Figure 9-43 goes from 0 to 250V in 50ms. This means that the gradient is as follows:

$$Gradient = \frac{dV_{IN}}{dt} = \frac{250}{50E^{-3}} = 5000$$

Therefore, substituting this into the expression for the output voltage, we get

$$V_{OUT} = -C_{RIN} R_F \frac{dV_{IN}}{dt} = -1E^{-6} \times 1E^{3} \times 5000 = -5V$$

If we examine the output voltage trace, Out1 in Figure 9-43, we see that apart from some ringing at the beginning of the trace the output voltage is the expected –5V.

We can further test the analysis of the differentiating Opamp by applying a sawtooth waveform that swings between 1 and –1 volts at a frequency of 50Hz. This means that the periodic time for the waveform will be 20ms. This in turn means that the voltage would go from –1V to 1V, a change of 2V, in 10ms. Then go from 1V to –1V in the same time of 10ms. Therefore, we can calculate the gradient as follows:

$$Gradient = \frac{2}{10E^{-3}} = 200$$

In the first 10m, the gradient is positive, then in the second 10ms, the gradient is negative.

We can then use this to calculate the output voltage as follows:

$$V_{OUT} = -C_{RIN} R_F \frac{dV_{IN}}{dt} = -1E^{-6} \times 1E^{3} \times 200 = -0.2V$$

This would be during the first 10ms, then during the second 10ms, the output would swing to +0.2V. When we simulate the circuit, we can use the transient analysis to display the input and output voltage over the first 50ms as shown in Figure 9-43.

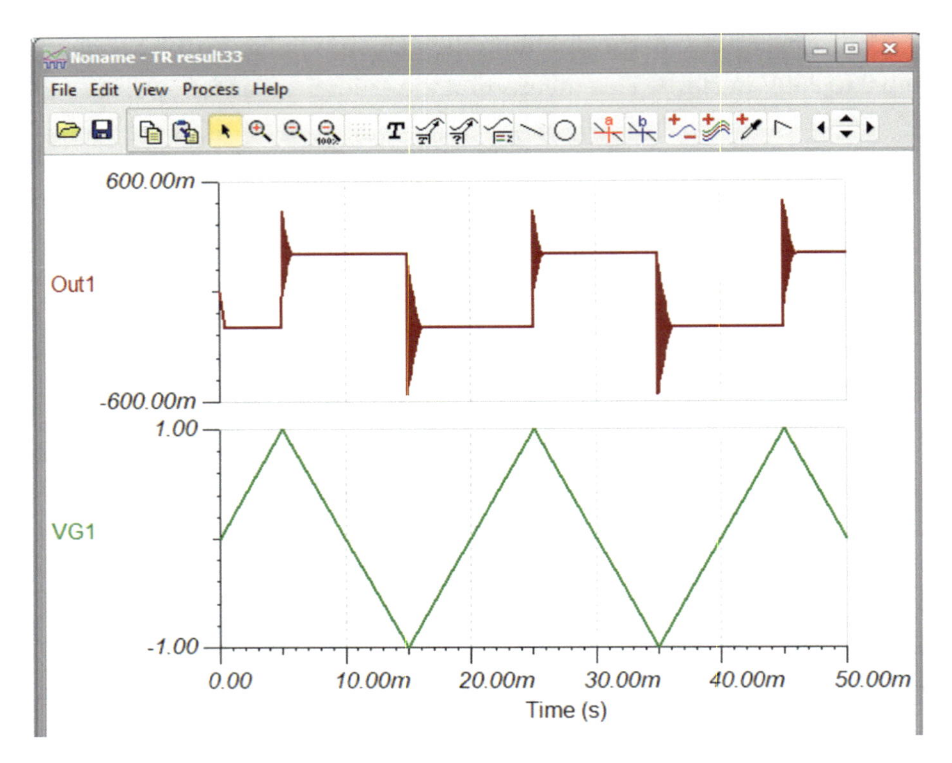


***Figure 9-44.*** *The Output of the Differentiating Opamp to a Sawtooth Input Voltage*

Figure 9-44 shows that we do get a square wave output when the input is a sawtooth waveform. Also, we can see that when the gradient of the input is positive, the output is –198.94mV. When the input gradient is negative, then the output is +201.06mV. This is very close to what we expect and so confirms our analysis of the differentiating Opamp. It also agrees with the statement that differentiation is the opposite of integration. There is some ringing on the output, and that is something we can deal with, but that's for a different book.

The main use of the integrating and differentiating Opamp is with analog control circuits that use a PID controller: proportional, integral, and differentiating control circuit. The proportional control can be implemented using the basic inverting Opamp.

## The Gain of the Differentiating Opamp

As we are using a capacitor in the circuit, then the expression for the voltage gain will have some frequency-dependent element in it. As always, we can say

$$V_{Gain} = \frac{V_{OUT}}{V_{IN}}$$

This is the type of expression we want to end up with. As the circuit is basically an inverting Opamp, we can start by stating

$$I_{IN} = \frac{V_{IN}}{Z_{IN}} \text{ and } I_F = \frac{-V_{OUT}}{Z_F}$$

With the differentiating Opamp, we can say

$$Z_{IN} = \frac{1}{j\omega C} \text{ and } Z_{OUT} = R_F$$

This means we can express the two currents as

$$I_{IN} = \frac{V_{IN}}{\frac{1}{j\omega C}} = V_{IN} j\omega C \text{ and } I_F = \frac{-V_{OUT}}{R_F}$$

Therefore, we can say

$$V_{IN} j\omega C = \frac{-V_{OUT}}{R_F}$$

This means we can rearrange this for the voltage gain as

$$\frac{V_{OUT}}{V_{IN}} = R_F\left(j\omega C\right)$$

This means that as the frequency changes from zero to infinity, the voltage gain will change from zero to infinity. However, that would only happen if the Opamp was ideal in that the Opamp itself does not have its own frequency dependence. We have seen that a realistic Opamp works, to some extent, as a low pass filter. This action will then counteract the gain response of the differentiator, and the actual Bode plot we get will look like that shown in Figure 9-45.

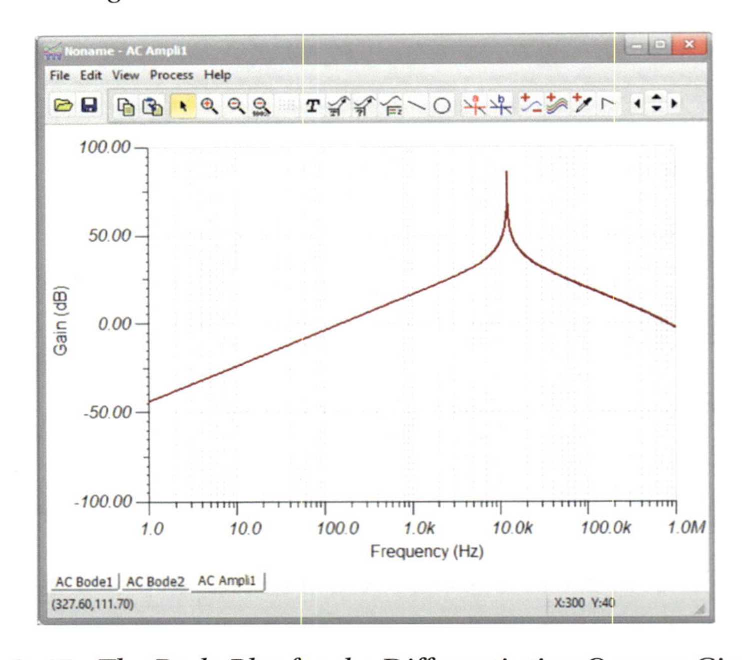


***Figure 9-45.*** *The Bode Plot for the Differentiating Opamp Circuit Shown in Figure 9-42*

During the low frequencies, from 0 to around 10kHz, the response follows the expression for the voltage gain. However, after that we can see how the low filter aspect of the Opamp takes over and the gain falls.

The differentiating Opamp is not normally used to add any gain; it is used to speed up the response of a system to a sudden change in the input. In this way, the output is proportional to the rate of change at the input. This means that if the input changed from say 0V to 1V in a nanosecond, then the output should shoot to a maximum, ideally in an instant, and as the input stayed at 1V, that is, the input did not change, the output should return to zero. Well, we know no voltage can change in an instant, and the change will be controlled by the time constant of the circuit, which, with the differentiator, is the product of CR seconds. We can best test this concept by applying a square wave to the input of the differentiator. The square wave is shown in Figure 9-46.

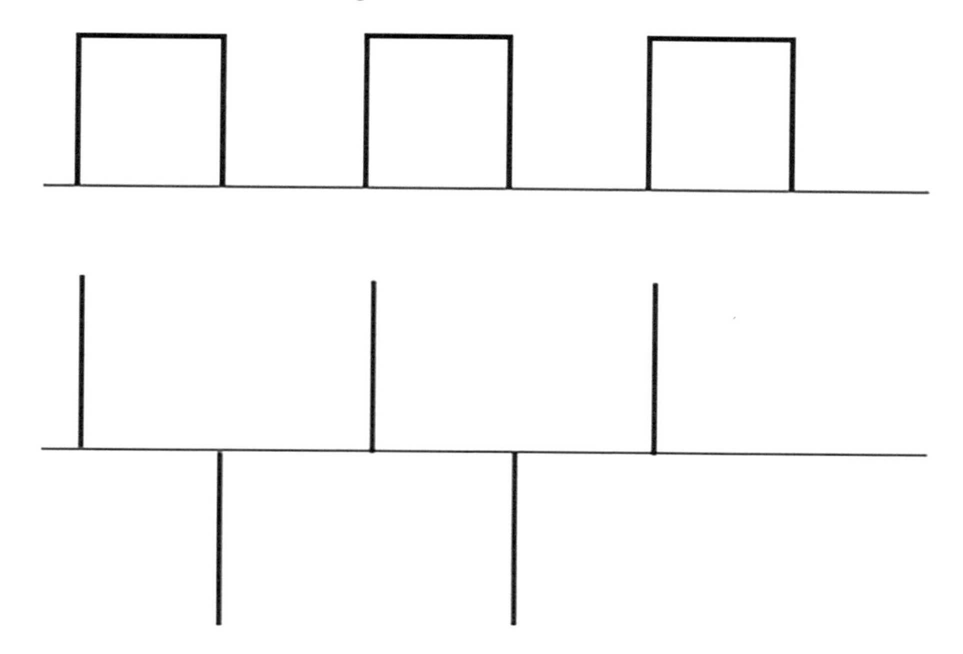

***Figure 9-46.*** *The Square Wave Input and the Ideal Response*

We are using a square wave with 50/50 duty cycle at a frequency of 50Hz. Ideally, we want the output to be a maximum when the input went from 0 to 1V as this is the point of maximum change. Then it should fall to zero, ideally at an instant, as the input is now undergoing zero change. Then, some 10ms later, we want the output to go to a maximum again as the input is again undergoing a maximum rate of change. However, it should be a negative maximum to differentiate from the first change. The output should then return to zero waiting for the next change at the input. This means the ideal output voltage would look like that shown in Figure 9-46; this is to the square wave shown in Figure 9-46. This would suggest that we should have a time constant of zero seconds. However, that is impossible, and you should also appreciate that we would get ringing on the output. This is because if you imagine you are running at full pelt and you were told to stop, it would take time to stop, and you would have passed the point, so you would have to come back. The output voltage would do the same sort of thing. One thing we must do is ensure the voltage had returned to zero before the input undergoes the next sudden change. With the 50Hz 50/50 square wave, this means the voltage must have fallen to zero in less than 10ms. From our work with electrical analysis, we know that it will take approximately 5Tau, that is, 5 time constants, for the voltage to settle down. This can be seen if we look at the waveforms from the simulation of the circuit shown in Figure 9-42, as shown in Figure 9-47.

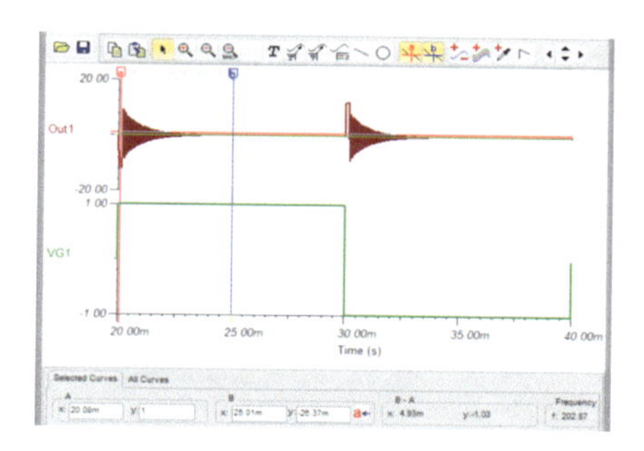


***Figure 9-47.** The Waveforms for the Differentiating Opamp Shown in Figure 9-42*

Using the Out1 trace, which is the output voltage of the Opamp, shows that when the input went from 0 to 1V, the output went initially to -12V. The negative is due to the inverting action of the Opamp. The voltage then starts to ring until it settles down close to 0V after 5ms. The time constant for the circuit is CR sec = 1ms. This means that it does take approximately 5 time constants, as expected, to settle down. Note the capacitor will eventually charge up to the DC of the input voltage. This means in the first 10ms, the output voltage would settle down to -1V. In the second 10ms, it would settle down to +1V.

The only way we can improve the response of the Opamp is to reduce the time constant. This can be done by reducing the value of either the capacitor or the resistor. However, we need to be aware of the current flowing through the components of the circuit. The resistor will control how much current is flowing through the components, so we may need to keep that rather high. This means we would normally change the value of the capacitor.

# The Instability of the Differentiating Opamp

This basic differentiating Opamp suffers from instability problems. This can be seen if we simulate the circuit while supplying it with a 1v sine wave at 15kHz. The output of the differentiator is shown in Figure 9-48.

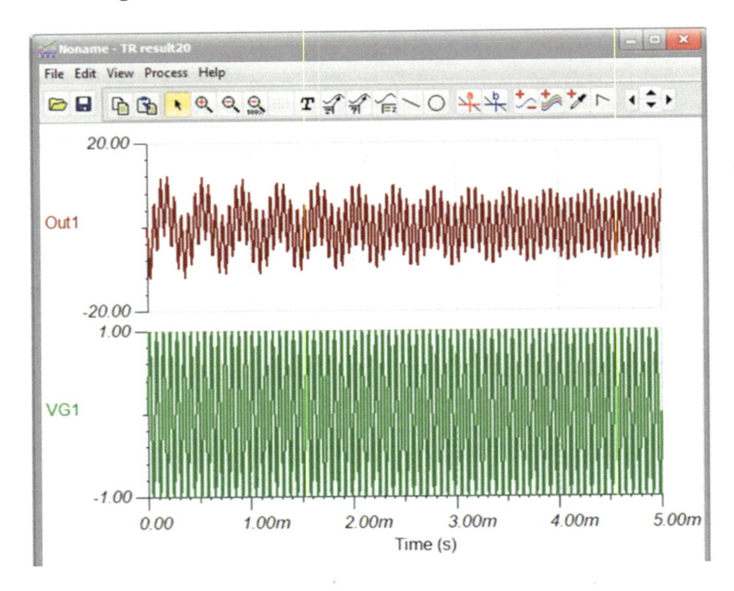


***Figure 9-48.*** *The Instability of the Basic Differentiator*

The output voltage is shown as trace Out1. The instability is quite clear to see. A circuit that combats this instability is shown in Figure 9-49.

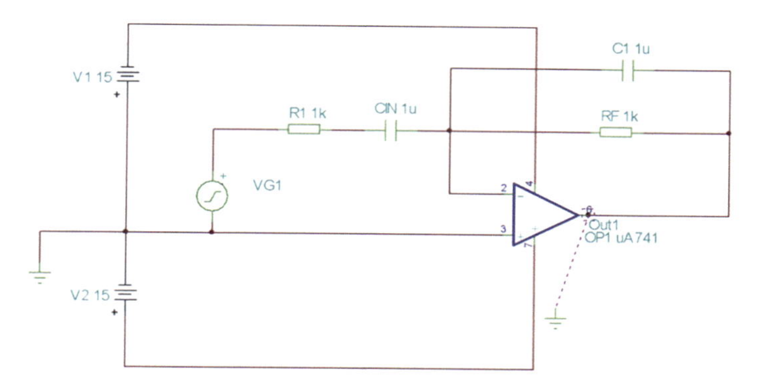


***Figure 9-49.*** *The Improved Differentiator*

I will leave the analysis of how this circuit works to another book; that would be a book dedicated to just Opamps. When we simulate the circuit, the waveforms obtained are shown in Figure 9-50.

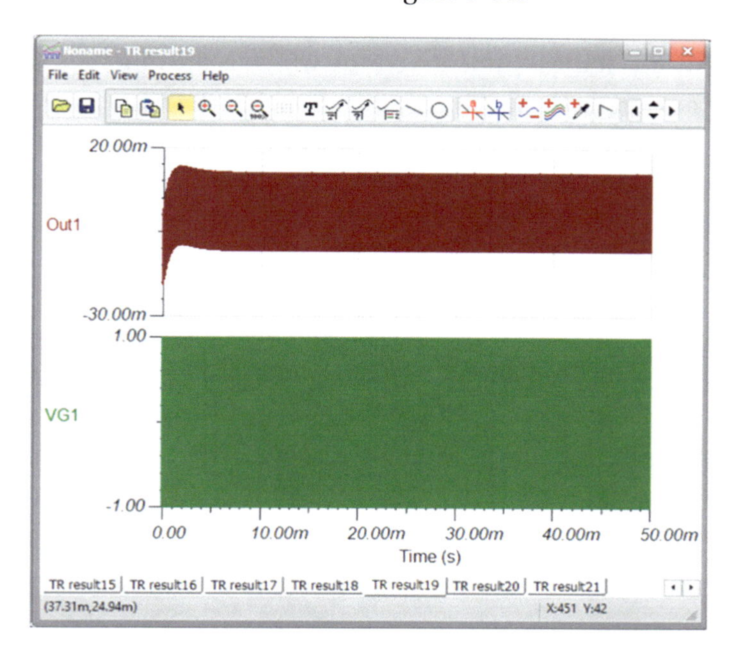


***Figure 9-50.*** *The Output of the Improved Differentiator*

It's not clear what the output voltage is, but it does show us how much more stable the output is. If we zoom in on just a few cycles, we can see the output more clearly. This is shown in Figure 9-51.

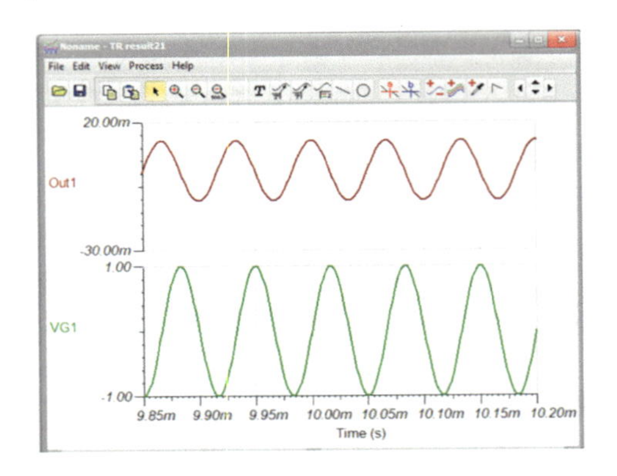


***Figure 9-51.*** *Just a Few Cycles of the Output*

This shows that the output is a cosine which is the differential of the input. There is attenuation, not amplification, but that analysis is for another book.

# The Voltage Comparator

When we started this look into Opamps, we said that there is one use of the Opamp when configured in its open-loop configuration. This is because we said the Opamp cannot distinguish between a 150mV and a 5V difference voltage; the output would be the same, that is, the voltage rails. Well, there is one application where we might be able to use the Opamp in its open-loop configuration, that is, as a comparator. The basic circuit for the comparator is shown in Figure 9-52.

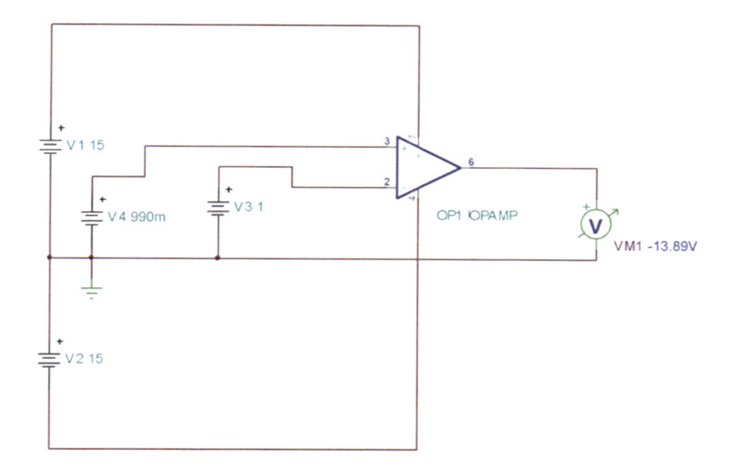


***Figure 9-52.*** *The Opamp As a Comparator*

In Figure 9-52, there is none of the output feedback to the input of the Opamp. Thus, the Opamp is configured in the open-loop configuration. The circuit will simply compare one of the inputs with the other. In this example, we are comparing $V_4$ with $V_3$, and as soon as $V_4$ is less than $V_3$, the output would swing to the negative rails at a maximum of –13.89. If $V_4$ went above $V_3$, then the output would swing from the –13.89 to +13.89. In this way, we can use the Opamp to compare a varying voltage with a set reference, and when it has reached or really just gone past it by around 150μV, the output voltage would swing from one voltage rail to the opposite rail.

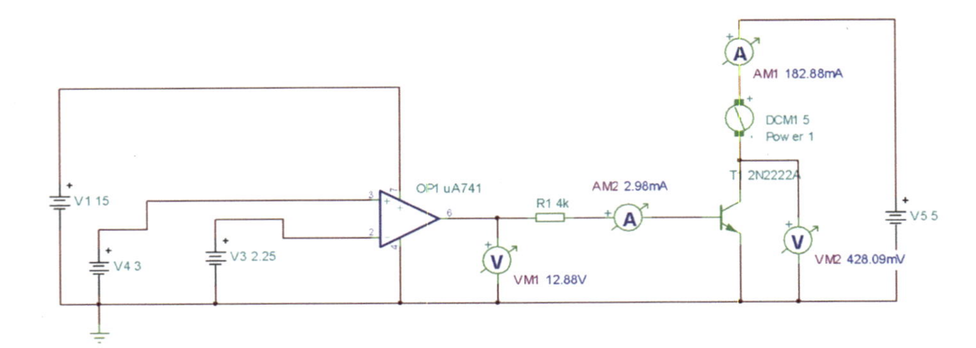


***Figure 9-53.*** *A Possible Application of the Opamp Comparator*

Figure 9-53 shows a possible application of the Opamp as a comparator. In this circuit, we are testing to see if the voltage V4 has become just that 150uV greater than V3. When that happens, the output voltage would swing from the negative voltage rail to the positive rail. This could be used to turn on a pump when the water level in a container has gone above a set point and start to empty it.

## Exercise 9.9

Briefly explain why the negative supply to the Opamp can be tied to 0V and briefly explain what the main considerations are in choosing the value for the resistor $R_1$.

# The Opamp As a Subtractor

Earlier, we looked at the summing Opamp where the output of the Opamp was related to the summing of the inputs. We will now complete this journey into studying the Operational Amplifier with a quick look at a subtractor circuit. The circuit is shown in Figure 9-54.

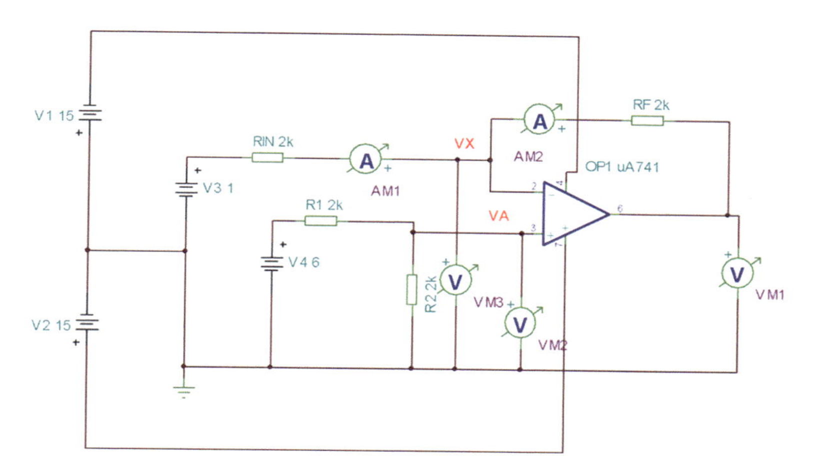


***Figure 9-54.*** *The Opamp Subtractor Circuit*

With this circuit, it is normal to make all the resistors to have the same value, 2k, in this case. In this circuit, the output will be equal to V4 – V3. Using the value shown in the circuit, this means that the output voltage $V_{OUT}$ will be 3V.

The way it works is that due to the virtual short circuit, $V_X$ will equal $V_A$. Using the voltage divider rule, we can say

$$V_A = \frac{V_4}{2} = \frac{6}{2} = 3V$$

This means $V_X$ = 3V. This means that the current flowing through $R_{IN}$, which we will call $I_{IN}$, is

$$I_{IN} = \frac{V_3 - V_X}{R_{IN}} = \frac{1-3}{2E^3} = -1mA$$

This means the current must be flowing out from pin 2 into V3. This means that the same current must be flowing out from Vout toward VX. This means that the output voltage $V_{OUT}$ can be calculated as

$$V_{OUT} = I_F R_F + V_X$$

But $I_F = I_{IN} = 1mA$. Therefore, we have

$$V_{OUT} = 1E^{-3}2E^{3} + 3 = 5V.$$

As all four resistors have the same value, then the voltage gain is simply "1." This is because the voltage output is simply the difference voltage:

$$V_{OUT} = V_4 - V_3$$

This assumes the voltages are numbered as in Figure 9-54.

# Summary

In this chapter, we have studied the Opamp and analyzed how it works in its five major configurations. We have used some basic maths to derive the expressions for the output voltage and the gain of the Opamp. We have also looked at how the Opamp can be used to carry out some mathematical operations and so show how it got its name. I hope you have found this chapter both interesting and informative.

In the next chapter, we will look at how we used the Opamp and also the BJT in two very useful applications such as oscillators and active filters.

CHAPTER 10

# Oscillators

In Chapter 9, we looked at the Operational Amplifier or Opamp. In this chapter, we will study oscillators starting with the multivibrator and ending with some Opamp oscillators. We will learn how the oscillators work and how we can analyze their operation using complex numbers.

## Multivibrators

There are many applications when electrical circuits use a signal that changes state at a regular interval. This has led to many different circuits that can be used to create this variety of signals. We will start our investigation of these types of circuits with a look at the multivibrator series of circuits. These types of circuits use resistor-capacitor combinations to create the time period between turning on and off transistors and so swing the output between states.

## The Monostable

The monostable was probably one of the first circuits that fell into the bracket of multivibrators. The circuit for the monostable is shown in Figure 10-1.


H. H. Ward, *Mastering Analog Electronics*, Maker Innovations Series,
https://doi.org/10.1007/979-8-8688-0245-4_10

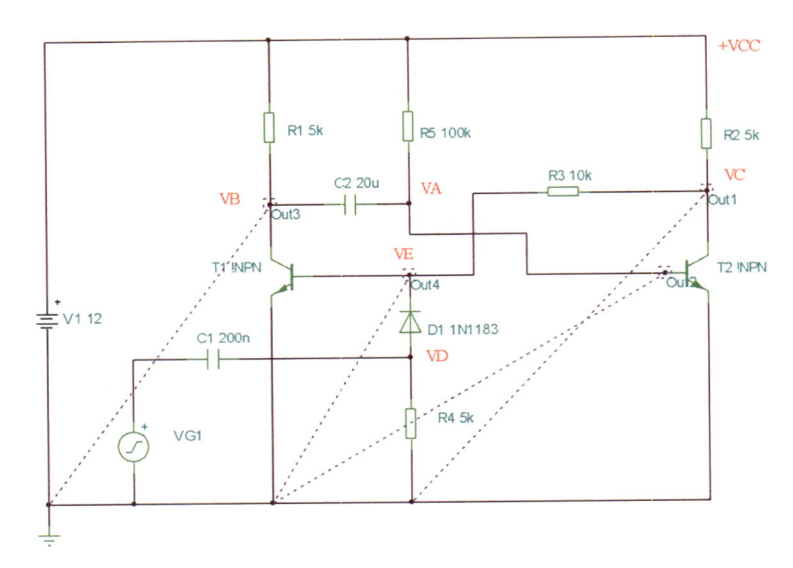


***Figure 10-1.*** *The Monostable Circuit in the Multivibrator Series*

We will use a transient analysis to show what is happening around the circuit over the first 500ms after triggering the circuit. As its name suggests, the monostable circuit has one stable state. This means we can force the output to change from its stable state, but it will eventually return to its stable state. The waveforms shown in Figure 10-2 should help us to see this action. We will investigate how the circuit works.

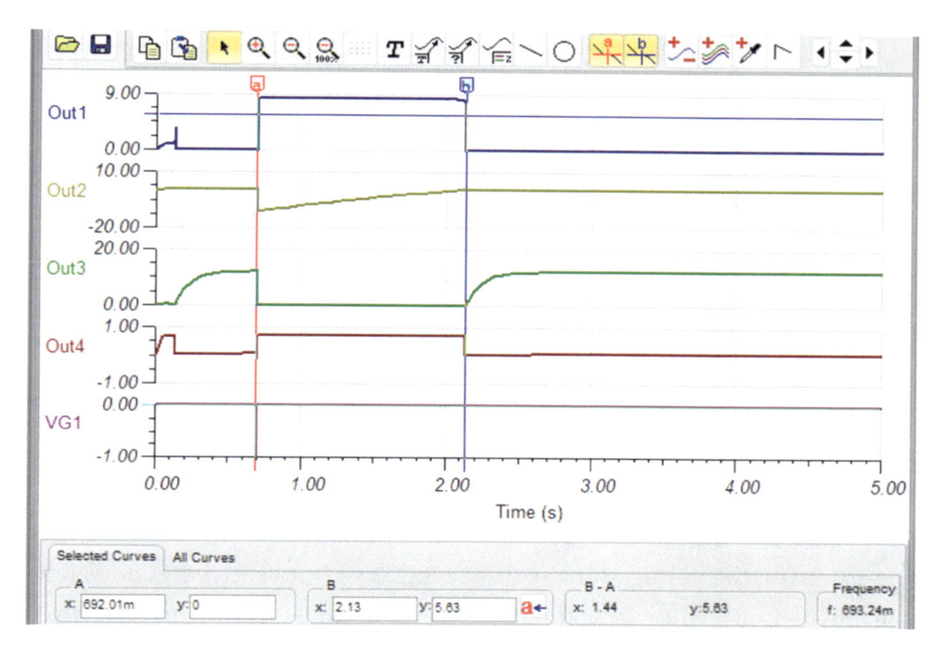


***Figure 10-2.** The Waveforms of the Circuit Shown in Figure 10-1*

At the initial switch-on, we can see that the base of $T_1$ is connected to $V_{CC}$ via the resistors $R_3$ and $R_2$. These two resistors will set the current flowing into the base of $T_1$. The voltage at the base of $T_1$ will be clamped to around 0.7V by the base emitter diode in $T_1$, which means $T_1$ will be turned on. This will connect the collector of T1 to around 0V. This would give a path for the capacitor $C_2$ to charge up via $T_1$ and $R_5$. However, one end of the capacitor is connected to the collector of $T_1$, and the other end is connected to the base of $T_2$. This means that when the voltage across $C_2$ reaches around 0.7V, $T_2$ turns on. This takes the collector of $T_2$ to around 0V, which in turn takes the base of $T_1$ to around 0V, which turns $T_1$ off.

This is how the circuit reaches the stable state. With the circuit shown in Figure 10-1, the stable state is when the transistor $T_2$ is turned on and $T_1$ is turned off. In this state, the voltage at Out1, the collector of $T_2$, will be close to 0V. The voltage at Out3, the collector of $T_1$, will be at $V_{CC}$, 12V in this case. The base of $T_2$ will be approximately 0.7V; with Tina, it will be around 0.655V. This means there will be 12 – 0.7 = 11.3V across the capacitor $C_2$.

Now if we force $T_1$ to turn on, this will force the collector of $T_1$ to go to around 0V. This is done 692ms after turning the circuit on. See the trace VG1 in Figure 10-2 when the voltage at VG1 goes down to –1V. This is confirmed by OUT3 going down to 0V, as we have just forced $T_1$ to turn on; see Figure 10-2. The circuit has been in its stable state for some 700ms, so there would have been 11.3V across the capacitor $C_2$, with the collector end of $C_2$ being 11.3V more positive than the other end. As we can't change the voltage across a capacitor instantly, as we have to either charge or discharge a capacitor to change the voltage across it, then this means that the base of $T_2$ will go to around –11.3V. This can be confirmed by measuring the voltage for the trace OUT2. We can see that it does fall to around –11.3V when $T_1$ is forced to turn on. This forces $T_2$ to turn off and connect the base of $T_1$ to VCC via $R_3$ and $R_2$ again keeping the base at around 0.7V, keeping T1 turned on. This will allow the capacitor $C_2$ to charge up toward $V_{CC}$ via the transistor $T_1$ and $R_5$. However, when the voltage across $C_2$ gets to around 0.7V, the transistor $T_2$ turns back on again which turns off $T_1$, and the circuit enters its stable state again. The trace OUT2 does confirm this; see Figure 10-2. It is the time it takes for the capacitor $C_2$ to charge up to around 0.7V from the starting voltage of –11.3V that dictates how long the output voltage, at the collector of $T_2$, stays high, that is, in the unstable state of the monostable.

One point worth noting is that when $T_1$ is turned on, the capacitor $C_2$ charges up via $T_1$ and $R_5$ toward $V_{CC}$ at the node "$V_A$"; the voltage at node "VB" will be 0V. However, when the voltage at "$V_A$" reaches 0.7V, $T_2$ turns on and $T_1$ turns off. This means the current path via $T_1$ has gone. The capacitor "$C_2$" now charges up through $R_1$ and the base emitter junction of $T_2$ toward $V_{CC}$. This will allow the voltage at node "$V_B$" to go to $V_{CC}$, 12V in this case, and the voltage at node "$V_A$" to stay at 0.7V. This means that the current flowing through $C_2$ must change direction, instantly. This can happen with capacitors but not with inductors.

The trigger circuit that is used to turn $T_1$ on and force the circuit out of its stable state is made up of $C_1$, $R_4$, and the diode $D_1$. The source VG1 is used to apply the pulse that triggers the circuit. The capacitor $C_1$ in series with $R_4$ creates what is termed a differentiator circuit. A test circuit for the differentiator is shown in Figure 10-3.

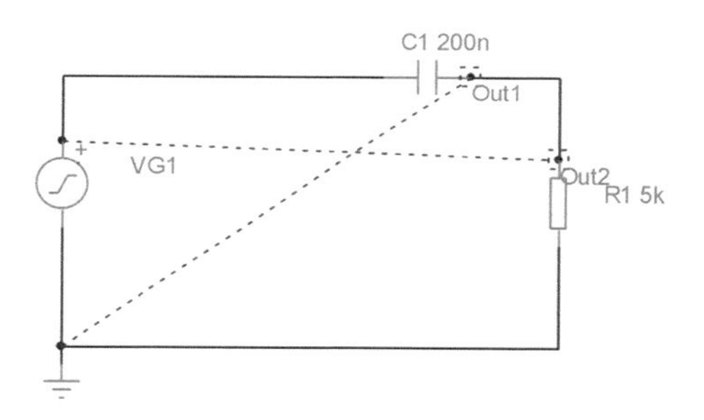


***Figure 10-3.*** *The CR Differentiator Circuit*

The circuit is really a simple series CR circuit, and the output voltage is taken from across the resistor. From a study of DC transients, which this trigger circuit relates to, we can use the following concepts.

We know that at all times we can say

$$V_{G1} = V_C + V_R$$

$$V_C = \frac{1}{C}\int i\,dt$$

$$V_R = iR$$

This means that using Kirchhoff's Voltage Law, we can say

$$V_{G1} = \frac{1}{C}\int i\,dt + iR$$

We can see that there are two terms on the RHS, and at first you might think that the current "i" is common to both terms. However, it is not common as one is simply "i," but the other is the integral of "i." Therefore, we need to change the equation if we are to transpose it. There are basically two approaches we can use to transpose this equation; one uses Laplace transforms, and the other uses calculus. Both approaches have their benefits, and, really, we as engineers should be able to use both methods. First, though, we should discuss what Laplace transforms are.

# Laplace Transforms

If we have an expression, as we do now, where the variable we are trying to look at is in both an algebraic term and a differential or integral term, then we will have problems trying to isolate it. This is where we can make use of Laplace transforms. Laplace transforms were initially invented by a French mathematician called Pierre Simon Laplace. They are used to help solve differential equations. The transforms consist of a table which shows how the common time equations have their own individual Laplace transforms. We will use them now to help analyze the triggering circuit shown in Figure 10-3 and derive an expression for the current "i."

The input voltage, applied to the series CR circuit shown in Figure 10-3, is a simple pulse that will go from 0 down to –1 and stay at –1 for 5ms. To transpose the expression, because there is an integral term in it, we need to move into the Laplace domain. This means taking the Laplace of the three terms.

To help with this analysis, it would be better to view the pulse voltage as two separate step inputs to the circuit. The first step input takes the voltage down from 0V to –1V. The second takes it up from –1V to 0V. This means that both steps are a 1V step, the first going negative and the second going positive. This would allow us to take the following Laplace transform of each of the terms. Using the standard Laplace transforms, shown in the appendix, we can say

$$L\{\ a\ Step\ input\} = \frac{V}{S}$$

$$L\left\{\frac{1}{C}\int i\,dt\right\} = \frac{1}{SC}\bar{i}$$

$$L\{iR\} = \bar{i}R$$

The term $\bar{i}$ is the Laplace of the variable we are transposing for, which is not the general term "x" but the current "i."

Therefore, the expression for the circuit is

$$V_{G1} = \frac{1}{C}\int i\,dt + iR$$

The Laplace expression is

$$\frac{V}{S} = \frac{1}{SC}\bar{i} + \bar{i}R$$

We can now treat this as an algebraic equation and take the $\bar{i}$ out as a common factor.

This gives

$$\frac{V}{S} = \bar{i}\left(\frac{1}{SC} + R\right)$$

Combining the terms in the bracket into a compound fraction, we have

$$\frac{V}{S} = \bar{i}\left(\frac{1+SCR}{SC}\right)$$

Now dividing both sides by the fraction in the bracket gives us

$$\bar{i} = \frac{SC}{1+SCR} \times \frac{V}{S}$$

This cancels down to

$$\bar{i} = \frac{VC}{1+SCR}$$

If we now divide the top and bottom of the RHS by the term "C," we get

$$\bar{i} = \frac{V}{\frac{1}{C} + SR}$$

Now we can divide the top and bottom of the RHS by the term "R," and we get

$$\bar{i} = \frac{\frac{V}{R}}{\frac{1}{CR} + S}$$

This can be written as

$$\bar{i} = \frac{V}{R} \times \frac{1}{S + \frac{1}{CR}}$$

We now have to take the inverse Laplace of the terms to go back to the time domain, which gives

$$L^{-1}\left\{\bar{i}\right\} = i$$

$$L^{-1}\left\{\frac{V}{R}\right\} = \frac{V}{R}$$

$$L^{-1}\left\{\frac{1}{S+\frac{1}{CR}}\right\} = e^{-\frac{t}{CR}}$$

This means that the time equation for the current in the circuit "i" is

$$i = \frac{V}{R}e^{-\frac{t}{CR}}$$

This is an exponential decaying expression in which the current starts off at its maximum value of $i = \frac{V}{R}$ and slowly decays away toward 0. The expression includes a time constant "tau" given the symbol "τ". The exponential term relates to the standard exponential decay expression:

$$e^{-\frac{t}{\tau}}$$

Relating this standard expression to the expression for the current "i," we can see that

$$\tau = CR\ seconds$$

It will normally take about five time constants for the circuit to settle down to its steady state conditions, either its maximum value or zero. In the CR circuit, shown in Figure 10-3, the value of the capacitance is 200nF, and the resistance is 5kΩ. This means that the time constant is

$$\tau = C \times R = 200^{-9} \times 5^{3} = 1ms$$

This means that the current in the circuit will take approximately 5ms to fall to 0Amps.

## The Output Voltage of Circuit 10.3

The output of this circuit, shown in Figure 10-3, is taken from across the resistor "$R_1$." Therefore, as we know the voltage across the resistor is simply iR, we have the expression for $V_R$ as

$$V_R = iR = \frac{V}{R} e^{-\frac{t}{CR}} \times R = Ve^{-\frac{t}{CR}}$$

This means that the expression for $V_R$ is

$$V_R = Ve^{-\frac{t}{CR}}$$

## The Voltage Across the Capacitor $V_C$

Again, applying KVL to the circuit shown in Figure 10-3, we can say

$$V = V_C + V_R$$

where "V" is the voltage applied to the circuit. Then, transposing this for $V_C$, we can state an expression for the voltage across the capacitor "$V_C$" as

$$V_C = V - V_R$$

We can use the previous expression for $V_R$, and substituting it gives

$$V_C = V - Ve^{-\frac{1}{CR}}$$

We can take the "V" out as a common factor to give the following expression:

$$V_C = V\left(1 - e^{-\frac{t}{CR}}\right)$$

This is the standard expression for the voltage across the capacitor as it is charged from a DC voltage source "V." This assumes there is no initial voltage across the capacitor.

If we now look at the expression for $V_R$ and if we call this the output of the circuit, we can then call this $V_{OUT}$. This means that the expression for $V_{OUT}$ is

$$V_{OUT} = Ve^{-\frac{t}{CR}}$$

We can use this expression to see how the circuit, shown in Figure 10-3, will respond to the pulse input. Firstly, we will see how it will respond to the negative edge of the pulse. This will be viewed as V = –1. Therefore, putting this into the expression for $V_{OUT}$ and using the values for C and R from that circuit, we get

$$V_{OUT} = -1e^{-\frac{t}{CR}} = -1e^{-1000t}$$

To enable us to see the full decay of the output voltage, I have set the pulse width to 5ms in time. This should enable us to see the output voltage fall to –1V, then climb back up to 0V, ending at 0V when time = 5ms.

Now the input voltage will itself swing back up from –1V to 0V. However, the circuit will simply see this as a step input of 1V. The output voltage would then follow this expression:

$$V_{OUT} = 1e^{-\frac{t}{CR}} = 1e^{-1000t}$$

This means the output voltage will swing up to 1V, then start to decay back to 0V, ending up at 0v after another 5ms, that is, when time now equals 10ms. The output voltage as well as the input voltage and the voltage across the capacitor are shown in Figure 10-4.

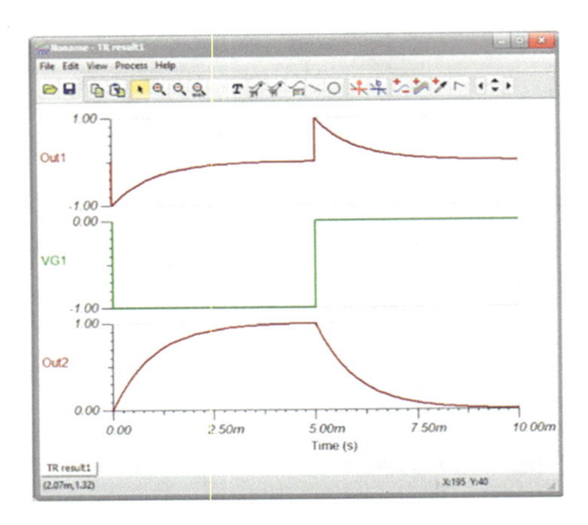


***Figure 10-4.*** *The Voltage Waveforms for the Circuit Shown in Figure 10-4*

We can see that the output voltage, shown as Out1 in Figure 10-4, does follow the two expressions for the $V_{OUT}$. Also, Out2 in Figure 10-4 does show the voltage across the capacitor. This does show the capacitor is charging up during the first 5ms and then discharging during the second 5ms. This is what we expect from the CR circuit shown in Figure 10-3,

## The Triggering of the Monostable

How, then, does this simple circuit, shown in Figure 10-3, trigger the monostable? The diode "D1" is there to ensure only the positive going voltage at the output of the CR circuit is actually passed onto the base of $T_1$, thus turning it on and forcing the output to change from its stable state.

The action from then on until the monostable returns to its stable state has already been described.

# The Time the Output Is High

We have stated that the output stays high as long as the transistor $T_2$ is turned off. The voltage at the base of $T_2$ is controlled by the charging circuit for $C_2$. This capacitor charges up toward $V_{CC}$ via the resistor $R_5$. We can analyze this charging time using the test circuit shown in Figure 10-5.

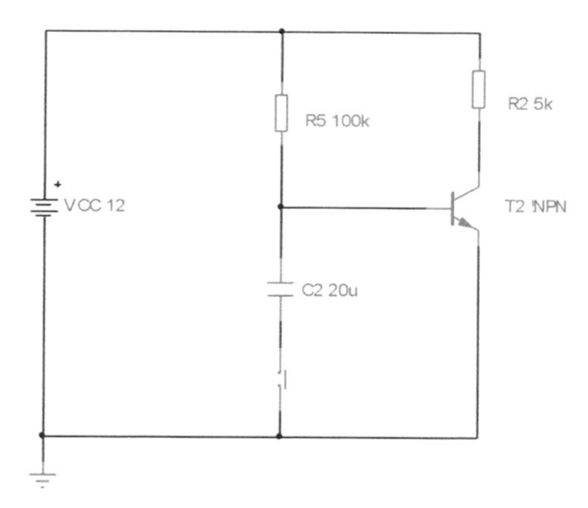


***Figure 10-5.*** *The Test Circuit for the C2R5 Charging Circuit*

The circuit in Figure 10-5 is a basic CR transient circuit, and this time we will use calculus, instead of Laplace transforms, to derive the expression for how the capacitor charges up. We know that at all times we can say

$$V_{CC} = V_C + V_R$$

We need to transpose the expression for the voltage across the capacitor. The following text will use calculus to transpose this equation.

From our work with the integrating Opamp in Chapter 9, we can say

$$CV_C = \int i \, dt$$

Therefore, if we differentiate both sides, we can say

$$i = C\frac{dV_C}{dt}$$

This current "i" is the current flowing through the resistor $R_1$ in the CR circuit shown in Figure 10-3. This means we can say

$$V_{CC} = C\frac{dV_C}{dt}R + V_C$$

Subtracting $V_C$ from both sides gives

$$V_{CC} - V_C = C\frac{dV_C}{dt}R$$

If we divide both sides by RC, we get

$$\frac{V_{CC} - V_C}{RC} = \frac{dV_C}{dt}$$

If we multiply both sides by "dt," we get

$$\frac{V_{CC} - V_C}{RC}dt = dV_C$$

If we now divide both sides by $V_{CC} - V_C$, we get

$$\frac{1}{RC}dt = \frac{1}{V_{CC} - V_C}dV_C$$

If we now integrate both sides, we get

$$\int \frac{1}{RC} dt = \int \frac{1}{V_{CC} - V_C} dV_C$$

These are two integrals, one integrated with respect to time "t" and the other integrated with respect to $V_C$. To make the integrals useful, we need to put in some limits of integration. The lower limit will be the initial values, which for time "t" is zero and for the capacitor voltage "$V_C$" is, in this case, also zero. The upper limits will be for time just "t" and for the capacitor voltage will be $V_{CAP}$. We will use $V_{CAP}$ to differentiate it from the variable of the integral $V_C$. Putting these variables in, we have

$$\int_0^t \frac{1}{RC} dt = \int_0^{V_{CAP}} \frac{1}{V_{CC} - V_C} dV_C$$

Using the standard integrals to carry out the integration, we get the following; see the table of standard integrals in the appendix.

$$\left[\frac{t}{RC}\right]_0^t = \left[-ln\left(V_{CC} - V_C\right)\right]_0^{V_{CAP}}$$

Putting the limits in, we get

$$\left(\frac{t}{RC}\right) - \left(\frac{0}{RC}\right) = \left(-ln\left(V_{CC} - V_{CAP}\right)\right) - \left(-ln\left(V_{CC} - 0\right)\right)$$

This simplifies to

$$\frac{t}{RC} = -ln\left(V_{CC} - V_{CAP}\right) + ln\left(V_{CC}\right)$$

If we multiply both sides by –1, we get

$$ln(V_{CC} - V_{CAP}) - ln(V_{CC}) = -\frac{t}{RC}$$

When subtracting logs, we can divide the terms as follows:

$$ln\left(\frac{V_{CC} - V_{CAP}}{V_{CC}}\right) = -\frac{t}{RC}$$

Taking the inverse ln of both sides, we get

$$\frac{V_{CC} - V_{CAP}}{V_{CC}} = e^{-\frac{t}{RC}}$$

Now multiplying both sides by $V_{CC}$, we get

$$V_{CC} - V_{CAP} = V_{CC}e^{-\frac{t}{RC}}$$

Now subtracting $V_{CC}$ from both sides, we get

$$-V_{CAP} = -V_{CC} + V_{CC}e^{-\frac{t}{RC}}$$

Now multiplying throughout by –1, we get

$$V_{CAP} = V_{CC} - V_{CC}e^{-\frac{t}{RC}}$$

Taking $V_{CC}$ out as a common factor, we get

$$V_{CAP} = V_{CC}\left(1 - e^{-\frac{t}{RC}}\right)$$

Now replacing the term $V_{CAP}$ with $V_C$ and $V_{CC}$ with V, we get

$$V_C = V\left(1 - e^{-\frac{t}{RC}}\right)$$

This is the same standard expression that shows how a capacitor charges up over time. When two approaches produce the same result, it should give you confidence the work is valid. The term $e^{-\frac{t}{RC}}$ will eventually decay down to zero, and so the voltage across the capacitor will settle down to the supply voltage "V." It will take approximately five time constants, that is, 5 tau, for the capacitor to charge up the supply voltage "V."

# The Charging of a Capacitor with an Initial Voltage

The expression we have derived earlier works if there was no initial voltage across the capacitor. However, with the monostable multivibrator shown in Figure 10-2, the capacitor $C_2$, which sets the time the output stays high, will be charging up toward $V_{CC}$ from an initial voltage of around −11.3V; see the preceding analysis of the circuit shown in Figure 10-1. Therefore, we need to modify the derivation as follows:

We can start from the same integral which states

$$\int \frac{1}{RC}\, dt = \int \frac{1}{V_{CC} - V_C}\, dV_C$$

However, we need to change the initial value for the capacitor integral to accommodate the fact that the capacitor is starting from some voltage that is not zero. We will call that initial voltage $V_0$. The integral now becomes

$$\int_0^{t_{CAP}} \frac{1}{RC} dt = \int_{V_0}^{V_{CAP}} \frac{1}{V_{CC} - V_C} dV_C$$

Using the standard integrals to carry out the integration, we get

$$\left[\frac{t}{RC}\right]_0^{t_{CAP}} = \left[-ln\left(V_{CC} - V_C\right)\right]_{V_0}^{V_{CAP}}$$

Putting the limits in, we get

$$\left(\frac{t}{RC}\right) - \left(\frac{0}{RC}\right) = \left(-ln\left(V_{CC} - V_{CAP}\right)\right) - \left(-ln\left(V_{CC} - V_0\right)\right)$$

Therefore, we have

$$-\frac{t}{RC} = ln\frac{V_{CC} - V_{CAP}}{V_{CC} - V_0}$$

Taking the inverse ln, we get

$$\frac{V_{CC} - V_{CAP}}{V_{CC} - V_0} = e^{-\frac{t}{RC}}$$

Multiplying both sides by $V_{CC} - V_0$, we get

$$V_{CC} - V_{CAP} = V_{CC} - V_0 e^{-\frac{t}{RC}}$$

From this, we can say

$$V_{CAP} = V_{CC} - \left(V_{CC} - V_0\right) e^{-\frac{t}{RC}}$$

Replacing the term $V_{CAP}$ with $V_C$ and $V_{CC}$ with V, we get

$$V_C = V - (V - V_0)e^{-\frac{t}{RC}}$$

This means that the capacitor $C_2$ will charge up toward the $V_{CC}$ according to the expression

$$V_C = V_{CC} - (V_{CC} - V_0)e^{-\frac{t}{RC}}$$

We can use this expression to determine the time the output voltage will stay high. This will be the time it takes the capacitor $C_2$ to charge up from around –11.3V to around 0.7V, as it will be then that T2 turns back on sending the output voltage back down to 0V, the stable output, and turning T1 off. We can transpose the expression for time "t" as follows:

$$V_C = V_{CC} - (V_{CC} - V_0)e^{-\frac{t}{RC}}$$

Subtract $V_{CC}$ from both sides:

$$V_C - V_{CC} = -(V_{CC} - V_0)e^{-\frac{t}{RC}}$$

Now multiplying throughout by –1 gives

$$V_{CC} - V_C = (V_{CC} - V_0)e^{-\frac{t}{RC}}$$

Now dividing both sides by $(V_{CC} - V_0)$ gives

$$\frac{V_{CC} - V_C}{V_{CC} - V_0} = e^{-\frac{t}{RC}}$$

Taking ln of both sides, we get

$$ln\left(\frac{V_{CC}-V_C}{V_{CC}-V_0}\right)=-\frac{t}{RC}$$

Now multiplying both sides by -RC, we get

$$t=-RCln\left(\frac{V_{CC}-V_C}{V_{CC}-V_0}\right)$$

Knowing the capacitor will charge up via the resistor R5 and putting the values in, we get

$$t=-100E^3 20^{-6} ln\left(\frac{12-0.7}{12--11.3}\right)=1.45s$$

If we look at Figure 10-2, we can see that the output voltage, trace Out1, stays high for approximately 1.44s; see cursor B – A. This does confirm that our workings are correct, and we can use the expressions we have derived.

I make no real apologies for taking you through these derivations. I feel it is essential that engineers realize that we can use maths to prove some of our concepts. I also want to show you that it is not that difficult a process, it just takes a bit of practice and a reason for doing it. I hope you can see that going through these types of mathematical proofs also reinforces some of the basic concepts of electrical analysis. That is another benefit we get from doing this type of work.

The analysis of the monostable has been very detailed, but there has been a lot that I wanted to show you. I hope you have found it informative and useful. The concepts we have learned will be used in the following analysis we go through.

## Exercise 10.1

1. Design a monostable that forces the output to stay high for 500ms when the value of C2 in Figure 10-1 is 1μF.
2. Calculate the value of C1 if we were using a value of 25k for R5 and we wanted the output to stay high for 1.5s.

# The Bistable Multivibrator

The next circuit we will look at is shown in Figure 10-6.

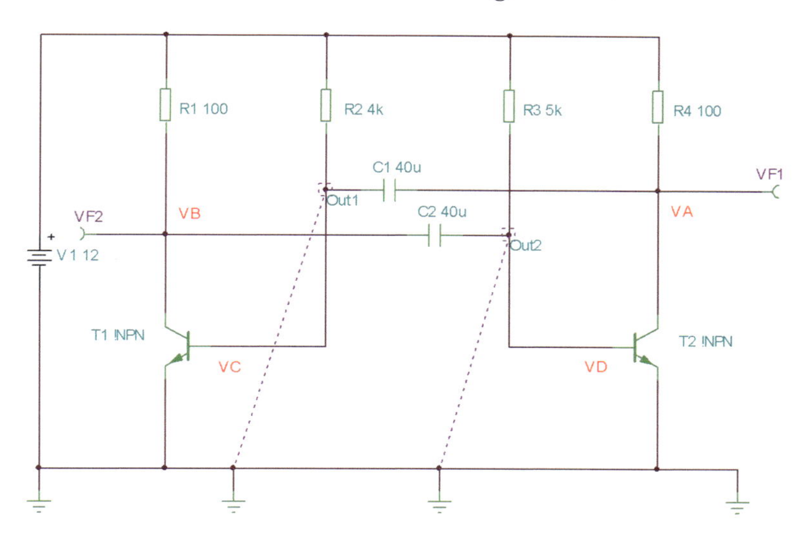


***Figure 10-6.*** *The Bistable Multivibrator*

As its name suggests, this circuit has two stable states, which basically means the output will switch between being high and low, and so it will produce a square wave type signal at the output. They are not ac type square waves as they do not change polarity, and that's why we don't call it an oscillator.

The main difference between this and the monostable circuit is that both transistors have a CR circuit controlling the voltage at their bases. Also, there is no need for a triggering input as it switches between stable states itself. It is possible to have two outputs taken from the circuit, one from each of the collectors, and they will be in antiphase with each other. The associated waveforms from the circuit are shown in Figure 10-7.

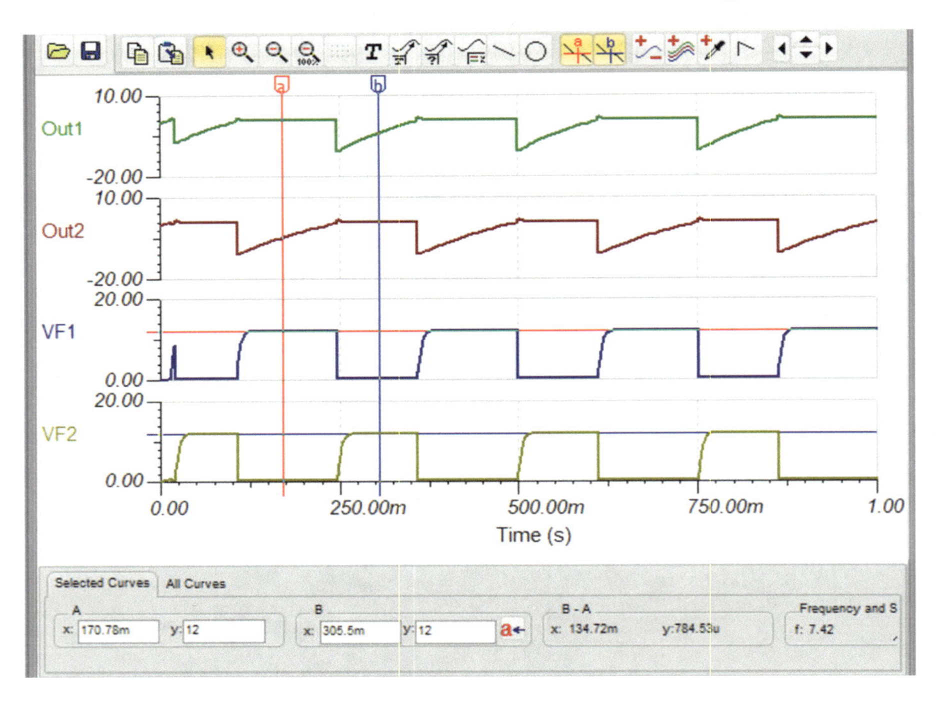


***Figure 10-7.** The Waveform for the Bistable Multivibrator*

The trace VF1 is the output from the transistor $T_2$, and VF2 is the output from $T_1$. The voltage at the base of $T_2$ is shown as trace Out2 and at the base of $T_1$ is shown as Out1. When $T_1$ is turned on, the capacitor $C_2$ charges up via the resistor R3 and $T_1$. When the voltage across $C_2$ reaches around 0.7V, $T_2$ turns on and so drives the base of $T_1$, that is, point VC on the circuit, to around –11V, thus ensuring $T_1$ is turned off. The reason why the base of $T_1$ goes to around –11V is that while $T_1$ was turned on the voltage at

VC went down to around 0.7V, while the voltage at VA went to 12V. This means there was around 11.3V across the capacitor $C_1$. We can see from the Out2 trace in Figure 10-3 that we cannot change the voltage across a capacitor instantly. Therefore, when $T_2$ turns on and forces the voltage at VA to go close to 0V, then to maintain this 11.3V across $C_1$, the voltage at VC must go lower than VA by approximately 11.3V. This ensures the voltage at VA, which has just gone down to 0V, is still 11.3V more positive than the voltage at VC.

Now that $T_1$ is turned off and $T_2$ has turned on, the capacitor $C_1$ now charges up from around -11.3V toward $V_{CC}$. However, when the voltage at VC reaches around 0.7V, $T_1$ turns back on and forces the voltage at VB to go to around 0V. Then, because the capacitor $C_2$ has now around 11.3V across it, the voltage at VD is driven to around -11V. This forces $T_2$ to turn off. The cycle starts to repeat automatically, and so we get the two stable states at the outputs of the circuit.

The time constant of $R_2$ and $C_1$ controls how long the output at VA, that is, VF1, is at 0V. Then the time constant of $R_3$ and $C_2$ controls how long the output at VA is at $V_{CC}$. We have derived an expression that describes how the voltage across a capacitor changes with time when the capacitor has an initial voltage across it. In this bistable circuit, we can see that we charge the capacitors up from around -11V, not 0V. The expression that describes how the voltage changes over time is

$$V_C = V_{CC} - \left(V_{CC} - V_0\right)e^{-\frac{t}{RC}}$$

The term $V_0$ is the initial voltage across the capacitor. The derivation of this expression has been done earlier in this chapter. The expression can be rearranged for time "t" to show

$$t = -RCln\left(\frac{V_{CC} - V_C}{V_{CC} - V_0}\right)$$

We should be able to use this expression to determine how long it will take for the CR circuit to reach the 0.7V, which is needed to turn the transistors on. Using the CR circuit made up of $R_3$ and $C_2$ which charges up when $T_1$ is turned on, we can calculate how long it will take to turn $T_2$ on and $T_1$ off. This will be the period of time during which the VF2 voltage at the collector of $T_1$ is low, the space time of the square wave. Putting these values in, we get

$$t = -40^{-6} \times 5^3 \times ln\left(\frac{(12-0.7)}{(12--11)}\right)$$

$$t = -0.2 \times ln(0.4913) = 142ms$$

If we look at the traces shown in Figure 10-8, we can see that the time period that VF2 is low is 138.52ms. This is pretty close to our calculated value. Note, this selecting the voltage at VF2 as the squarewave output.

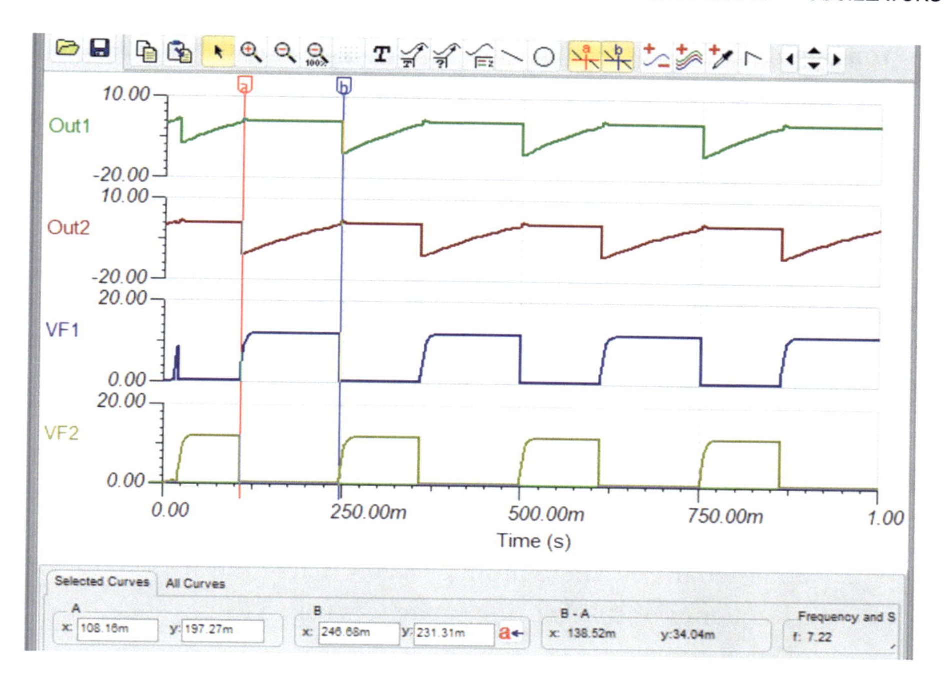


***Figure 10-8.** The Output Traces Showing the Outputs of the Bistable Circuit*

The time period that the voltage at VF2 is high, the mark time of the square wave, is controlled by the time constant of the CR circuit $C_1$ and $R_2$ as the capacitor charges up via $T_2$. We can use the same expression to calculate the time the voltage is high as

$$t = -40^{-6} \times 4^{3} \times ln\left(\frac{(12-0.7)}{(12--11)}\right)$$

$$t = -0.16 \times ln(0.4913) = 114ms$$

When we use the cursors to measure this time period, we get a value of 111ms. This is again close to what we have calculated. This does help to confirm the expression for the circuit.

You might think that creating a 50/50 duty cycle would be quite an easy task as all it requires is that the CR time constants made up of $R_3$, $C_2$ and $R_2$, $C_1$ are the same. However, when you make the two time constants the same, the circuit could crash. Therefore, using empirical design, I suggest the closest value to 5k value for $R_2$ would be 4.7k.

## Exercise 10.2

1. Design a bistable multivibrator to produce a 5kHz square wave using 10μF capacitors when the mark time was 75% and the space time was 25% of the total periodic time of the square wave.

2. Redesign the circuit to produce a 20% mark time at a frequency of 10kHz using capacitors with a value of 20μF.

## The Phase Shift Oscillator

This is the first of the real oscillator circuits that we will look at. An oscillator will produce an output that swings both positive and negative, and this differentiates it from the bistable we have just looked at. We will restrict the analysis of oscillators to those that use the Opamp. This is because it is easier to set the gain of the Opamp, and they produce a more reliable oscillator output.

## The Requirements of the Oscillator Circuit

To analyze the oscillator, we need to understand its requirements. These are identified as follows:

- The oscillator should keep itself going, and so the input must come from the output in some form of a feedback signal.
- This feedback signal must be fed back in such a way that it will help the circuit maintain the output at its steady state values.
- This means it must be fed back in phase with the output.
- This means the overall phase shift across the circuit must be 0 degrees.
- Normally, as we are using the inverting input of the Opamp, then the output would be 180° out of phase, that is, in antiphase with the input.
- This means we should add a further 180° phase shift in the feedback path to bring the feedback back in phase with the input.
- Also, as the feedback circuit which produces this 180° phase shift it will inevitably attenuate the signal before it gets to the input of the oscillator, the oscillator should have its own gain that recoups the loss from the feedback circuit.
- This concept of what the oscillator circuit should do can be represented in the block diagram shown in Figure 10-9.

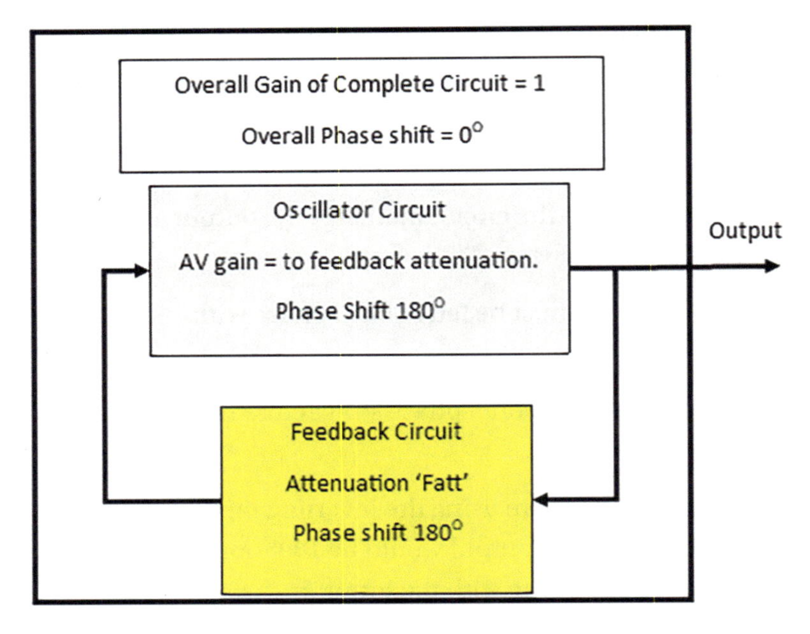


***Figure 10-9.*** *The Block Diagram of an Oscillator Circuit*

You should appreciate that there is no input to the oscillator, only an output. The input is simply some of the output fed back to the oscillator.

## The Phase Shift Oscillator

There are two main oscillator circuits we will look at. The first is the phase shift oscillator, and the basic circuit is shown in Figure 10-10. The second is the Wien Bridge oscillator, which we will look at later in this chapter.

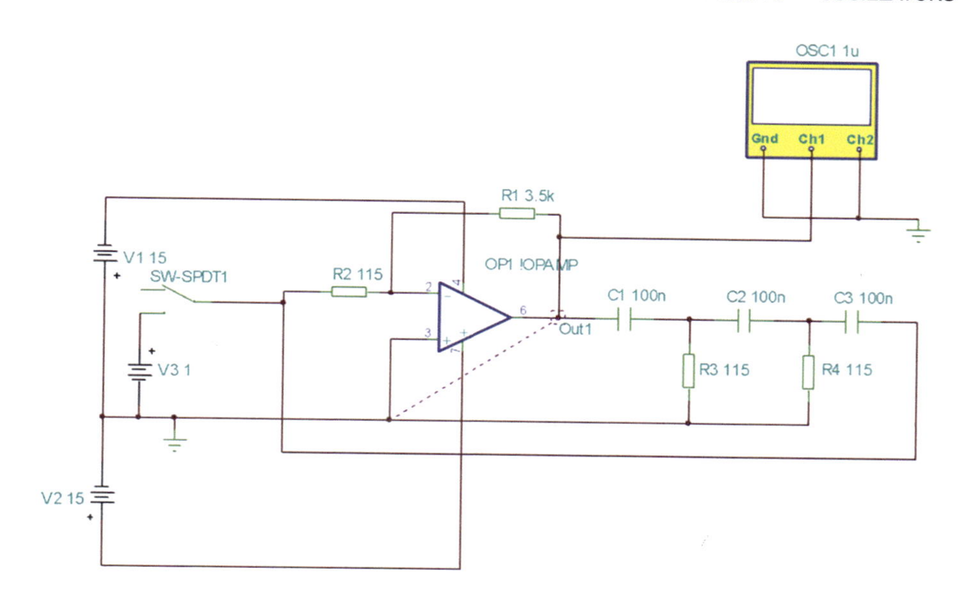


***Figure 10-10.*** *The Opamp Circuit for the Phase Shift Oscillator*

The basic inverting Opamp will normally produce a 180° phase difference between its input and its output. This means the feedback circuit must produce a further 180° to create a full 360° phase shift or 0° at the input. The feedback circuit is made up of the three CR combinations of $C_1$ with $R_3$, $C_2$ with $R_4$, and $C_3$ with $R_2$. Before we look at the three combinations in the feedback path, we will look at a single RC circuit to see how it brings about this phase shift. The basic CR circuit is shown in Figure 10-11.

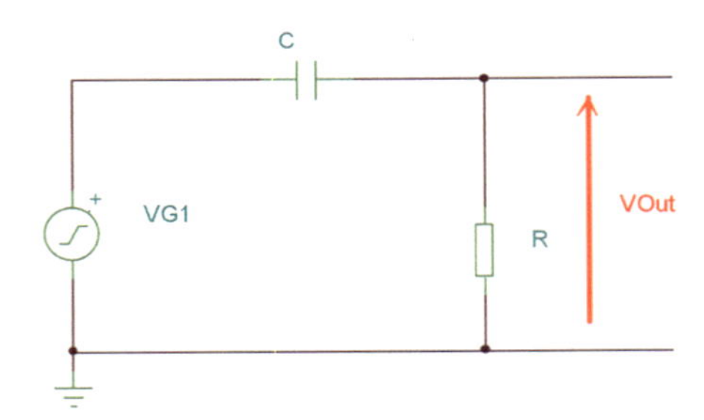


***Figure 10-11.*** *A Single CR Phase Shift Circuit*

This is actually the same as the high pass filter circuit we have looked at in Chapter 6 and indeed the differentiator circuit we looked at in Chapter 9. The circuit can be put to many different uses, and producing a phase shift is the one we will be looking at now.

The circuit will simply divide the input signal between the capacitor and resistor. However, because the capacitor is a reactive component, and its impedance varies with frequency, whereas the resistor is a passive component, and its resistance does not vary with frequency, then the division of the input signal will vary with the frequency of the signal. In Figure 10-11, the output is taken from across the resistor. The input voltage "$V_{IN}$" is the VG1 signal. We can create an expression that shows how the two components divide the input signal using the voltage divider rule as follows:

$$v_{out} = \frac{v_{in}R}{R - j\dfrac{1}{\omega C}}$$

I feel I should explain how I am referring to the impedance of the capacitor. I am using "complex numbers" as they can be used to represent phasor quantities. I am sure we use that name "complex numbers" to try and show how clever we are. However, they are not complex, and we will study more about how we can use them in Chapter 11. I am using the term "j" to show that the capacitive impedance appears on the vertical axis of a phasor diagram; it's on the negative part of the vertical axis. The term "ω" is just a shorthand way of writing 2πf, that is, ω = 2πf. The resistor goes on the positive horizontal axis. This means that there is a 90° difference between the capacitor and the resistor. However, this is the maximum phase difference we can have between them.

Really, we can use complex numbers to represent the phasor quantities that exist with most electrical quantities and components. This means that most quantities and components have a magnitude, their measured value, and an angle, their phase value, associated with them. With resistors, being the only nonreactive component, the angle is 0, and so we don't need to state it with them. This is because with resistor the current flowing through it is always in phase with voltgae across it. However, with all other values and components, we should always state the angle as well as the value, or magnitude, as they are all phasor quantities. I hope this explains the way I have stated the expression for $v_{out}$. Note, I am using lowercase symbols as these are ac values.

If we now divide both sides by $v_{in}$, we get

$$\frac{v_{out}}{v_{in}} = \frac{R}{R - j\frac{1}{\omega C}}$$

The term $\frac{v_{out}}{v_{in}}$ is known as the gain or in this case as the "transfer function (TF)" of the circuit. It can be used to show how the input is transferred by the circuit, or system, to become the output. We will use the TF to show how the circuit produces some phase shift.

It is normal in a transfer function to have a numerator that has a value of 1; we will see why later. Therefore, we must divide the top and bottom by R or multiply the top and bottom by $\frac{1}{R}$ which does the same job. This gives

$$TF = \frac{R \times \frac{1}{R}}{R \times \frac{1}{R} - j\frac{1}{\omega C} \times \frac{1}{R}}$$

Canceling out where we can gives

$$TF = \frac{1}{1 - j\dfrac{1}{\omega CR}}$$

This is the division of two complex numbers as we can say $1 = 1 + j0$:

$$TF = \frac{1 + j0}{1 - j\dfrac{1}{\omega CR}}$$

These two complex numbers can be converted into their polar format, which would make the division easier. Converting rectangular to polar can be shown as being

$$a + jb \; rectangular = \sqrt{a^2 + b^2} \;\; \langle Tan^{-1}\frac{b}{a}\rangle$$

where

$$\sqrt{a^2 + b^2} \; is \; the \; magnitude$$

and

$$\langle Tan^{-1}\frac{b}{a}\rangle \; is \; the \; associated \; angle$$

With respect to the numerator of our transfer function, we can say that "a" = 1 and "b" = 0. With respect to the denominator, we can say "a" = 1 and "b" $= -\dfrac{1}{\omega CR}$.

Changing the transfer function in this way gives

$$TF = \frac{\sqrt{1^2 + 0^2}\ \langle Tan^{-1}\frac{0}{1}\rangle}{\sqrt{1^2 + \left(-\frac{1}{\omega CR}\right)^2}\ \langle Tan^{-1}\frac{-\frac{1}{\omega CR}}{1}\rangle} = \frac{1\langle 0\rangle}{\sqrt{1^2 + \left(-\frac{1}{\omega CR}\right)^2}\ \langle Tan^{-1} - \frac{1}{\omega CR}\rangle}$$

To carry out this division, we must divide the two magnitude parts and subtract the two angle parts. This means that the transfer function now becomes

$$TF = \frac{1}{\sqrt{1^2 + \left(-\frac{1}{\omega CR}\right)^2}} - \langle Tan^{-1} - \frac{1}{\omega CR}\rangle$$

We have stated that the maximum phase shift we can produce is 90°. However, this can only happen if the resistive value was 0. If R was 0, the ωCR would be 0, and the angle would become

$$-Tan^{-1} - \frac{1}{0} = -Tan^{-1} - \infty = 90$$

It is not possible for the resistor to have a value of 0, so we cannot create a CR circuit that has a phase shift of 90°. If we could, then we would only need two such phase shift CR circuits. The only other shift we could use would be a phase shift of 60° as we want the phase shift for each CR circuit to be the same throughout. This means we use three such CR combinations as 3 × 60 = 180.

This means the angle part of the transfer function must equate to 60°. The Tan of 60 is 1.732 radians. We must use radians and not degrees as we are using "w" to describe the angle. This means that the term

$$\frac{1}{\omega CR}\ must\ equal\ 1.732$$

For this to happen, we must know the values of ω, C, and R. The value of ω depends upon the frequency of oscillation we want. In this example, we will choose a frequency of 8kHz. This means that

$$\omega = 2 \times \pi \times 8^3 = 50,265.5$$

The units for this value would be Radians or Rads.

If we choose an arbitrary value of 100nF for the capacitor, then we have

$$\frac{1}{\omega CR} = 1.732$$

which means

$$R = \frac{1}{1.732 \times \omega C} = \frac{1}{1.723 \times 50265.5 \times 100^{-9}} = 114.86$$

Let R = 115Ω.

If we simulate the CR circuit, shown in Figure 10-11, with the values calculated, we get the traces of the input and output waveforms as shown in Figure 10-12.

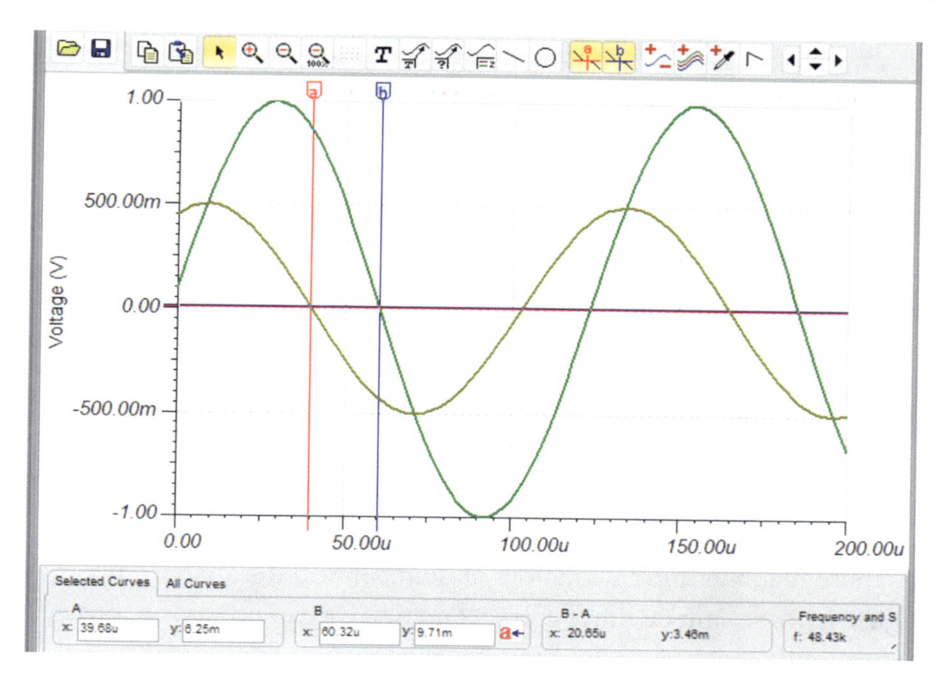


***Figure 10-12.*** *The Input and Output Waveforms of the CR Phase Shift Circuit*

The two cursors are being used to measure the time difference between the output, which is the smaller of the two waves, and the input. The time difference is shown as being approximately 20.65μs. This time difference can be used to calculate the phase difference as follows:

$$Phase\ Difference = time\ difference \times 360 \times Frequency$$

Therefore, we have

$$PD = TD \times 360 \times F = 20.65^{-6} \times 360 \times 8^{3} = 59.472^{\circ}$$

This is very close to the expected phase shift of 60°. Also, because the output waveform reaches zero before the input, then the output waveform is leading the input waveform, which gives the required phase shift of +60°. Note, the two negative signs in the expression for the angle make the angle a positive value.

One other parameter we can see from the display in Figure 10-12 is the voltage gain that can be calculated as

$$v_{gain} = \frac{v_{out} Peak\ to\ Peak}{v_{in} Peak\ to\ Peak} = \frac{1}{2} = 0.5$$

This shows that the CR phase shift circuit actually attenuates the input signal. Using the magnitude part of the transfer function, we can determine the magnitude of the TF as follows:

$$|TF| = \frac{1}{\sqrt{1^2 + \left(-\frac{1}{\omega CR}\right)^2}} = \frac{1}{\sqrt{1^2 + \left(-\frac{1}{50265.5 \times 100^{-9} \times 115}\right)^2}} = \frac{1}{\sqrt{1 + 2.993}} = \frac{1}{2}$$

These readings and calculations should confirm the analysis of the single CR phase shift circuit is correct and that we can now go on and apply a similar analysis to the complete feedback circuit.

The complete feedback circuit is shown in Figure 10-13.

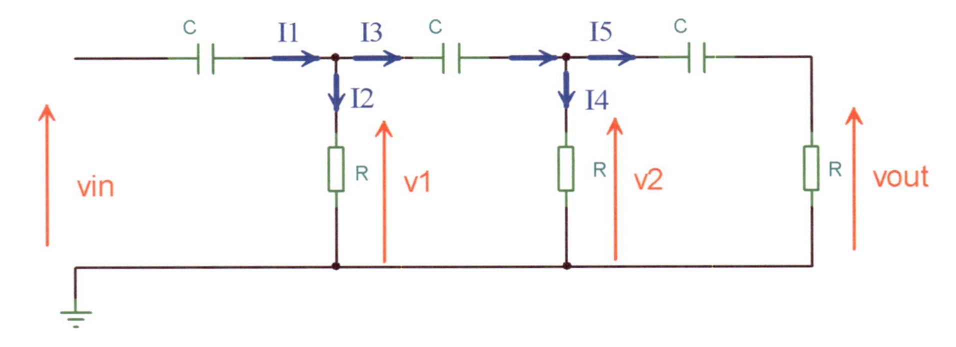


***Figure 10-13.*** *The Three-Stage Feedback Circuit*

You might be forgiven for saying that the circuit shown in Figure 10-13 is not the same as the feedback circuit shown in Figure 10-10. This is because the third resistor, which is $R_2$ in Figure 10-10, is not physically connected to ground, as it is in Figure 10-13. However, we need to remember that there is a virtual short circuit across the input terminals of the Opamp. This means that pin 2 of the Opamp is, to all intents and purposes, connected to ground which connects $R_2$ to ground. Also, it is the current flowing through $R_2$ that flows around the feedback path of the Opamp, that is, it flows through $R_1$. That is what controls the output voltage of the Opamp. The current that flows through $R_1$ is the same current as shown as I5 in the circuit shown in Figure 10-13. Therefore, the two feedback circuits are the same.

We can show that the transfer function for the feedback circuit is

$$TF = \frac{1}{1 - \frac{5}{(\omega CR)^2} + j\left(\frac{1}{(\omega CR)^3} - \frac{6}{\omega CR}\right)}$$

The derivation of this expression is shown in the appendix.

We can use this expression to determine the phase and the magnitude of the transfer function. We will look at the phase of the transfer function first. We are using this feedback circuit to introduce a further 180° phase shift to ensure that the feedback is fed back in phase with the original input to the Opamp. If we examine the graph of tan (θ), we can see that at 180° the value of tan (180) = 0. This means that the "j" term of the transfer function must equal 0. This means that

$$\frac{1}{(\omega CR)^3} - \frac{6}{\omega CR} = 0$$

Therefore

$$\frac{1}{(\omega CR)^3} = \frac{6}{\omega CR}$$

Therefore, we can say

$$1 = \frac{6(\omega CR)^3}{\omega CR} = 6(\omega CR)^2$$

Therefore, we can say

$$(\omega CR)^2 = \frac{1}{6} \; or \; \frac{1}{(\omega CR)^2} = 6$$

This means that we can say

$$\omega CR = \sqrt{\frac{1}{6}}$$

From this, we can say

$$f_o = \frac{1}{2\pi CR}\sqrt{\frac{1}{6}}$$

This can be rewritten as

$$f_o = \frac{1}{2\pi CR} \times \frac{\sqrt{1}}{\sqrt{6}}$$

Therefore, the frequency at which the feedback circuit will produce the required phase shift of 180° would be

$$f_o = \frac{1}{2\pi CR\sqrt{6}}$$

To determine the magnitude of the feedback circuit, we can again use the concept that with a phase shift of 180° we can say

$$\frac{1}{(\omega CR)^3} - \frac{6}{\omega CR} = 0$$

Putting this into the expression for the transfer function, we can say

$$TF = \frac{1}{1 - \frac{5}{(\omega CR)^2} + j0}$$

This would produce a magnitude of

$$|TF| = \frac{1}{\sqrt{\left(1 - \frac{5}{(\omega CR)^2}\right)^2 + 0^2}}$$

From before, we know

$$\frac{1}{(\omega CR)^2} = 6$$

Substituting this in, we get

$$|TF| = \frac{1}{\sqrt{(1 - 5 \times 6)^2}} = \frac{1}{\sqrt{(-29)^2}} = \frac{1}{29}$$

This shows that the feedback circuit will attenuate the signal by a factor of 29. This means that to ensure the overall gain of the oscillating circuit is unity, then the gain of the actual Opamp must be 29 or really just a little bit larger than 29. The gain of the Opamp in the circuit shown in Figure 10-10 is

$$v_{gain} = \frac{R_f}{R_{in}} = \frac{R_1}{R_2} = \frac{3.5k}{115} = 30.43$$

If we use the values of the components in Figure 10-10, we will be able to calculate the frequency that the circuit will oscillate at. The values from that circuit mean that C = 100nF and R = 115. Therefore, we can calculate the frequency of oscillation as

$$f_o = \frac{1}{2\pi CR\sqrt{6}} = \frac{1}{2\times\pi\times100^{-9}\times115\times\sqrt{6}} = 5.65kHz$$

If we simulate the circuit and measure the display of the output voltage, we should be able to determine the frequency of the oscillator. The output voltage waveform is shown in Figure 10-14.

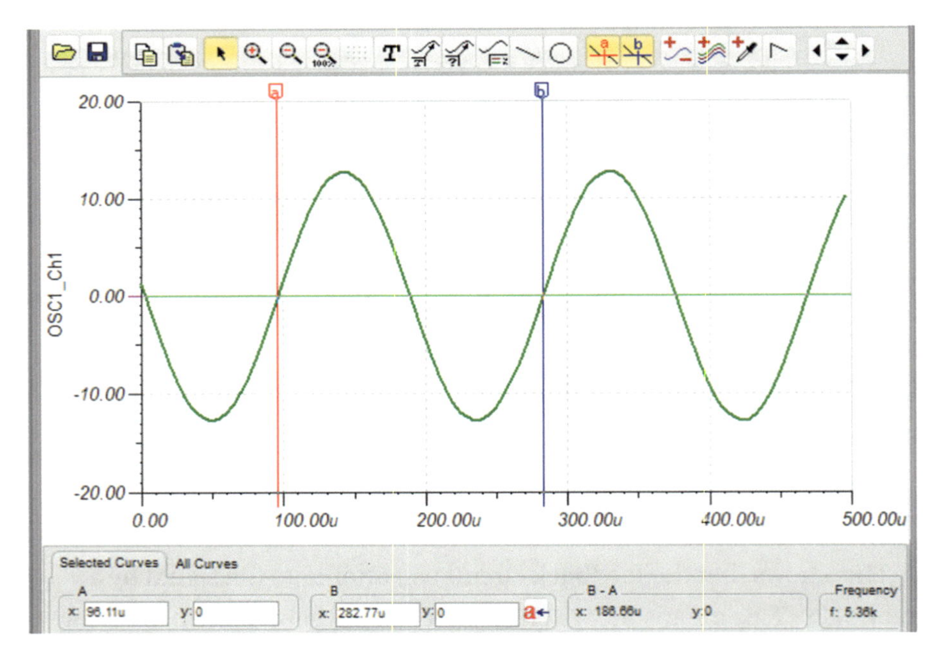


***Figure 10-14.*** *The Waveform of the Output Voltage of the RC Phase Shift Oscillator*

We can use the two cursors to measure the periodic time for the waveform, and from the display, we can see that the periodic time "T" = 186.66µs. Using this value, we can calculate the frequency as follows:

$$f_o = \frac{1}{T} = \frac{1}{186.67E^{-6}} = 5.357kHz$$

This is very close to the calculated value. This should give us some confidence in the theory explained here of how the CR phase shift oscillator works. One thing I should mention is that in real life the oscillator should get enough of an input voltage from stray noise that could be induced in the wiring of the circuit. However, with the simulation, it may be necessary to provide the noise by switching the switch down to the 1V battery momentarily so as to trigger the oscillator into action.

## Exercise 10.3

1. With respect to the phase shift circuit shown in Figure 10-11, design the circuit to produce a phase shift of 75° using a resistor value of 2.2k.

2. Design an RC phase shift oscillator that has a frequency of 1200Hz using the existing 100nF capacitors.

# The Wien Bridge Oscillator

This is another oscillator that uses the Opamp to provide some gain in the oscillator. The circuit for the Wien Bridge oscillator is shown in Figure 10-15.

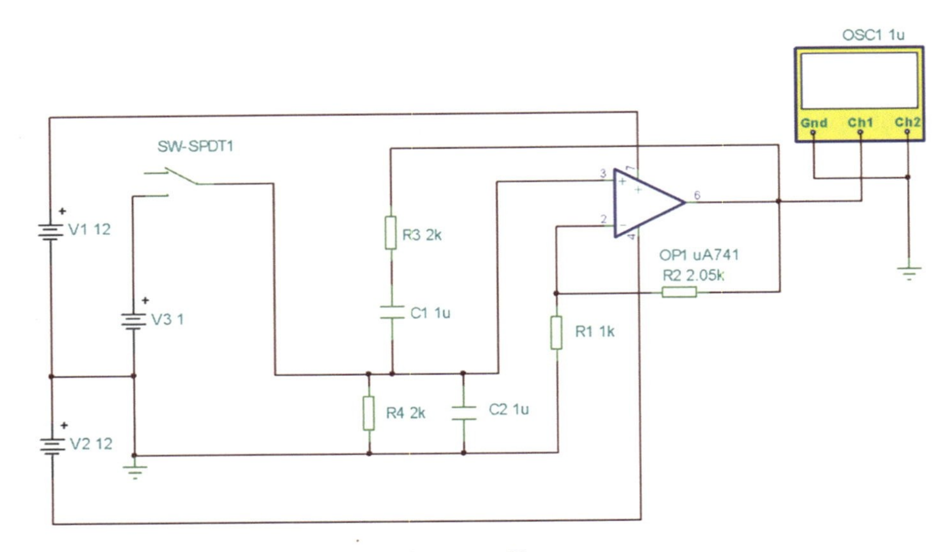


***Figure 10-15.*** *The Wien Bridge Oscillator*

The principle of operation for the oscillator is the same as the phase shift oscillator in that some of the output is fed back to the input, and the overall gain of the oscillator circuit should be unity or 1. However, the Opamp is being operated in its non-inverting configuration, and the phase shift in the feedback circuit should be 0°. To see how this is brought about and what the gain of the Opamp must be set to, we could analyze the feedback circuit shown in Figure 10-16.

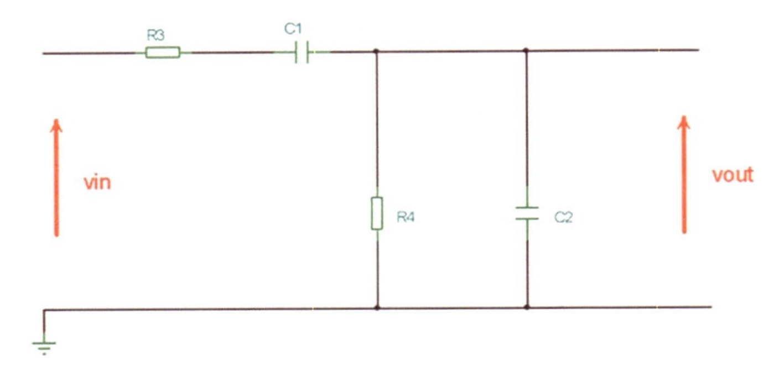


***Figure 10-16.*** *The Feedback Circuit for the Wien Bridge Oscillator*

Using the voltage divider rule, it can be shown that the transfer function is

$$TF = \frac{0 + j\omega CR}{1 - \omega^2 C^2 R^2 + 3j\omega CR}$$

# Exercise 10.4

As an exercise, see if you can derive the expression for the transfer function. My solution is in the appendix. As an aid to get you started, you can represent the feedback circuit as shown in Figure 10-17.

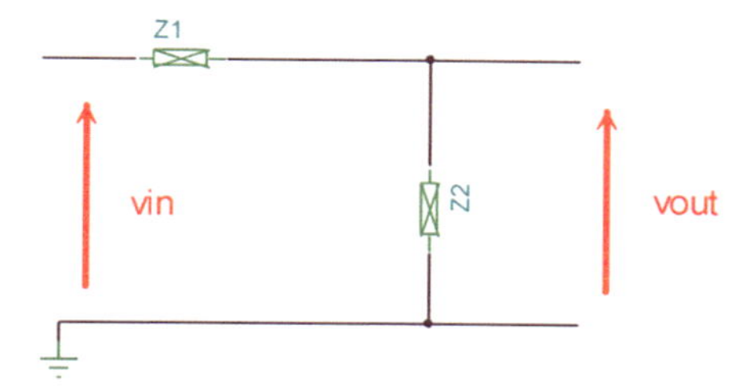


***Figure 10-17.*** *The Representation of the Feedback Circuit*

The impedance $Z_1$ is

$$Z_1 = R_3 - j\frac{1}{\omega C_1}$$

The impedance $Z_2$ is the parallel combination of $R_4$ with $C_2$:

$$Z_2 = \frac{R_4 \times -j\dfrac{1}{\omega C_2}}{R_4 - j\dfrac{1}{\omega C_2}}$$

The purpose of asking you to practice this sort of work is not to torture you, but to try and help develop your ability to use mathematics as a way of proving your theories and ensuring your analysis stand up to investigation. This is a skill that is essential for good engineers. So do try it, take your time, and you might just surprise yourself.

## The Attenuation and Phase Shift of the Feedback Path

Using the expression for the transfer function, we could develop an expression for the attenuation and phase shift of the feedback path. We know the feedback path will attenuate the signal from our work with the phase shift oscillator and also because the circuit uses only passive components.

To help develop this expression, it might be best to change the format of the complex numbers from their current rectangular format to their polar format. This would give us

$$TF = \frac{\sqrt{0^2 + (\omega CR)^2}\ \langle Tan^{-1}\left(\frac{\omega CR}{0}\right)\rangle}{\sqrt{\left(1-\omega^2C^2R^2\right)^2 + (3\omega CR)^2}\ \langle Tan^{-1}\left(\frac{3\omega CR}{1-\omega^2C^2R^2}\right)\rangle}$$

If we consider the angle part of the complex number on the numerator, we have

$$angle = \langle Tan^{-1}\left(\frac{\omega CR}{0}\right)\rangle$$

I hope you can appreciate that any number divided by "0" will produce a result of infinity, ∞. If you are not sure, then try this. Let $0 = 1E^{-100}$, a very, very small value. Then infinity = $1E^{100}$, a very, very large value. Then use your calculator to divide $100/1E^{-100}$. You should get $1^{102}$, a very, very large value.

That being the case, then if we examine the graph of Tan (θ) we should see that when (θ) = 90°, then Tan (θ) = infinity. This means that the angle associated with the numerator must be 90°. Knowing that when we divide complex numbers in their polar format, we simply subtract the angles. Then this means that the angle associated with the denominator must also equate to 90° to give 90 – 90 = 0. This means that

$$\frac{3\omega CR}{1-\omega^2C^2R^2}\ must = infinity\ \infty$$

This will happen when the denominator equates to 0. Therefore, we can say

$$1-\omega^2C^2R^2 = 0$$

This means that

$$1=\omega^2C^2R^2=\left(\omega CR\right)^2$$

Therefore, taking the square root of both sides, we can say

$$1=\omega CR$$

From this, we can say

$$f_o=\frac{1}{2\pi CR}$$

This then is the expression we can use to determine the frequency of oscillation if we know the values of C and R.

If we now consider the magnitude of the transfer function, we can say

$$|TF| = \frac{\sqrt{0^2 + (\omega CR)^2}}{\sqrt{(1-\omega^2C^2R^2)^2 + (3\omega CR)^2}}$$

## Exercise 10.5

Complete the simplification of the transfer function to show that the magnitude of the transfer function will result in an attenuation of 1/3. As a hint, you should remember

$$1-\omega^2C^2R^2 = 0$$

It only requires three more steps. Good luck.

The solution is in the appendix.

# The Gain of the Opamp

Knowing that the feedback circuit attenuates the signal by a 1/3, then this means that the gain of the Opamp should be just a bit more than "3." Knowing that the gain of the non-inverting Opamp can be calculated as

$$v_{gain} = 1 + \frac{R_1}{R_2}$$

Then using the values from the circuit, we have

$$v_{gain} = 1 + \frac{2.05^3}{1^3} = 3.05$$

Also, using the values from the circuit, we can calculate the frequency of oscillation as

$$f_o = \frac{1}{2\pi CR} = \frac{1}{2\times\pi\times 1^{-6}\times 2^3} = 79.58Hz$$

If we simulate the circuit shown in Figure 10-15 and use the oscilloscope to measure the output waveform, we will get the display as shown in Figure 10-18.

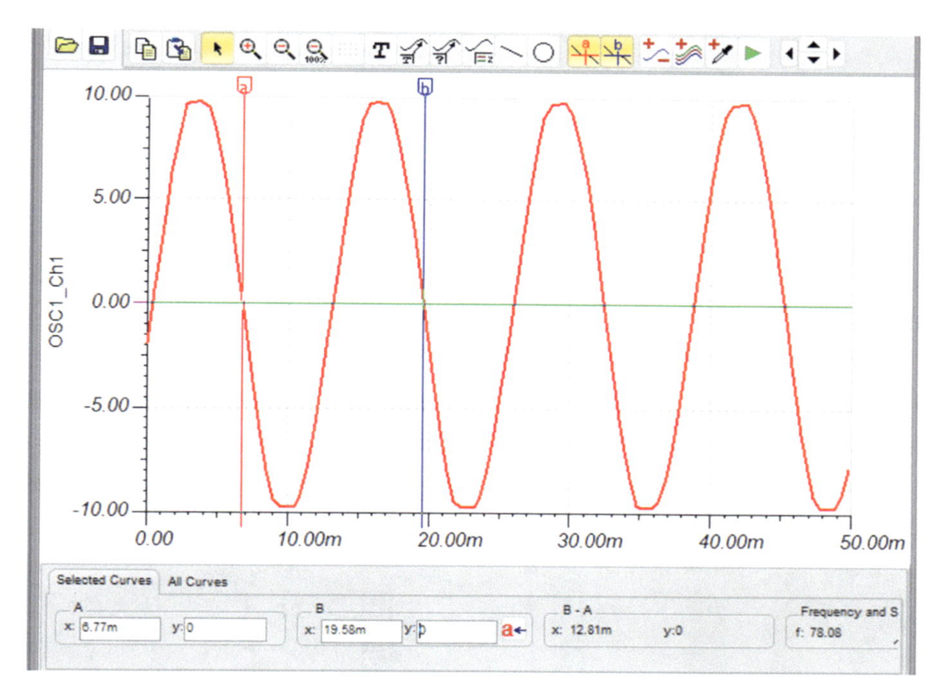


***Figure 10-18.*** *The Output Voltage of the Wien Bridge Oscillator*

Using the two cursors, we can measure the periodic time which was 12.81ms. Using this value, we can calculate the frequency as

$$f_o = \frac{1}{12.81E^{-3}} = 78.06Hz$$

This does agree with our calculations.

## Exercise 10.6

1. Design a Wien Bridge oscillator that has a frequency of 1kHz using a 50nF capacitor.
2. Design a Wien Bridge oscillator that has a frequency of 10kHz using a 500nF capacitor.

# The 555 Timer

The next device we will look at is the 555 timer. It does not actually produce a waveform that goes both positive and negative, so it might be wrong to call it an oscillator; however, we looked at the multivibrator at the beginning of this chapter, so we should look at the 555 timer.

This is an integrated circuit or IC, which can be used as a monostable, that is, a device which has one stable state at its output, either low or high, and as an astable, that is, a device which produces a free running square wave at its output. To appreciate how it works, a study of what the eight pins are used for would be helpful. The pin out is shown in Figure 10-19.

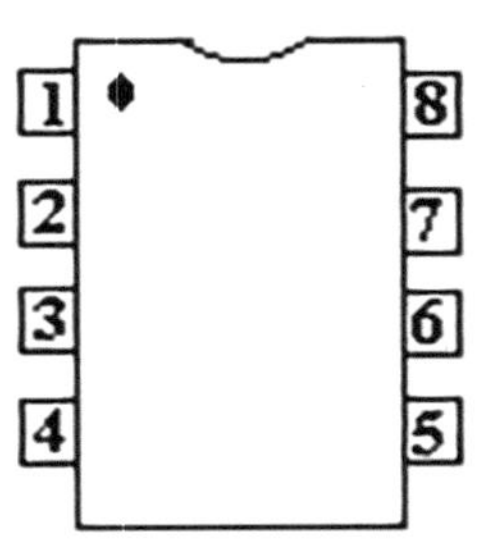


***Figure 10-19.*** *The 555 Timer IC*

# The Pins of the 555 Timer

- Pin 1: This is used to connect the IC to the ground.
- Pin 2: This is the trigger input. The output at pin 3 of the chip is normally held at ground or logic low, usually between 0v and 0.4v. The output will go high, that is, to $V_{CC}$ - 1.7V, when the input at the trigger, or pin 3, goes down from $V_{CC}$ to approximately 1/3 of $V_{CC}$. For example, if $V_{CC}$ was 12 volts, then when the voltage at this pin drops to 4V or less, the chip is triggered, and the output at pin 3 will go high, that is, $V_{CC}$ - 17v, therefore 10.3V, if $V_{CC}$ is 12V. Note if the trigger input is held low for a time longer than a period determined by the external CR components, then the output will stay low until the trigger input is driven high again.
- Pin 3: This is the output of the chip, and it can be made to switch high, for a set period of time, before returning to its stable state of logic low. Or it can be made to produce a square wave output at a frequency and mark to space ratio determined by the CR components around the chip. Note the output is normally capable of sourcing 200mA, that is, delivering 200mA to a load, or sinking 200mA, that is, providing a path for up to 200mA to flow through the chip to ground.
- Pin 4: This is the reset pin, and it can be used to reset the output by driving the voltage at this pin low, that is, to a voltage between 0v and 0.4v. This pin can be used to reset the output regardless of the states at any of the other pins. If this pin is not to be used as a reset pin, then it can be held high by connecting to $V_{CC}$.

- Pin 5: This is the control voltage input pin. By applying a voltage between 45% and 90% of $V_{CC}$, it is possible to vary the timing of the device independent of the external CR network. This controls the chip in the monostable mode and so controls the width of the pulse independent of the CR network. When using the chip in the astable mode, this control voltage can be varied from 1.7V to $V_{CC}$ to enable control of the frequency of the square wave at the output. In most cases, this pin is not used, and so it should be grounded normally via a capacitor to prevent noise affecting the chip.
- Pin 6: This is the threshold input. It is used to reset the output back to logic low when the voltage at pin 6 rises to 2/3 of $V_{CC}$. Therefore, if $V_{CC}$ was 12V, then when this pin rose to 8V, the output would reset to a logic low. It should be noted that for the timer to work at all, there must be a minimum current of 0.1μA flowing into this pin. This concept sets the maximum value of resistance connected between $V_{CC}$ and this pin.
- Pin 7: This is the discharge pin. There is an internal NPN transistor, and when this is turned on, it is used to provide a discharge path to ground for the timing capacitor connected to this pin.
- Pin 8: This is the $V_{CC}$ pin. This pin can be connected to a voltage supply from 4.5V to 16V. This makes the timer very adaptable to a range of supply voltages.

# The Timer Used As a Monostable

The stable state of the 555 timer is logic low. When triggered, the output goes high and stays high until the voltage at the discharge pin, which is connected to the threshold pin, reaches 2/3 of $V_{CC}$. In this case, when $V_{CC} = 5V$, the output would stay high until the threshold voltage reached 3.33V. Figure 10-20 shows a circuit we can use to test the basic operation of the timer when configured as a monostable.

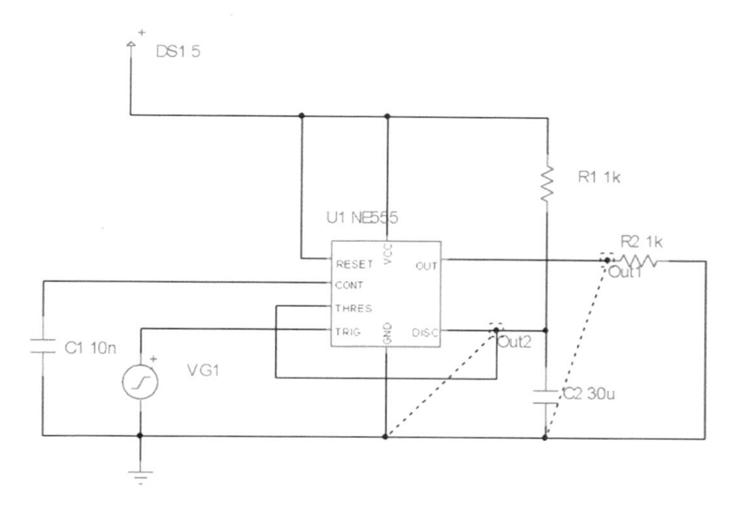


***Figure 10-20.*** *The Basic Monostable*

The basic idea of the monostable circuit is that it has one state that it wants to stay stable in, which is with the output on PIN3 at around 0V. It can be forced out of that state temporarily by driving the voltage at the “trigger input” down to around 0V. In this circuit, we are using the VG1 voltage source to drive the trigger input down to 0V. To achieve this, the VG1 source starts off at 5V. It stays at 5V for 48ms, then changes to 0V. The voltage stays at 0V for the next 2ms. The voltage then repeats the cycle forever. Applying this voltage to the trigger input would force the monostable out of its stable state, which is at around 0V, and send it to around $V_{CC}$, that is, 5V. The monostable stays in its stable state of around 0V until the trigger pulse goes down to around 0V. This then forces the output out of its stable state and sends it to around +5V.

The capacitor $C_2$ is connected to the disc and threshold pins of the monostable. It tries to charge up toward $V_{CC}$ via $R_1$. When the capacitor reaches around 3.3V, that is, 2/3 of $V_{CC}$, the output resets and returns to its stable state of around 0V. The output will remain in this stable state until the trigger input is again forced to go to 0V. In this case, this is at 50ms later.

Figure 10-3 shows the voltage waveforms we are interested in:

- OUT1 is the output on pin 3 of the timer. We can see that, because the trigger input, VG1, starts off at 5V, the output stays low as the 555 timer has not been triggered.
- OUT2 is the voltage at the threshold and disc pins of the timer, pins 6 and 7. The internal transistor at the disc pin is currently turned off, and so the capacitor $C_1$ is allowed to charge up via the resistor $R_1$ toward $V_{CC}$. This means the voltage at the threshold pin starts to rise as the capacitor charges up. When this voltage gets to 3.32V, as shown in trace OUT2 in Figure 10-3, the output of the timer switches back to its stable state, as shown by the OUT1 trace in Figure 10-3. At the same time, the internal transistor, in the timer, turns on to provide a path to earth so that the capacitor $C_1$ can discharge. When the transistor turns on, there is very little resistance between the capacitor and ground, and so the capacitor discharges almost instantly.
- VG1 is the voltage applied to the trigger input.

Using this triggering action, we can see that the timer does want to stay in the one stable state of 0V at its output. We can force the output to change from this stable state for a set period of time. After which it does return to its stable state, waiting for the trigger to go to 0V once again. We can see that the timing of this set period is set by the time it takes for the capacitor, $C_1$, to charge up to 3.32V, that is, 2/3 of the $V_{CC}$. The charging of the capacitor is controlled by the time constant set by the CR combination of $C_1$ and $R_1$. With Figure 10-20, this is set to 1k × 30μF = 30ms. From our work with analog electronics, we know the voltage across the capacitor, $C_1$, and so the input to the threshold pin follows the expression

$$V_C = V\left(1 - e^{-\frac{t}{CR}}\right)$$

Putting the values in, we have

$$V_C = 5\left(1 - e^{-\frac{t}{30ms}}\right)$$

We can transpose this expression to determine the time this voltage will take to reach the 3.32V required to send the output back to its stable state of 0V.

The expression for this time is

$$t = -CRln\left(1 - \frac{V_C}{V}\right)$$

Putting the values in, we get

$$t = -30E^{-3}\ln\left(1 - \frac{3.32}{5}\right) = 32.7ms$$

This means that 32.7ms after the output voltage rose to 5V, it would return to the stable state of around 0V.

If we look at Figure 10-21, we should see that all the preceding predictions are confirmed by all the traces displayed. Also, using the readings of the two cursors, we can confirm that the output voltage stays out of its stable state, at around 5V, for 32.77ms. This is very close to the time period we have calculated.

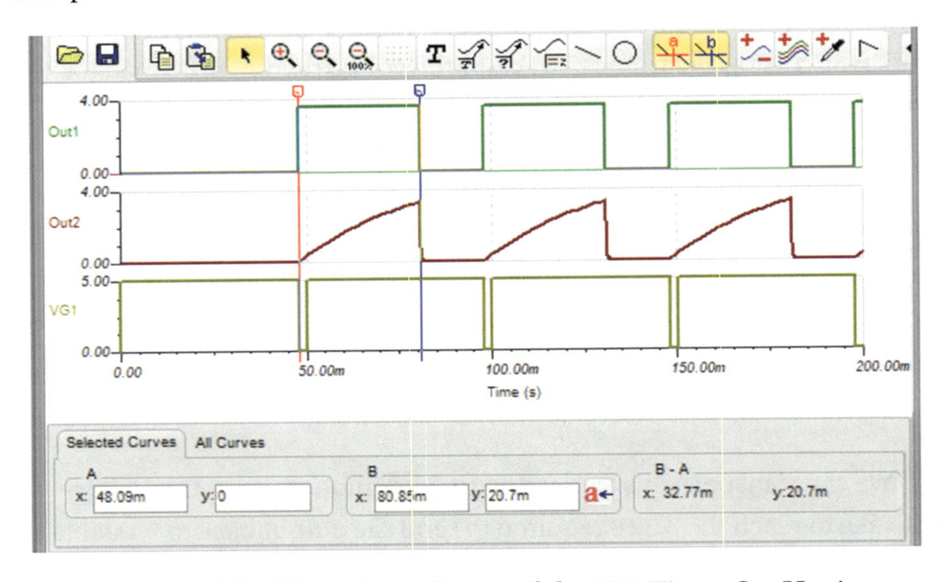


***Figure 10-21.** The Three Waveforms of the 555 Timer Set Up As a Monostable when the CR Setting Equals 30ms*

To further confirm that the circuit works in this way, we can change the capacitor $C_1$ to 10μF. This would change the CR constant to 10ms. Therefore, using the expression to determine the time the output voltage remains high, we get

$$t = -10E^{-3}\ln\left(1 - \frac{3.32}{5}\right) = 10.98ms$$

Figure 10-22 shows that this calculation is correct, that is, the two cursors show that "b" – "a" = 11.06ms.

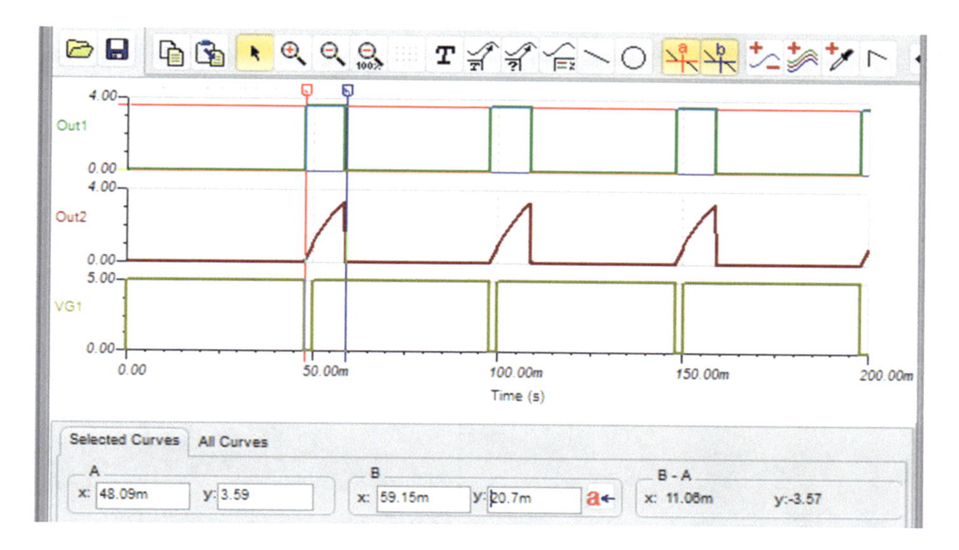


***Figure 10-22.*** *The Display of the Waveforms with C = 10μF and R = 1kΩ*

# Exercise 10.7

Design a monostable using the 555 timer that can be forced out of its stable state for a period of 150ms. Your company wants to use the 50uF capacitors it has a surplus off.

# The Basic Astable

To change the monostable into an astable circuit, which will use the 555 timer to create a square wave at its output, we simply need to keep triggering the 555 timer, that is, keep sending pin 2 to below 1/3 $V_{CC}$. The simplest way of doing this is to connect the varying voltage across the capacitor $C_1$ to the trigger as well as to the threshold pin. The test circuit for this arrangement is shown in Figure 10-23.

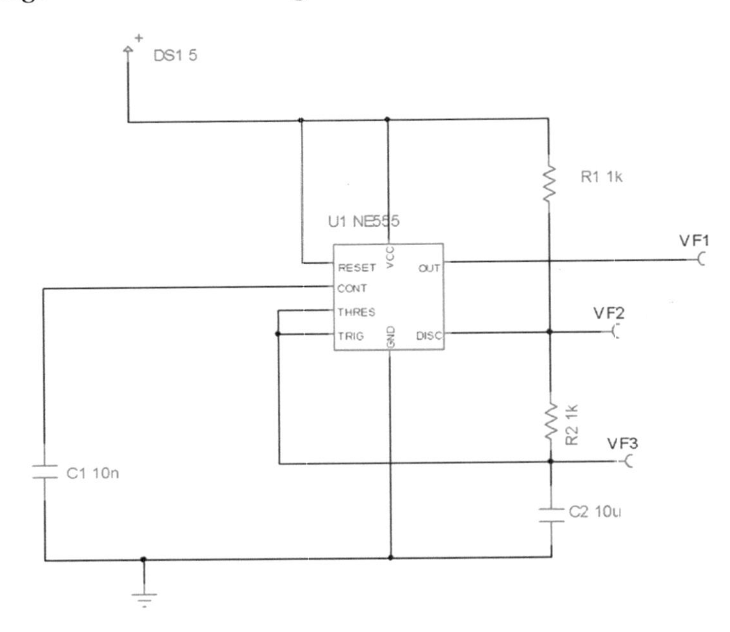


***Figure 10-23.*** *The Basic Astable Test Circuit*

There are now two resistors controlling the charging and discharging of the capacitor. As the capacitor discharges via $R_2$ and the internal transistor that provides the capacitor with a path to ground, then $R_2$ controls the discharge time. Once discharged, the internal transistor turns off and allows the capacitor to charge up toward $V_{CC}$ via $R_1$ and $R_2$. Therefore, $R_1$ and $R_2$ control the charge up time for the capacitor.

The charge up time of the capacitor controls how long the output is high, that is, the mark time of the square wave. The discharge time of the capacitor controls how long the output takes to return to its stable state, the logic low or space time of the square wave. Therefore, $R_1$ and $R_2$ control the mark time, and $R_2$ controls the space time.

Note the capacitor $C_2$ is just to prevent any noise from upsetting the control voltage input pin.

The frequency of the output is set by the following expression taken from the datasheet for the 555 timer:

$$f_o = \frac{1.44}{(R1 + 2R2)C}$$

Therefore, with the circuit values shown in Figure 10-23, the frequency of oscillation is

$$f_O = \frac{1.44}{\left(1E^3 + 2E^3\right)10E^{-6}} = 48Hz$$

The datasheet for the 555 timer states that the mark time can be calculated using

$$Marktime = 0.69\left(R1 + R2\right)C1$$

Putting the values for the circuit into the expression gives us

$$Marktime = 0.69\left(1E^3 + 1E^2\right)10E^{-6} = 13.8ms$$

This is confirmed by the trace VF1 as shown in Figure 10-24. With this simulation, we are using the voltage pins within TINA to allow us to measure the voltages in question. We will study how to use TINA in more detail in Chapter 12.

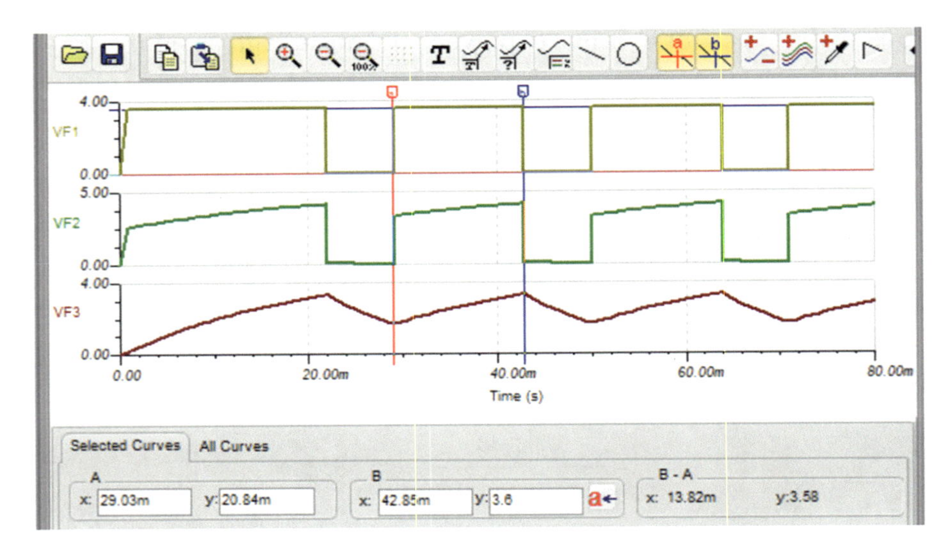


***Figure 10-24.*** *Trace Showing the Mark Time of 13.8ms*

With respect to the space time, using the datasheet, the space time is set by

$$Spacetime = 0.69R2C$$

Putting the values in, we get

$$Spacetime = 0.69 \times 1E^{3} \times 10E^{-6} = 6.9ms$$

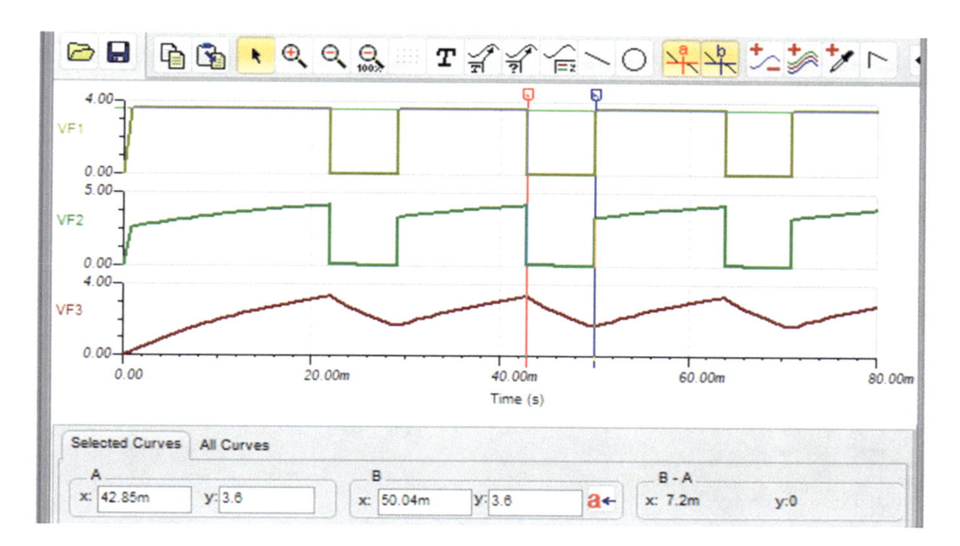


***Figure 10-25.*** *Trace Showing the Space Time*

Figure 10-25 confirms the setting of the space time. Knowing that the periodic time can be determined by adding the mark and space time together, this gives a periodic time of

$$Periodic\ Time = M + S = 13.8ms + 6.9ms = 20.7ms$$

This agrees closely to the measurement in Figure 10-26.

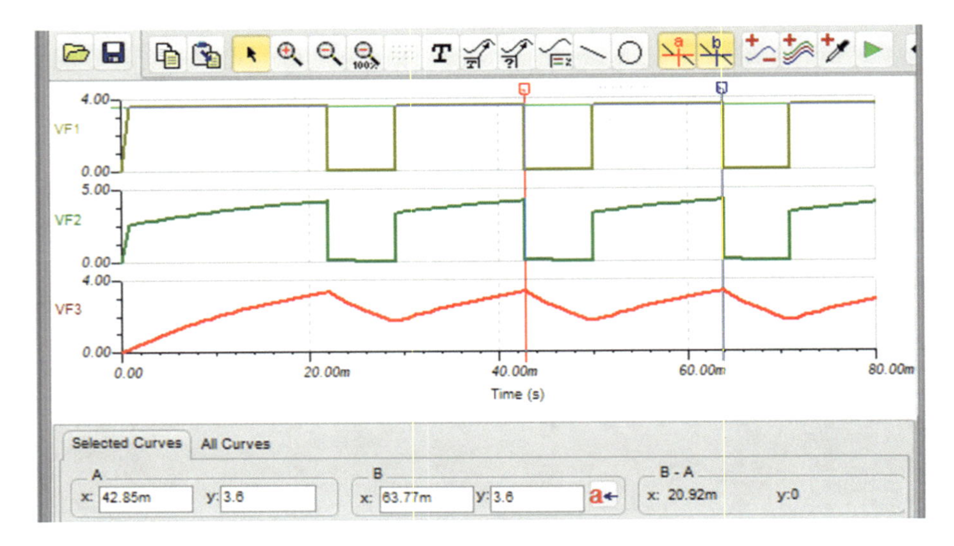


***Figure 10-26.*** *Trace Showing the Periodic Time for the Square Wave Output*

Figure 10-26 shows the periodic time for the square wave to be 20.92ms. Therefore, using the expression to determine the frequency of a waveform knowing the periodic time "T," we can calculate the frequency to be

$$F = \frac{1}{T} = \frac{1}{20.92E^{-3}} = 47.8Hz$$

This agrees closely to the frequency calculated using the expression for the frequency taken from the datasheet.

# Creating a 50/50 Duty Cycle Square Wave

It is not essential that the duty cycle of a square wave should be 50/50, but it would be useful if we could do that. To create a 50/50 duty cycle square wave, we need to make the mark time the same as the space time. If we keep the value of the capacitor constant, then it is the resistor values that control these periods. We can rearrange the expressions for mark time and space time to derive expressions for the resistors $R_1$ and $R_2$ as follows:

$$Marktime = 0.69\left(R1 + R2\right)C$$
$$\therefore R1 = \frac{Marktime - 0.69R2C}{0.69C}$$

$$Spacetime = 0.69R2C$$
$$\therefore R2 = \frac{Spacetime}{0.69C}$$

Now, substituting the expression for $R_2$ into the expression for $R_1$, we can

$$R1 = \frac{Marktime = \dfrac{0.69 \times Spacetime \times C}{0.69C}}{0.69C}$$
$$\therefore R1 = \frac{Marktime - Spacetime}{0.69C}$$

This means that it is impossible to get a true 50/50 duty cycle as to make the mark time the same as the space time would result in $R_1$ being 0Ω.

A compromise that is given in the datasheet is to get close to 50/50, we must let $R_1$ be between R2/8 and R2/5. At lower frequencies, let $R_1$ be closer to R2/5, and at higher frequencies, let $R_1$ be closer to R2/8. Note this is without changing the value of the capacitor.

Example 1:

When f = 500Hz, let space = 1ms, that is, half the periodic time. Therefore, when C = 1μF, we have

$$R2 = \frac{Spacetime}{0.69C} = \frac{1E^{-3}}{0.69 \times 1E^{-6}} = 1449.27$$

Therefore, using

$$R1 = \frac{R2}{5} = \frac{1449.27}{5} = 289.86$$

Figure 10-27 shows the trace of the output from the 555 timer using the component values calculated earlier. The mark time is shown as 1.21ms, and if you used the simulation to measure the space time, you should get a value of around 1.24ms. This does confirm that we can achieve close to a 50/50 duty cycle, but the output of the 555 timer is not very accurate. However, if your application does not require a very accurate frequency, then this simple timer can be quite useful and easy to set up.

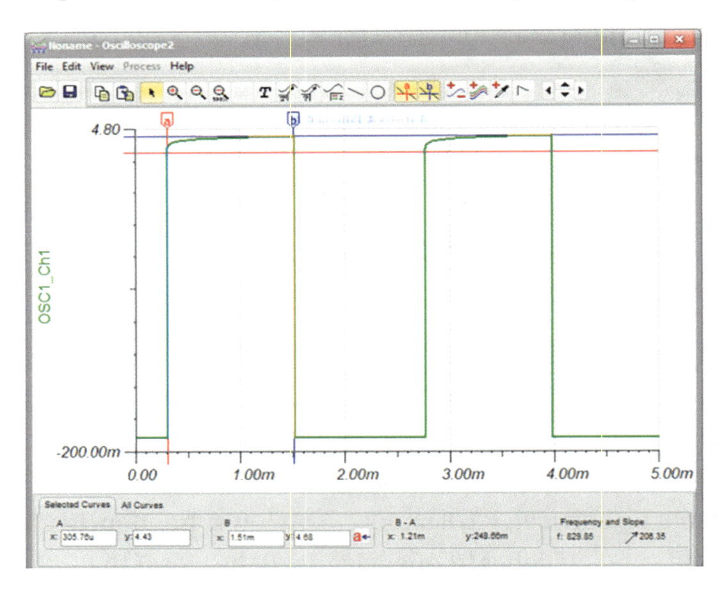


***Figure 10-27.*** *The Trace Showing the Mark Time of the 50/50 500Hz Square Wave*

Example 2:

When f = 50kHz, let space = 10μs. Therefore, when C = 1nF, we have

$$R2 = \frac{Spacetime}{0.69C} = \frac{10E^{-6}}{0.69 \times 1E^{-9}} = 14.49k$$

Therefore, using

$$R1 = \frac{R2}{8} = \frac{14.49}{8} = 1.81k$$

The display of the oscilloscope is shown in Figure 10-28.

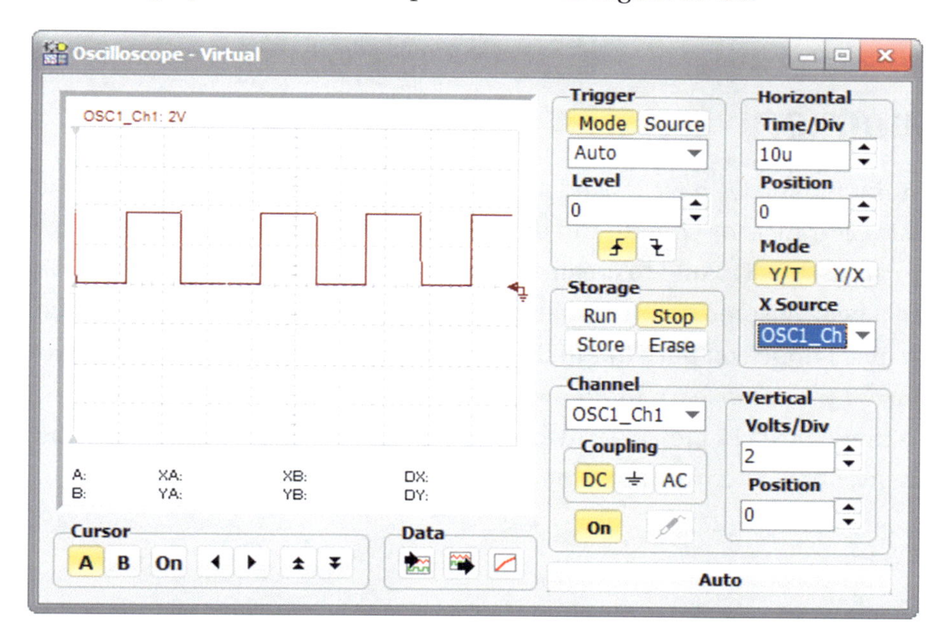


***Figure 10-28.*** *The 50kHz Oscilloscope Trace*

I have changed the capacitor value to 1nF as if we kept it at 1uF, the value of the resistors would be too low, that is, R2 = 14.49Ω and $R_1$ = 1.81Ω. This would suggest that at high frequencies, use low values for capacitor $C_1$, and for low frequencies, use high values of capacitance.

## Exercise 10.8

1. Design a 555 timer that can produce a 1Hz square wave with a mark time of 666.67ms, using a 10μF capacitor for $C_2$.
2. Design a 555 timer that can produce a 50kHz square wave with a mark time of 11μs, using a 1nF capacitor for $C_2$.

## Summary

In this chapter, we have looked at some multivibrator circuits, such as the basic CR transistor circuits and the 555 timer. We have also looked at how we can use the Opamp to create two common types of oscillators. We also had a peak at some of the challenging mathematics we use to analyze the circuits. I hope you have found this interesting and informative and it has spurred you on to learn more.

In the next chapter, we will look at filter circuits. We will start with the passive filters and then move on to the active filters.

CHAPTER 11

# Filters

In Chapter 10, we looked at multivibrator circuits and oscillator circuits. In this chapter, we will look at filters. We will study both passive and active filters, learning how they work. We will restrict the analysis to the basic low and high pass filters as filters can be a book on its own.

We will use the response of filters to show what complex numbers are and how they can be related to the argand or phasor diagram and circuit impedances, explaining how we can use them to analyze electrical circuits.

We will look at Bode plots and how we can create an approximation to a Bode plot, called "an asymptotic Bode plot."

I feel I should warn you that there will be a lot of work, using complex numbers, to derive the transfer function for the filters. If engineers are going to understand how circuits work, so that they can design them, then the ability to be able to derive the transfer function, for any system, is an essential skill they need to learn. When it comes to any phasor quantity, and really all electrical quantities are phasor quantities, then it is essential that we can use complex numbers to help describe the phasors.

I hope that after studying the derivations in this chapter, you will appreciate that complex numbers are not overly complex and that you can use them, along with some basic rules of electrical theory, to analyze the circuits you come across. Of course, you can skip the derivations and just use the transfer functions, but I hope you do stick with it and learn this useful analytical skill. You never know, you might find that you can do it yourself and enjoy it.


H. H. Ward, *Mastering Analog Electronics*, Maker Innovations Series,
https://doi.org/10.1007/979-8-8688-0245-4_11

# Filters and Passive Filters

A filter is a circuit that prevents signals of unwanted frequencies from being passed onto a load, that is, a circuit, which wants to use the signal. For example, a hi-fi system would not want any noise, picked at the input, to be passed onto the amplifier and so out at the speakers. In this case, as noise is normally of a higher frequency than the normal audio range of frequencies, we humans will not hear it, but electric noise can damage electrical signals. Then a low pass filter could be used to filter out the noise.

A passive filter circuit is one which does not use power from the supply to add power to the signal the circuit is using, whereas the active filters, which we will look at later in this chapter, use amplifiers that add power to the signal. Passive filters are circuits made up with passive components which are resistors, capacitors, and inductors. The resistor can also be classified a nonreactive component, whereas capacitors and inductors are reactive components. Capacitors and inductors cause a reaction when a voltage is applied across them to force current to flow through them. This means they cause a reaction to the current flowing through them. With a capacitor, the current leads the voltage, ideally by 90°. With the inductor, the current lags the voltage, ideally by 90°. However, with the resistor, the voltage and current are in phase with each other. We will find this concept useful when we analyze the passive filters.

# The CR Passive Filters

The circuit for the first filter we will look at is shown in Figure 11-1.

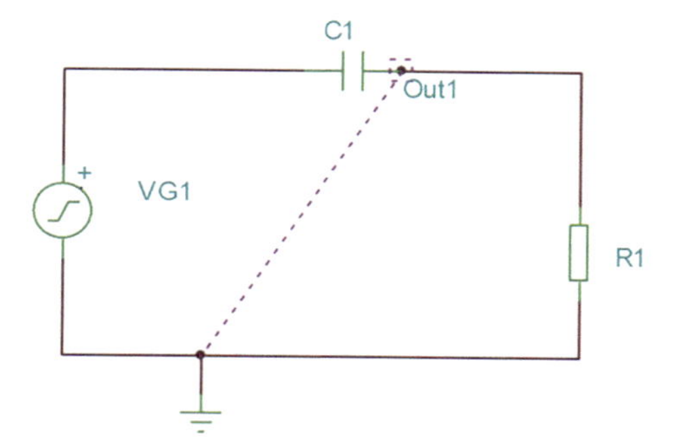


***Figure 11-1.*** *The First CR Filter Circuit*

We need to develop the expression for the transfer function as it will help describe how the system, in this case, the CR filter circuit, will transfer the input to becoming the output. To derive the transfer function, we can use the voltage divider rule. However, before we do that, it might be useful to represent the circuit as two impedances. Indeed, we could do that for the first four passive filters we will look at. This configuration is shown in Figure 11-2.

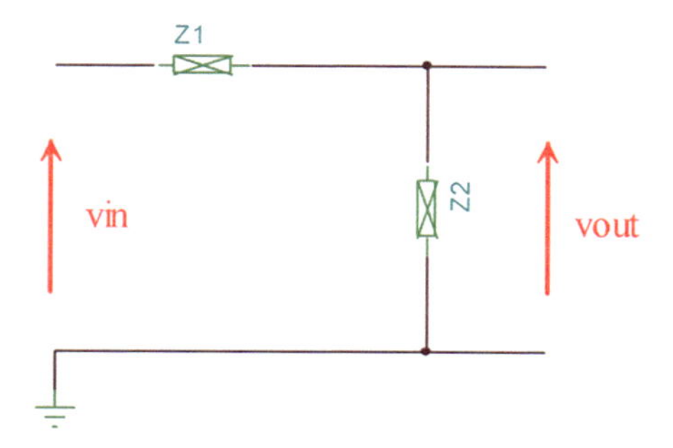


***Figure 11-2.*** *Representing the Passive Filter with Two Impedances*

Using the voltage divider rule, we can say

$$v_{out} = \frac{v_{in} Z_2}{Z_1 + Z_2}$$

Now dividing both sides by $v_{in}$, we get

$$\frac{v_{out}}{v_{in}} = TF = \frac{Z_2}{Z_1 + Z_2}$$

## Complex Numbers and Phasor Quantities

It might be useful to explain how we can express the impedances using complex numbers. To do that, we need to appreciate that all impedances, including resistance, are phasor quantities. You should appreciate that phasor quantities have magnitude and direction or magnitude and an angle describing the direction. This is the difference between "speed" and velocity. Speed is a scalar quantity as it does not have a direction; it just has a value. However, velocity is a phasor quantity as it has a value, magnitude, and direction.

We can draw phasor quantities on a phasor diagram, called an argand diagram, and describe them using complex numbers, which are not really overly complex; believe me, they are not. A typical argand diagram, showing all three impedances of R, XL, and XC, is shown in Figure 11-3.

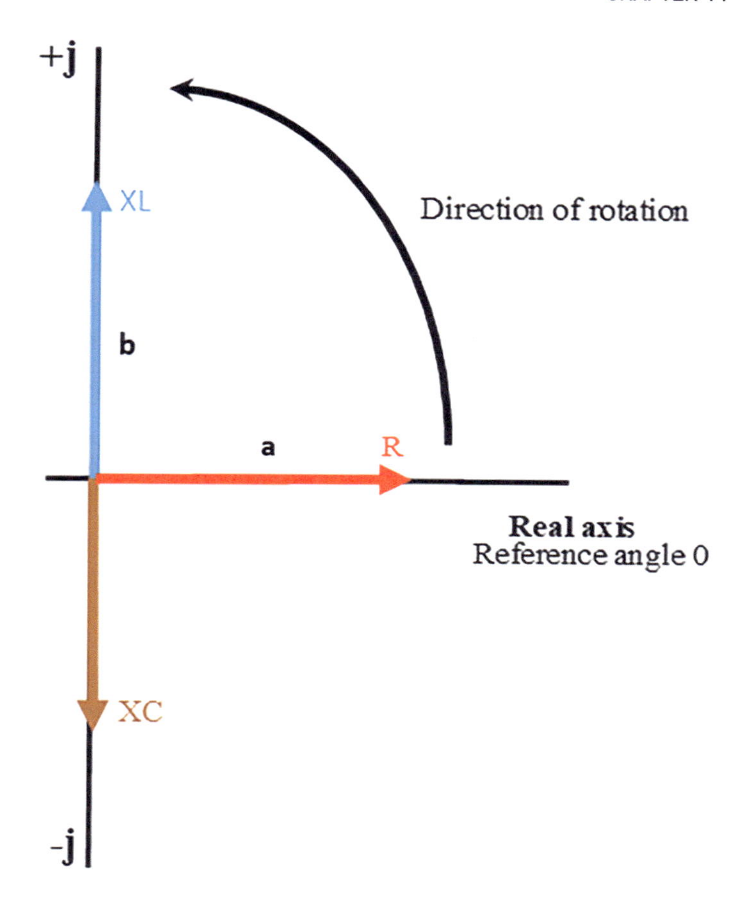


***Figure 11-3.*** *A Typical Argand Diagram of the Three Passive Components*

Figure 11-3 is an attempt to try and show the phase relationship between the three main passive components and how we can represent them on a phasor or argand diagram. The two axes of the phasor diagram are the "real axis" and the "j or imaginary" axis.

The length of the arrow depicts the magnitude of the impedance, and their placing on the phasor diagram depicts their phase relationship with respect to the reference, that is, the real axis.

The resistance is shown as being in line with the real axis as the angle associated with resistance is normally "0". That is why we do not normally consider the angle when dealing with only resistive circuits. This agrees with the concept that the current flowing through the resistor is in phase with the voltage across it.

The inductive impedance, represented by the "XL" term, is on the +j axis, which means a pure inductance impedance is 90° ahead of the reference. Note, we are talking about impedance, not current or voltage. We will see later that this agrees with the principle that current flowing through the inductor lags the voltage across it.

The capacitive impedance, represented by the "XC" term, is shown as being on the -j axis, which means a pure capacitive impedance is 90° behind or lagging the reference. We will see later that this agrees with the principle that current flowing through the capacitor leads the voltage across it.

The relationship between voltage and current and impedance can be explained using Ohm's Law as

$$i = \frac{v}{z} = \frac{|v|\langle\theta_v\rangle}{|z|\langle\theta_z\rangle}$$

With this complex division, we simply divide the two magnitudes, identified by the term within the two lines, |*Magnitude*|, and simply subtract the two angles. This means the current can be expressed as

$$i = \frac{|v|}{|z|}\langle\theta_v - \theta_z\rangle$$

If we apply this to a pure inductor with an impedance of

$$|X_L|\langle 90\rangle$$

then the current flowing through can be calculated using

$$i_L = \frac{|v|}{|X_L|}\langle 0 - 90 \rangle = \frac{|v|}{|X_L|}\langle -90 \rangle$$

Hence, the current lags the voltage by 90°. This assumes that the voltage is the reference given the angle "0". This is normal in electrical analysis as we get nothing without a voltage.

If we apply this to a capacitor with an impedance of

$$|X_C|\langle -90 \rangle$$

then the current flowing through can be calculated using

$$i_C = \frac{|v|}{|X_C|}\langle 0 - -90 \rangle = \frac{|v|}{|X_C|}\langle 90 \rangle$$

Hence, the current in a capacitor leads the voltage by 90°.

It is normal to use complex numbers to represent any impedance or phasor quantity, and all electrical quantities are phasor quantities. That being the case, it would be useful to represent the three passive components as phasor quantities using their complex number representation.

There are two ways of writing complex numbers.

One is in rectangular or coordinate format, which in general is

$$a + jb$$

that is, "a" along the real axis and "b" along the j axis as shown in Figure 11-3.

The other format is in polar format, which in general is

$$r\langle \theta \rangle$$

where "r" is the magnitude or length of the phasor and "θ" is the associated angle which positions the phasor on the argand diagram. The angle is sometimes referred to as the argument of a complex number.

If we consider the resistance "R," we can see it has a length of "R," which means its magnitude in polar format is "R." As it is sitting on the real axis, which is the reference, and has an angle of "0", then the angle associated with the resistance is "0". Therefore, in polar format we have

$$Resistance = R\langle 0 \rangle$$

If we consider the coordinate format, we can see that the end of the phasor is at a distance "R" from the origin, and it has a "0" coordinate on the "j" axis. Therefore, in coordinate or rectangular format, we can say

$$Resistance = R + j0$$

I prefer to use the term "coordinate format" as it describes the format better than rectangular. This is because the coordinate format states the coordinates of the very end of the phasor on the phasor diagram.

If we consider how we can convert from coordinate format to polar and polar to coordinate format, you might be able to see that the two formats equal each other as they should do. Using the general format, we can convert from coordinate format to polar using

$$a + jb = \sqrt{a^2 + b^2} \ \langle Tan^{-1} \frac{b}{a} \rangle$$

Really, this is simply using Pythagoras to determine the magnitude and trigonometry to determine the angle.

Then converting from polar to coordinate format, we have

$$r\langle \theta \rangle = rCos(\theta) + jrSin(\theta)$$

We can again use trigonometry to prove the conversion from polar to coordinate format.

Using these conversions, we can show that

$$R + j0 = \sqrt{R^2 + 0^2} \ \langle Tan^{-1}\frac{0}{R}\rangle = R\langle 0\rangle$$

Now if we consider the pure inductive phasor, we can see, using coordinate format, the real component "a" is zero. Also, the "j" component is +XL. Therefore, for the pure inductor, we can say

$$\textit{Pure Inductive Impedance } X_L = 0 + jX_L$$

In polar format, this would be

$$\textit{Pure Inductive Impedance } X_L = X_L \ \langle 90\rangle$$

Intuitively, we can see that the length of the phasor is "XL," and its position is on the +j axis, which is 90° ahead of the reference. Also, we can convert the coordinate format to polar using

$$0 + jX_L = \sqrt{0^2 + X_L{}^2}\langle Tan^{-1}\left(\frac{X_L}{0}\right)\rangle = X_L\langle 90\rangle$$

We should remember anything divided by "0" equals infinity and $Tan^{-1}$(Infinity) = 90°.

Therefore, bearing all that in mind, I hope you can appreciate how we can express the impedance of the capacitor "XC."

$$X_C = 0 - jX_C = X_C \ \langle -90\rangle$$

Hopefully, this explanation has given you some insight into using complex numbers. Apart from some small challenging aspects, which we might have to consider later, complex numbers are not overly complex as we will see when we use them to describe the action of these passive filters.

We should now continue with the analysis of the passive filter shown in Figure 11-1. The impedance $Z_1$ is that of the capacitor, which means we can express $Z_1$ as

$$Z_1 = 0 - j\frac{1}{\omega C}$$

The impedance $Z_2$ is that of the resistor, which means we can express $Z_2$ as

$$Z_2 = R + j0$$

If we now consider the addition of the two impedances, that is, $Z_1 + Z_2$, this can be expressed as

$$Z_1 + Z_2 = \left(0 - j\frac{1}{\omega C}\right) + (R + j0) = R - j\frac{1}{\omega C}$$

We simply add all the real terms together and all the imaginary terms together separately.

Therefore, the transfer function for Figure 11-1 is

$$TF = \frac{R}{R - j\frac{1}{\omega C}}$$

It is normal to create a transfer function that has a numerator of unity, that is, 1. It will become clear why we do this as we progress. Therefore, in this case, we must divide the top and bottom of the transfer function by "R." Another way of doing this would be to multiply both top and bottom by $\frac{1}{R}$, which is mathematically the same as dividing by "R." This would give

$$TF = \frac{R \times \frac{1}{R}}{\left(R - j\frac{1}{\omega C}\right) \times \frac{1}{R}} = \frac{1}{R \times \frac{1}{R} - j\frac{1}{\omega C} \times \frac{1}{R}}$$

This now results in

$$TF = \frac{1}{1 - j\frac{1}{\omega CR}}$$

This is now a division of two complex numbers, especially if we rewrite it as

$$TF = \frac{1 + j0}{1 - j\frac{1}{\omega CR}}$$

We need to convert this into polar format as it's easier to divide in polar. Therefore, we get

$$TF = \frac{\sqrt{1^2 + 0^2}\ \langle Tan^{-1}\left(\frac{0}{1}\right)\rangle}{\sqrt{1^2 + \left(-\frac{1}{\omega CR}\right)^2}\ \langle Tan^{-1}\left(\frac{-\frac{1}{\omega CR}}{1}\right)\rangle}$$

which gives

$$TF = \frac{1\langle 0\rangle}{\sqrt{1^2 + \left(-\frac{1}{\omega CR}\right)^2}\ \langle Tan^{-1}\left(-\frac{1}{\omega CR}\right)\rangle}$$

Note, when dividing two complex numbers, we simply divide the two magnitudes and subtract the angles.

This gives

$$\frac{1}{\sqrt{1^2+\left(-\frac{1}{\omega CR}\right)^2}}\langle 0-Tan^{-1}\left(-\frac{1}{\omega CR}\right)\rangle$$

This simplifies to

$$TF=\frac{1}{\sqrt{1^2+\left(-\frac{1}{\omega CR}\right)^2}}-\langle Tan^{-1}\left(-\frac{1}{\omega CR}\right)\rangle$$

We can separate the transfer function into its two parts of magnitude and angle as follows.

The magnitude is

$$|TF|=\frac{1}{\sqrt{1^2+\left(-\frac{1}{\omega CR}\right)^2}}$$

The angle is

$$\theta=-\langle Tan^{-1}\left(-\frac{1}{\omega CR}\right)\rangle$$

# An Asymptotic Bode Plot

We can produce a Bode plot that will show how the circuit reacts to a range of different frequencies. This could be done if we had a frequency analyzer or with a simulation using an ECAD package such as TINA. However, if we do not have access to a practical circuit or an ECAD package, we could use some values to produce a straight-line approximation and that might be enough for what we need. The posh name for such an approximation is "an asymptotic Bode plot," and we will go through producing one now.

We will need some values, namely, two or three values for the frequency; we will use three to draw the asymptote. However, we cannot actually produce any values for the magnitude or phase, as they both depend upon the value of "$\omega$", which is a shorthand way of writing $2\pi f$, and so we need to know the frequency. Except, if we choose some obvious values for the frequency, we will be able to draw the plot quite quickly and accurately enough for our needs. The first two values we will choose are the two extremes of the frequency spectrum, that is, 0Hz and a frequency of infinity. The third value is a little less obvious until you realize what we are trying to do with these filters. We are trying to filter out unwanted frequencies, and the basic idea of these filters is that we filter out frequencies that are lower than a set value, as in a high pass filter, or are above a set value, as in a low pass filter. For example, a high pass filter will only allow frequencies that are higher than a set value and reject all frequencies that are lower than that set value. However, you must appreciate that the circuit cannot simply block the frequencies that are lower than this set frequency. They will still be passed onto the output, but the power level will be too low for them to be of any use. This begs the question: What power level is too low? Well, that is where the benchmark of the "half power point," which we have mentioned in Chapter 6, comes in. We say that any frequency that has a power level that is less than half the maximum power will be deemed useless and will not be of any use at the output of the filter, even though it will be passed on. It will be as though it has been filtered out. So, before we can look at the transfer function using all three frequencies, we need to understand this half power point. The frequency at which we reach this half power point is called the frequency of cutoff, that is, $f_C$.

# The Half Power Point Benchmark and the Frequency of Cutoff

This half power point uses the ratio of power out divided by power in. However, when applying it to a Bode plot, we use dBs as the unit for this ratio. The power ratio in dBs can be expressed as

$$Power\ ratio = 10log\left(\frac{P_{out}}{P_{in}}\right)$$

Then knowing at this cutoff frequency the power ratio will be ½, we can say

$$Power\ ratio = 10log\left(\frac{1}{2}\right) = 10 \times -0.3010 = -3\ dbs$$

We can relate this half power point to the voltage gain of a system or current gain using

$$Power\ ratio = 20log\left(\frac{v_{out}}{v_{in}}\right)$$

Or

$$Power\ ratio = 20log\left(\frac{i_{out}}{i_{in}}\right)$$

This is because power can be calculated as

$$Power = v \times i = \frac{v^2}{R} = \frac{i^2}{R}$$

Therefore, using the third rule of logs, the power of "2" can be brought down and multiplied into the log. Hence, we get 2x10 = 20 in the 20Log.

When the ratio of

$$\frac{v_{out}}{v_{in}} = \frac{1}{\sqrt{2}}$$

then we have

$$Power\ ratio = 20log\left(\frac{1}{\sqrt{2}}\right) = 20 \times -0.1505 = -3.01\ \text{dBs}$$

This is the same as the half power point, and so the frequency of cutoff will be when the magnitude of the transfer function is

$$|TF| = \frac{1}{\sqrt{2}}$$

As we have ensured, the numerator of the transfer function was "1," and the square root in the denominator, for this transfer function, was

$$\sqrt{1^2 + \left(-\frac{1}{\omega CR}\right)^2}$$

Then the term in the square root will equal two, which is what we want for the half power point, when we have

$$|TF| = \frac{1}{\sqrt{1+1}}$$

This will be when

$$\left(-\frac{1}{\omega CR}\right)^2 = 1$$

This will be when

$$\omega CR = 1$$

From this, we can say

$$2\pi f_C CR = 1$$

This then means that the cutoff frequency "$f_C$" is when

$$f_c = \frac{1}{2\pi CR}$$

For the asymptotic Bode plot, we do not need to calculate the actual frequency; we just need to know that the magnitude of the transfer function will be at the half power point when we are at the frequency when the term

$$\omega CR = 1$$

We can now calculate the magnitude as

$$|TF| = \frac{1}{\sqrt{1+1}} = \frac{1}{\sqrt{2}} = 0.707$$

The angle "θ" can be calculated as

$$\theta = -\langle Tan^{-1}\left(-\frac{1}{1}\right)\rangle = -\langle Tan^{-1}(-1)\rangle = 45°$$

Therefore, at the cutoff frequency, the transfer function is

$$TF = 0.707\,\langle 45\rangle$$

We now have our three frequencies we will use to draw the asymptote for this CR filter shown in Figure 11-1; they are as follows:

1. When f = 0, the lowest value
2. When f = infinity, the highest value
3. When f is the cutoff frequency, the frequency at which the output reaches the benchmark of the half power point

We will calculate both the magnitude and angle at these frequencies starting with f = 0Hz.

As f = 0, then

$$\omega = 2\pi f = 2\pi \times 0 = 0$$

This means that

$$\omega CR = 0$$

Therefore, the magnitude is

$$|TF| = \frac{1}{\sqrt{1^2 + \left(-\frac{1}{\omega CR}\right)^2}} = \frac{1}{\sqrt{1 + \left(-\frac{1}{0}\right)^2}} = \frac{1}{\sqrt{1+\infty}} = \frac{1}{\infty}$$

which gives

$$|TF| = 0$$

We should remember that anything divided by "0" will result in a value of infinity. Also, one divided by infinity will produce a value of 0. This is a little difficult to prove, but if you use your calculator and define the value of "0" as $0 = 1^{-99}$ and infinity as $\infty = 1^{99}$ and then try the calculations out on your calculator, you should get the correct appropriate results.

If we now consider the angle "θ", we have

$$\theta = -\langle Tan^{-1}\left(-\frac{1}{\omega CR}\right)\rangle = -\langle Tan^{-1}\left(-\frac{1}{0}\right)\rangle = -\langle Tan^{-1} - (\infty)\rangle$$

Therefore, we have

$$\theta = 90°$$

Therefore, when f = 0Hz the transfer function is

$$TF = 0\,\langle 90\rangle$$

Next, we will let f = infinity.

# Exercise 11.1

As an exercise, see if you can calculate the magnitude and angle of the transfer function when f = infinity and so show that when f = infinity

$$TF = 1\langle 0\rangle$$

Finally, at the cutoff frequency, the transfer function is

$$TF = 0.707\,\langle 45\rangle$$

Using these three values, we can create the asymptote of the Bode plot except that we must express the magnitude in dBs. This is done using

$$20log\left(|TF|\right)$$

When f = 0, the magnitude is 0 and so we have

$$20Log(0)$$

To do this, let 0 = $1^{-99}$ and infinity = $1^{99}$.

$$20Log\left(1^{-99}\right) = -100\, dBs$$

Really, the log of "0" is -infinity.

When f = infinity, the magnitude is "1," so we have

$$20Log\left(1\right) = 0dBs$$

We have already shown that at the cutoff frequency the magnitude in dBs is -3dbs.

This means we can now create the Bode plot. The typical Bode plot for the passive CR filter shown in Figure 11-1 is shown in Figure 11-4.

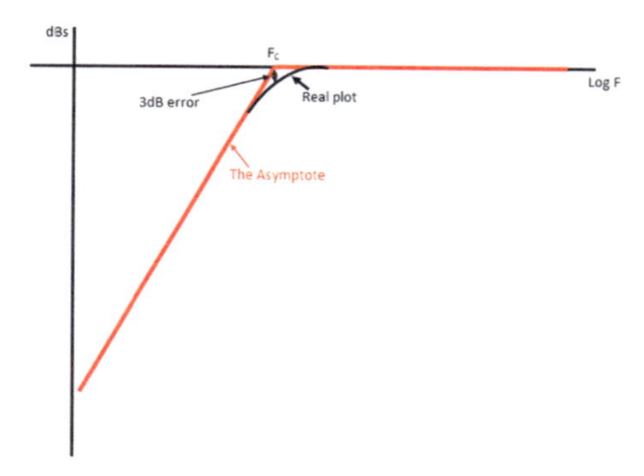


***Figure 11-4.** The Asymptotic Bode Plot for the CR Filter in Figure 11-1*

The asymptote is drawn as the two straight lines shown in red in Figure 11-1. There is a small section of curve drawn in black. This is to try to show how the real Bode plot would vary from the approximation plot. I hope you can see that the approximation only varies from the real plot around the frequency of cutoff "$f_C$." Around that frequency, the approximation is in error by about the value of -3dBs. This is because the approximation shows that at fc the gain is 0dBs, whereas we know it should be –3dBs. That is why we call this error the 3dB error. However, apart from that, the approximation does show us what we need to know, and that is that this filter is a high pass filter, as any frequencies below the cutoff frequency "$f_c$" will be of no use as their power content will be below the benchmark of the half power point.

One thing we should appreciate is that the gradient of the slope in the approximation is constant at 20dBs per decade. If we could show it on the plot, the slope would be starting at minus infinity as that is the result of the log of infinity.

If we give the CR filter shown in Figure 11-1 some value such as C = 100nF and R = 1kΩ, we should be able to calculate the frequency of cutoff, $f_C$. Then we can simulate the circuit in TINA and see if the frequency response does follow the approximation as shown in Figure 11-4. The frequency of cutoff can be calculated as follows:

$$f_c = \frac{1}{2\pi CR} = \frac{1}{2\pi \times 100E^{-9} \times 1E^{3}} = 1.591kHz$$

When we simulate the circuit and conduct a frequency analysis, the Bode plot can be created as shown in Figure 11-5.

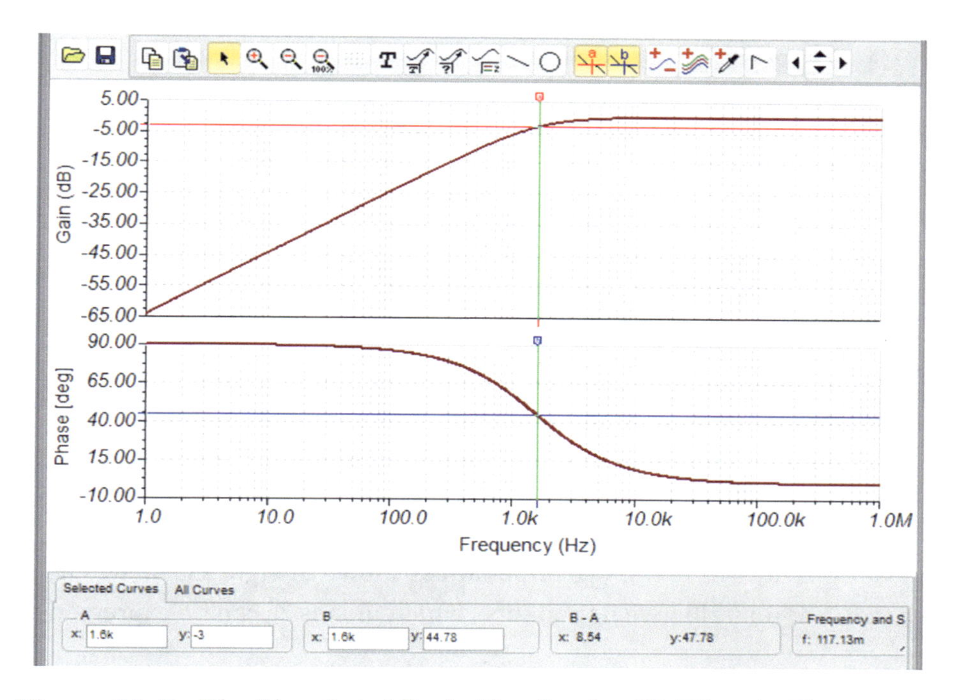


***Figure 11-5.** The Simulated Bode Plot for the CR Filter in Figure 11-1*

Figure 11-5 shows the results of the simulated Bode plot. From it, we can see that the maximum gain of the CR filter is 0dBs, that is, a gain of one. This must be correct as there is no amplification in the circuit as it just uses passive components. Also, it shows that the gain falls by 3dBs to be –3dBs at a frequency of 1.6kHz. This is close to the calculated cutoff frequency of 1.591kHz. If we examine the phase plot, we also see that at the cutoff frequency the phase is 45° again, close to what we have calculated.

The overall shape of the Bode plot in Figure 11-5 does agree closely to the approximated Bode plot shown in Figure 11-4. If we examine the gradient of the simulated Bode plot, we can see that it is close to the estimated 20dBs per decade as stated for the asymptote we have drawn up.

These measurements should suggest that the theory we are using is correct.

# The RL Filter

The circuit for the next filter is shown in Figure 11-6.

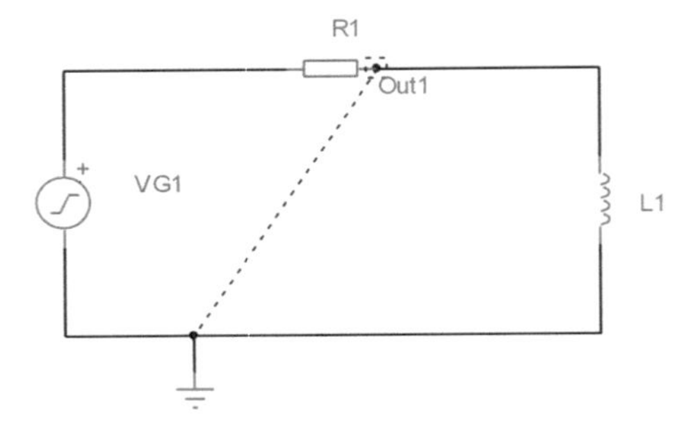


***Figure 11-6.*** *The RL Passive Filter*

Comparing this to the two impedances shown in Figure 11-2, we can see that $Z_2$ is an inductive impedance, which means $Z_2$ can be expressed as

$$Z_2 = 0 + j\omega L$$

Also, we can see that $Z_1$ is simply the resistor. Therefore, we can express $Z_1$ as

$$Z_1 = R + j0$$

Adding the two impedances together, we can say

$$Z_1 + Z_2 = (R + j0) + (0 + j\omega L) = (R + 0) + j(0 + \omega L) = R + j\omega L$$

Using these expressions, we can express the transfer function as

$$TF = \frac{0 + j\omega L}{R + j\omega L}$$

Now dividing the top and bottom by $j\omega L$ or multiplying the top and bottom by $\frac{1}{j\omega L}$, we get

$$TF = \frac{\frac{0}{j\omega L} + \frac{j\omega L}{j\omega L}}{\frac{R}{j\omega L} + \frac{j\omega L}{j\omega L}}$$

Canceling out where we can gives

$$TF = \frac{1}{1 + \frac{R}{j\omega L}}$$

We need to appreciate that

$$\frac{R}{j\omega L} = \frac{1}{j} \times \frac{R}{\omega L}$$

However, it can be shown that

$$\frac{1}{j} = -j$$

The proof of this involves the use of the complex conjugate, and this is shown in the appendix.

This now means the transfer function becomes

$$TF = \frac{1}{1 - j\frac{R}{\omega L}}$$

Converting this to polar so that we can easily carry out the division gives

$$TF = \frac{\sqrt{1^2 + 0^2}\ \langle Tan^{-1}\left(\frac{0}{1}\right)\rangle}{\sqrt{1^2 + \left(-\frac{R}{\omega L}\right)^2}\ \langle Tan^{-1}\left(\frac{-\frac{R}{\omega L}}{1}\right)\rangle}$$

Therefore, carrying out the division and separating it into magnitude and phase, we have

$$|TF| = \frac{1}{\sqrt{1^2 + \left(-\frac{R}{\omega L}\right)^2}}$$

$$Phase = -\langle Tan^{-1}\left(-\frac{R}{\omega L}\right)\rangle$$

We can use the expression for the magnitude to derive an expression for the cutoff frequency as we know, at the cutoff frequency, the magnitude must be

$$|TF| = \frac{1}{\sqrt{2}}$$

This would happen when

$$\frac{R}{\omega L} = 1$$

The expression for the cutoff frequency $f_C$ is

$$f_C = \frac{R}{2\pi L}$$

We can now evaluate the transfer function when f = 0 and when f = infinity so that we can draw an asymptotic Bode plot.

When f = 0, we have

$$\omega L = 0$$

Therefore, we have

$$|TF| = \frac{1}{\sqrt{1^2 + \left(-\frac{R}{0}\right)^2}} = \frac{1}{\sqrt{1^2 + \infty^2}} = 0$$

The phase shift is

$$Phase = -\langle Tan^{-1}\left(-\frac{R}{\omega L}\right)\rangle = -\langle Tan^{-1}\left(-\frac{R}{0}\right)\rangle = 90°$$

When f = infinity, the magnitude of the transfer function will be

$$|TF| = \frac{1}{\sqrt{1^2 + \left(-\frac{R}{\infty}\right)^2}} = \frac{1}{\sqrt{1^2 + 0^2}} = 1$$

The phase of the transfer function will be

$$Phase = -\langle Tan^{-1}\left(-\frac{R}{\infty}\right)\rangle = 0°$$

We now have all three values, and we can draw the Bode plot as shown in Figure 11-7.

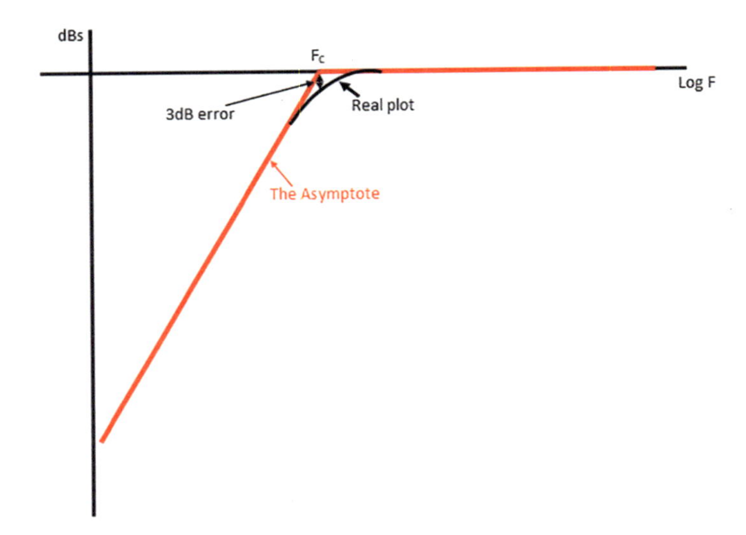


***Figure 11-7.*** *The Asymptotic Bode Plot for the RL Filter in Figure 11-6*

This is the same as the asymptote shown in Figure 11-4. That is because, just like the filter circuit shown in Figure 11-2, the RL filter shown in Figure 11-6 is a high pass filter. Using the expression for the cutoff frequency and knowing the value of R was 1kΩ and L was 1mH, then the frequency of cutoff "$f_C$" can be calculated as follows:

$$F_C = \frac{R}{2\pi L} = \frac{1E^3}{2\pi \times 1E^{-3}} = 159.154kHz$$

We can calculate the phase of the filter at the cutoff frequency as follows:

$$Phase = -\langle Tan^{-1}\left(-\frac{R}{\omega L}\right)\rangle$$

Then knowing at the cutoff frequency, we have

$$\frac{R}{\omega L} = 1 \; then \; phase = -\langle Tan^{-1}(-1)\rangle = 45°$$

Simulating the circuit and creating the Bode plot gives us the display shown in Figure 11-8.

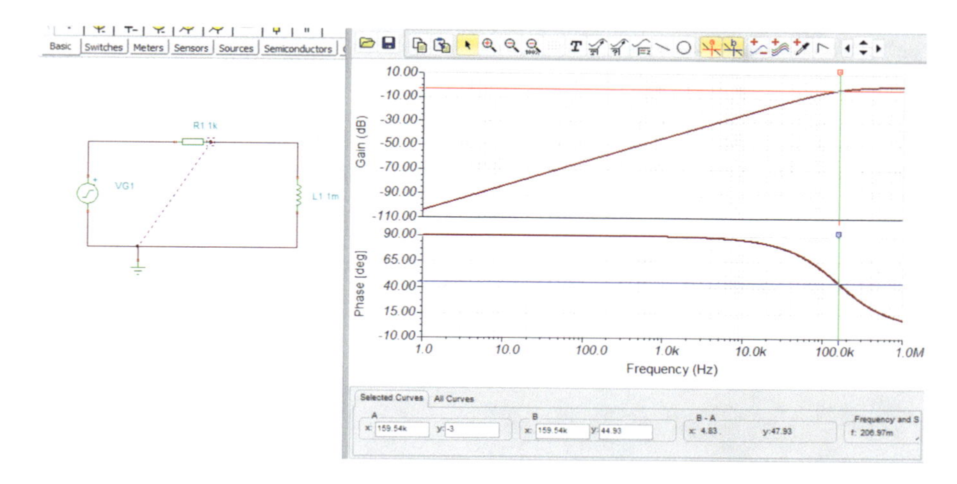


***Figure 11-8.*** *The Bode Plot for the RL High Pass Filter*

From the Bode plot shown in Figure 11-8, we can see that at a frequency of 159.54kHz the gain of the filter is –3dBs, which is the expected benchmark figure. Also, the phase has come down from 90° at 0Hz to 44.93° at the cutoff frequency. These two measurements do confirm our calculations and theory.

# Exercise 11.2

As an exercise, derive the transfer function for the two passive filter circuits shown in Figure 11-9 and so confirm they are as stated. Then using a resistor value of 2.2kΩ for each circuit, calculate the value of capacitance in circuit "a" and inductance in circuit "b" so that they both have a cutoff frequency of 10kHz. State what type, low pass or high pass filter, they are.

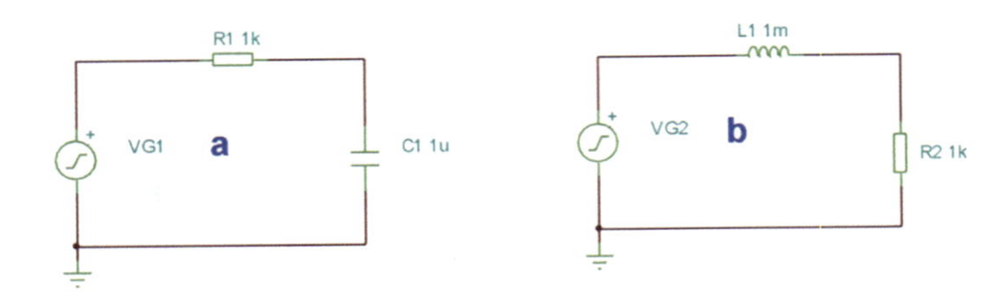


***Figure 11-9.** The Two Passive Filter Circuits for Exercise 11.2*

Circuit a

Use the following hint:

$$-j\frac{1}{\omega C} = \frac{1}{j\omega C}$$

$$\frac{1}{j} = -j$$

You should be able to show that the transfer function is

$$TF = \frac{1}{1 + j\omega CR}$$

$$\frac{1}{\omega CR} = 1$$

$$Phase = -\langle Tan^{-1}(\omega CR)\rangle$$

Circuit b

You should be able to show that the transfer function is

$$TF = \frac{1}{1 + j\frac{\omega L}{R}}$$

# Active Filters

There are many more passive filters that we could look at such as the following:

Band stop

Band pass

The notch filter

Symmetrical M filters

However, it would require a book devoted to filters to cover them all. So, we will leave the passive filters now and start our investigation into active filters.

The main disadvantage of passive filters is that they attenuate the frequencies we want to use. Active filters use an amplifier, usually an Opamp, so they can retrieve the signal level we lose with passive filters and even add some gain to them.

As with passive filters, there are a range of responses we can get from the design of an active filter, and they are

- Low pass
- High pass
- Band pass
- Band stop

However, as this is only an introduction into filters, we will restrict our investigation into low and high pass filters.

There are three distinctive designs of active filters, and they are

- Butterworth
- Chebyshev
- Bessel

Again, we will restrict our analysis to the Butterworth filter, as it would take a book just to look at all the filters. The first circuit we will look at is shown in Figure 11-10.

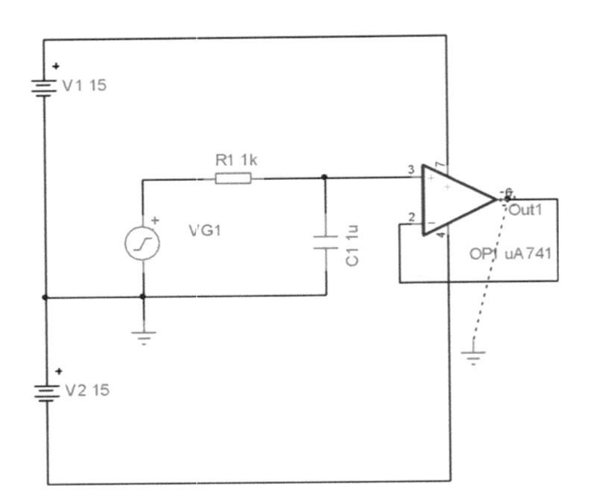


***Figure 11-10.*** *A First-Order Low Pass Filter*

I hope you can see that this uses an Opamp that is configured in the non-inverting configuration and that the gain is unity or "1". Really, this is a passive filter with an amplifier added to it, but with this circuit the gain of the amplifier is "1".

I hope you can see that the transfer function of the passive components is

$$TF = \frac{1}{1 + j\omega CR}$$

The expression for the cutoff frequency is

$$f_C = \frac{1}{2\pi CR}$$

The expression for the phase would be

$$Phase = -\langle Tan^{-1}(\omega CR)\rangle$$

Putting the RC values from the circuit into the expression for fc and the phase gives us a cutoff frequency of 159.15Hz with a phase of –45°. If we simulate the circuit, we will get the Bode plot as shown in Figure 11-11.

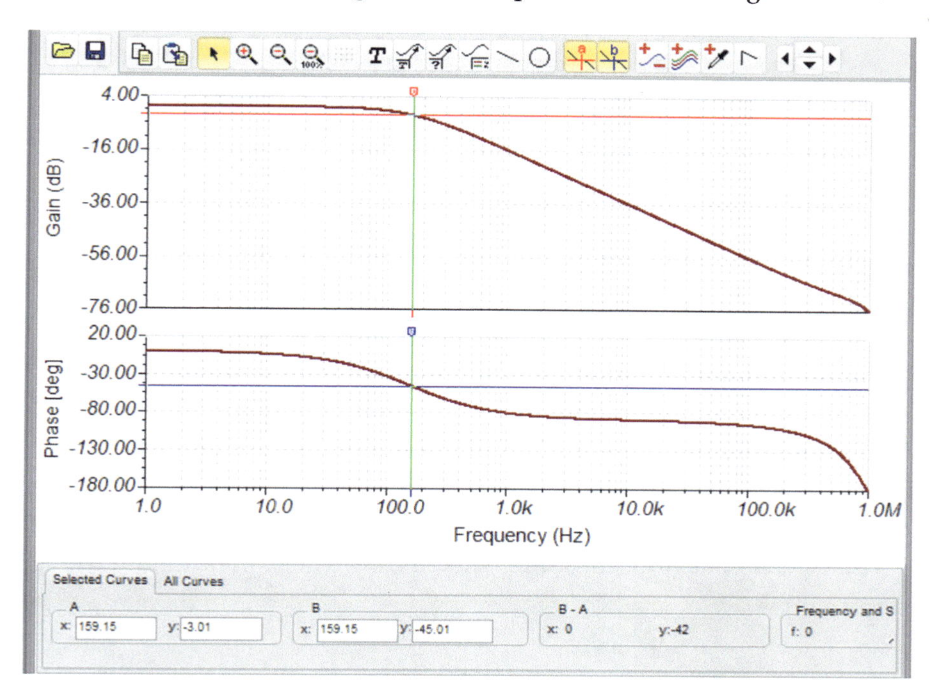


***Figure 11-11.*** *The Simulated Bode Plot for the Circuit in Figure 11-10*

We can see from the Bode plot that the –3.01dB point is at a frequency of 159.15Hz. The phase shift at that frequency is –45.01°. These values both agree closely with our calculations.

Now we will see what happens now when we apply some gain to the non-inverting Opamp. The circuit to evaluate this is shown in Figure 11-12.

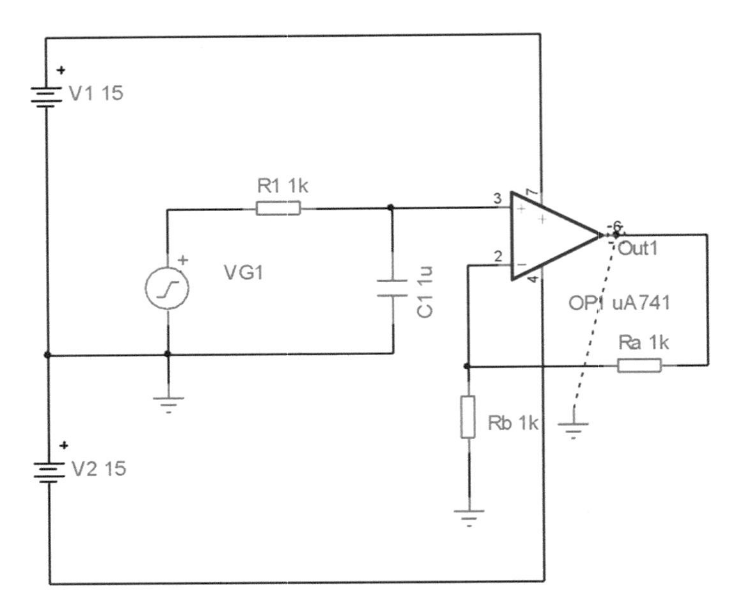


***Figure 11-12.*** *The First-Order Low Pass Filter with Gain*

The Opamp has now got some voltage gain. The expression for the voltage gain is

$$v_{gain} = 1 + \frac{R_a}{R_b}$$

I am using the letters “a” and “b” to try and avoid confusion with the actual resistors in the circuit. I hope you can appreciate that the expression for the voltage gain of the Opamp is the same as the transfer function for the Opamp. This means that this active filter is made up of two sections, the passive filter circuit and the Opamp. When determining the transfer function of a circuit with more than one section, we can simply multiply the functions together. This would give a transfer function for the active filter in Figure 11-12 as

$$TF = \left(\frac{1}{1 + j\omega CR}\right) \times \left(1 + \frac{R_a}{R_b}\right) = \frac{1 + \frac{R_a}{R_b}}{1 + j\omega CR}$$

If we put the values for Ra and Rb into the transfer function, it will now become

$$TF = \frac{2}{1 + j\omega CR}$$

Now we can add some gain to the filter circuit; we will see what happens when we have a wide band of frequencies that we want to pass onto the output. For example, we will design a low pass filter that has a cutoff frequency of 15kHz. This is close to the upper end of the audio range, especially my upper limit. If we use a typical resistor value of 2.2kΩ, then we can calculate the value of the capacitor as follows:

$$C = \frac{1}{2\pi fR} = \frac{1}{2\pi \times 15E^3 \times 2.2E^3} = 4.83nF$$

Changing $R_1$ and $C_1$ to those values and simulating the circuit gives us the Bode plot shown in Figure 11-13.

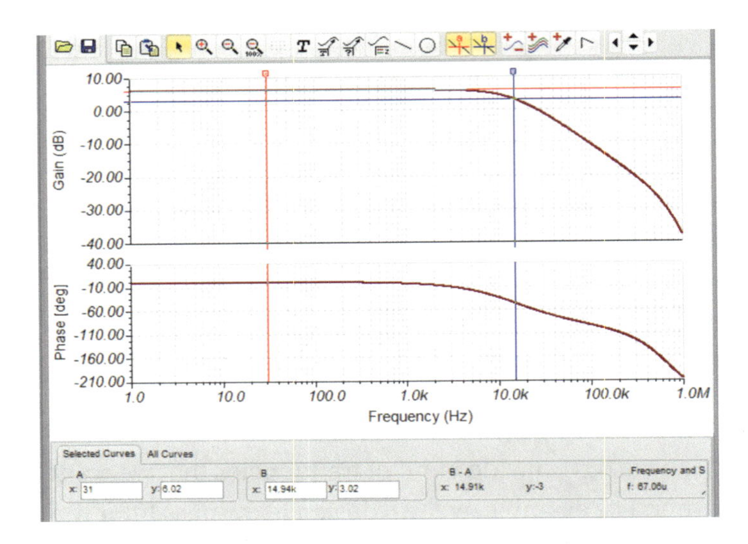


***Figure 11-13.** The Bode Plot for the First-Order Low Pass Filter with a Gain of Two and Cutoff Frequency of 15kHz*

When using a Bode plot, we need to remember the gain is plotted in dBs. To express voltage gain in dBs, we use the expression

$$v_{gain} = 20Log(gain)\ in\ dBs$$

The maximum dBs on the Bode plot would be

$$20Log(2) = 6.02\ dBs$$

Cursor "a" shows the maximum gain on the Bode plot is 6.02 dBs. We can now use cursor "b" to measure the frequency at which the gain reduces by 3dBs. Cursor "b" shows that the frequency when the gain falls to 3.02dBs is 14.94kHz. This is close to the desired cutoff frequency of 15kHz we have designed the filter to have. One thing that differs from this Bode plot and that for the passive filters is that the roll-off at the high frequency does not show a constant gradient of 20dBs/decade. This is because the Opamp has its own natural high-end roll-off, and this increases the gradient at the higher frequencies.

## Exercise 11.3

Design a low pass active filter that has a cutoff frequency of 10kHz when the resistor R is 4.7kΩ. Let $R_a$ and $R_b$ both be set at 1kΩ.

# The First-Order High Pass Filter

The circuit shown in Figure 11-14 is a high pass first-order active filter.

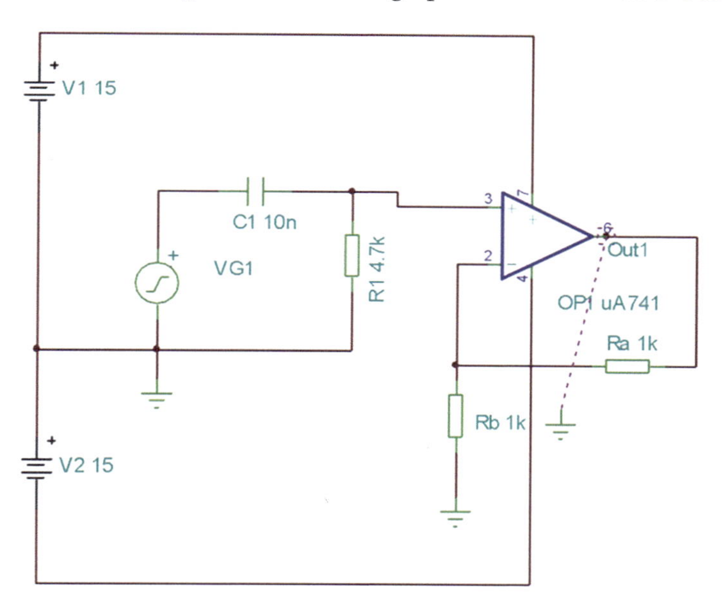


***Figure 11-14.*** *The High Pass First-Order Active Filter*

The gain of the Opamp is still set to 2 using resistors $R_a$ and $R_b$. That being the case, we can show that the transfer function of the filter is

$$TF = \frac{1}{1 - j\frac{1}{\omega CR}} \times 2 = TF = \frac{2}{1 - j\frac{1}{\omega CR}}$$

The cutoff frequency fc can be calculated using

$$f_c = \frac{1}{2\pi CR} = \frac{1}{2\pi \times 10E^{-9} \times 4.7E^3} = 3.386kHz$$

When we simulate the circuit, the Bode plot created is as shown in Figure 11-15.

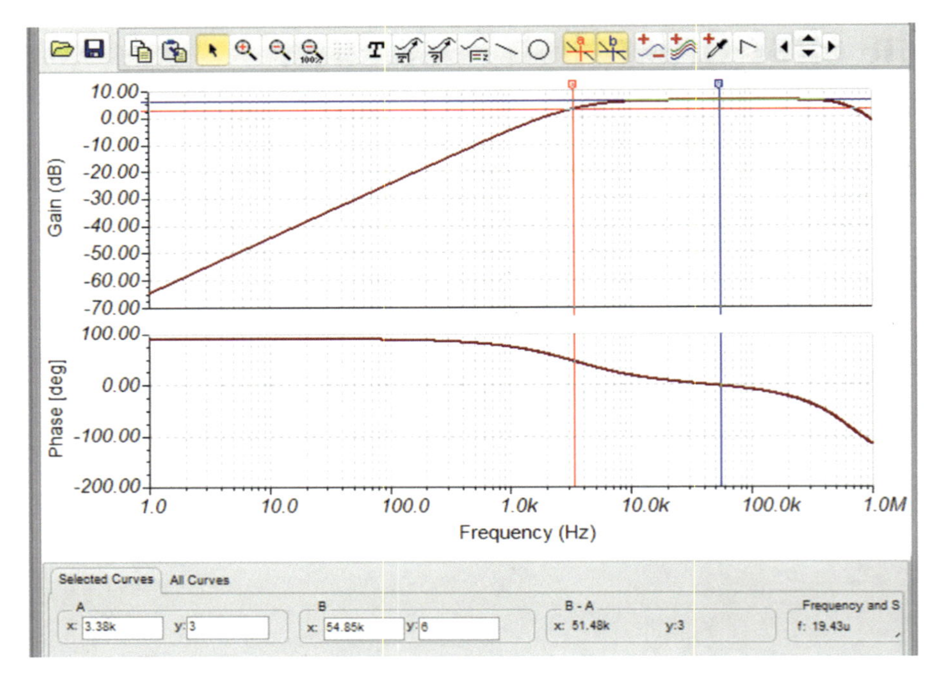


***Figure 11-15.*** *The Bode Plot for the High Pass Filter*

Cursor "b" shows that the maximum gain is the 6dBs we expect. Cursor "a" shows that when the gain has fallen down the 3dBs, that is, 3dBs down from the maximum, the frequency is 3.38kHz. This is close to the calculated cutoff frequency.

Figure 11-15 shows one issue with using the basic 741 Opamp as an active filter. We can see that when the frequency reaches around the 500kHz mark, the gain starts to fall off. This is because the Opamp itself can be viewed as a low pass filter. Due to the internal circuitry, the gain will fall off at higher frequencies. Some datasheets show this as an FT figure, which is the frequency of transition. This is the frequency at which the Opamp undergoes a transition, as it stops being an amplifier and starts being an attenuator. This is similar to what happens with the BJT amplifier, and it does restrict the overall bandwidth of the Opamp.

# High-Order Active Filters

I have described the active filters we have looked at so far as being first order. It is now time to explain what I mean by this. If you consider what you would expect from a filter, such as a low pass filter, you could say that the filter should amplify all frequencies up to the cutoff frequency with the same gain. Then all frequencies above the cutoff would be stopped completely. The ideal response of the filter would be as shown in Figure 11-16.

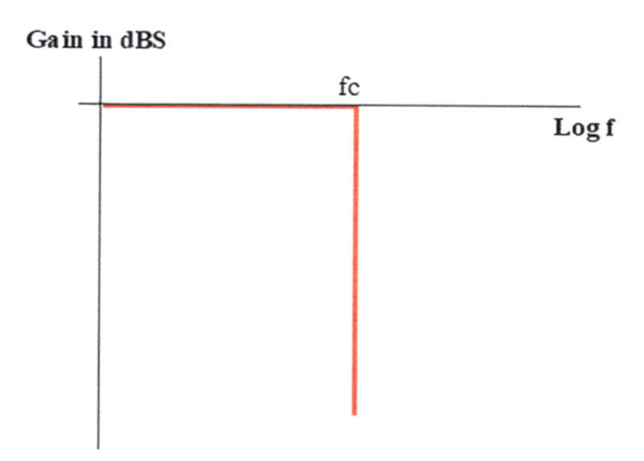


***Figure 11-16.*** *The Ideal Low Pass Filter Response*

We have seen that the response of the first-level filters is not vertical, as it is with the ideal frequency response shown in Figure 11-16. The simulated Bode plot response shown in Figure 11-11 has a gradient of 20dBs per decade of frequency. So, a first-order filter does not come close to the ideal frequency response. As an attempt to improve the filter's drop-off rate, we could add more filter sections to the filter. The first attempt was to add another section to the passive filter as shown in Figure 11-17.

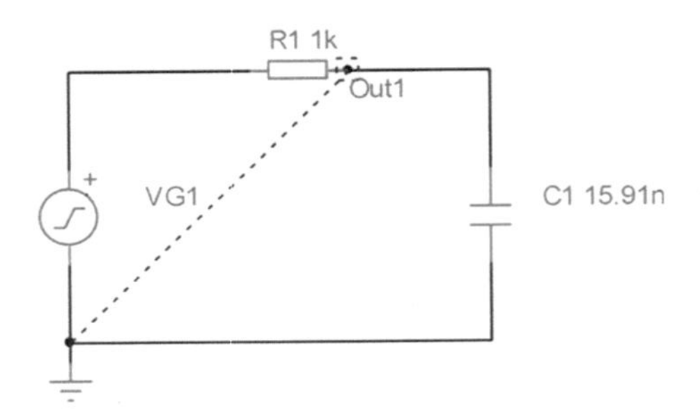


***Figure 11-17.*** *A Low Pass Passive First-Order Filter*

I hope you have been able to derive the transfer function for the single RC low pass passive filter shown in Figure 11-9. We will look at the filter again here. The transfer for the circuit shown in Figure 11-17 is

$$TF = \frac{1}{1 + j\omega CR}$$

$$TF = \frac{1}{\sqrt{1 + (\omega CR)^2}} - \langle Tan^{-1}(\omega CR) \rangle$$

Just to confirm that the transfer function is correct, we can evaluate the transfer function using the following values for the components shown in Figure 11-17. We could then simulate the circuit and use the Bode plot to confirm our calculations. We will evaluate the transfer function when f = 5kHz, C = 15.91nF, and R = 1kΩ.

$$TF = \frac{1}{\sqrt{1 + \left(2\pi \times 5E^3 \times 15.91E^{-9} \times 1E^3\right)^2}} - \langle Tan^{-1}\left(\left(2\pi \times 5E^3 \times 15.91E^{-9} \times 1E^3\right)\right) \rangle$$

$$TF = \frac{1}{\sqrt{1+0.25}} - \langle Tan^{-1}(0.5) \rangle$$

$$TF = 0.8944\langle -26.57 \rangle$$

We need to express the magnitude of the transfer function in dBs as this is the unit the Bode plot uses. As we are using a voltage gain to derive the transfer function, then we convert the magnitude to dBs using

$$20Log\left(\frac{v_{out}}{v_{in}}\right) = 20Log(0.8944) = -0.969\ dBs$$

# Exercise 11.4

Calculate the magnitude of the transfer function when f = 12kHz and so show that it would be –3.87 dBs.

After simulating the circuit shown in Figure 11-17, we could produce the Bode plot shown in Figure 11-18. The gain at 5kHz and 12kHz is shown as being the same as those calculated earlier.

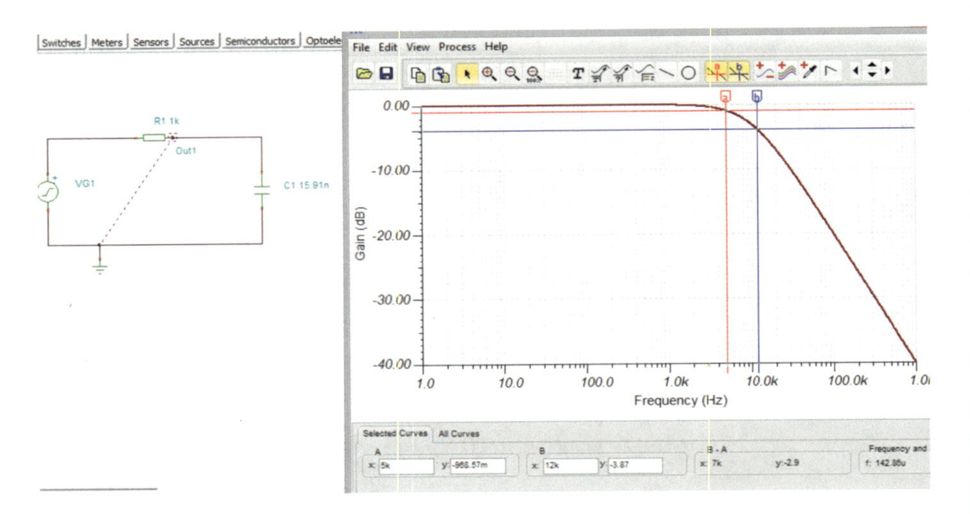


***Figure 11-18.*** *The Bode Plot for the First-Order Low Pass Passive Filter*

# The Second-Order Filter

We can change the circuit shown in Figure 11-17 to a second-order filter by adding another CR filter to the circuit. This arrangement is shown in Figure 11-19.

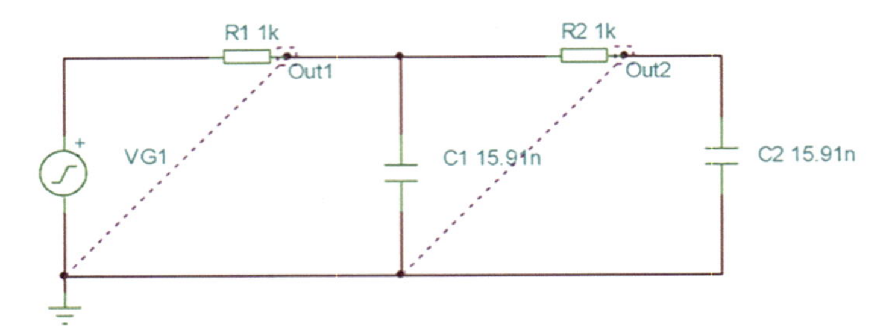


***Figure 11-19.*** *The Second-Order Low Pass RC Filter*

To derive the transfer function for the two-stage passive filter shown in Figure 11-19, we need to extend the analysis to accommodate the loading effect of the second stage on the first. To help with that, we will use the circuit shown in Figure 11-20.

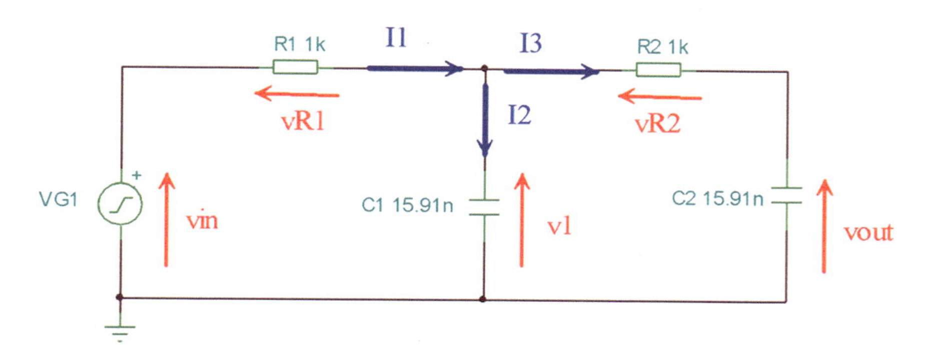


***Figure 11-20.*** *The Two-Stage Passive Low Pass Filter*

This is a second-order filter, and so hopefully we should see an improvement of the filter's response. The approach to deriving the transfer function is to use Ohm's Law to develop relationships between the currents, voltages, and impedances. When trying to derive a transfer function for an electrical system, we want to end up with the expression

$$TF = \frac{v_{out}}{v_{in}}$$

This means that we want to end up with expressions that have only the terms $v_{out}$ and $v_{in}$ in them. This means we must relate all the interim variables to the terms of $v_{out}$ and $v_{in}$. This will be our first in-depth analysis of an electrical circuit. You should take your time reading it and try to appreciate all the steps. I will try to explain all the steps, but it is a difficult balance between writing too much and leaving some steps out. I hope I have got the balance right.

It is usually best to start the furthest end away from the input, so we will look at the output voltage, $v_{out}$, and the current, $i_3$, in the circuit shown in Figure 11-20.

The volt drop across a component can be calculated by multiplying the impedance of the component by the current flowing through it. The current "i3" flows through the capacitor $C_2$. The impedance "XC" of the capacitor can be calculated as

$$XC = -j\frac{1}{\omega C} = \frac{1}{j\omega C}$$

We can replace the term *jω with'S'* as this is an accepted approach that takes us into the Laplace domain.

Using that replacement, we can say

$$v_{out} = i_3\frac{1}{SC}$$

If we divide both sides by $\frac{1}{SC}$, we can say

$$i_3 = v_{out} \div \frac{1}{SC}$$

When dividing by a fraction, we simply invert it and multiply:

$$i_3 = v_{out}SC$$

We could have simply multiplied both sides by "SC," and we would have the same result.

We can also express "$i_3$" in terms of the volt drop across $R_2$. Note, in the circuit we have made $R_1 = R_2 = R$. The volt drop across a component is simply the voltage at one end minus the voltage at the other end.

Using this approach, we can say the volt drop across $R_2$ is $v_1 - v_{out}$. Using Ohm's Law, we can say

$$I = \frac{V}{R}$$

This means that the current flowing through a component is equal to the volt drop across the component divided by the impedance of the component. Using this concept, we can say

$$i_3 = \frac{v_1 - v_{out}}{R} = \frac{v_1}{R} - \frac{v_{out}}{R}$$

We now have two expressions for "$i_3$", so we can equate them to each other as follows:

$$\frac{v_1}{R} - \frac{v_{out}}{R} = v_{out} SC$$

$$\frac{v_1}{R} = v_{out} SC + \frac{v_{out}}{R}$$

Taking $v_{out}$ as a common factor on the RHS, we get

$$\frac{v_1}{R} = v_{out}\left(SC + \frac{1}{R}\right)$$

Now multiplying both sides by "R" gives

$$v_1 = v_{out}\left(SC + \frac{1}{R}\right)R$$

Therefore, we get

$$v_1 = v_{out}\left(SCR + 1\right)$$

which we can rewrite as

$$v_1 = v_{out}\left(1 + SCR\right)$$

Now that we have an expression for $v_1$ in terms of $v_{out}$, we can now move on to consider the current "$i_2$". We can say

$$i_2 = v_1 \div \frac{1}{SC}$$

When dividing by a fraction, we simply invert it and multiply, which gives

$$i_2 = v_1 SC$$

Now we can substitute for $v_1$ in this expression with the expression for $v_1$ from before:

$$i_2 = v_{out}\left(1 + SCR\right)SC$$

$$i_2 = v_{out}\left(SC + S^2C^2R\right)$$

Now we can derive an expression for the current $i_1$ as follows:

$$i_1 = \frac{v_{in} - v_1}{R} = \frac{v_{in}}{R} - \frac{v_1}{R}$$

Now using Kirchhoff's Current Law, we can say

$$i_1 = i_2 + i_3$$

Substituting with the expressions for the three currents, we can say

$$\frac{v_{in}}{R} - \frac{v_1}{R} = \left(v_{out}\left(SC + S^2C^2R\right)\right) + \left(v_{out}SC\right)$$

If we add the term $\frac{v_1}{R}$ to both sides, we get

$$\frac{v_{in}}{R} = \left(v_{out}\left(SC + S^2C^2R\right)\right) + \left(v_{out}SC\right) + \frac{v_1}{R}$$

If we now multiply both sides by "R," we get

$$v_{in} = \left[\left(v_{out}\left(SC + S^2C^2R\right)\right) + \left(v_{out}SC\right) + \frac{v_1}{R}\right]R$$

We can now expand the "R" into the outer bracket, which gives

$$v_{in} = \left(v_{out}\left(SC + S^2C^2R\right)\right)R + \left(v_{out}SC\right)R + v_1$$

Now expand the "R" into the smaller brackets, which gives

$$v_{in} = v_{out}\left(SCR + S^2C^2R^2\right) + v_{out}SCR + v_1$$

We need to replace the term $v_1$ with the expression for $v_1$ in terms of $v_{out}$ from before. This gives

$$v_{in} = v_{out}\left(SCR + S^2C^2R^2\right) + v_{out}SCR + v_{out}\left(1 + SCR\right)$$

We now have an expression that relates all the interim variables to the terms of $v_{in}$ and $v_{out}$. We can now use this to derive the expression for the transfer function.

We can take the term $v_{out}$ out as a common factor from the RHS. This gives

$$v_{in} = v_{out}\left[\left(SCR + S^2C^2R^2\right) + SCR + \left(1 + SCR\right)\right]$$

Collecting all the like terms gives

$$v_{in} = v_{out}\left[1 + 3SCR + S^2C^2R^2\right]$$

Now we can substitute back by replacing the "S" with $j\omega$. This gives

$$v_{in} = v_{out}\left[1 + 3j\omega CR + (j\omega)^2 C^2R^2\right]$$

We know that

$$(j\omega)^2 = j^2\omega^2 \text{ and } j^2 = -1 \text{ therefore } (j\omega)^2 = -\omega^2$$

This uses the fact that

$$j = \sqrt{-1} \text{ therefore } j^2 = -1$$

This is a difficult concept to appreciate, but using "j" in this way allows us engineers and mathematicians to determine the square root of a negative number, which is something we do come across especially in control engineering.

Therefore, collecting the real terms and the j terms gives

$$v_{in} = v_{out}\left[1 - \omega^2C^2R^2 + 3j\omega CR\right]$$

We can now finally arrive with an expression for $\frac{v_{out}}{v_{in}}$, which is the transfer function:

$$\frac{v_{out}}{v_{in}} = \frac{1}{1 - \omega^2C^2R^2 + 3j\omega CR}$$

We can convert the denominator to polar format using the following relationships:

$$a + jb = \sqrt{a^2 + b^2}\ \langle Tan^{-1}\left(\frac{b}{a}\right)\rangle$$

$$a + jb \quad a = 1 - \omega^2 C^2 R^2 \quad b = 3\omega CR$$

Putting these values in, we get

$$TF = \frac{1}{1 - \omega^2 C^2 R^2 + 3j\omega CR} = \frac{1}{\sqrt{\left(1 - \omega^2 C^2 R^2\right)^2 + \left(3\omega CR\right)^2}\ \langle Tan^{-1}\left(\frac{3\omega CR}{1 - \omega^2 C^2 R^2}\right)\rangle}$$

$$TF = \frac{1}{\sqrt{\left(1 - \omega^2 C^2 R^2\right)^2 + \left(3\omega CR\right)^2}} - \langle Tan^{-1}\left(\frac{3\omega CR}{1 - \omega^2 C^2 R^2}\right)\rangle$$

Therefore, the transfer function for the second-order passive RC filter circuit shown in Figure 11-20 is

$$TF = \frac{1}{\sqrt{\left(1 - \omega^2 C^2 R^2\right)^2 + \left(3\omega CR\right)^2}} - \langle Tan^{-1}\left(\frac{3\omega CR}{1 - \omega^2 C^2 R^2}\right)\rangle$$

So, we have gone through a lot of work to get to the transfer function. However, I hope that after reading through it carefully, you will see that it is not too complicated, and you should be able to do this kind of work yourselves; you just need to take your time and check your work as you progress.

We can now use some different frequencies to try and confirm the expression is correct. First, we will use a frequency of 5kHz.

## When f = 5kHz

$$\omega CR = 0.4998$$

$$(\omega CR)^2 = 0.2498$$

$$TF = \frac{1}{\sqrt{(1-0.2498)^2 + (3\times 0.4998)^2}} - \langle Tan^{-1}\left(\frac{3\times 0.4998}{1-0.2498}\right)\rangle$$

$$TF = \frac{1}{\sqrt{0.5628 + 2.248}} - \langle Tan^{-1}(1.999)\rangle$$

$$TF = 0.5965\langle -63.435\rangle$$

Expressing the magnitude in dBs gives

$$20Log(0.5965) = -4.49\ dBs$$

## When f = 10kHz

$$\omega CR = 0.99965$$

$$(\omega CR)^2 = 0.99931$$

$$TF = \frac{1}{\sqrt{(1-0.99931)^2 + (3\times 0.99965)^2}} - \langle Tan^{-1}\left(\frac{3\times 0.99965}{1-0.99931}\right)\rangle$$

$$TF = 0.3333\langle -90\rangle$$

In dBs, we have

$$200.3333 = -9.54\ dBs$$

# Exercise 11.5

Calculate the magnitude of the transfer function when f = 15kHz.

There is a problem when calculating the angle of the phasor with the calculator as once you have gone past the 90° mark, the calculating gives you the wrong angle.

When you completed Exercise 11.5, the expression for the angle would have produced

$$-Tan^{-1}\left(\frac{4.4985}{1-2.248}\right)$$

This would give an angle of 74.495° or 1.3 Rads. We should have some idea that this would be the wrong angle as at f = 0 the angle would be 0°. Then when f = 5kHz, the angle was -63.435°, and when f = 10kHz, the angle was -90°. This should suggest that as the frequency increased, the angle went further negative. Also, if we simulate the circuit and examine the Bode plot produced, as shown in Figure 11-22, we see the angle does increase negatively as the frequency increases. Indeed, at a frequency of 15kHz, the angle from the Bode plot is -105.5°; see Figure 11-22.

To understand why the angle should be -105.5 and not the +74.495, we need to convert the polar format of the transfer function to the rectangular or coordinate format. This is done using the following relationship:

$$r\langle\theta\rangle = rCos(\theta) + jrSin(\theta)$$

Therefore, as an example, we will convert the complex number in polar format to coordinate format:

$$TF = 0.214\langle-(105.5)\rangle$$

$$0.214\langle-(105.5)\rangle = 0.214Cos(-105.5) + j0.214Sin(-105.5)$$

Therefore

$$0.214\langle -(105.5)\rangle = -0.0572 - j0.2062$$

Whenever the complex number has a negative real term, we must add 180 to the calculated angle that $-Tan^{-1}\left(\frac{b}{a}\right)$ results in. Also, you should note that 180 + 74.5 = 254.5, and that this is the same as –105.5.

Really, it would be useful to sketch the complex number as a phasor on the argand diagram. This is shown in Figure 11-21.

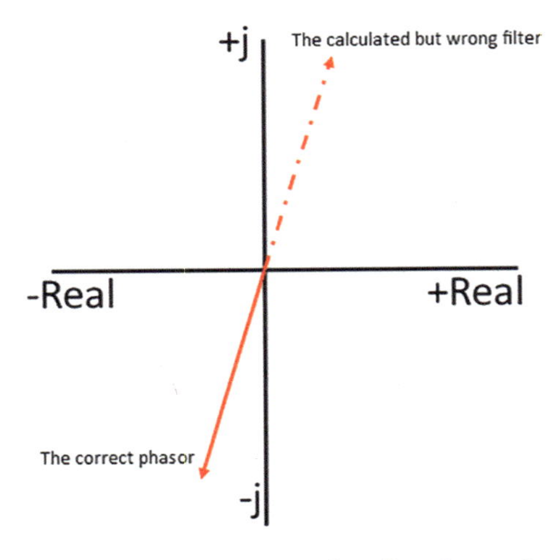


***Figure 11-21.*** *The Argand Diagram for the Complex Number*

$$TF = 0.214\langle -(105.5)\rangle$$

You do really need to practice using complex numbers and phasor quantities. This aspect will require some experience with drawing the phasors on an argand diagram. Some calculators allow you to conduct a complex division by expressing them in coordinate or polar format. If you can use your calculator in this way, your calculator will solve this issue for you. However, not all calculators have this ability, and the way to use your calculator in this way varies too much. The method I describe in the chapter works on any calculator, but you need to be aware of the issues as described here.

We can now simulate the circuit shown in Figure 11-17 and use the Bode plot to confirm the calculations. The Bode plot is shown in Figure 11-22.

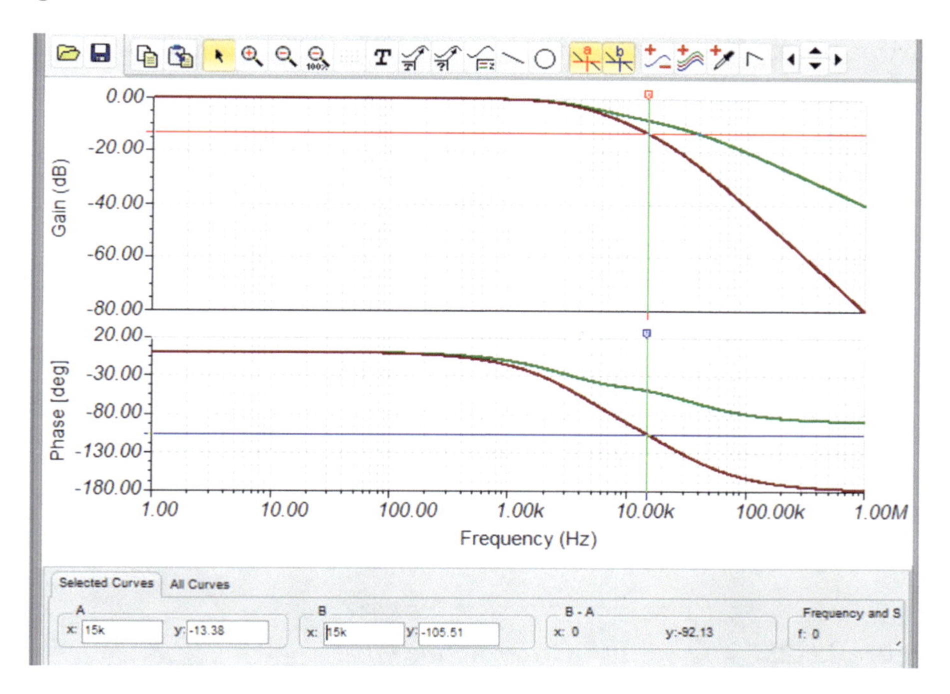


***Figure 11-22.*** *The Bode Plot for the Second-Order Passive Circuit*

We can see that at 15kHz the reading from the Bode plot does agree with our calculations. An important response we can see from the Bode plot is that the gradient of the first section, shown as the green trace, has a gradient of 20dBs per decade, which is what is expected, but the gradient of the second section, shown in brown, is 40dBs per decade. This is a steeper response and shows that the higher-order filter is an improvement. However, the disadvantage is that the attenuation is greater. Therefore, we could counter this disadvantage by using an active filter with some gain added.

# The Second-Order Active Filter

We can use the Opamp to provide some amplification to counteract the attenuation of the higher-order filter, and the first circuit we will look at is shown in Figure 11-23.

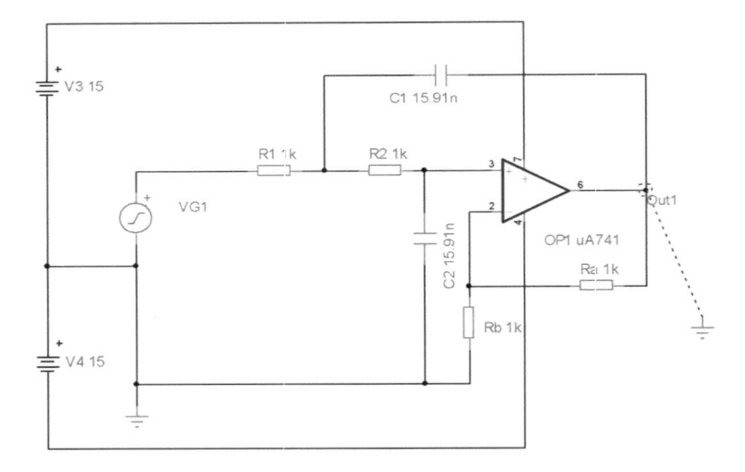


***Figure 11-23.*** *The Second-Order Active Filter*

The circuit uses a non-inverting Opamp, and the gain is set using

$$v_{gain} = 1 + \frac{R_a}{R_b} = 1 + \frac{1^3}{1^3} = 2$$

We can use the circuit shown in Figure 11-24 to try and derive the transfer function for the second-order filter.

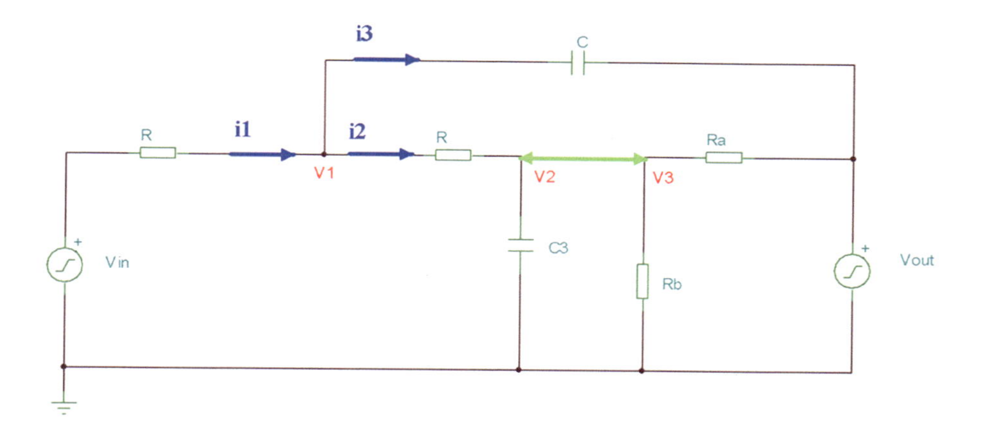


***Figure 11-24.*** *The Filter Circuit with Feedback Around the Non-inverting Opamp*

To begin with, we should understand that the double-headed arrow, shown in light green in Figure 11-24, is to signify the virtual short circuit across the input terminals of the Opamp. The resistors $R_a$ and $R_b$ are what are used to set the voltage gain of the Opamp. If we assume $R_a$ and $R_b$ have the same value, then we can say

$$v_3 = v_{out}\frac{R_b}{R_a + R_b} = \frac{v_{out}}{2}$$

This means that because of the virtual short circuit across the terminals of the Opamp, then we can say

$$v_2 = v_3 = \frac{v_{out}}{2}$$

Now we can look at the currents flowing around the circuit, and we start by applying Kirchhoff's Current Law (KCL). Applying this to the node at $V_1$, we can say

$$i_1 = i_2 + i_3$$

We can now apply Ohm's Law to express these currents in terms of their associated volt drops. Therefore, with respect to $i_1$, we can say

$$i_1 = \frac{v_{in} - v_1}{R} = \frac{v_{in}}{R} - \frac{v_1}{R}$$

$$i_1 = \frac{v_{in}}{R} - \frac{v_1}{R}$$

Similarly, with respect to $i_2$, we can say

$$i_2 = \frac{v_1 - v_2}{R} = \frac{v_1}{R} - \frac{v_2}{R}$$

$$i_2 = \frac{v_1}{R} - \frac{v_2}{R}$$

Finally, looking at $i_3$, we can say

$$i_3 = \frac{v_1 - v_{out}}{XC}$$

Knowing that, we can express XC as

$$XC = \frac{1}{SC}$$

Putting this into the expression for "$i_3$", we get

$$i_3 = \left(v_1 - v_{out}\right) \div \frac{1}{SC}$$

When dividing by a fraction, we can simply invert and multiply, which gives

$$i_3 = (v_1 - v_{out}) \times SC = v_1 SC - v_{out} SC$$

Therefore, we have

$$i_3 = v_1 SC - v_{out} SC$$

We can replace all references to $v_2$ in the expression for the currents with $\frac{v_{out}}{2}$. The only current expression with a reference to $v_2$ is the current $i_2$. Therefore, we can say

$$i_2 = \frac{v_1}{R} - \frac{v_{out}}{2} \div \frac{R}{1}$$

Therefore, we have

$$i_2 = \frac{v_1}{R} - \frac{v_{out}}{2} \times \frac{1}{R}$$

which gives

$$i_2 = \frac{v_1}{R} - \frac{v_{out}}{2R}$$

Now we need to remove the term $v_1$ from the current expressions. If we look at the small section of the circuit, as shown in Figure 11-25 we should be able to derive an expression for $V_1$ in terms of $V_{OUT}$.

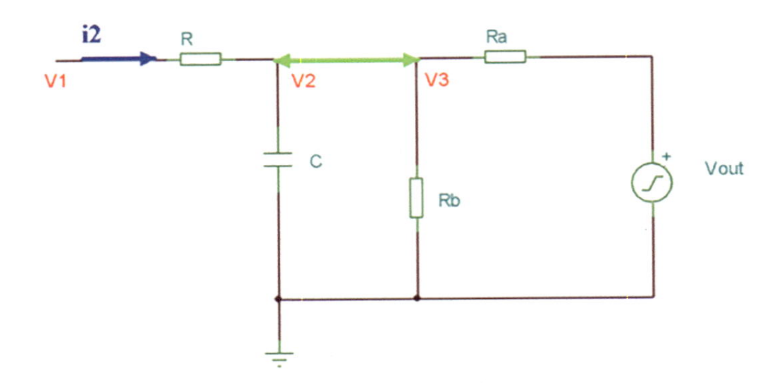


***Figure 11-25.*** *The Small Section of the Filter Circuit*

We should be able to see that the voltage $v_1$ is divided by the RC circuit to produce the voltage $v_2$. That being the case, we can say

$$v_2 = v_1 \times \frac{\frac{1}{SC}}{R + \frac{1}{SC}}$$

If we multiply the top and bottom of the fraction by *SC*, we get

$$v_2 = v_1 \times \frac{1}{1 + SCR}$$

This then means we can say

$$v_1 = V_2 \left(1 + SCR\right)$$

However, we have already shown that

$$v_2 = \frac{v_{out}}{2}$$

Therefore, we can say

$$v_1 = \frac{v_{out}}{2}(1 + SCR)$$

This means we can replace all references to $v_1$ in the expression for the currents. In this way, we can say

$$i_1 = \frac{v_{in}}{R} - \frac{v_{out}}{2}(1 + SCR) \div R$$

Therefore, we have the following expressions for the three currents with reference to $v_{in}$ and $v_{out}$ only:

$$i_1 = \frac{v_{in}}{R} - \frac{v_{out}}{2R}(1 + SCR)$$

and

$$i_2 = \frac{v_{out}}{2R}(1 + SCR) - \frac{v_{out}}{2R}$$

and

$$i_3 = \frac{v_{out}}{2}(1 + SCR)SC - v_{out}SC$$

We have now removed all references to the intermediate variables, $v_1$ and $v_2$, and are left with just $v_{in}$ and $v_{out}$, which is what we wanted.

So, using the expressions for the three currents and applying them to KCL, we can say

$$\frac{v_{in}}{R} - \frac{v_{out}}{2R}(1 + SCR) = \frac{v_{out}}{2R}(1 + SCR) - \frac{v_{out}}{2R} + \frac{v_{out}}{2}(1 + SCR)SC - v_{out}SC$$

OK, we need to take a deep breath and just check what we have written is correct. Then when we are ready, we can first expand the brackets. This will give

$$\frac{v_{in}}{R}-\frac{v_{out}}{2R}-\frac{v_{out}}{2}SC=\frac{v_{out}}{2R}+\frac{v_{out}}{2}SC-\frac{v_{out}}{2R}+\frac{v_{out}}{2}SC+\frac{v_{out}}{2}(SC)^2R-v_{out}SC$$

Now, after checking this expression, we can collect all like terms on the RHS. This reduces the expression to

$$\frac{v_{in}}{R}-\frac{v_{out}}{2R}-\frac{v_{out}}{2}SC=\frac{v_{out}}{2}(SC)^2R$$

Now add $\frac{v_{out}}{2R}+\frac{v_{out}}{2}SC$ to both sides and simplify the expression to

$$\frac{v_{in}}{R}=\frac{v_{out}}{2}(SC)^2R+\frac{v_{out}}{2R}+\frac{v_{out}}{2}SC$$

Now multiply the whole expression by 2R, which will remove all the denominators in the expression as follows:

$$2v_{in}=v_{out}(SCR)^2+v_{out}SCR+v_{Out}$$

Now take the term $v_{out}$ out of the RHS as a common factor. This gives

$$2v_{in}=v_{out}\left[(SCR)^2+SCR+1\right]$$

Now divide both sides by the term in the bracket. This simplifies the expression to

$$\frac{2v_{in}}{(SCR)^2+SCR+1}=v_{out}$$

Now divide both sides by $v_{in}$, which gives

$$\frac{v_{out}}{v_{in}} = \frac{2}{(SCR)^2 + SCR + 1}$$

This means we have

$$\frac{v_{out}}{v_{in}} = \frac{2}{S^2C^2R^2 + SCR + 1}$$

If we now replace the term "S" with jω, we go back into real time, and this gives

$$\frac{v_{out}}{v_{in}} = \frac{2}{(j\omega)^2 C^2R^2 + j\omega CR + 1}$$

As $j^2 = -1$, we can say

$$\frac{v_{out}}{v_{in}} = \frac{2}{-\omega^2C^2R^2 + j\omega CR + 1}$$

Therefore, we have

$$\frac{v_{out}}{v_{in}} = \frac{2}{1 - \omega^2C^2R^2 + j\omega CR}$$

In its full complex form, we have

$$\frac{v_{out}}{v_{in}} = \frac{2 + j0}{1 - \omega^2C^2R^2 + j\omega CR}$$

This is the transfer function for the circuit shown in Figure 11-23, the second-order low pass active filter. Therefore, we can say

$$TF = \frac{2 + j0}{1 - \omega^2C^2R^2 + j\omega CR}$$

Converting this into polar format gives

$$TF = \frac{2\langle 0\rangle}{\sqrt{\left(1-\omega^2C^2R^2\right)^2+\left(\omega CR\right)^2}\ \langle Tan^{-1}\left(\frac{\omega CR}{1-\omega^2C^2R^2}\right)\rangle}$$

If we carry out the division by dividing the magnitude parts and subtracting the angles, this then equates to

$$TF = \frac{2}{\sqrt{\left(1-\omega^2C^2R^2\right)^2+\left(\omega CR\right)^2}}\ \langle -\langle Tan^{-1}\left(\frac{\omega CR}{1-\omega^2C^2R^2}\right)\rangle\rangle$$

We can check the expression by using some different values for the frequency "f" as follows.

## When f = 5kHz

$$\omega CR = 0.4998$$

$$\left(\omega CR\right)^2 = 0.2498$$

$$1-\left(\omega CR\right)^2 = 0.75017$$

$$\left(1-\left(\omega CR\right)^2\right)^2 = 0.56276$$

$$\sqrt{\left(1-\omega^2C^2R^2\right)^2+\left(\omega CR\right)^2} = 0.9014$$

We can now evaluate the magnitude of the transfer function as follows:

$$|TF| = \frac{2}{\sqrt{\left(1-\omega^2C^2R^2\right)^2+\left(\omega CR\right)^2}} = 2.219$$

Converting this into dBs gives

$$|TF| = 20Log(2.219) = 6.923dBs$$

We can calculate the phase using

$$Phase = -\langle Tan^{-1}\left(\frac{\omega CR}{1-\omega^2 C^2 R^2}\right)\rangle = -Tan^{-1}\left(\frac{0.499}{0.75017}\right) = -33.63°$$

If we simulate the circuit and create the Bode plot as shown in Figure 11-26, we can see that the simulated result at f = 5kHz agrees with the calculated results.

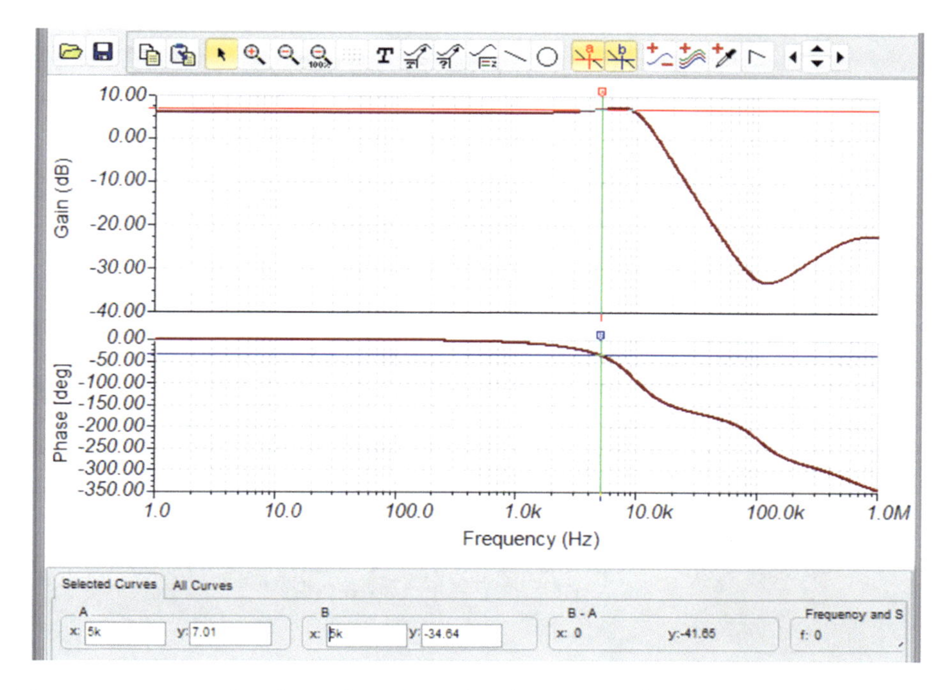


**Figure 11-26.** *The Simulated Bode Plot at f = 5kHz*

We can repeat the calculations at a frequency of f = 10kHz:

$$\omega CR = 0.99965$$

$$(\omega CR)^2 = 0.99931$$

$$1-(\omega CR)^2 = 6.9032^{-4}$$

$$\left(1-(\omega CR)^2\right)^2 = 4.7654^{-7}$$

$$\sqrt{\left(1-\omega^2C^2R^2\right)^2+(\omega CR)^2} = 0.99966$$

$$|TF| = \frac{2}{\sqrt{\left(1-\omega^2C^2R^2\right)^2+(\omega CR)^2}} = 2$$

In dBs, it gives

$$20Log(2) = 6.02dBs$$

We can calculate the phase using

$$Phase = -\langle Tan^{-1}\left(\frac{\omega CR}{1-\omega^2C^2R^2}\right)\rangle = Tan^{-1}\left(\frac{0.99965}{6.9032^{-4}}\right) = -89.97^\circ$$

We can repeat the calculations at a frequency of f = 30kHz:

$$\omega CR = 2.99896$$

$$(\omega CR)^2 = 8.9938$$

$$1-(\omega CR)^2 = -7.99379$$

$$\left(1-(\omega CR)^2\right)^2 = 63.9$$

$$\left(1-\omega^2C^2R^2\right)^2 + (\omega CR)^2 = 72.8944$$

$$\sqrt{\left(1-\omega^2C^2R^2\right)^2 + (\omega CR)^2} = 8.5378$$

$$|TF| = \frac{2}{\sqrt{\left(1-\omega^2C^2R^2\right)^2 + (\omega CR)^2}} = 0.23425$$

In dBs, this is

$$20Log(0.23425) = -12.61dBs$$

We can calculate the phase using

$$Phase = -\langle Tan^{-1}\left(\frac{\omega CR}{1-\omega^2C^2R^2}\right)\rangle = -Tan^{-1}\left(\frac{2.99896}{-7.99379}\right) = 20.56^\circ$$

However, because the rectangular format has a negative real term, then we must subtract 180° from this, which gives an angle of

$$20.56 - 180 = -159.44^\circ$$

This is another side to the issue of simply using the inverse Tan to calculate the phase angle. It is always a good idea to express the complex number in coordinate format and sketch the phasor on a phasor diagram. This will allow you to know how the angle you have just calculated relates to the correct angle of the phasor. It will take some practice to gain some experience.

Again, if we use the Bode plot from the simulated circuit, we see that the simulated results are close to the calculated results.

## Exercise 11.6

As an exercise, use the expression for the transfer function to calculate the magnitude, in dBs, and the phase of the transfer function when f = 7kHz and f = 20kHz.

$$TF = \frac{2}{\sqrt{\left(1-\omega^2C^2R^2\right)^2+\left(\omega CR\right)^2}} \langle -\langle tan^{-1}\left(\frac{\omega CR}{1-\omega^2C^2R^2}\right)\rangle\rangle$$

# The High Pass Second-Order Active Filter

The circuit we will use to analyze the high pass second-order active filter is shown in Figure 11-27.

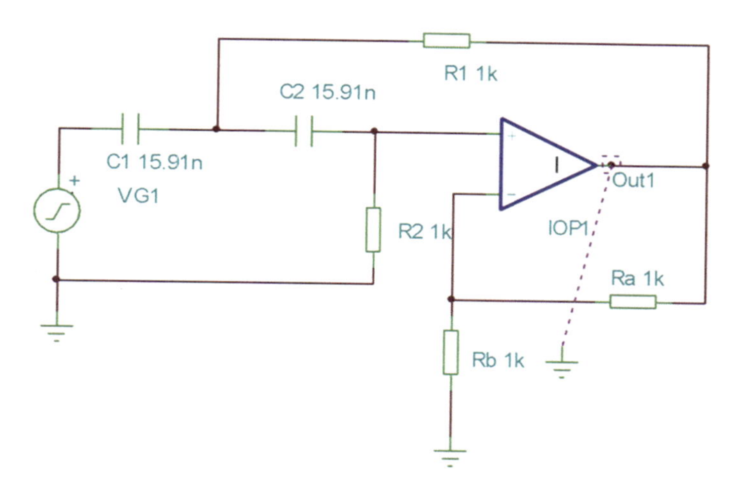


***Figure 11-27.** The High Pass Second-Order Active Filter*

We are using the ideal Opamp available in TINA, as it is a simpler circuit to build. Also, our mathematical model will be closer to that of TINA's, as TINA's model for the non-ideal Opamp is much more accurate than our transfer function, as it accounts for the imperfections of the Opamp.

# Exercise 11.7

Using the feedback model of the high pass filter, shown in Figure 11-28, derive the transfer function for the high pass second-order active filter. Assume that $R_a = R_b$.

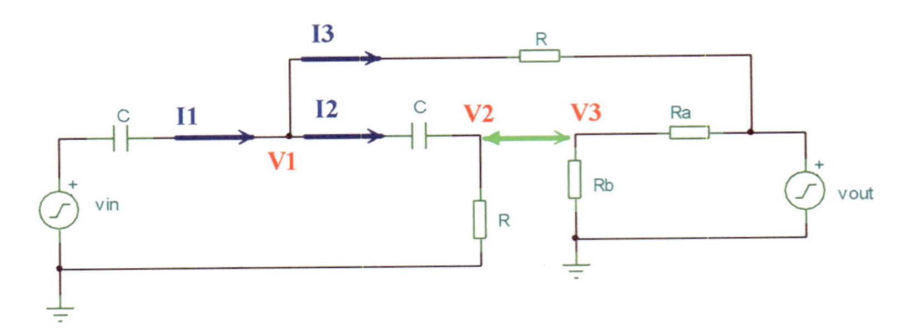


***Figure 11-28.*** *The Feedback Circuit for the High Pass Second-Order Active Filter*

You should be able to show that the transfer function is

$$TF = \frac{2}{\sqrt{\left(1 - \frac{1}{(\omega CR)^2}\right)^2 + \left(-\frac{1}{\omega CR}\right)^2}} - \langle Tan^{-1}\left(\frac{-\frac{1}{\omega CR}}{1 - \frac{1}{(\omega CR)^2}}\right)\rangle$$

We can evaluate the transfer function when f = 5kHz as follows. The magnitude of the transfer function is

$$\frac{2}{\sqrt{\left(1 - \frac{1}{(\omega CR)^2}\right)^2 + \left(\frac{1}{\omega CR}\right)^2}} = 0.5547$$

Converting this to dBs gives

$$20Log(0.5547) = -5.119dBs$$

We can calculate the phase using

$$Phase = \langle -\langle Tan^{-1}\left(\frac{\frac{1}{\omega CR}}{1-\frac{1}{(\omega CR)^2}}\right)\rangle\rangle = -Tan^{-1}\left(\frac{-2}{-3}\right) = -33.69°$$

If we changed the expression for the transfer function into its coordinate format, the transfer function would be

$$TF = \frac{2-\frac{2}{(\omega CR)^2}+j\frac{2}{\omega CR}}{\left(1-\frac{1}{(\omega CR)^2}\right)^2+\left(-\frac{1}{\omega CR}\right)^2}$$

The process to express the transfer function in this rectangular format involves using the complex conjugate. The process is shown in the appendix.

If we let f = 5kHz, then evaluating the transfer function would give

$$TF = -0.46194 + j0.3077$$

At this frequency, the phasor would be in the second quadrant, which means we need to add 180 to the calculated value of –33.69, giving us the correct phase of 146.32°.

When F = 30kHz, we have

$$\frac{2}{\sqrt{\left(1-\frac{1}{(\omega CR)^2}\right)^2+\left(\frac{1}{\omega CR}\right)^2}} = 2.1069$$

Converting this to dBs gives

$$20Log(2.1069) = 6.473dBs$$

We can calculate the phase using

$$Phase = \langle -\langle Tan^{-1}\left(\frac{\frac{1}{\omega CR}}{1-\frac{1}{(\omega CR)^{2}}}\right)\rangle\rangle = -Tan^{-1}\left(\frac{-0.33345}{0.8888}\right) = 20.56°$$

If we let f = 30kHz, then evaluating the transfer function in rectangular format would give

$$TF = 1.9725 + j0.74$$

This would be in the first quadrant, and so the angle of 20.56° is correct.

These calculations can be confirmed by creating the Bode plot from the simulation as shown in Figure 11-29. The 5kHz frequency on the magnitude, cursor "a," confirms that the gain is –5.13 dBs. The 30kHz frequency on the phase, cursor "b," confirms that the phase angle is 20.56. The Bode plot also confirms that the circuit is a high pass filter.

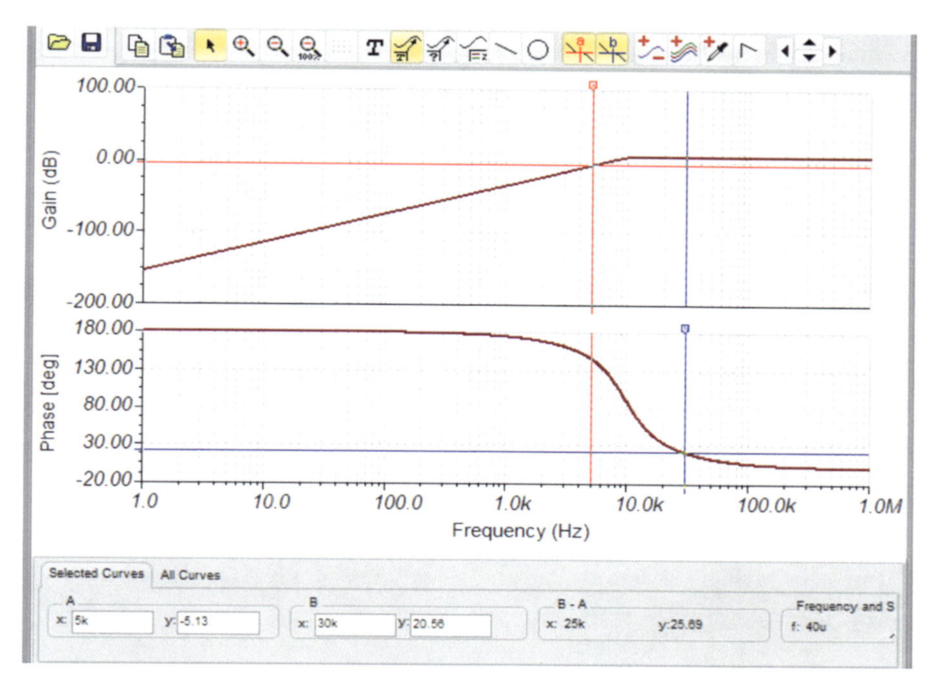


***Figure 11-29.*** *The Bode Plot for the High Pass Active Filter As Shown in Figure 11-26*

# Exercise 11.8

As an exercise, use the expression for the transfer function to calculate the magnitude, in dBs, and the phase of the transfer function when f = 7kHz and f = 20kHz.

# The Butterworth Filter

If we simulate the circuit shown in Figure 11-27, we can display the Bode plot as shown in Figure 11-30.

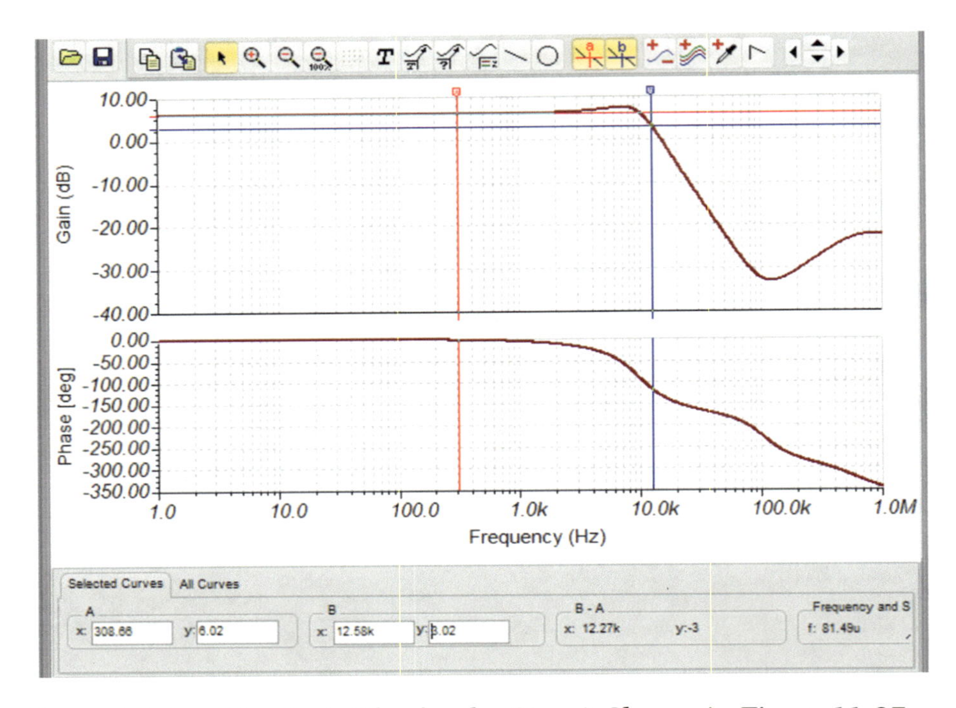


***Figure 11-30.** The Bode Plot for the Circuit Shown in Figure 11-27*

There are some interesting things we can see from the Bode plot shown in Figure 11-30. Firstly, the gain of the low frequencies, just before the cutoff frequency, has an unusual rise for those between 1.1k and 10k. Ideally, the gain should be flat over the frequencies below the cutoff frequency. Also, if we look at the frequency when the gain had fallen to 3.02dBs, which is 3dBs down from the maximum, we see it is a frequency of 12.58kHz. However, if we use the expression for this cutoff frequency, we do not get that value.

$$f_C = \frac{1}{2\pi \times C \times R} = \frac{1}{2\pi \times 15.91^{-9} \times 1^{3}} = 10kHz$$

We can see that there is some gain as at the low frequencies the Bode plot has a value of 6.02dBs. We can confirm this agrees with a gain of two, which is what the Opamp has been set to as

$$20Log\left(v_{gain}\right) = 20Log\left(2\right) = 6.021dBs$$

We need to do some work with this active filter as we need to get a flat response for the Bode plot, and we do need to be able to calculate the frequency of cutoff. Thankfully, there has been a lot of work conducted by engineers with a lot more experience than me, and we now have an active filter termed “the Butterworth filter.” This was first designed by an engineer named Stephen Butterworth in 1930. This is based on his work involving using the polynomials that can be derived for the filters of order “n.” The expression for determining the gain of the Opamp using these polynomials is

$$V_{gain} = 3 - \alpha$$

The term α is the coefficient of the “S” term in the appropriate polynomial. Some of the polynomials are given in Table 11-1.

***Table 11-1.*** *The Polynomials for Some of the Order of Filters*

| Filter Order "n" | Polynomial |
|---|---|
| 1 | $S + 1$ |
| 2 | $S^2 + 1.414S + 1$ |
| 3 | $(S + 1)(S^2 + 1S + 1)$ |
| 4 | $(S^2 + 1.848S + 1)$ $(S^2 + 0.7456S + 1)$ |
| 5 | $(S + 1)(S^2 + 1.618S + 1)$ $(S^2 + 0.618S + 1)$ |

Stephen Butterworth realized that the gain of the Opamp had a dramatic effect on the response of the filter. We can use this table to calculate the required gain of the Opamp to gain the flattest response of the filter in relation to the order of the filter. We are designing a second-order filter, and from the table, we can see that the value for "α" is 1.414, that is, the coefficient of the "S" term in the quadratic equation. Using this value for "α," we can calculate the required gain value that we need to set the Opamp to. This is done using

$$v_{gain} = 3 - \alpha = 3 - 1.414 = 1.586$$

The expression for the gain of the non-inverting Opamp is

$$v_{gain} = 1 + \frac{R_a}{R_b}$$

Therefore, we can say

$$1.586 = 1 + \frac{R_a}{R_b}$$

Therefore

$$\frac{R_a}{R_b} = 1.586 - 1 = 0.586$$

If we let Rb = 1k, then the value for Ra is

$$R_a = R_b \times 0.586 = 1E^3 \times 0.586 = 586\Omega$$

If we now change $R_a$ to 586Ω and simulate the circuit, we will get the Bode plot as shown in Figure 11-31.

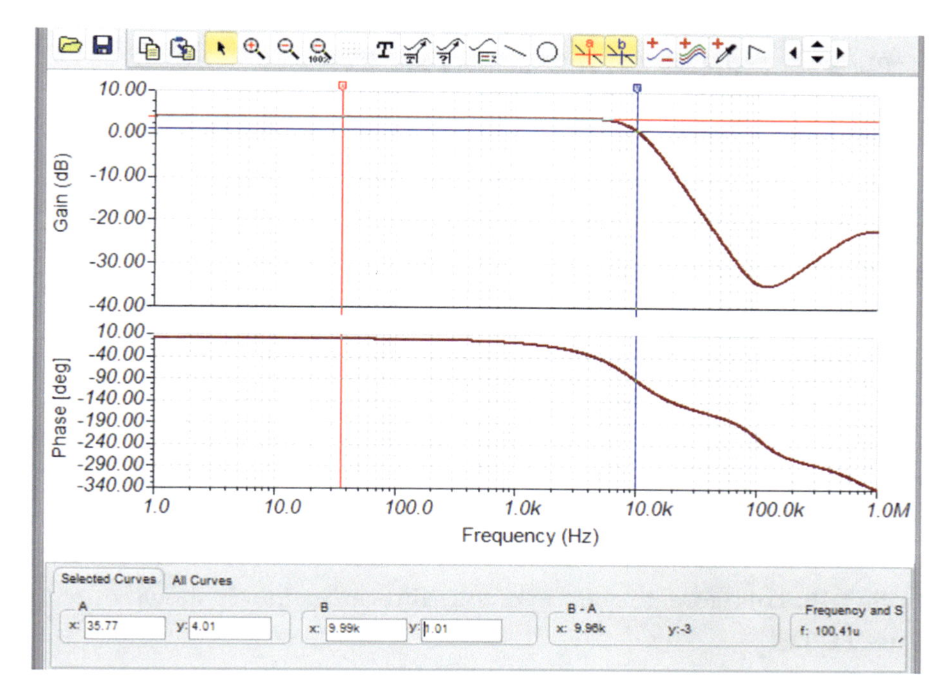


***Figure 11-31.*** *The Bode Plot for the Filter with the Gain Corrected*

We can see from Figure 11-28 that the response is very flat over the required frequency range. The maximum gain of 4.01dBs does agree with the calculation:

$$v_{gain} = 20Log(1.586) = 4dBs$$

The frequency when the gain fell by 3dBs to 1.01dBs is at 9.99k. This agrees with the calculated cutoff frequency from before. We can also use the Bode plot to confirm the gradient of the roll-off has been improved to give a roll-off that is now 40dBs per decade.

This does suggest that the Butterworth filter does work well. There is still the problem of the low transition frequency, FT, of the 741 Opamp. It does mean we need to use an Opamp that has a very high transition frequency.

## A Third-Order Butterworth Filter

We will design a third-order filter, and just to show you that a better Opamp would help with the high-frequency roll-off, we will use the ideal Opamp from within TINA. This will reduce the schematic as we do not need to add the supply rails. If we look at Table 11-1, we can see that the third-order filter uses a first-order expression multiplied into a second order. This means we can use two Opamps, the first organized as a first-order filter and the second as a second-order filter. We know that for the second-order filter, the gain must be set to 1.586. We need to use the table to set the gain of the first-order filter. From the table, we can see that the coefficient of "S" is 1, which means that "α" = 1. This means that the gain of the Opamp must be set to 2, that is, 3–1, which means we need to make $R_a = R_b = 1k$. Therefore, the circuit of the third-order filter is shown in Figure 11-32.

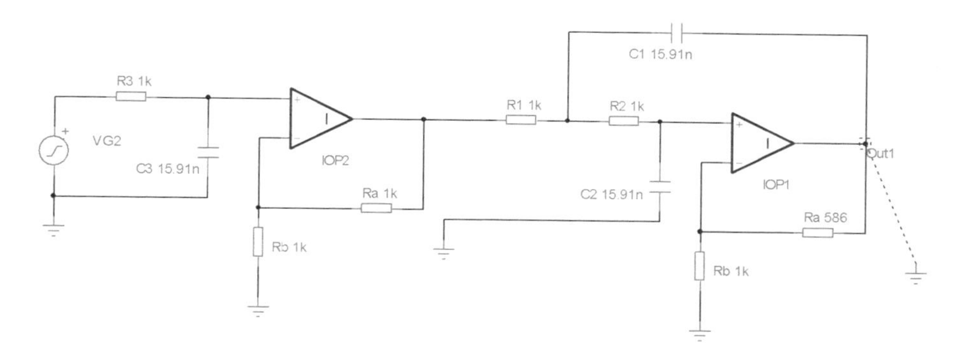


***Figure 11-32.*** *A Third-Order Low Pass Active Filter*

To allow us to change the gain of the second-order filter, to achieve a flatter frequency response, we need to modify the transfer function for the second-order filter and then the third-order filter by including the variable "A" to represent the ratio of the two gain resistors Ra and Rb. In the previous derivation of the second-order transfer function, we simply let Ra = Rb = 1k. Introducing this variable aspect into the design, we have

$$A = \frac{R_b}{R_a + R_b}$$

Doing this, we can derive the following transfer functions.

For the second-order filter, we have

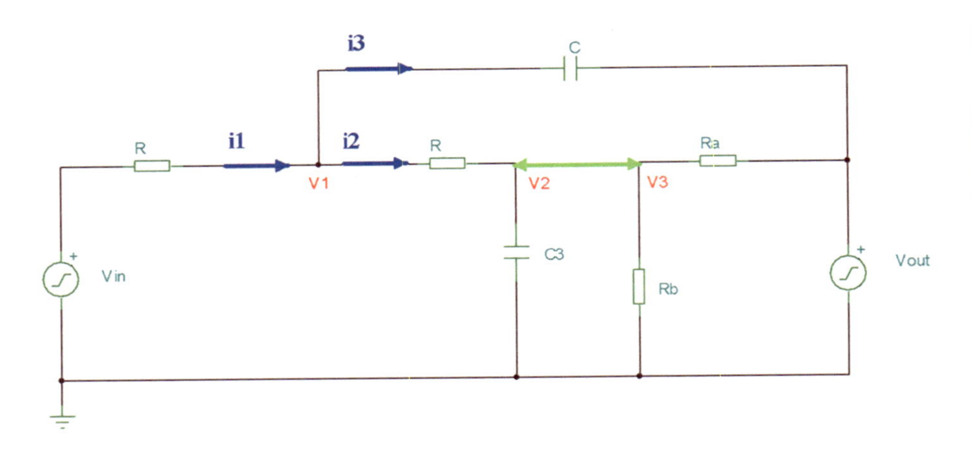


First, express the three currents in terms of $v_{in}$ and $v_{out}$.

$$i_2 = \frac{v_2}{\frac{1}{SC}} = v_2 SC$$

$$v_2 = v_3 = v_{out} \frac{R_b}{R_a + R_b}$$

$$A = \frac{R_b}{R_a + R_b}$$

$$v_2 = Av_{out}$$

$$i_2 = Av_{out} SC$$

$$i_3 = \frac{v_1 - v_{out}}{\frac{1}{SC}} = \frac{v_1}{\frac{1}{SC}} - \frac{v_{out}}{\frac{1}{SC}} = SCv_1 - SCv_{out}$$

$$v_1 = v_2\left(1+SCR\right)$$

$$v_1 = Av_{out}\left(1+SCR\right)$$

$$i_3 = Av_{out}\left(1+SCR\right)SC - v_{out}SC$$

$$i_1 = \frac{v_{in} - v_1}{R} = \frac{v_{in}}{R} - \frac{v_1}{R}$$

$$i_1 = \frac{v_{in}}{R} - \frac{Av_{out}\left(1+SCR\right)}{R}$$

Now apply Kirchhoff's Current Law, which means

$$i_1 = i_2 + i_3$$

$$\frac{v_{in}}{R} - \frac{Av_{out}\left(1+SCR\right)}{R} = Av_{out}SC + Av_{out}\left(1+SCR\right)SC - v_{out}SC$$

Now multiply throughout by "R" to remove the denominator on the LHS:

$$v_{in} - Av_{out}\left(1+SCR\right) = Av_{out}SCR + Av_{out}\left(1+SCR\right)SCR - v_{out}SCR$$

Now add the term $Av_{out}(1 + SCR)$ to both sides:

$$v_{in} = Av_{out}SCR + Av_{out}\left(1+SCR\right)SCR - v_{out}SCR + Av_{out}\left(1+SCR\right)$$

Now expand the brackets:

$$v_{in} = Av_{out}SCR + Av_{out}SCR + +Av_{out}\left(SCR\right)^2 - v_{out}SCR + Av_{out} + Av_{out}SCR$$

Now collect all the like terms:

$$v_{in} = Av_{out}\left(SCR\right)^2 + 3Av_{out}SCR - v_{out}SCR + Av_{out}$$

Now take $v_{out}$ as a common factor on the RHS:

$$v_{in} = v_{out}\left[A\left(SCR\right)^2 + 3ASCR - SCR + A\right]$$

Now factorize the terms $3ASCR - SCR$:

$$v_{in} = v_{out}\left[A\left(SCR\right)^2 + (3A-1)\left(SCR\right) + A\right]$$

Now divide both sides by the term in the bracket:

$$\frac{v_{in}}{A\left(SCR\right)^2 + (3A-1)\left(SCR\right) + A} = v_{out}$$

Now divide both sides by $v_{in}$:

$$\frac{v_{out}}{v_{in}} = \frac{1}{A\left(SCR\right)^2 + (3A-1)\left(SCR\right) + A}$$

Now divide the top and bottom of the fraction by "A":

$$\frac{v_{out}}{v_{in}} = \frac{\frac{1}{A}}{\left(SCR\right)^2 + \left(3 - \frac{1}{A}\right)\left(SCR\right) + 1}$$

Now replace the Laplace "S" term with jω:

$$\frac{v_{out}}{v_{in}} = \frac{\frac{1}{A}}{(j\omega CR)^2 + \left(3 - \frac{1}{A}\right)(j\omega CR) + 1}$$

Knowing $j^2 = -1$, we have

$$\frac{v_{out}}{v_{in}} = \frac{\frac{1}{A}}{-(\omega CR)^2 + \left(3 - \frac{1}{A}\right)(j\omega CR) + 1}$$

Now collect the real terms and the j terms:

$$\frac{v_{out}}{v_{in}} = \frac{\frac{1}{A}}{1 - (\omega CR)^2 + j\left(3 - \frac{1}{A}\right)(\omega CR)}$$

Therefore, the transfer function is

$$TF = \frac{\frac{1}{A}}{1 - (\omega CR)^2 + j\left(3 - \frac{1}{A}\right)(\omega CR)}$$

For the third-order filter, we have

$$TF = \frac{\frac{2}{A}}{1 - \left(4 - \frac{1}{A}\right)(\omega CR)^2 + j\left[\left(4 - \frac{1}{A}\right)\omega CR - (\omega CR)^3\right]}$$

This transfer function assumes that the two gain resistors, $R_a$ and $R_b$, for the first-order filter have the same value, giving the numerator of the first-order filter the value "2."

# Exercise 11.9

If you feel confident, see if you can derive the expression for the general third-order filter using the variable "A."

This means that we can derive the transfer function for the circuit shown in Figure 11-32.

This circuit uses the polynomials shown in Table 11-1 where the first-order coefficient is "1" and the second-order coefficient is 1.414. This means the gains of the two Opamps can be set as follows.

The first-order filter:

$$gain = 3 - \alpha = 3 - 1 = 2$$

The gain of the non-inverting Opamp is

$$v_{gain} = 1 + \frac{R_a}{R_b}$$

We need to let $R_a = R_b$, so let both = 1kΩ.

For the second-order filter, we have

$$gain = 3 - \alpha = 3 - 1.414 = 1.586$$

To achieve this, let $R_a$ = 586Ω and Rb = 1kΩ.

The transfer function for the third-order filter shown in Figure 11-31 is

$$TF = \frac{3.1721}{1 - 2.41396(\omega CR)^2 + j\left[2.41396\omega CR - (\omega CR)^3\right]}$$

The expression for the magnitude is

$$\lceil TF \rceil = \frac{3.1721}{\sqrt{\left[\left(1 - 2.41396(\omega CR)^2\right)^2 + \left(2.41396\omega CR - (\omega CR)^3\right)^2\right]}}$$

The expression for the phase is

$$Phase = -Tan^{-1}\left(\frac{2.41396\omega CR - (\omega CR)^3}{1 - 2.41396(\omega CR)^2}\right)$$

## Exercise 11.10

Evaluate the expression for the magnitude and phase of the third-order filter when f = 5kHz and 15kHz. Hint, you will need to be very accurate in your calculations.

## Adding More Gain

When we use the Butterworth polynomials to try and achieve the maximally flat frequency response for the active filters, we cannot apply any value of gain with the actual filters. However, we can achieve a larger overall gain for the signals by adding an Opamp just to amplify the output of the active filter. An example of this is shown in Figure 11-33.

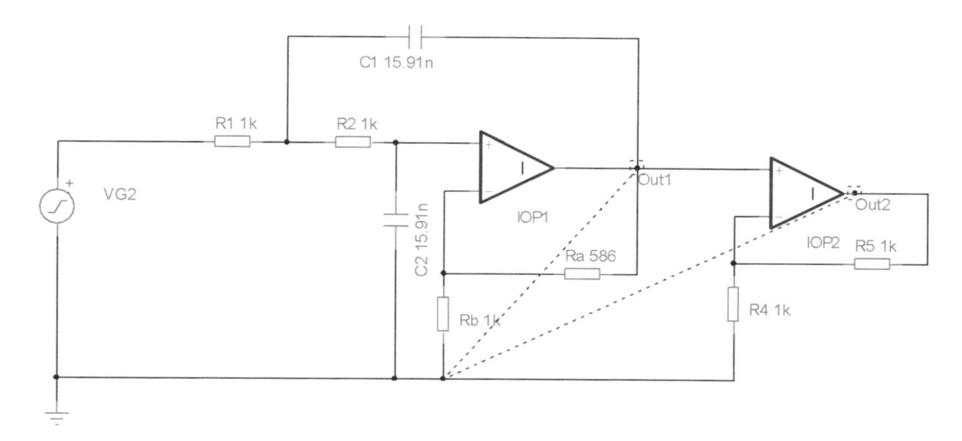


***Figure 11-33.*** *A Second-Order Active Filter with Added Gain*

This circuit has the basic Butterworth second-order filter with a gain set by

$$gain = 3 - \alpha = 3 - 1.414 = 1.586$$

The final Opamp is just there to add more gain of a factor of two. This means the overall gain of the filter and amplifier is

$$Overall\ gain = 1.586 \times 2 = 3.172$$

In dBs, this is

$$20Log(3.172) = 10.027dBs$$

This can be confirmed from the Bode plot for the circuit.

## Exercise 11.11

Design the following active filter circuits:

1. A first-order circuit with an overall gain of four
2. A third-order circuit with an overall gain of six

# Higher-Order Filters

The circuit shown in Figure 11-34 is that of a basic fourth-order filter using the Butterworth polynomials to set the gain of the two Opamps. The coefficient of the first Opamp is 1.848, which means the gain for that Opamp is set as follows:

$$gain = 3 - \alpha = 3 - 1.848 = 1.152$$

The gain of the non-inverting Opamp is

$$v_{gain} = 1 + \frac{R_a}{R_b}$$

Therefore

$$1 + \frac{R_a}{R_b} = 1.152$$

Therefore

$$\frac{R_a}{R_b} = 0.152$$

The simplest way to achieve this is to let $R_a = 152\Omega$ and $R_b = 1k\Omega$.

The coefficient for the second Opamp is 0.7456, which means the gain would be

$$gain = 3 - \alpha = 3 - 0.7456 = 2.2544$$

Therefore

$$\frac{R_a}{R_b} = 2.2544 - 1 = 1.2544$$

The simplest way to achieve this is to let $R_a = 1.2544k\Omega$ and $R_b = 1k\Omega$. That is how the gain resistors of the Opamp were set.

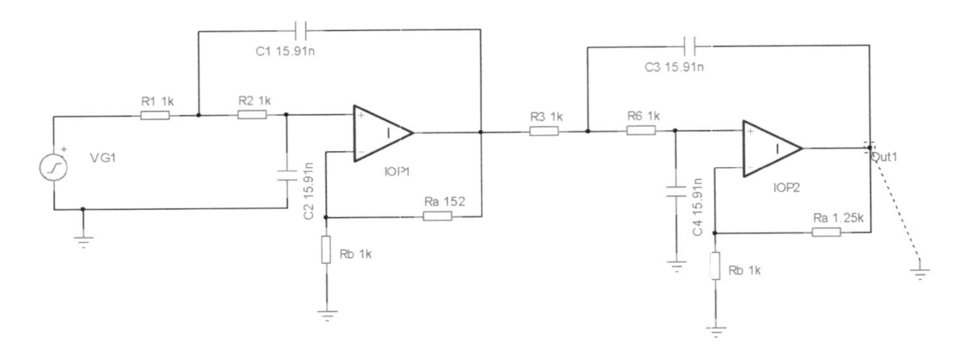


***Figure 11-34.*** *A Fourth-Order Active Filter*

When we simulated the circuit, we could confirm that the gradient of the Bode plot was around 80dBs per decade.

# Exercise 11.12

Use the Butterworth polynomials to design a basic fifth-order filter.

# Summary

In this chapter, we have introduced passive and active filters. We have learned how we can use complex numbers to derive the important transfer functions for the circuits we have looked at. We have studied how to create higher-order filters and apply the Butterworth polynomials to achieve a maximally flat frequency response. In the next chapter we will look at some basic aspects of how to use the ECAD software TINA.

CHAPTER 12

# Using TINA 12

In this chapter, we will look at how to use the ECAD software TINA 12. As we work through the book, we will use this software to simulate the circuits so that we can test the theories that are covered in this book. This chapter will help you learn how to use some of the basic aspects of this software, and we will also look at some of the more specialized aspects that we might use in this book. This chapter is not intended to be a manual for using TINA, but after reading this chapter, you should be able to carry out all the simulations discussed in this book.

## What Is ECAD and TINA 12

ECAD stands for Electronic Computer-Aided Design, and TINA 12 is just one of many pieces of software you can buy to use in the design of electrical and electronic circuits. I am not saying it is the best software, as there will be pros and cons for any of the software available to you. However, I have used TINA while teaching electrical and electronic engineering, and I found it more than sufficient for every aspect of what I was trying to teach. At around £100 for the basic software, it is one of the more affordable pieces of software while giving you more than enough to start your career as an engineer. Even the demo version, which is free to download, will be enough for you to use alongside most, if not all, the simulations in this book. With the demo version, you are allowed 31 runs of the software, on separate days, before you need to buy it. With this demo


H. H. Ward, *Mastering Analog Electronics*, Maker Innovations Series,
https://doi.org/10.1007/979-8-8688-0245-4_12

version, there are no major restrictions in what it can do except there is a limitation on the number of nodes in the circuits you can simulate. This will mean that some of the larger circuits won't run in the demo software, and also you can't save your files. However, you should be able to simulate most of the circuits and get a good appreciation of what the book is about.

Having TINA 12 at your disposal is simply like having a full electrical and electronics lab at your disposal. You can have almost any component you want at any value. You also have any piece of test equipment you could ever want at your disposal. You even have some specialist analysis tool that would be difficult to have anywhere else. However, to be fair, this is only what you would expect from any industrial standard ECAD software.

The other ECAD software programs I have used are Proteus and Multisim.

These are very good ECAD software, but they do come at a price. It is up to you what software you eventually end up buying. However, as this book uses TINA 12, to simulate all the circuits in this book, I think it would be useful if I showed you how to use the most common aspects of the software.

# Running the Software

The software can be downloaded from `www.tina.com`. You will have to register your request for the demo version, but that should not put you off. Assuming you have successfully downloaded the software, you need to click the Tina icon as shown in Figure 12-1.

***Figure 12-1.*** *The Program Icon for Tina*

You may be presented with a registration window or run as a demo. You may have to choose to run the software as administrator. The program should then open up with the main editing window as shown in Figure 12-2.

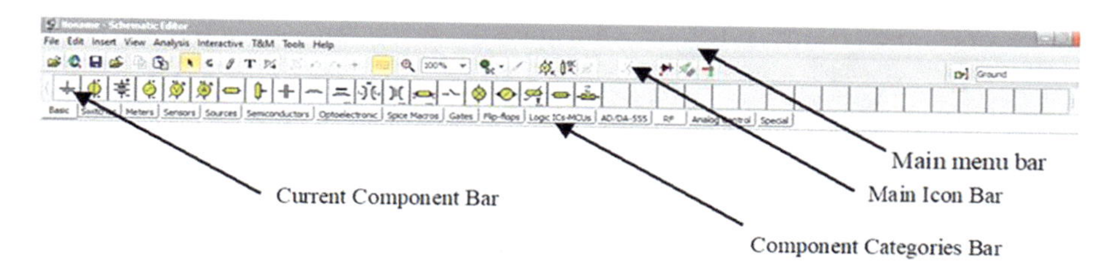


Main Drawing Area

***Figure 12-2.*** *The Opening Screen*

There are four toolbars on the main working window of TINA. These are identified in Figure 12-2 as

- The Main Menu Bar: This is where we can select the following options.
    - File, Edit, Insert, View, Analysis, Interactive, T&M Tools, and Help
- The Main ICON Bar: This has some graphical icons for the open, save, and other commands.
- The Current Component Bar: TINA splits the different components that you can use into different categories. When a category has been selected, the components in that category are shown in this menu bar.

- The Component Categories Bar: This is where TINA allows you to select one of the many different component categories that TINA uses.

Note, these are my names that I have given to the menu bars. This is so that I can refer to them in the following text.

# Creating Our First Test Circuit

The first circuit we will look at creating is the simple CR high pass filter. We will use this test circuit to show you how to run an ac analysis to produce a Bode plot. The first thing we will do is find the two components. This is in the "Basic" component category for TINA. Therefore, we must click the mouse of the word "Basic" on the component categories bar. The current category bar should change to that shown in Figure 12-3.

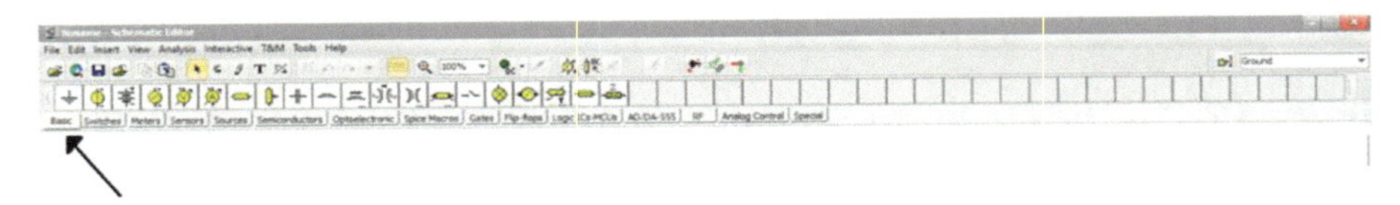

***Figure 12-3.*** *The Basic Current Category Menu Bar*

We are looking for the resistor symbol; this is the seventh icon on the current category bar.

I feel I should point out that TINA allows you to choose from two sets of symbols for the components you use. You can choose either the USA ANSII or European DIN. I prefer the European DIN set of symbols. To choose the set of symbols you want, you need to click the "View" tab from the main menu bar. When you do that, a menu will drop down, and you must select the "Options" tab from it as shown in Figure 12-4.

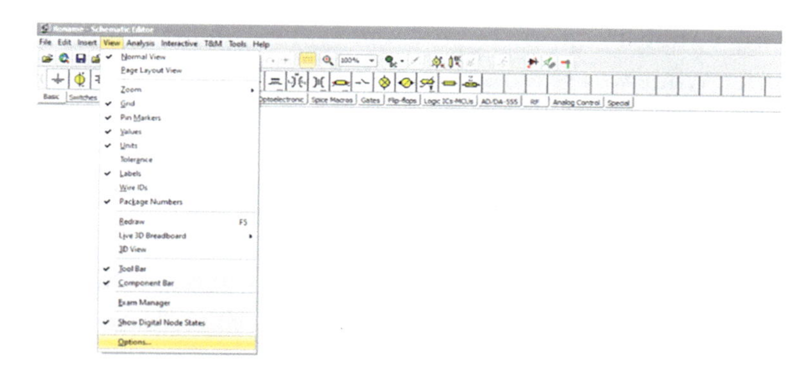


***Figure 12-4.** The Drop-Down Menu from the View Tab*

When you select the Options tab, the "Editor Options" window will appear as shown in Figure 12-5.

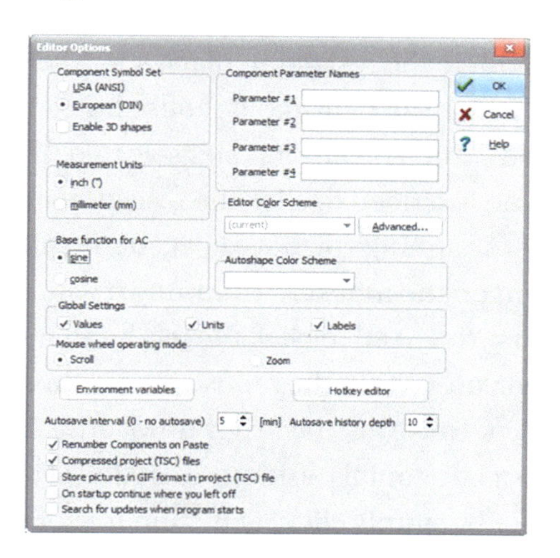


***Figure 12-5.** The Editor Options Window*

You will be able to see the Component Symbol Set in the top left-hand corner. While we are looking at this window, we should set the base function for AC. It defaults to the mathematical base of cosine. However, I use the sine function as the base function for my ac analysis, so I suggest you change it to sine here.

If you click the mouse on that resistor icon, you will select the resistor symbol from the current component menu bar. If you move the mouse – note, you don't have to click and drag the mouse – you will see that the symbol for the resistor follows the mouse around the main drawing area. Now, with the mouse in a suitable position, simply click the mouse and the resistor will be dropped in the place the mouse is pointing to when you clicked the mouse. You should see that the resistor is still red, which means it is still selected. If the resistor turned green, that is, it is deselected, don't worry; it just means you clicked the mouse too many times.

Now we want to rotate the resistor to place it in a vertical orientation. To do this, we simply click the right-hand mouse button while the resistor is still highlighted red, which means it is still selected. If the resistor is not red, but green, you just need to click the resistor using the left-hand button of the mouse. Assuming you have clicked the selected resistor with the right-hand button on the mouse, a drop-down window should appear with multiple options. You need to click the left-hand mouse button on the "Rotate Right." When you do this, the mouse moves through 90 degrees into a vertical position. However, the text is now vertical, and you may want to rotate it into a horizontal position. To do this, you must deselect the resistor altogether by simply clicking the mouse somewhere on the main drawing area of the screen. The resistor symbol will now turn green to show that it is deselected. Now click the right-hand mouse button on the text only, not the resistor symbol. You should now be able to rotate the text to the right through 90 degrees.

We will now change the value of the resistor to show you how this can be done. We can either double-click the mouse button on the whole resistor symbol or just the text for the component. When you do either of those procedures, the component flyout window appears as shown in Figure 12-6.

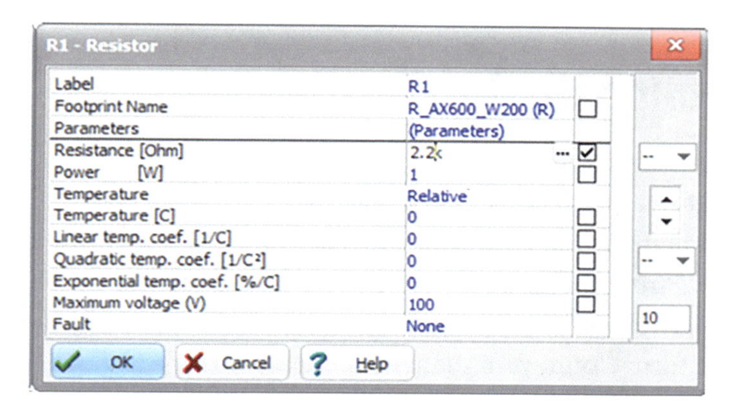


***Figure 12-6.*** *The Component Value Window*

With this window, you can simply change the value of the component. You can also change the label if you want to. There are various other parameters you can change for a more detailed simulation. You can also add a fault such as an open circuit or a short circuit. We will just change the value to 2.2k as shown in Figure 12-6.

Now we need to add the capacitor. This is done in the same way as adding the resistor except that the capacitor is the ninth symbol on the current component menu bar. We can keep the default horizontal orientation and the default value of 1μF.

Now we need to add the earth or ground. We must do this for all our circuits as there has to be a reference as voltage needs a reference. The normal reference for any circuit is ground or earth or 0V. The earth symbol is the first one on the current component menu bar. I normally place this on the bottom left-hand corner of the circuit. You will see this when we look at Figure 12-8.

The next item we need to add is the signal generator. This is used to apply a variety of input voltage to the circuit. The signal generator is called the voltage generator in Tina, and it is the fourth symbol on the current component menu bar. I normally place this above the ground symbol; see Figure 12-8.

## Connecting Up the Circuit

Now we have all the components we need to wire them together. TINA does provide us with a grid to help align the components we use. If the grid is not present in the main drawing area, then if you click the mouse on the word "View," from the main menu bar, you should see a drop-down menu appear. If you now select the word "GRID," the tick should appear, and the grid will be turned onto your main drawing area. The grid can help you align the wire and components as you place them.

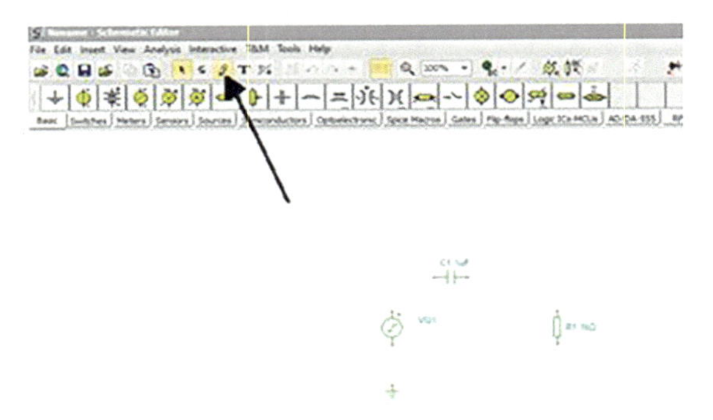

***Figure 12-7.*** *The Wire Tool*

To start the wire once you have selected the wire tool as shown in Figure 12-7, move the wire symbol, which looks like a soldering iron, to where you want to start the wire. Then click the mouse button to connect the tool and drag it around the circuit, while keeping the mouse button pressed, to where you want to end the wire. Then click the mouse button again to fasten the end of the wire at the required position. You should be able to move to the end of another component and start wiring again. However, you may now need to select the wire tool again to complete the wiring of the components.

Now we have wired the circuit completely, it should look like the circuit shown in Figure 12-8.

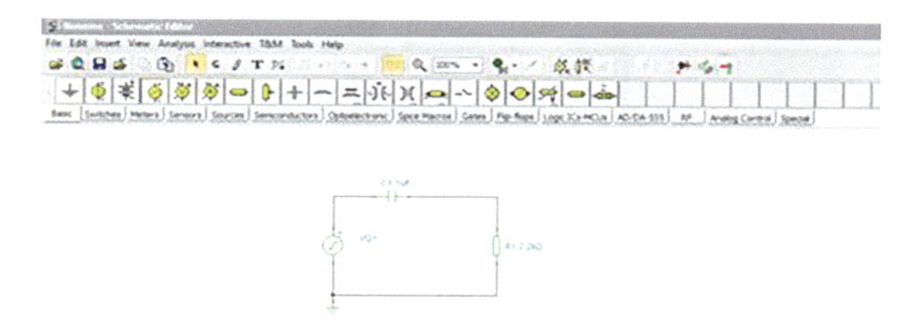

***Figure 12-8.** The Circuit Wired Completely*

We need to identify where we want to monitor what the circuit will do in the analysis we will simulate. There are a variety of ways we can identify the point of the circuit. I have inserted outputs as shown in Figure 12-9.

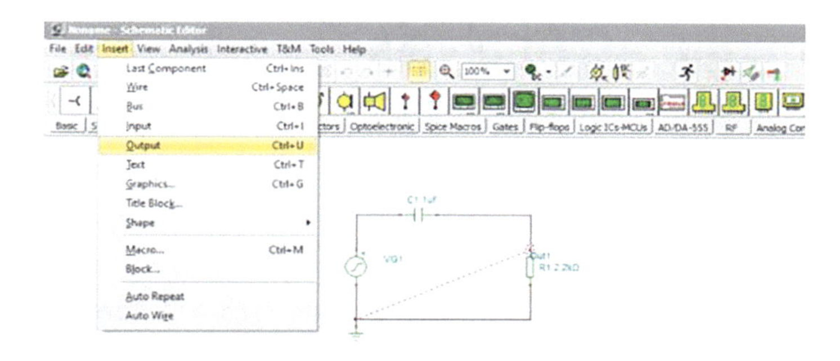


***Figure 12-9.** Using the Insert Output to Add a Monitoring Point on the Circuit*

However, I find that you sometimes have to move the out wire slightly to ensure that it makes a connection. An alternative approach would be to add a voltage pin as shown in Figure 12-10.

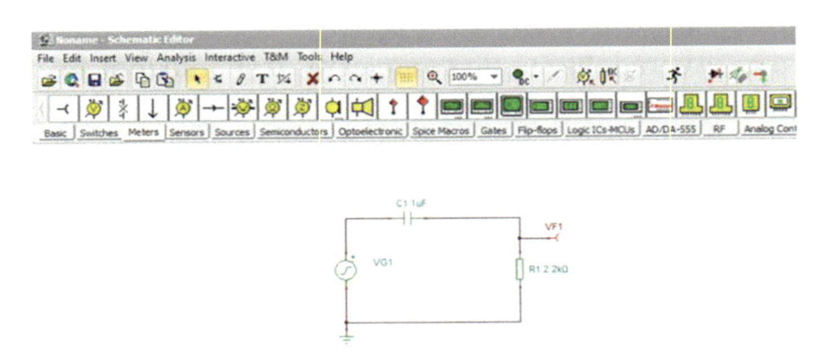


***Figure 12-10.*** *Using the Voltage Pin*

To do this, you must first change the component categories tab to "Meters" by simply clicking the mouse button on the tab. The voltage pin is the first symbol on the now changed current component bar. We simply place the voltage pin in the same way as any component and use the wire tool to connect it to where we want to monitor the voltage.

# Running the AC Analysis

We are now ready to simulate the circuit and run the analysis we want. This will be an AC Transfer Characteristics. This will allow us to create a Bode plot to show how the gain of the circuit varies with a change in frequency. To select this type of analysis, we firstly click the Analysis tab on the main menu bar. Then we select the AC Analysis option from the drop-down menu that appears. Finally, we select the AC Transfer Characteristics from the flyout menu that appears. This is shown in Figure 12-11.

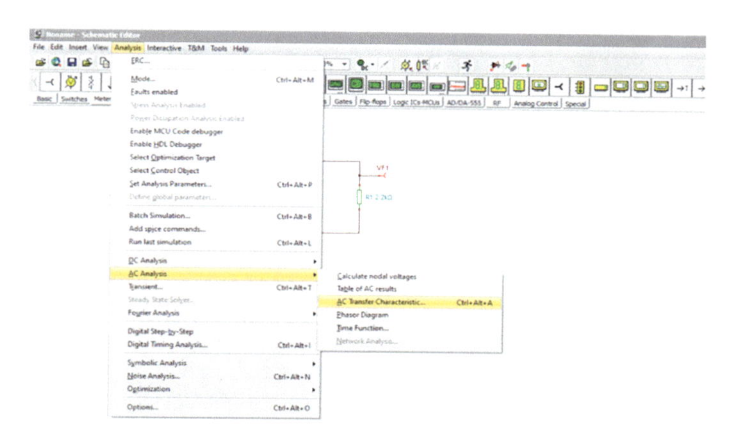


***Figure 12-11.*** *The AC Transfer Characteristics Option*

When we select the AC Transfer Characteristics, the AC Transfer Characteristics window appears. I have changed the number of points to 1000 but kept everything else the same as the default. This is shown in Figure 12-12.

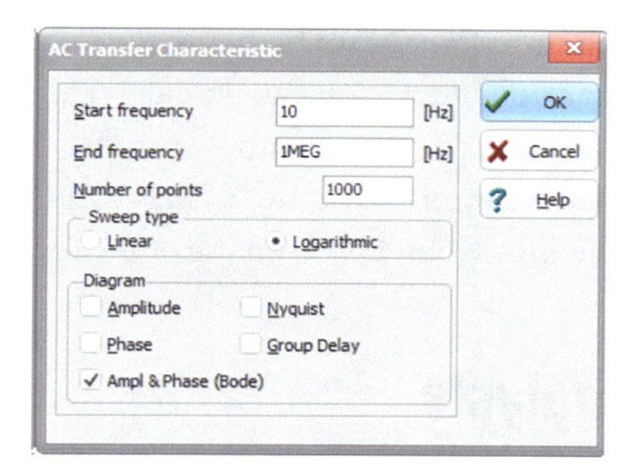


***Figure 12-12.*** *The AC Transfer Characteristics Window*

Once you are happy with the settings, simply click the OK button. The circuit will simulate, and the Bode plot will appear on the screen. This is shown in Figure 12-13.

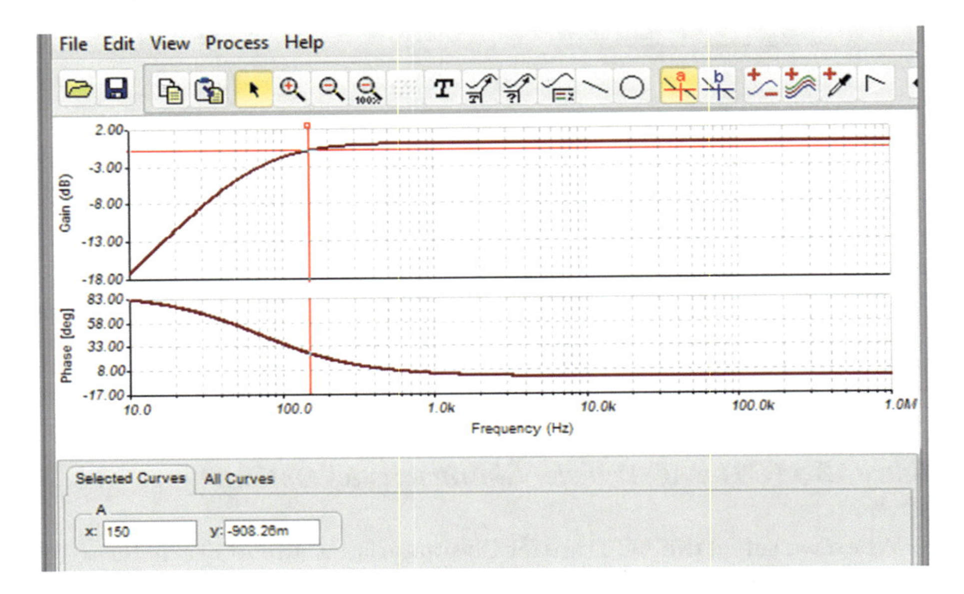


***Figure 12-13.*** *The Bode Plot*

There are two cursors we can use to highlight the plot at a particular value. I have selected the cursor "a" to highlight the gain value, in dBs, of the plot at a particular frequency. You can actually type the value that you are interested in into one of the small windows on the display. In this case, I have entered the frequency of 150 in the "x" axis window, and the cursor has gone to that frequency when I pressed the enter key.

# Transient Analysis

Another useful analysis that we have used in the book is the transient analysis. This looks at how the voltage or current at different points in the circuit changes over time. This is especially useful to see what happens in the first few milliseconds after a change has occurred within the circuit, such as being switched on. It enables the software to act like a storage scope. We will use this type of analysis to examine the voltage rise across a capacitor and so confirm the time constant of the DC CR circuit. We will firstly create the circuit which is similar to the first circuit we created. We will select the components in the same way and wire them up as before to create the circuit shown in Figure 12-14.

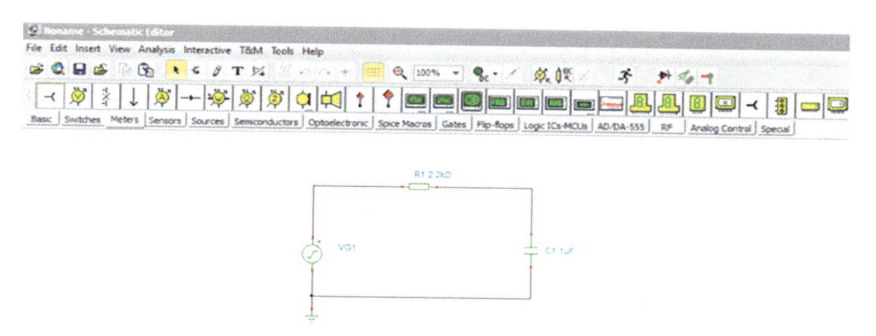


***Figure 12-14.** The RC Circuit for the DC Transient Analysis*

We need to set up the input signal to apply a DC wave that will stay high for enough time for the capacitor to charge up fully and then go to 0V where it will stay low long enough for the capacitor to discharge. This means you must have an idea of how long it will take the capacitor to charge up and discharge. The theory states that it will take 5 Tau, that is, 5 time constants, for this to happen. With the RC circuit, one time constant is equal to RC seconds. In this circuit, the time constant Tau is

$$Tau = \tau = CR\ seconds = 1E^{-6} \times 2.2E^{3} = 2.2ms$$

Therefore, five time constants will equal 11ms. This means that we will need to create a voltage that stays high for 11ms and then low for 11ms. It will then repeat the cycle. In Tina, this is a general waveform, and to create it, we need to click the mouse button on the VG1, voltage generator symbol in the circuit. This will open up the voltage generator window as shown in Figure 12-15.

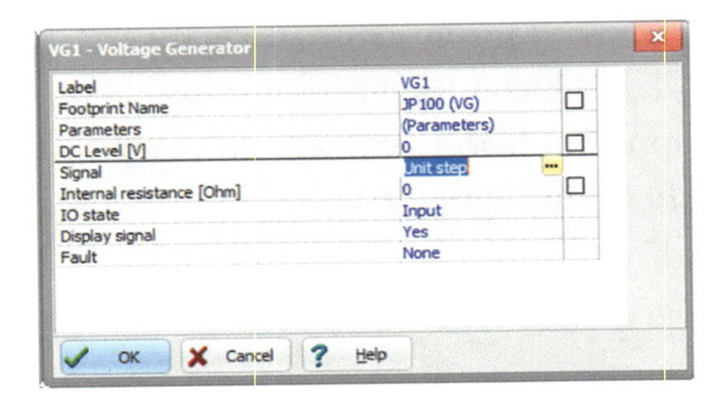


***Figure 12-15.*** *The Voltage Generator Window*

The default signal is the Unit Step. To change this, we need to click the mouse button on the small three dots. This will open up the signal editor window as shown in Figure 12-16.

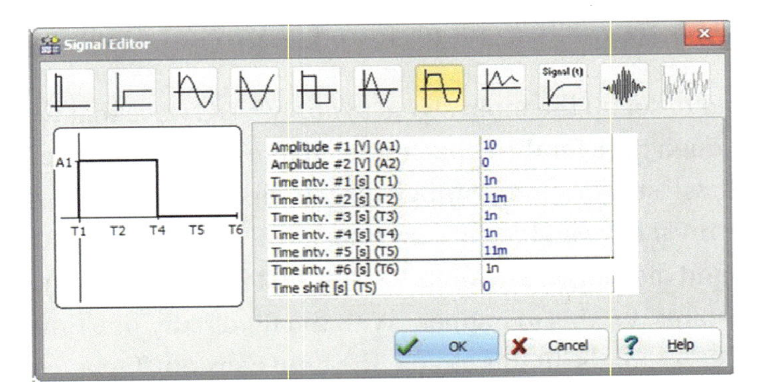


***Figure 12-16.*** *The Signal Editor Window*

The signal type we want is the seventh which has been selected here. There are seven time settings we need to set, and they are

1. The amplitude the signal should rise to. I have set it to 10V.
2. The amplitude the signal should fall to. I have set it to 0V.

3. The (T1) which is the rise time of the signal. I have left this at 1ns.

4. The (T2) which is the time the signal stays high. This has been set to 11ms.

5. The (T3) which is the fall time of the signal. I have left this at 1ns.

6. The (T4) which is the second rise time of the signal. I have left this at 1ns.

7. The (T5) which is the time the signal stays low. This is set to 11ms.

8. The (T6) which is the second fall time. This has been left at 1ns.

9. The (Ts) which is a delay time you might want to introduce into the starting of the signal waveform. This has been left at 0.

As you enter these settings, the waveform will be shown in the preview window. When you are happy with the setting, simply click the mouse button on the OK button, and the window will close. Then click the OK button to close the voltage generator window.

We now need to add a voltage pin and place it above the top of the capacitor. This will allow us to monitor the voltage across the capacitor over time.

We now need to set up the transient analysis. To do this, click the mouse button on the Analysis tab on the main menu bar and then select "Transient" from the drop-down menu that appears. The Transient Analysis Window will appear as shown in Figure 12-17.

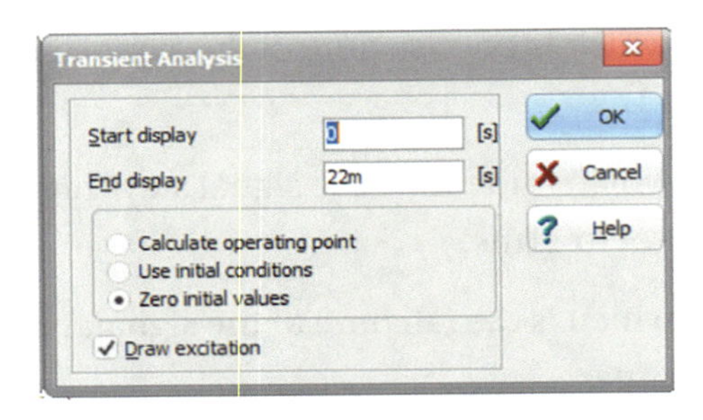


*Figure 12-17.* *The Transient Window*

We need to change the setting to those as shown in Figure 12-17. Make sure you select the Zero initial values option. When you are happy with the settings, click the OK button.

The default view shows the curves on top of each other. To separate them, we can click the mouse button on the "View" option on the menu bar, then select separate curves. The display will change to that shown in Figure 12-18.

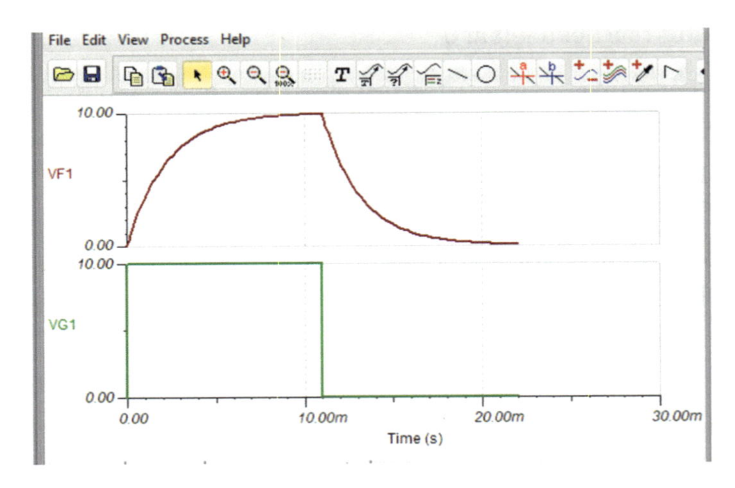


*Figure 12-18.* *The TR Results Window*

# Using the Oscilloscope

We will use the oscilloscope to monitor the input and output of amplifiers in the book. We will create a class A stabilized amplifier now to show you how to use the oscilloscope. The circuit uses four resistors, one capacitor, and voltage generator which we have shown how to put on the circuit. The circuit will need a DC supply for the VCC, and the simplest would be a battery. This is the third symbol on the basic component category bar. Placing the battery in the circuit and changing the voltage to 15V gives us the circuit so far as shown in Figure 12-19.

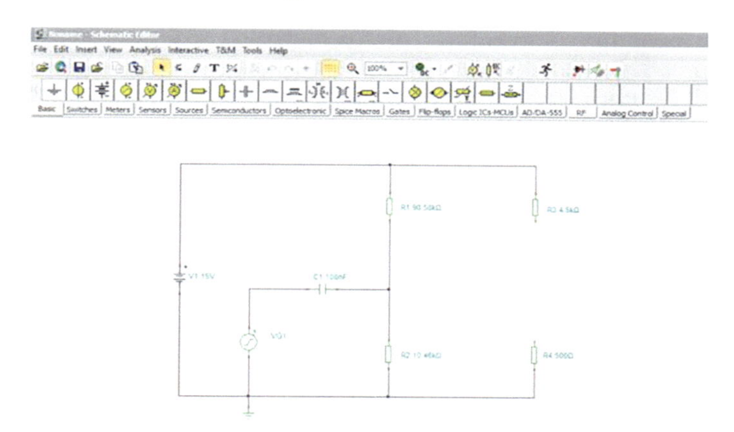


***Figure 12-19.*** *The Amplifier Circuit So Far*

We now need to add the BJT transistor. This can be found on the "semiconductor menu," and it is the eighth symbol on that menu bar, the NPN transistor, that we want. You should place it on the circuit in a suitable position, then when you double-click the mouse button on the transistor, the NPN Bipolar Transistor window will appear as shown in Figure 12-20.

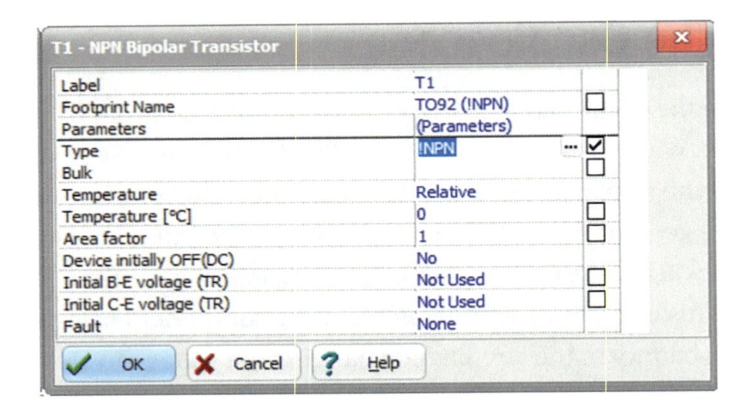


***Figure 12-20.*** *The NPN Bipolar Transistor Window*

Again, if we click the mouse button on the three dots, the Catalog Editor window will appear, as shown in Figure 12-21.

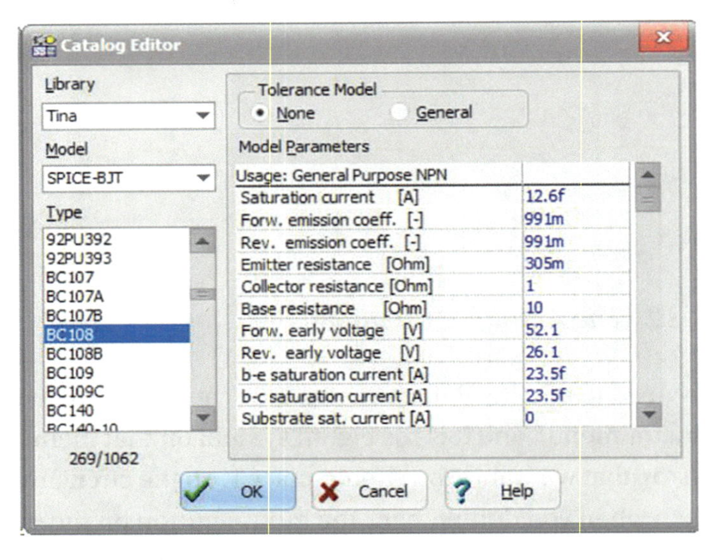


***Figure 12-21.*** *The Transistor Catalog Editor Window*

We can use this window to change the transistor type we want to use. I have selected the BC108 transistor here. We can use this window to change the various parameters of the transistor, such as the Beta and internal capacitance values. Once you are happy with your settings, click OK to close this window and OK again to close the first window.

We must change the input signal to a sine wave, and to do this, we need to open up the signal editing window and select the sine wave. We can set the parameters up as shown in Figure 12-22.

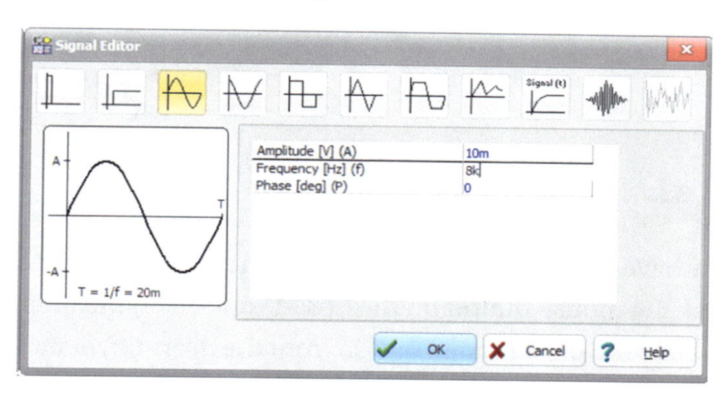


***Figure 12-22.*** *Setting Up a Sine Wave Voltage at the Input*

Now we will add the oscilloscope. This can be found on the "Meters" tab; it is the fourteenth symbol on that menu bar. When you click the oscilloscope, select the three-terminal option from those given and place it somewhere convenient on the screen. We should connect the oscilloscope up as shown in Figure 12-23.

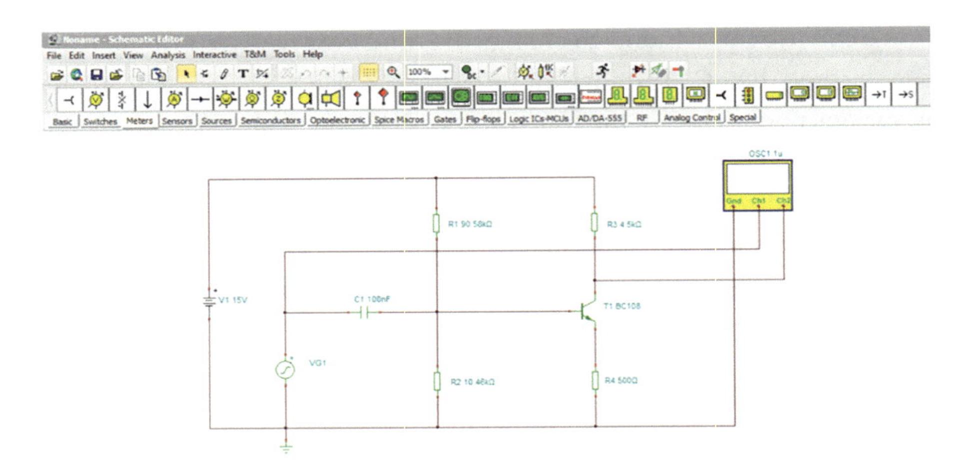


*Figure 12-23.* *Connecting the Oscilloscope*

We now have to set the oscilloscope up and run the simulation. We need to click the mouse button on the "T&M" option on the main menu bar. Then select the oscilloscope option from the drop-down menu that appears. This will open up the Oscilloscope window as shown in Figure 12-24.

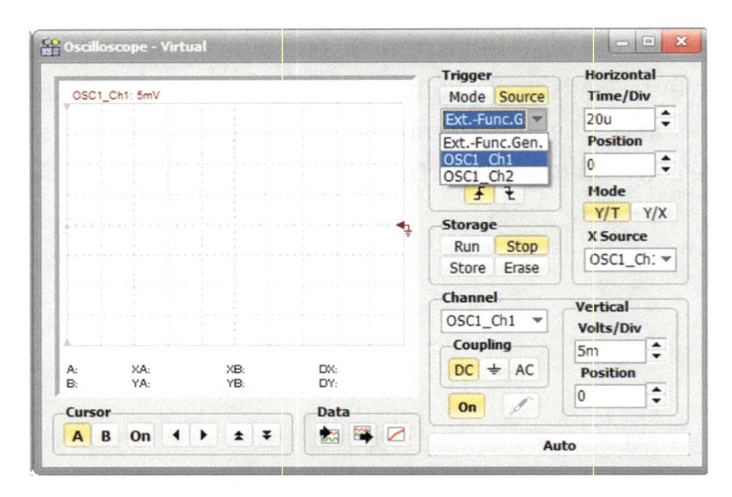


*Figure 12-24.* *The Oscilloscope Window*

We need to set the Time/Div option and the Volts/Div option. With respect to the Time/Div, we really should try and display two cycles of the waveform we are measuring. This means we need to display two times the periodic time on the screen. We know that the horizontal width of the screen has ten divisions. Knowing this, we can set the Time/Div using

$$\frac{Time}{Div} = \frac{2}{10f}$$

When f = 8kHz, we have

$$\frac{Time}{Div} = \frac{2}{10 \times 8E^{3}} = 25E^{-6} = 25\mu s$$

We do not have this setting on the oscilloscope, so we will set the Time/Div to 50µs.

As to the volts/Div, we know the peak of the input is 10mv, and we have four divisions to display this on the vertical. Therefore, we must set the volts/div to 5mV as this is the best resolution we have in that order.

Now we need to select the trigger input for the oscilloscope. This is done by clicking the "Source" option on the oscilloscope and selecting channel 1 as shown in Figure 12-24.

Finally, set the source option to "Normal."

This is really what you should do before you use any oscilloscope; however, most of us just twiddle the knobs until they get a steady picture not very good. Once you have set up the oscilloscope correctly, press the Run button, and the display will change to that as shown in Figure 12-25.

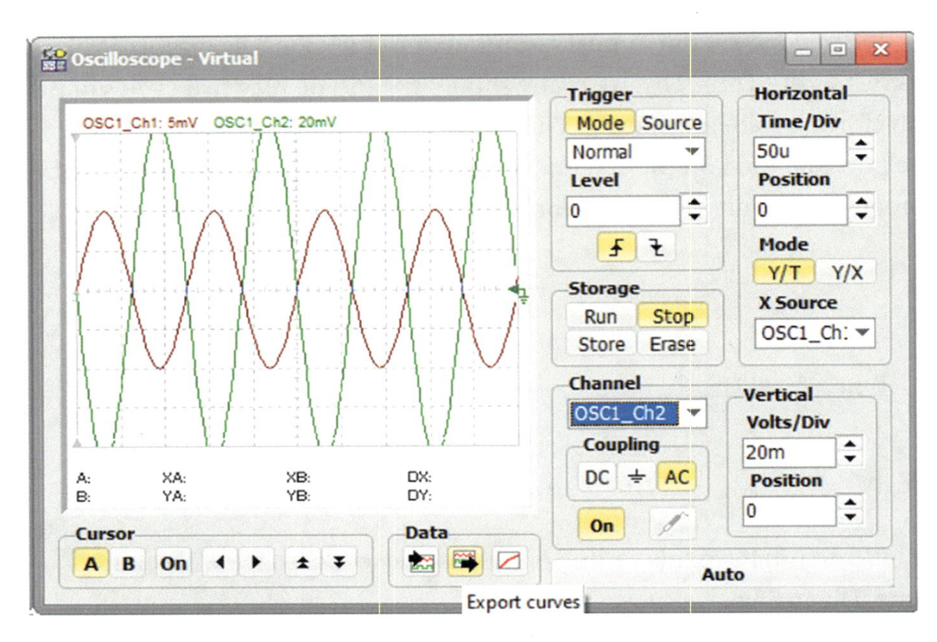


***Figure 12-25.*** *The Oscilloscope Display*

Once you are happy with the display, you should press the stop button, and the display should freeze. Now you can click the second data arrow to export the display to a graph you can use and save for later.

# Summary

In this chapter, we have looked at using some options available to you within the Tina 12 ECAD software. We will go through these and some more in the book as we use them. I hope you do find this software useful in your design work and that you do enjoy using it.

This chapter was not intended to provide you with a manual on how to use TINA. It was only meant as an introduction to how you can use TINA to help create and simulate the circuits in this book.

## APPENDIX 1

# The Average Voltage of a Half-Wave Rectifier

To determine the average of a range of values, we must add all the values up in the range and divide the total by the number of values in that range. When it comes to a waveform, we can use integration to add all the values up and then divide them by the period of the waveform. We need only do this over one cycle as the complete waveform is a repetition of this one cycle. Therefore, we can use the following expression to calculate the average of a waveform:

$$Average = \frac{1}{T}\int_0^T f_x \, dx$$

This means we need to develop the expression $f_x$ for the waveform. The waveform of the output of a half-wave rectifier is shown in Figure A-1.


H. H. Ward, *Mastering Analog Electronics*, Maker Innovations Series,
https://doi.org/10.1007/979-8-8688-0245-4

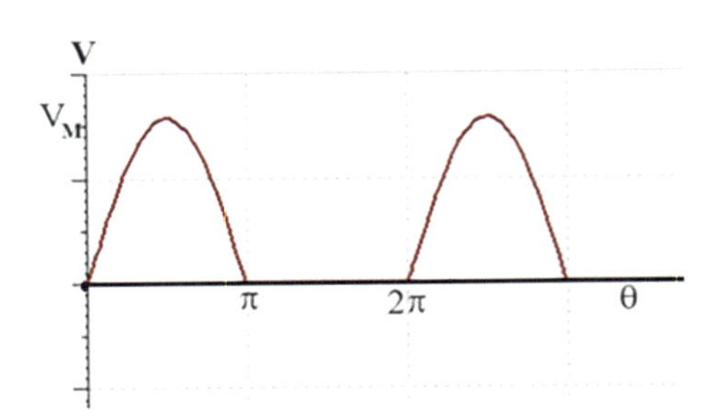


***Figure A-1.*** *The Waveform of a Half-Wave Rectifier*

We can generalize this waveform by saying that it will rise to a maximum voltage of $V_M$. The vertical axis is in V for volts, and the horizontal axis is in θ for angles. We can express the angle in degrees, but we will use radians. This means the period, which is cyclic, will be 2π. This is shown in Figure A-1.

We can see that the waveform in part can be expressed using the sin function. However, this cannot be used for the complete waveform. Therefore, we must split the waveform up into parts. In this case, we can split the waveform up into just two parts, which we will call part "A" and part "B." This means that the expression for the average voltage, which we will call "$V_{AVE}$," is

$$V_{AVE} = \frac{1}{2\pi}\int_0^{2\pi} PartA + PartB \, d\theta$$

We can see that part A exists for θ = 0 and π. The expression for part A is

$$V = V_M \, Sin(\theta)$$

We can see that part B exists for θ = π and 2π. The expression for part B is

$$V = 0$$

This means that the expression for the average voltage is

$$V_{AVE} = \frac{1}{2\pi}\left[\int_0^{\pi} V_M \, Sin(\theta)\, d\theta + \int_{\pi}^{2\pi} 0 \, d\theta\right]$$

Using the standard integrals, we know

$$\int Sin = -Cos$$

Therefore, we have

$$V_{AVE} = \frac{1}{2\pi}\left[\left[-V_M \, Cos(\theta)\right]_0^{\pi} + 0\right]$$

Putting the limits in gives

$$V_{AVE} = \frac{1}{2\pi}\left[\left(-V_M \, Cos(\pi)\right) - \left(-V_M \, Cos(0)\right)\right]$$

which gives

$$V_{AVE} = \frac{1}{2\pi}\left[\left(V_M\right) - \left(-V_M\right)\right] = \frac{1}{2\pi}\left[2V_M\right]$$

Therefore, we can say

$$V_{AVE} = \frac{V_M}{\pi}$$

This is the expression for the average voltage of the output of a half-wave rectifier.

# Exercise 3.1

1. Using the expression for the voltage across the capacitor in the CR phase shift circuit shown in Figure 3-7, determine the time when the SCR would turn on if the resistor $R_1$ was changed to 15kΩ. Determine the delay in degrees and radians as well as time. Calculate the peak of the voltage across the capacitor. This is with the Opamp buffer included in the circuit.

2. Repeat the calculation but now with $R_1$ changed to 22k.

Using the general expression for time "t," we have

$$t = \frac{Sin^{-1}\left(\frac{v_{out}\sqrt{1^2 + (100\pi CR)^2}}{339}\right) + Tan^{-1}(100\pi CR)}{100\pi}$$

Using C = $1E^{-6}$ and R = $15E^3$, we have

$$CR = 1E^{-6} \times 15E^3 = 15E^{-3}$$

Therefore, we can say

$$100\pi CR = 100\pi \times 15E^{-3} = 4.7124$$

Therefore, we can say

$$\sqrt{1^2 + (100\pi CR)^2} = \sqrt{1 + 4.7124^2} = 4.8173$$

We can also say

$$Tan^{-1}\left(100\pi CR\right) = Tan^{-1}\left(4.7124\right) = 1.362$$

We can also say

$$\frac{v_{out}\sqrt{1^2 + \left(100\pi CR\right)^2}}{339} = \frac{0.7 \times 4.8173}{339} = 9.947ms$$

Finally, we can say that time "t" is

$$t = \frac{Sin^{-1}\left(9.946^{-3}\right) + 78.02}{100\pi} = \frac{9.936E^{-3} + 1.362}{100\pi} = 4.367ms$$

This time will be 4.367ms after the start of the cycle of the supply voltage.

In radians, the delay angle would be 1.362 Radians.

In degrees, this would be

$$Degrees = Rads\frac{180}{\pi} = 1.362 \times \frac{180}{\pi} = 78.04$$

When R = 20kΩ, we get the following results.

Using C = $1E^{-6}$ and R = $20E^{3}$, we have

$$CR = 1E^{-6} \times 20E^{3} = 20E^{-3}$$

Therefore, we can say

$$100\pi CR = 100\pi \times 20E^{-3} = 6.2832$$

Therefore, we can say

$$\sqrt{1^2 + (100\pi CR)^2} = \sqrt{1 + 6.2832^2} = 6.3623$$

We can also say

$$Tan^{-1}(100\pi CR) = Tan^{-1}(6.3623) = 1.4149$$

We can also say

$$\frac{v_{out}\sqrt{1^2 + (100\pi CR)^2}}{339} = \frac{0.7 \times 6.3623}{339} = 13.137ms$$

Finally, we can say that time "t" is

$$t = \frac{Sin^{-1}(13.137^{-3}) + 1.4149}{100\pi} = \frac{13.137E^{-3} + 1.4149}{100\pi} = 4.5456ms$$

This time will be 4.5456ms after the start of the cycle of the supply voltage.

In radians, the delay angle would be 1.4149 Radians.

In degrees, this would be

$$Degrees = Rads\frac{180}{\pi} = 1.4149 \times \frac{180}{\pi} = 81.07$$

# Exercise 3.2

Determine the average voltage across the load if the resistor $R_1$ in the phase shift circuit was changed to 10k and then 20k if the capacitor was set at 1μF.

When R1 = 10k, we use

$$ave = \frac{v_{Peak}}{2\pi}\left(1 + cos\left(\propto\right)\right)$$

$$\alpha = Tan^{-1}\left(100\pi CR\right)$$

The vpeak is the peak of the UK mains at 339v assuming an rms of 240v. Then

$$\alpha = Tan^{-1}\left(314.16 \times 1E^{-6} \times 10E^{3}\right) = 1.2626$$

Therefore, we get

$$ave = \frac{339}{2\pi}\left(1 + Cos\left(1.2626\right)\right) = 70.32v$$

When R1 = 20k, we get

$$\alpha = 1.413\ Rads$$

$$ave = 124.86v$$

# Exercise 3.3

1. If the inductance value in the test circuit shown in Figure 3-15 was increased to 50mH, calculate the new angle "β" and the average voltage across the RL load.

2. Repeat the calculations if the resistor in the CR phase shift circuit was increased to 20k, and the inductance value in the load was reduced to 40mH.

Using

$$\beta = -Tan^{-1}\left(\frac{\omega_L}{R}\right)$$

Knowing f = 50 Hz, R = 10Ω, and L = 50mH, we get

$$\beta = -Tan^{-1}\left(\frac{2\pi \times 50E^{-3}}{10}\right) = -1.004\ Rads$$

We need to calculate the angle "α" using

$$\propto = \mathrm{Tan}^{-1}(\omega CR)$$

where the values of C and R are from the phase shift circuit. Therefore, C = 1mF and R = 5kΩ. Therefore

$$\propto = Tan^{-1}\left(2\pi \times 50 \times 1E^{-6} \times 5E^{3}\right) = 1\ Rad$$

We can now calculate the average of the voltage across the load using

$$ave = \frac{v_{Peak}}{2\pi}\int_{\propto}^{\pi+\beta} Sin(\theta)d\theta = \frac{339}{2\pi}\int_{1}^{\pi+1.004} Sin(\theta)d\theta$$

Integrating, we get

$$ave = \frac{v_{Peak}}{2\pi}\left[-Cos(\theta)\right]_{1}^{\pi+1.004}$$

This gives

$$ave = \frac{v_{Peak}}{2\pi}\left[\left(-Cos\left(\pi + 1.004\right)\right) - \left(-Cos\left(1\right)\right)\right]$$

This gives

$$ave = \frac{339}{2\pi}\left[\left(- - 0.537\right) - \left(-0.5403\right)\right]$$

$$ave = \frac{339}{2\pi}\left[1.077\right] = 58.11v$$

Solution for part 2

$$\beta = -Tan^{-1}\left(\frac{2\pi \times 40E^{-3}}{10}\right) = -0.8986\ Rads$$

The angle "a":

$$\propto = Tan^{-1}\left(2\pi \times 50 \times 1E^{-6} \times 20E^{3}\right) = 1.4129\ Rad$$

$$ave = \frac{v_{Peak}}{2\pi}\int_{\propto}^{\pi+\beta} Sin\left(\theta\right)d\theta = \frac{339}{2\pi}\int_{1.4129}^{\pi+0.8986} Sin\left(\theta\right)d\theta$$

Integrating, we get

$$ave = \frac{v_{Peak}}{2\pi}\left[-Cos\left(\theta\right)\right]_{1.4129}^{\pi+0.8986}$$

This gives

$$ave = \frac{v_{Peak}}{2\pi}\left[\left(-Cos\left(\pi + 0.8986\right)\right) - \left(-Cos\left(1.4129\right)\right)\right]$$

This gives

$$ave = \frac{339}{2\pi}\left[\left(- - 0.623\right) - \left(-0.157\right)\right]$$

$$ave = \frac{339}{2\pi}\left[0.7802\right] = 42.97v$$

# Exercise 5.1

As an exercise, using the following specification of a $V_{CC}$ of 22V and a base current $I_B$ of 10μA, assuming the "β" was 100, design a class A stabilized amplifier using the preceding expressions. My circuit to meet this specification is shown in Figure A-2.

The expressions we will use are

$$V_{CEQ} = \frac{V_{CC}}{2}$$

$$I_{CQ} = \beta I_B = 100 I_B$$

$$R_L + R_E = \frac{1}{Gradient} = \frac{V_{CEQ}}{I_{CQ}} = \frac{V_{CC}}{2I_{CQ}}$$

$$V_E = I_{CQ} \times R_E$$

$$V_B = 0.7 + V_E$$

$$I_B = \frac{I_{Bleed}}{10}$$

$$I_{Bleed} = 10 \times I_B$$

$$V_{R1} = V_{CC} - V_B$$

$$R_1 = \frac{V_{R1}}{I_{Bleed}}$$

$$R_2 = \frac{V_B}{I_{Bleed} - I_B}$$

Putting the values in, we get

$$V_{CQ} = \frac{V_{CC}}{2} = \frac{22}{2} = 11$$

$$I_C = \beta I_B$$

Therefore

$$I_C = 100 \times 10^{-6} = 1mA$$

$$R_L + R_E = \frac{1}{Gradient} = \frac{V_{CEQ}}{I_{CQ}} = \frac{V_{CC}}{2I_{CQ}} = \frac{22}{2 \times 1^{-3}} = 11k$$

Let $R_L$ = 10k and $R_E$ = 1k:

$$V_E = I_{CQ} \times R_E = 1E^{-3} \times 1E^{3} = 1V$$

$$V_B = 0.7 + V_E = 0.7 + 1 = 1.7V$$

$$I_{Bleed} = 10 \times I_B = 10 \times 10E^{-6} = 100\mu A$$

$$V_{R1} = V_{CC} - V_B = 22 - 1.7 = 20.3V$$

$$R_1 = \frac{V_{R1}}{I_{Bleed}} = \frac{20.3}{100^{-6}} = 203k$$

$$R_2 = \frac{V_B}{I_{Bleed} - I_B} = \frac{1.7}{100^{-6} - 10^{-6}} = 18.889k$$

The circuit is shown in Figure A-2.

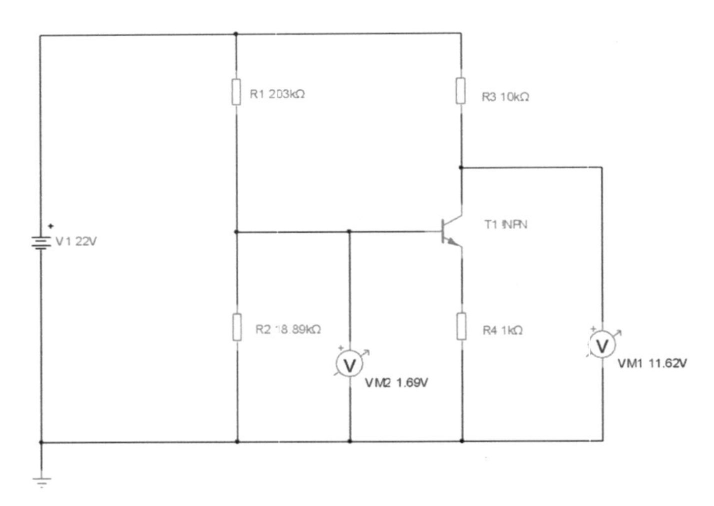


***Figure A-2.** The Class A Amplifier*

Your circuit might change slightly as you might split the 11k differently.

# Exercise 5.2

$$R_{IN} = \frac{1}{\frac{1}{R_1} + \frac{1}{R_2} + \frac{1}{101 \times R_E}}$$

$$C_1 = \frac{1}{2 \times \pi \times F_C \times R_{IN}}$$

Use the preceding expressions to calculate the high pass filter capacitor "$C_1$" for the amplifier in the first exercise if the cutoff frequency was 30Hz. The solution is shown here:

$$R_{IN} = \frac{1}{\frac{1}{R_1} + \frac{1}{R_2} + \frac{1}{101 \times R_E}} = \frac{1}{\frac{1}{203^3} + \frac{1}{18.889^3} + \frac{1}{101 \times 1^3}} = 14.756k$$

$$C_1 = \frac{1}{2 \times \pi \times F_C \times R_{IN}} = \frac{1}{2 \times \pi \times 30 \times 14.756^3} = 359.5nF$$

# Exercise 6.1

Design a class A amplifier to the following specification:

The VCC should be 15V.

The DC quiescent collector current should be 1.5mA.

The DC quiescent emitter voltage should be 1V.

The typical beta value of 100 should be assumed.

The base current Ib:

$$I_b = \frac{I_c}{beta} = \frac{1.5E^{-3}}{100} = 15\mu A$$

The emitter current IE:

$$I_E = I_b + I_c = 15E^{-6} + 1.5E^{-3} = 1.515mA$$

The collector resistor RE.

The collector voltage $V_C$

The emitter resistor RC:

$$R_E = \frac{V_E}{I_E} = \frac{1}{1.515^{-3}} = 660.07$$

$$V_C = \frac{V_{CC} - V_E}{2} + V_E = \frac{15-1}{2} + 1 = 8V$$

$$R_C = \frac{V_{CC} - V_C}{I_C} = \frac{15-8}{1.515E^{-3}} = 4.62k$$

The base voltage VB: The currents $IR_2$ and $IR_1$.

$$V_B = V_{BE} + V_E = 0.655 + 1 = 1.655V$$

$$I_{R2} = 9 \times I_B = 9 \times 15E^{-6} = 135uA$$

$$I_{R1} = 10 \times I_B = 10 \times 15F^{-6} = 150uA$$

Using this value for $I_{R2}$, we can calculate the value for $R_2$ as follows:

$$R_2 = \frac{V_B}{I_{R2}} = \frac{1.655}{135E^{-6}} = 12.259k$$

Similarly, we can determine the value for $R_1$ knowing it is the volt drop across $R_1$ divided by the current flowing through it. This can be done as follows:

$$R_1 = \frac{V_{CC} - V_B}{I_{Bleed}} = \frac{15 - 1.655}{150E^{-6}} = 88.967k$$

The completed circuit with the simulated results is shown in Figure A-3.

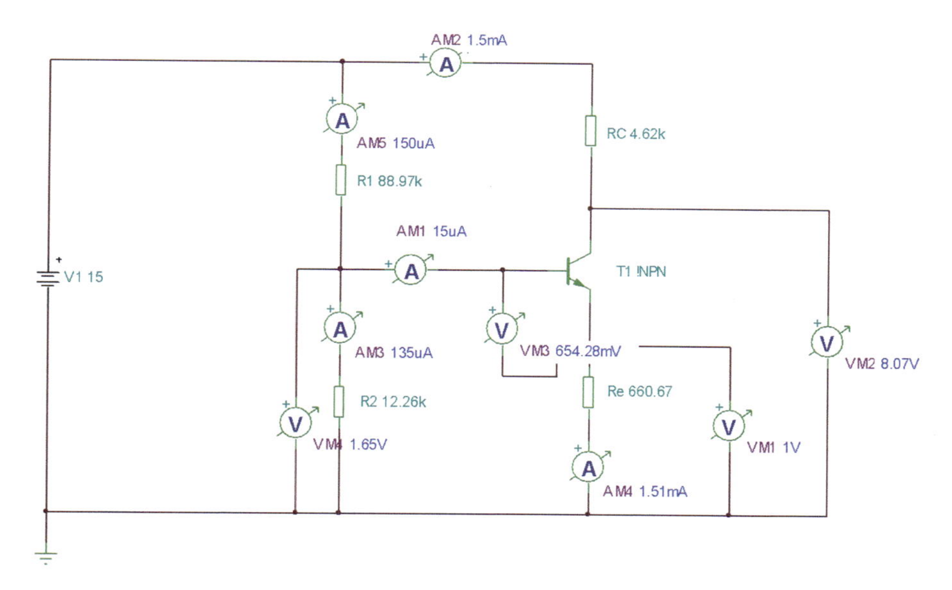


***Figure A-3.*** *The Simulated Circuit for Exercise 6.1*

Of course, these are nonstandard values for the resistors, but it is more to show you that the process and the expressions can be used to design a stabilized class A amplifier.

# Exercise 6.2

With respect to the class A amplifier circuit shown in Figure 6-24, calculate the following knowing that the DC quiescent $V_E$ voltage is to be around 2V and the beta is 100:

1. The emitter current
2. The base current
3. The base voltage
4. The output impedance
5. The input impedance
6. The voltage gain

Using Ohm's Law

$$I = \frac{V}{R} = \frac{2}{1E^3} = 2mA$$

Using

$$I_E = (\beta + 1) I_B$$

Then

$$I_B = \frac{I_E}{\beta + 1} = \frac{2E^{-3}}{101} = 19.8E^{-6} = 19.8\mu A$$

Using

$$V_B = V_E + V_{BE} = 2 + 0.7 = 2.7V$$

The Output Impedance

To ac, the capacitor $C_2$ is a short circuit, which means the resistor $R_3$ is in parallel with $R_L$. Note due to the capacitance across the output of the power supply for $V_{CC}$, the supply rail $V_{CC}$ is the same as the ground rail. These two resistors make up the output impedance, which means the output impedance is

$$R_{OUT} = \frac{R_L \times R_3}{R_L + R_s} = \frac{3.75E^3 \times 500E^3}{3.75E^3 + 500E^3} = 3.722k\Omega$$

The Input Impedance

Again, because of the various capacitances being a short circuit to ac, the input impedance is made up of the parallel combinations of $R_1$, $R_2$, (Beta + 1)$R_E$.

$$R_{IN} = \frac{1}{\frac{1}{R_1} + \frac{1}{R_2} + \frac{1}{(\beta+1)R_E}} = \frac{1}{\frac{1}{61.5E^3} + \frac{1}{15E^3} + \frac{1}{(100+1)1E^3}}$$

$$= \frac{1}{16.26E^{-6} + 66.67^{-6} + 9.901E^{-6}} = 10.772k\Omega$$

The Voltage Gain

This can be calculated using the expression

$$V_{Gain} = \frac{R_L}{R_E} = \frac{3.75E^3}{1E^3} = 3.75$$

# Exercise 6.3

With respect to the circuit shown in Figure 6-25, determine the input and output impedances of both transistors as well as the voltage gain of each transistor and the overall circuit.

Starting with the transistor T2 in the two-stage amplifier circuit.

The Output Impedance

This is simply the parallel combination of R7 and R5.

Therefore, we have

$$R_{OUT} = \frac{R_5 \times R_7}{R_5 + R_7} = \frac{3.75E^3 \times 500^3}{3.75E^3 + 500E^3} = 3.722k\Omega$$

The Input Impedance

This will be the parallel combination of R3, R4, and R6. We can ignore the effect of re as there is no emitter bypass capacitor, and so re compared to the 1k R6 can be ignored if you want to.

$$\frac{1}{R_{in}} = \frac{1}{R_3} + \frac{1}{R_4} + \frac{1}{(\beta+1)R_6} = \frac{1}{61.5E^3} + \frac{1}{15E^3} + \frac{1}{101 \times 1E^3} = 9.283E^{-5}$$

This means the input resistance $R_{in}$ is

$$R_{in} = \frac{1}{9.283E^{-5}} = 10.773k\Omega$$

Now consider the Transistor T1

The Output Impedance

This is the parallel combination of the input resistance of T2 and the load resistance of the first transistor. Therefore, we have

$$R_{OUTT1} = \frac{R_{LT1} \times R_{INT2}}{R_{LT1} + R_{INT2}} = \frac{3.75E^3 \times 10.773E^3}{3.75E^3 + 10.773E^3} = 2.782k$$

The Input Impedance of T1

As the second amplifier stage is a replica of the first amplifier stage, then the input impedance of the first amplifier will be the same as the second amplifier. This means that the input impedance of T1 is

$$10.773k\Omega$$

The Voltage Gain

To some extent, we can ignore the effect of the internal resistance re as there is no emitter bypass capacitor. Therefore, the voltage gain of the T1 amplifier is

$$V_{Gain} = -\frac{R_{OUT}}{R_E} = -\frac{3.722E^3}{1E^3} = -3.722$$

The voltage gain of T2 is

$$V_{Gain} = -\frac{R_{OUT}}{R_E} = -\frac{2.782E^3}{1E^3} = -2.782$$

The overall gain will be the product of the two gains:

$$V_{Gainoverall} = -3.722 \times -2.782 = 10.355$$

The Low Cutoff Frequency of the Amplifier

The capacitor will make a high pass filter with the input resistance of the first amplifier. This means that we can calculate the cutoff frequency using

$$f_C = \frac{1}{2\pi CR_{IN}} = \frac{1}{2\pi \times 400E^{-9} \times 10.773E^3} = 36.94Hz$$

# Exercise 6.4

To try and confirm that the Darlington transistor does allow the speaker to be added to the amplifier without loading the amplifier, the circuit shown in Figure 6-43 was simulated. The voltage gain was measured using the oscilloscope and the pk-to-pk output voltage was measured at 574.6μV. The input voltage was a 60mv pk-to-pk sine wave at 8kHz. This means the circuit actually attenuated the input signal. Use the theories you have studied to determine the ac impedances of the circuit and so confirm that this attenuation would happen.

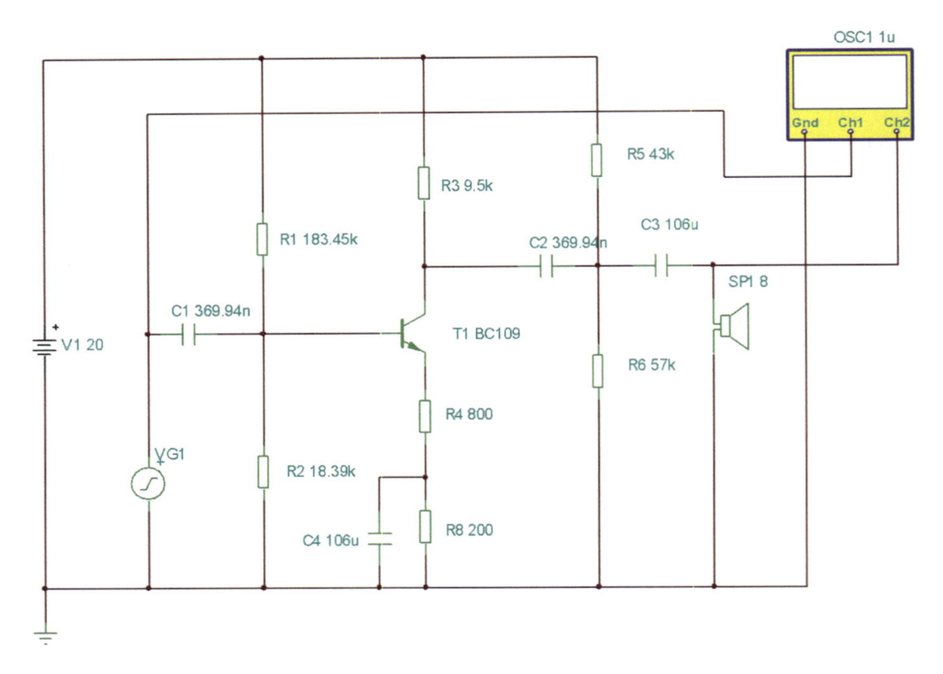


To ac, the capacitors are a short circuit. This means that starting at the output, the resistor R6 is in parallel with the 8Ω speaker. This makes a parallel $R_{COMB1}$ of

$$R_{COMB1} = \frac{R_6 \times R_{Speaker}}{R_6 + R_{Speaker}} = \frac{57E^3 \times 8}{57E^3 + 8} = 8$$

This now means that the output impedance of the transistor is the parallel combination of $R_L$, $R_5$, and $R_{COMB1}$. This means that the output impedance of the transistor is

$$R_{OUT} = \frac{1}{\frac{1}{R_L} + \frac{1}{R_5} + \frac{1}{R_{COMB1}}} = \frac{1}{\frac{1}{9.5E^3} + \frac{1}{43^3} + \frac{1}{8}} = 7.99$$

The voltage gain of the amplifier can be calculated using the expression

$$V_{Gain} = \frac{R_{OUT}}{(r_e + R_4)}$$

Assuming re = 26Ω, then we get

$$V_{Gain} = \frac{7.99}{(26 + 800)} = 9.67E^{-3}$$

This means that the output voltage can be calculated as

$$v_{out} = v_{in} \times v_{gain} = 60E^{-3} \times 9.67E^{-3} = 580.42^{-6} = 580.2\mu v$$

# Exercise 7.1

Design a MOSFET amplifier with the following specification:

VDD = 15V

ID = 5mA

The minimum gate voltage is 3.19 as we are using the 2N6755 MOSFET.

Calculate the voltage gain using a value of 3Ω for RDSON.

Then add a bypass capacitor if the lower frequency of interest was 25Hz.

Recalculate the voltage gain of the improved amplifier.

# Exercise 8.1

Design a current mirror circuit to source a 3mA current with a VCC of 15V for the control circuit and a VCC of 9V for the mirror circuit. Determine the maximum value of the load resistor for the mirror circuit. Assume the beta of the transistors was 100.

$$R_{LMAX} = \frac{V_{CC} - V_{BE}}{I_{C1}} = \frac{15 - 0.65}{3E^{-3}} = 4.783k$$

$$R_{LMAX} = \frac{V_{CC} - V_{BE}}{I_{C1}} = \frac{9 - 0.65}{3E^{-3}} = 2.783k$$

# Exercise 9.1

Design an inverting Opamp for the two following specifications:

1. The input impedance is 15kΩ, and the gain is –4.
2. The gain is –5, and the feedback resistor $R_F$ is 15kΩ.

Solution for part 1 $R_{IN} = 15k$

$$V_{Gain} = \frac{R_F}{R_{IN}}\ therefore, R_F = R_{IN} \times V_{Gain} = 15k \times 4 = 60k$$

Solution for part 2

$$V_{Gain} = \frac{R_F}{R_{IN}} \ therefore, R_{IN} = \frac{R_F}{V_{Gain}} = \frac{15E^3}{5} = 3k$$

# Exercise 9.2

Design a non-inverting Opamp that has a gain of 5 if the value of the resistor in the position of $R_2$ (see Figure 9-19) was 2kΩ.

Redesign the Opamp if we wanted to use a 15kΩ resistor in the position of $R_1$ (see Figure 9-19), and the gain required was a gain of 4.

Using

$$V_{GAIN} = 1 + \frac{R_1}{R_2}$$

Transposing for $R_1$, we have

$$R_1 = R_2\left(V_{Gain} - 1\right) = 2E^3\left(5 - 1\right) = 8k$$

Transposing the expression for the gain for R2, we have

$$R_2 = \frac{R_1}{V_{Gain} - 1} = \frac{15k}{4 - 1} = 5k$$

# Exercise 9.3

Design a summing Opamp to conform to the following specifications. The supply rail for the Opamps was the normal $\mp 15V$.

1. A summing Opamp has the following three input voltages: V1 = 2V, V2 = 0.5V, and V3 = 1V. The input impedance for each of the inputs should be 5k, and the voltage gain of the Opamp should be –3.

2. A weighted summing Opamp had the same three inputs as those in specification 1. The feedback resistor was chosen to be 9k. The gain for V1 was to be –1, the gain for V2 was to be –3, and the gain for V3 was to be –2.

3. In the design for specification 1, explain what the issue would be if the V1 input voltage was changed to 5V.

Using

$$V_{Gain} = -\frac{R_3}{R_{IN}}$$

Transposing for R3, we get

$$R_3 = V_{Gain} \times R_{IN} = 3 \times 5k = 15k$$

The individual gain will be set using the following relationship:

$$V_{Gain1} = \frac{R_F}{R_1}, V_{Gain2} = \frac{R_F}{R_2}, V_{Gain3} = \frac{R_F}{R_3}$$

Using this relationship, we can say that

$$R_1 = \frac{R_F}{V_{Gain1}}, R_2 = \frac{R_F}{V_{Gain2}}, R_3 = \frac{R_F}{V_{Gain3}},$$

Putting the values in, we get

$$R_1 = \frac{9k}{1} = 9k, R_2 = \frac{9k}{3} = 3k, R_3 = \frac{9k}{2} = 4.5k$$

If the input at V1 went to 5V and the others added up to 1.5, then with a gain of –3 the output would try to go to –3 times 6.5, i.e., 19.5V, which is above the maximum of the supply voltage, and so we would get an incorrect result.

# Exercise 9.4

Design a differential amplifier that has a gain of 6 when the two input resistors $R_1$ and $R_2$ were 3k. Calculate what the expected output voltage would be if $V_2$ was 250mv and $V_1$ was 100mV. Calculate the current $I_F$.

Knowing the gain can be calculated using the expression

$$V_{Gain} = \frac{R_4}{R_2}$$

The value of the resistor R4 would be

$$R_4 = R_2 \times V_{Gain} = 3k \times 6 = 18k$$

$$V_{OUT} = \left(V_2 - V_1\right)\frac{R_4}{R_2} = \left(250mv - 100mv\right)\frac{18E^3}{3E^3} = 900mV$$

Using Ohm's Law, the current IF would be

$$I_F = \frac{V_{OUT} - V_X}{R_F} = \frac{450E^{-3}}{18E^3} = 25\mu A$$

$$V_X = V_2 \times \frac{R_4}{R_2 + R_4} = 250^{-3} \times \frac{18E^3}{3E^3 + 18E^3} = 214.29mv$$

Therefore

$$I_F = \frac{900E^{-3} - 214.29E^{-3}}{18E^3} = 38.095\mu A$$

# Exercise 9.5

Using the same value for the PT100 and the components in the circuit shown in Figure 9-28, determine the output voltage of the amplifier when

1. T = 12°C.
2. T = 75°C.

$$R = R_0\left(1 + \alpha T\right)$$

When T = 12°C and $R_0$ = 100, we get

$$R = 100\left(1 + 0.025 \times 12\right) = 130$$

The difference voltage $V_B$ is

$$V_B = \frac{V_1 R_8}{R_6 + R_8}$$

Putting the values in, we get

$$V_B = \frac{5 \times 100}{130 + 100} = 2.174V$$

We can now calculate the difference voltage $V_{DIFF}$:

$$V_{DIFF} = V_A - V_B = 2.5 - 2.174 = 326.1mV$$

$$V_{Out} = V_{Diff} \times 10 = 326.1E^{-3} \times 10 = 3.261V$$

When T = 75OC

$$R = 100(1 + 0.025 \times 75) = 287.5$$

The difference voltage $V_B$ is

$$V_B = \frac{V_1 R_8}{R_6 + R_8}$$

Putting the values in, we get

$$V_B = \frac{5 \times 100}{387.5 + 100} = 1.026V$$

We can now calculate the difference voltage $V_{DIFF}$:

$$V_{DIFF} = V_A - V_B = 2.5 - 1.026 = 1.474mV$$

$$V_{Out} = V_{Diff} \times 10 = 1.474 \times 10 = 14.74V$$

# Exercise 9.6

Show that if $R_2$ was 1k and $R_4$ was 10k, which gives the same voltage gain, remember we must make $R_2 = R_1$ and $R_3 = R_4$, the voltage $V_A$ would reduce to 2.489V. Not much difference, but then an error of 11.31mV would produce an error in the temperature reading of 0.363°C.

This would put a 1k resistor in parallel with the resistor R8 in the Wheatstone bridge. This would load the resistor and change its value to

$$R_8 = \frac{100 \times 11k}{100 + 11k} = 99.1\Omega$$

This would mean that at 0°C the voltage VB would be

$$V_B = \frac{5 \times 99.1}{100 + 99.1} = 2.486v$$

# Exercise 9.7

Determine the gradient and the output voltage after 250µs and 300µs if the following changes to the circuit shown in Figure 9-30 were made

VIN changed to 7.5V

R1 changed to 4.7k

C1 changed to 100nF

$$V_{OUT} = -\frac{V_{IN}}{C_F R_{IN}} t = -\frac{7.5}{100E^{-9} \times 4.7E^3} t = -15.957E^3 t$$

The gradient is $-15.957E^3$.
When t = 250us

$$V_{OUT} = -15.957E^3 t = -15.957E^3 \times 250E^{-6} = -3.989v$$

When t = 300us

$$V_{OUT} = -15.957E^{3}t = -15.957E^{3} \times 300E^{-6} = -4.787v$$

# Exercise 9.8

Design a sawtooth generator that uses a 5kHz square wave with a peak voltage of 5V. The pk-to-pk voltage of the sawtooth should be 6V. The capacitor must be 100nF as the company has hundreds of them spare. The design should employ a resistor in the feedback path to ensure the output centers around 0V.

$$C_{F}R_{IN} = \frac{D_{C1}}{(pk - to - pk)2F}$$

Transposing this for RIN, we have

$$R_{IN} = \frac{D_{C1}}{C_{F}(pk - to - pk)2F} = \frac{5}{100E^{-9}(6)2 \times 5E^{3}} = 833.33$$

To lift the output voltage, we need to add a feedback resistor. The value of this feedback resistance must be ten times the impedance of the capacitor at the frequency of the waveform. Therefore, we can calculate the value of this feedback resistor using

$$R_{F} = \frac{10}{2\pi fC_{f}} = \frac{10}{2\pi \times 5E^{3} \times 100E^{-9}} = 3.183k$$

# Exercise 9.9

Briefly explain why the negative supply to the Opamp can be tied to 0V and briefly explain what the main considerations are in choosing the value for the resistor $R_1$.

There is no requirement for the output voltage to go negative. The main purpose of the resistor $R_1$ is to set the base current flowing into the transistor.

# Exercise 10.1

1. Design a monostable that forces the output to stay high for 500ms when the value of C2 in Figure 10-1 has a value of 50µF.
2. Calculate the value of C1 if we were using a value of 25k for R5, and we wanted the output to stay high for 1.5s.

Note you should increase the delay period for VG1 to ensure the capacitor $C_2$ has been able to charge up the $V_{CC}$.

$$t = -RCln\left(\frac{V_{CC} - V_C}{V_{CC} - V_0}\right)$$

This can be transposed for R or C. Transposing for R, we get

$$R = \frac{t}{-Cln\left(\frac{V_{CC} - V_C}{V_{CC} - V_0}\right)}$$

Therefore

$$R = \frac{500E^{-3}}{-50E^{-6} ln\left(\frac{12-0.7}{12--11.3}\right)} = 13.818k$$

$$C = \frac{t}{-Rln\left(\frac{V_{CC}-V_C}{V_{CC}-V_0}\right)} = \frac{2}{-25E^3 ln\left(\frac{12-0.7}{12--11.3}\right)} = 110.55\mu F$$

As this is a large value capacitor, we need to increase the delay time to 2.8s. The time constant for R1 and C2 is

$$CR = 5E^3 \times 110.55E^{-6} = 0.553s$$

Therefore, it will take approximately 2.76s for $C_2$ to charge up to 11.3V.

# Exercise 10.2

1. Design a bistable multivibrator to produce a 5kHz square wave using 10μF capacitors when the mark time was 75% and the space time was 25% of the total periodic time of the square wave.

2. Redesign the circuit to produce a 20% mark time at a frequency of 10kHz using capacitors with a value of 20μF.

The expression that shows how the space and mark times are controlled is

$$t = -RCln\left(\frac{V_{CC}-V_C}{V_{CC}-V_0}\right)$$

We can transpose this for R as

$$R = \frac{t}{-Cln\left(\frac{V_{CC} - V_C}{V_{CC} - V_0}\right)}$$

We can calculate the 75% mark time at 5Hz, and we can calculate the value of time "t" as

$$t = 0.75 \times \frac{1}{5E^3} = 150\mu s$$

Using this time, we can calculate the value of $R_2$ as

$$R_2 = \frac{150E^{-6}}{-10E^{-6}ln\left(\frac{12 - 0.7}{12 - 11.3}\right)} =$$

The space time will be

$$t = 0.25 \times \frac{1}{5E^3} = 50\mu s$$

Using this time, we can calculate the value of $R_5$ as

$$R_5 = \frac{50E^{-6}}{-10E^{-6}ln\left(\frac{12 - 0.7}{12 - 11.3}\right)} =$$

When the frequency is 10kHz and the mark time was 20%, we have

$$mark\ time\, t = 0.2 \times \frac{1}{10E^3} = 20\mu s$$

Using this time, we can calculate the value of $R_2$ as

$$R_2 = \frac{20E^{-6}}{-20E^{-6}\, ln\left(\frac{12-0.7}{12-11.3}\right)} =$$

The space time will be

$$t = 0.8 \times \frac{1}{10E^3} = 80\mu s$$

Using this time, we can calculate the value of $R_5$ as

$$R_5 = \frac{80E^{-6}}{-10E^{-6} ln\left(\frac{12-0.7}{12-11.3}\right)} =$$

# Exercise 10.3

1. With respect to the phase shift circuit shown in Figure 10-11, design the circuit to produce a phase shift of 75° using a resistor value of 2.2k.
2. Design an RC phase shift oscillator that has a frequency of 1200Hz using the existing 100nF capacitors with a phase shift of 60°.

This means the angle part of the transfer function must equate to 75°. The Tan of 75 is 3.732. This means that the term

$$\frac{1}{\omega CR} \; must \;\; equal \; 3.732$$

Transposing this for R, we have

$$R = \frac{1}{3.732 \times 2\pi fC} = \frac{1}{3.732 \times 2\pi \times 2.2E^{3} \times 100E^{-9}} = 193.84\Omega$$

Tan 60 = 1.732.

Therefore

$$R = \frac{1}{1.732 \times 2\pi fC} = \frac{1}{1.732 \times 2\pi \times 1.2E^{3} \times 100E^{-9}} = 765.73\Omega$$

# Exercise 10.4

As an exercise, see if you can derive the expression for the transfer function. My solution is shown here. As an aid to get you started, you can represent the feedback circuit as shown in Figure 10-17.

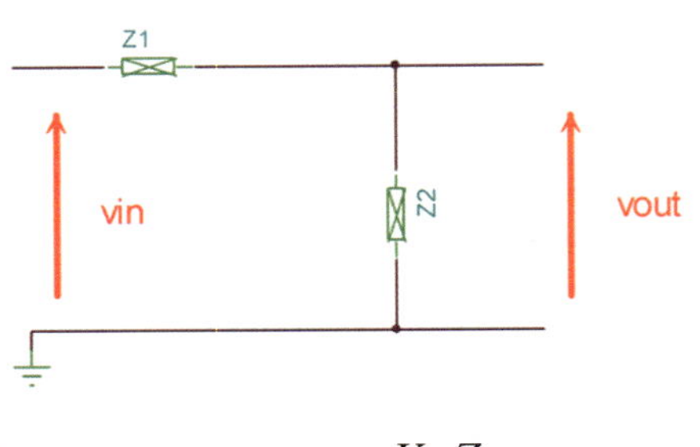


$$V_{OUT} = \frac{V_{IN} Z_2}{Z_1 + Z_2}$$

Therefore

$$\frac{V_{OUT}}{V_{IN}} = \frac{Z_2}{Z_1 + Z_2}$$

$$Z_1 = R_3 + \frac{1}{j\omega C_1}$$

$$Z_2 = R_4 \text{ in parallel with } C_2 = \frac{R_4 \dfrac{1}{j\omega C_2}}{R_4 + \dfrac{1}{j\omega C_2}} = \frac{\dfrac{R_4}{j\omega C_2}}{R_4 + \dfrac{1}{j\omega C_2}}$$

$$\frac{V_{OUT}}{V_{IN}} = \frac{\dfrac{\dfrac{R_4}{j\omega C_2}}{R_4 + \dfrac{1}{j\omega C_2}}}{\left(R_3 + \dfrac{1}{j\omega C_1}\right) + \left(\dfrac{\dfrac{R_4}{j\omega C_2}}{R_4 + \dfrac{1}{j\omega C_2}}\right)}$$

R3 = R4 = R and C1 = C2 = C also.

$$\frac{V_{OUT}}{V_{IN}} = TF$$

$$TF = \frac{\dfrac{\dfrac{R}{j\omega C_2}}{R + \dfrac{1}{j\omega C}}}{\left(R + \dfrac{1}{j\omega C}\right) + \left(\dfrac{\dfrac{R}{j\omega C_2}}{R + \dfrac{1}{j\omega C}}\right)}$$

If we multiply the top and bottom by $R + \frac{1}{j\omega C}$, we get

$$TF = \frac{\frac{R}{j\omega C}}{\left(R + \frac{1}{j\omega C}\right)\left(R + \frac{1}{j\omega C}\right) + \frac{R}{j\omega C}}$$

Now multiplying the top and bottom by $j\omega C$, we get

$$TF = \frac{R}{\left(R + \frac{1}{j\omega C}\right)\left(R + \frac{1}{j\omega C}\right)(j\omega C) + R}$$

Expanding one of the brackets in the denominator, we get

$$TF = \frac{R}{\left(R + \frac{1}{j\omega C}\right)(j\omega CR + 1) + R}$$

Now expanding the remaining brackets, we get

$$TF = \frac{R}{j\omega CR^2 + R + R + \frac{1}{j\omega C} + R}$$

Collecting the terms gives

$$TF = \frac{R}{j\omega CR^2 + 3R + \frac{1}{j\omega C}}$$

If we now multiply the top and bottom by $j\omega C$, we get

$$TF = \frac{j\omega CR}{1 - \omega^2 C^2 R^2 + 3j\omega CR}$$

$$\therefore TF = \frac{0 + j\omega CR}{1 - \omega^2 C^2 R^2 + 3j\omega CR}$$

Converting this to polar format, we get

$$TF = \frac{\omega CR \angle 90}{\sqrt{\left(1 - \omega^2 C^2 R^2\right)^2 + \left(3\omega CR\right)^2} \angle \left(\frac{3\omega CR}{1 - \omega^2 C^2 R^2}\right)}$$

# Exercise 10.5

Complete the simplification of the transfer function to show that the magnitude of the transfer function will result in an attenuation of 1/3. As a hint, you should remember

$$1 - \omega^2 C^2 R^2 = 0$$

It only requires three more steps. Good luck.

The solution is shown here.

The magnitude of the transfer function is

$$|TF| = \frac{\omega CR}{\sqrt{\left(1 - \omega^2 C^2 R^2\right)^2 + \left(3\omega CR\right)^2}}$$

Knowing

$$1-\omega^2C^2R^2=0$$

then

$$|TF|=\frac{\omega CR}{\sqrt{(3\omega CR)^2}}=\frac{\omega CR}{3\omega CR}=\frac{1}{3}$$

# Exercise 10.6

1. Design a Wien Bridge oscillator that has a frequency of 1kHz using a 50nF capacitor.
2. Design a Wien Bridge oscillator that has a frequency of 10kHz using a 500nF capacitor.

$$f_o=\frac{1}{2\pi CR}$$

This can be transposed for R as

$$R=\frac{1}{2\pi Cf_o}=\frac{1}{2\pi\times 50E^{-9}\times 1E^3}=3.183k$$

The solution for part 2

$$R=\frac{1}{2\pi Cf_o}=\frac{1}{2\pi\times 500E^{-9}\times 10E^3}=31.83\Omega$$

# Exercise 10.7

Design a monostable using the 555 timer that can be forced out of its stable state for a period of 150ms. Your company wants to use the 50uF capacitors it has a surplus of.

The expression for the time period is

$$t = -CRln\left(1 - \frac{V_C}{V}\right)$$

This can be transposed for R as

$$R = \frac{t}{-Cln\left(1 - \frac{V_C}{V}\right)}$$

Putting the values in, we get

$$R = \frac{150E^{-3}}{-50E^{-6} \times ln\left(1 - \frac{3.32}{5}\right)} = 2.75k$$

# Exercise 10.8

1. Design a 555 timer that can produce a 1Hz square wave with a mark time of 666.67ms, using a 10μF capacitor for $C_2$.

2. Design a 555 timer that can produce a 50kHz square wave with a mark time of 11μs, using a 1nF capacitor for $C_2$.

$$Marktime = 0.69(R1 + R2)C1$$

We can calculate the value of $R_2$ using

$$R_2 = \frac{Spacetime}{0.69C}$$

We need the value of the space time. We know that at a frequency of 1Hz, the periodic time will be one second. Therefore, knowing that the periodic time is the mark time + the space time, that is, T = M + S, then we can say

$$Spacetime\ S = 1 - 0.6667 = 0.33333$$

Now we can calculate the value of R2:

$$R_2 = \frac{0.33333}{0.69 \times 10E^{-6}} = 48.309k$$

We can transpose the expression for mark time to an expression for R1 as

$$Marktime = 0.69(R1 + R2)C1$$

Therefore

$$R_1 = \frac{Marktime}{0.69C_1} - R_2$$

Putting the values in, we get

$$R_1 = \frac{0.66667}{0.69 \times 10E^{-6}} - 48.309E^3 = 48.3k$$

The solution for part 2

When f = 50kHz, the periodic time T is

$$T = \frac{1}{f} = \frac{1}{50E^3} = 20\mu s$$

Therefore

$$Space\ time\ S = 20E^{-6} - 11E^{-6} = 9\mu s$$

Therefore

$$R_2 = \frac{9E^{-6}}{0.69 \times 1E^{-9}} = 13.043k$$

Therefore

$$R_1 = \frac{11E^{-6}}{0.69 \times 1E^{-9}} - 13.043E^3 = 2.899k$$

# Exercise 11.1

As an exercise, see if you can calculate the magnitude and angle of the transfer function when f = infinity and so show that when f = infinity

$$TF = 1\langle 0 \rangle$$

Finally, at the cutoff frequency, the transfer function is

$$TF = 0.707\,4\langle 5 \rangle$$

$$|TF| = \frac{1}{\sqrt{1^2 + \left(-\frac{1}{\omega CR}\right)^2}}$$

When f = infinity, $\omega CR = \infty$.

$$|TF| = \frac{1}{\sqrt{1^2 + \left(-\frac{1}{\infty}\right)^2}} = \frac{1}{\sqrt{1^2 + (0)^2}} = 1$$

$$\theta = -\langle Tan^{-1}\left(-\frac{1}{\omega CR}\right)\rangle = -\langle Tan^{-1}\left(-\frac{1}{\infty}\right)\rangle = -\langle Tan^{-1} - (0)\rangle = 0$$

# Exercise 11.2

As an exercise, derive the transfer function for the two passive filter circuits as shown in Figure 11-9 and so confirm they are as stated. Then using a resistor value of 2.2kΩ for each circuit, calculate the value of capacitance in circuit "a" and inductance in circuit "b" so that they both have a cutoff frequency of 10kHz. State what type, low pass or high pass filter, they are.

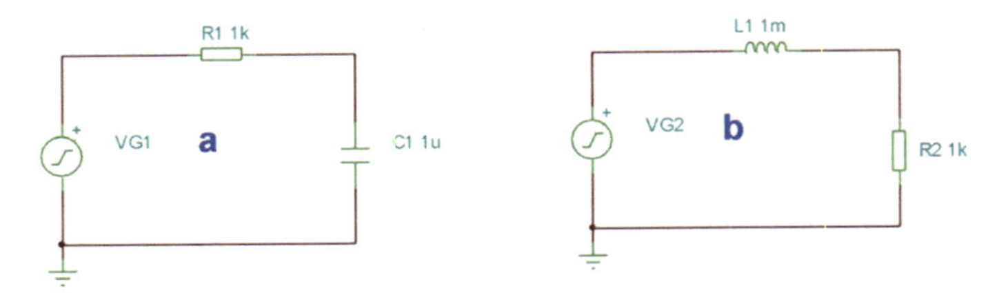


Circuit "a":

Using

$$TF = \frac{Z_2}{Z_1 + Z_2} = \frac{\frac{1}{j\omega C}}{R + \frac{1}{j\omega C}}$$

Multiplying the top and bottom by $j\omega C$, we get

$$TF = \frac{1}{1 + j\omega CR}$$

Converting to polar format, we get

$$TF = \frac{1}{\sqrt{1^2 + (\omega CR)^2} \langle Tan^{-1}\left(\frac{\omega CR}{1}\right)\rangle}$$

This gives

$$TF = \frac{1}{\sqrt{1^2 + (\omega CR)^2}} - \langle Tan^{-1}(\omega CR)\rangle$$

This splits to

$$|TF| = \frac{1}{\sqrt{1^2 + (\omega CR)^2}}$$

The phase is

$$\theta = -tan^{-1}(\omega CR)$$

# Exercise 11.3

Design a low pass active filter that has a cutoff frequency of 10kHz when the resistor R is 4.7kΩ. Let $R_a$ and $R_b$ both be set at 1kΩ.

We can calculate the value of the capacitor using

$$C = \frac{1}{2\pi fR} = \frac{1}{2\pi \times 10E^3 \times 4.7E^3} = 3.39nF$$

# Exercise 11.4

Calculate the magnitude of the transfer function when f = 12kHz and so show that it would be –3.87dBs.

Using

$$|TF| = \frac{1}{\sqrt{1+(\omega CR)^2}}$$

Putting the values in, we get

$$|TF| = \frac{1}{\sqrt{1+\left(2\pi \times 12E^3 \times 15.91E^{-9} \times 1E^3\right)^2}} = 0.6403$$

In dBs, this is

$$dBs = 20log\left(0.6403\right) = -3.87dBs$$

# Exercise 11.5

Calculate the magnitude of the transfer function when f = 15kHz:

$$TF = \frac{1}{\sqrt{\left(1-\omega^2C^2R^2\right)^2 + (3\omega CR)^2}} \langle -Tan^{-1}\left(\frac{3\omega CR}{1-\omega^2C^2R^2}\right)\rangle$$

The magnitude can be calculated using

$$|TF| = \frac{1}{\sqrt{\left(1-(\omega CR)^2\right)^2 + (3\omega CR)^2}}$$

Putting the values in, we get

$$|TF| = \frac{1}{\sqrt{\left(1-\left(2\pi \times 15E^{3} \times 15.19E^{-9} \times 1E^{3}\right)^{2}\right)^{2} + \left(3 \times 2\pi \times 15E^{3} \times 15.19E^{-9} \times 1E^{3}\right)^{2}}}$$

$$|TF| = \frac{1}{\sqrt{1.5586 + 20.236}} = 0.2142$$

In dBs, it is

$$dBs = 20log\left(0.2142\right) = -13.3836dBs$$

# Exercise 11.6

As an exercise, use the expression for the transfer function to calculate the magnitude, in dBs, of the transfer function when f = 7kHz and f = 20kHz.

$$TF = \frac{2}{\sqrt{\left(1-(\omega CR)^{2}\right)^{2} + (\omega CR)^{2}}} \langle -\langle Tan^{-1}\left(\frac{\omega CR}{1-(\omega CR)^{2}}\right)\rangle\rangle$$

When f = 7kHz

$$\omega CR = 2\pi \times 7E^{3} \times 15.91E^{-9} \times 1E^{3} = 0.6998$$

$$\left(\omega CR\right)^{2} = 0.48966$$

$$1-\left(\omega CR\right)^{2} = 0.51034$$

$$\left(1-(\omega CR)^2\right)^2 = 0.26045$$

$$|TF| = \frac{2}{\sqrt{0.26045+0.48966}} = \frac{2}{0.8661} = 2.309$$

In dBs, this is

$$dBs = 20\log(2.309) = 7.27dBs$$

When f = 20kHz

$$\omega CR = 2\pi \times 20E^3 \times 15.91E^{-9} \times 1E^3 = 1.9993$$

$$(\omega CR)^2 = 3.9972$$

$$1-(\omega CR)^2 = -2.9972$$

$$\left(1-(\omega CR)^2\right)^2 = 8.9834$$

$$|TF| = \frac{2}{\sqrt{8.9834+3.9972}} = \frac{2}{3.6029} = 0.55511$$

In dBs, this is

$$dBs = 20log(0.55511) = -5.112dBs$$

# Exercise 11.7

Using the feedback model of the high pass filter, shown in Figure 11-28, derive the transfer function for the high pass second-order active filter. Assume that $R_a = R_b$.

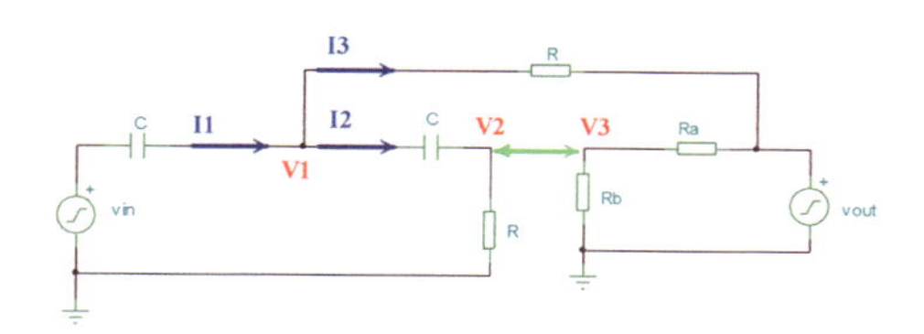


$$v_3 = v_{out} \frac{R_b}{R_a + R_b}$$

$$A = \frac{R_b}{R_a + R_b}$$

If Ra = Rb

$$A = \frac{1}{2}$$

$$v_3 = \frac{v_{out}}{2}$$

$$v_2 = v_3 = \frac{v_{out}}{2}$$

$$i_1 = i_2 + i_3$$

$$i_1 = \frac{v_{in} - v_1}{XC} = \frac{v_{in}}{XC} - \frac{v_1}{XC}$$

Knowing we can express XC as

$$XC = \frac{1}{SC}$$

$$i_1 = v_{in}SC - v_1SC$$

Similarly, with respect to $i_2$, we can say

$$i_2 = v_1SC - v_2SC$$

$$i_2 = v_1SC - \frac{v_{out}}{2}SC$$

Finally, looking at $i_3$, we can say

$$i_3 = \frac{v_1 - v_{out}}{R}$$

$$i_3 = \frac{v_1}{R} - \frac{v_{out}}{R}$$

Using KCL, we can say

$$i_1 = i_2 + i_3$$

$$v_{in}SC - v_1SC = \left(v_1SC - \frac{v_{out}}{2}SC\right) + \left(\frac{v_1}{R} - \frac{v_{out}}{R}\right)$$

Now multiply throughout by R and remove the brackets as we don't need them:

$$v_{in}SCR - v_1SCR = v_1SCR - \frac{v_{out}}{2}SCR + v_1 - v_{out}$$

Now adding $v_1SCR$ to both sides gives

$$v_{in}SCR = v_1SCR - \frac{v_{out}}{2}SCR + v_1 - v_{out} + v_1SCR$$

Now collect the like terms:

$$v_{in}SCR = v_1SCR + v_1 + v_1SCR - \frac{v_{out}}{2}SCR - v_{out}$$

Now take $v_1$ out as a common factor where we can:

$$v_{in}SCR = v_1\left(2SCR + 1\right) - \frac{v_{out}}{2}SCR - v_{out}$$

We need to remove the $v_1$ term from the expression. From the circuit, we can see that

$$v_2 = v_1 \frac{R}{R + \frac{1}{SC}}$$

Therefore, we can say

$$v_1 = v_2 \frac{R + \frac{1}{SC}}{R} = \frac{SCR + 1}{SC} \times \frac{1}{R} = v_2 \frac{SCR + 1}{SCR}$$

However

$$v_2 = \frac{v_{out}}{2}$$

Therefore, we have

$$v_1 = \frac{v_{out}}{2}\frac{SCR+1}{SCR}$$

$$v_{in}SCR = \frac{v_{out}}{2}\frac{SCR+1}{SCR}(2SCR+1) - \frac{v_{out}}{2}SCR - v_{out}$$

Multiplying throughout by SCR gives

$$v_{in}(SCR)^2 = \frac{v_{out}}{2}(SCR+1)(2SCR+1) - \frac{v_{out}}{2}(SCR)^2 - v_{out}SCR$$

Expanding the bracket gives

$$v_{in}(SCR)^2 = \frac{v_{out}}{2}\left[(2SCR)^2 + 3SCR + 1\right] - \frac{v_{out}}{2}(SCR)^2 - v_{out}SCR$$

Now multiply throughout by 2:

$$v_{in}2(SCR)^2 = v_{out}\left[(2SCR)^2 + 3SCR + 1\right] - v_{out}(SCR)^2 - v_{out}2SCR$$

Now take vout out as a common factor:

$$v_{in}2(SCR)^2 = v_{out}\left[\left[(2SCR)^2 + 3SCR + 1\right] - (SCR)^2 - 2SCR\right]$$

Now collect like terms:

$$v_{in} 2(SCR)^2 = v_{out}\left[(SCR)^2 + SCR + 1\right]$$

Now divide by the bracket on the RHS:

$$v_{out} = \frac{v_{in} 2(SCR)^2}{(SCR)^2 + SCR + 1}$$

Now divide by $v_{in}$:

$$\frac{v_{out}}{v_{in}} = \frac{2(SCR)^2}{(SCR)^2 + SCR + 1}$$

Now divide by $(SCR)^2$:

$$\frac{v_{out}}{v_{in}} = \frac{2}{1 + \frac{1}{SCR} + \frac{1}{(SCR)^2}}$$

Now replace S with jw:

$$\frac{v_{out}}{v_{in}} = \frac{2}{1 + \frac{1}{j\omega CR} + \frac{1}{(j\omega CR)^2}}$$

Knowing $\frac{1}{j} = -j$ and $j^2 = -1$

$$\frac{v_{out}}{v_{in}} = \frac{2}{1 - j\frac{1}{\omega CR} - \frac{1}{(\omega CR)^2}} = \frac{2}{1 - \frac{1}{(\omega CR)^2} - j\frac{1}{\omega CR}}$$

Converting to polar format gives

$$TF = \frac{2\langle 0\rangle}{\sqrt{\left(1-\frac{1}{(\omega CR)^2}\right)^2+\left(-\frac{1}{\omega CR}\right)^2}\langle Tan^{-1}\left(\frac{-\frac{1}{\omega CR}}{1-\frac{1}{(\omega CR)^2}}\right)\rangle}$$

Dividing the complex numbers gives

$$TF = \frac{2}{\sqrt{\left(1-\frac{1}{(\omega CR)^2}\right)^2+\left(-\frac{1}{\omega CR}\right)^2}}\langle -Tan^{-1}\left(\frac{-\frac{1}{\omega CR}}{1-\frac{1}{(\omega CR)^2}}\right)\rangle$$

# Exercise 11.8

As an exercise, use the expression for the transfer function to calculate the magnitude, in dBs, of the transfer function when f = 10kHz and f = 20kHz.

$$TF = \frac{2}{\sqrt{\left(1-\frac{1}{(\omega CR)^2}\right)^2+\left(-\frac{1}{\omega CR}\right)^2}}\langle -Tan^{-1}\left(\frac{-\frac{1}{\omega CR}}{1-\frac{1}{(\omega CR)^2}}\right)\rangle$$

When f = 10kHz

$$\omega CR = 2\pi \times 10E^3 \times 15.91E^{-9} \times 1E^3 = 0.99965$$

$$(\omega CR)^2 = 0.99931$$

$$\frac{1}{\omega CR} = \frac{1}{0.99965} = 1.00035$$

$$\frac{1}{(\omega CR)^2} = \frac{1}{0.99931} = 1.00069$$

$$1 - \frac{1}{(\omega CR)^2} = 1 - 1.00069 = -0.00069$$

$$\left(1 - \frac{1}{(\omega CR)^2}\right)^2 = (-0.00069)^2 = 476.76E^{-9}$$

$$\left(-\frac{1}{\omega CR}\right)^2 = (-1.0035)^2 = 1.00701$$

$$|TF| = \frac{2}{\sqrt{\left(1 - \frac{1}{(\omega CR)^2}\right)^2 + \left(-\frac{1}{\omega CR}\right)^2}}$$

$$|TF| = \frac{2}{\sqrt{476.76E^{-9} + 1.00701}} = 1.99302$$

In dBs, this is

$$dBs = 20\log(1.99302) = 5.99dBs$$

When f = 30kHz

$$\omega CR = 2\pi \times 30E^3 \times 15.91E^{-9} \times 1E^3 = 3$$

$$(\omega CR)^2 = 9$$

$$\frac{1}{\omega CR} = \frac{1}{3} = 0.3333$$

$$\frac{1}{(\omega CR)^2} = \frac{1}{9} = 0.111111$$

$$1 - \frac{1}{(\omega CR)^2} = 1 - 0.11111 = 0.88889$$

$$\left(1 - \frac{1}{(\omega CR)^2}\right)^2 = (0.88889)^2 = 0.79$$

$$\left(-\frac{1}{\omega CR}\right)^2 = (-0.33333)^2 = 0.11111$$

$$|TF| = \frac{2}{\sqrt{\left(1 - \frac{1}{(\omega CR)^2}\right)^2 + \left(-\frac{1}{\omega CR}\right)^2}}$$

$$|TF| = \frac{2}{\sqrt{0.79 + 0.11111}} = 2.1069$$

In dBs, this is

$$dBs = 20\,log(2.1069) = 6.47dBs$$

# Exercise 11.9

This is an exercise in multiplying fractions, expanding brackets and factorising terms. We start with the transfer functions for the first order and second order active filters that are cascaded together to make the third order filter.

The transfer function for the first order active filter which has been derived already is.

$$TF = \frac{2}{1 + j\omega CR}$$

The transfer function for the second order filter with the general term.

$$A = \frac{R_b}{R_a + R_b}$$

Is.

$$\frac{\frac{1}{A}}{1 - (\omega CR)^2 + j\left(3 - \frac{1}{A}\right)(\omega CR)}$$

We can now simply multiply the two transfer functions together to derive the transfer function for the third order active filter. This is.

$$TF = \left(\frac{2}{1 + j\omega CR}\right)\left(\frac{\frac{1}{A}}{1 - (\omega CR)^2 + j\left(3 - \frac{1}{A}\right)(\omega CR)}\right)$$

When multiplying fractions, we simply multiply the numerators together and the denominators together. Staring with the numerators we have.

$$(2)\left(\frac{1}{A}\right)=\frac{2}{1}\times\frac{1}{A}=\frac{2}{A}$$

Therefore, the numerator of the transfer function is.

$$\frac{2}{A}$$

There are two terms in the denominator which can be written as.

$$(1+j\omega CR)\left(1-(\omega CR)^2+j\left(3-\frac{1}{A}\right)(\omega CR)\right)$$

I hope this makes it easier to see that we are now expanding two brackets each with two terms in. To try and make it easier to see this we could write the two brackets out as.

$$(a+b)(c+d)$$

To expand the brackets, we simply multiply everything in the second bracket by everything in the first bracket. This expands to.

$$(ac)+(ad)+(bc)+(bd)$$

Where.

$$a=1$$

$$b=j\omega CR$$

$$c = 1 - (\omega CR)^2$$

$$d = j\left(3 - \frac{1}{A}\right)(\omega CR)$$

$$\left(1-(\omega CR)^2\right)+\left(j\left(3-\frac{1}{A}\right)(\omega CR)\right)+\left(j\omega CR - j(\omega CR)^3\right)+\left(j^2\left(3-\frac{1}{A}\right)(\omega CR)^2\right)$$

(ac) (ad) (bc) (bd)

We can replace $j^2$ with -1 which is in the fourth bracket. This will give.

To help with the next stage we can move the 'j' in the second bracket next to the ($\omega CR$), which is legal as 2x3 is the same as 3x2. This now gives.

$$\left(1-(\omega CR)^2\right)+\left(\left(3-\frac{1}{A}\right)j(\omega CR)\right)+\left(j\omega CR - j(\omega CR)^3\right)-\left(\left(3-\frac{1}{A}\right)(\omega CR)^2\right)$$

Next, we can multiply the $\left(3-\frac{1}{A}\right)$ in the second and fourth bracket by the terms linked to them. This gives.

$$1-(\omega CR)^2 + 3j\,\omega CR - \frac{1}{A}j\,\omega CR + j\omega CR - j(\omega CR)^3 - \left(3(\omega CR)^2 - \frac{1}{A}(\omega CR)^2\right)$$

Now, we can arrange the whole denominator expression in its real and 'j' terms which gives.

$$1-(\omega CR)^2 - \left(3(\omega CR)^2 - \frac{1}{A}(\omega CR)^2\right) + 3j\,\omega CR - \frac{1}{A}j\,\omega CR + j\omega CR - j(\omega CR)^3$$

(real terms) ('j' terms)

Now, collect the whole numbers of the $-(\omega CR)^2$ in the real terms and collect the whole numbers of $\omega CR$ in the 'j' terms. This gives.

$$1-\left(4(\omega CR)^2-\frac{1}{A}(\omega CR)^2\right)+4j\,\omega CR-\frac{1}{A}j\,\omega CR-j(\omega CR)^3$$

Now we can take the $(\omega CR)^2$ out as a common factor in the real terms and take the $j\,\omega CR$ in the 'j' terms. This gives.

$$1-\left(4-\frac{1}{A}\right)(\omega CR)^2+\left(4-\frac{1}{A}\right)j\,\omega CR-j(\omega CR)^3$$

Finally, we can take the 'j' out as a common factor in the 'j' terms. This gives.

$$1-\left(4-\frac{1}{A}\right)(\omega CR)^2+j\left[\left(4-\frac{1}{A}\right)\omega CR-(\omega CR)^3\right]$$

This is the denominator of the transfer function. Therefore, combining it with the numerator of the transfer function gives.

$$TF=\frac{\frac{2}{A}}{1-\left(4-\frac{1}{A}\right)(\omega CR)^2+j\left[\left(4-\frac{1}{A}\right)\omega CR-(\omega CR)^3\right]}$$

# Exercise 11.10

Evaluate the expression for the magnitude of the third-order filter when f = 5kHz and 15kHz. Hint, you will need to be very accurate in your calculations.

$$TF = \frac{3.1721}{\sqrt{\left[\left(1-2.41396(\omega CR)^2\right)^2+\left(2.41396\omega CR-(\omega CR)^3\right)^2\right]}}$$

When F = 5kHz

$$\omega CR = 2\pi \times 5E^3 \times 15.91^{-9} \times 1^3 = 0.4998274$$

$$(\omega CR)^2 = 0.2498274$$

$$(\omega CR)^3 = 0.124871$$

$$2.41396\omega CR = 1.206563$$

$$2.41396(\omega CR)^2 = 0.603073$$

$$1-2.41396(\omega CR)^2 = 0.39693$$

$$2.41396\omega CR-(\omega CR)^3 = 1.08169$$

$$\left(1-2.41396(\omega CR)^2\right)^2 = 0.15755$$

$$\left(2.41396\omega CR-(\omega CR)^3\right)^2 = 1.17005$$

$$\left(1-2.41396(\omega CR)^2\right)^2+\left(2.41396\omega CR-(\omega CR)^3\right)^2 = 1.3276$$

$$TF = \frac{3.1721}{\sqrt{[1.3276]}} = 2.75304$$

In dBs, this is

$$dBs = 20 log(2.75304) = 8.796 dBs$$

When F = 15kHz

$$\omega CR = 2\pi \times 15E^3 \times 15.91^{-9} \times 1^3 = 1.49948$$

$$(\omega CR)^2 = 2.248447$$

$$(\omega CR)^3 = 3.371506$$

$$2.41396\omega CR = 3.619685$$

$$2.41396(\omega CR)^2 = 5.42766$$

$$1 - 2.41396(\omega CR)^2 = -4.42766$$

$$2.41396\omega CR - (\omega CR)^3 = 0.248179$$

$$\left(1 - 2.41396(\omega CR)^2\right)^2 = 19.6042$$

$$\left(2.41396\omega CR - (\omega CR)^3\right)^2 = 0.061593$$

$$\left(1 - 2.41396(\omega CR)^2\right)^2 + \left(2.41396\omega CR - (\omega CR)^3\right)^2 = 19.66577$$

$$TF = \frac{3.1721}{\sqrt{[19.66577]}} = 0.7153053$$

In dBs, this is

$$dBs = 20log(0.7153053) = -2.91dBs$$

# Exercise 11.11

Design the following active filter circuit:

1. A first-order circuit with an overall gain of four
2. A third-order circuit with an overall gain of six

A first-order circuit has a gain of two. This means that we need to add an amplifier with a gain of two. The circuit for this active filter is shown as follows.

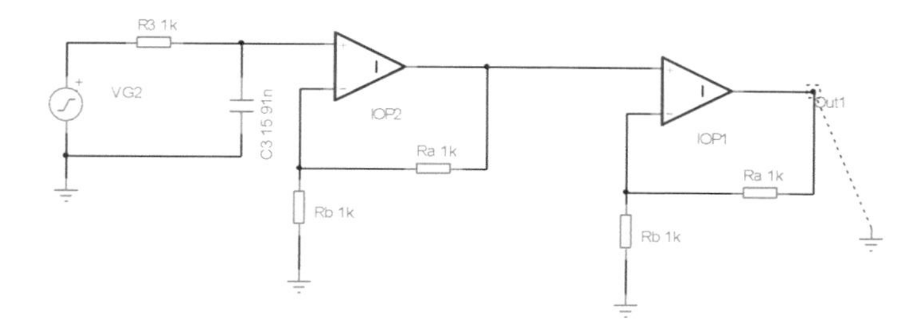


A third-order filter has a gain of 3.1721; to change this to a gain of 6, we need to add an non-inverting Opamp with a gain of

$$gain = \frac{6}{3.1721} = 1.8915$$

The gain of the non-inverting Opamp is

$$Vgain = 1 + \frac{R_a}{R_b} = 1.8915$$

Therefore

$$\frac{R_a}{R_b} = 0.8915$$

Let Ra = 891.5Ω and Rb = 1k.

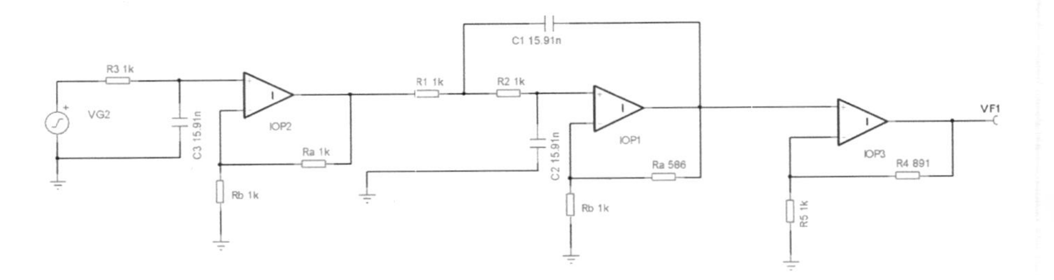


# Exercise 11.12

Use the Butterworth polynomials to design a basic fifth-order filter.

A fifth-order filter can be created from two second-order filters and one first-order filter. The circuit is shown as follows.

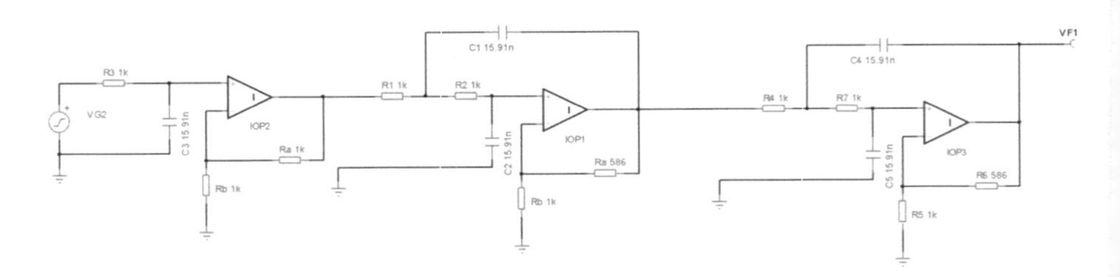

Complex Number Proofs

This proof shows that

$$\frac{1}{j} = -j$$

This can be written as

$$\frac{1 + j0}{0 + j}$$

We can multiply anything by 1, and it won't change. Also, we know that anything divided by itself equals 1. For example:

$$\frac{10}{10} = 1, \quad \frac{2m+1}{2m+1} = 1$$

We will choose a very particular way of representing the number 1. We will use something that is called "the complex conjugate." This sounds very complicated, but it's not too complex. If we look at the general complex number

$$a + jb$$

its complex conjugate is

$$a - jb$$

For example, the complex conjugate of

$$2 + j4 \; is \; 2 - j4$$

We simply change the sign of the j term.

Therefore, as it is normal to use the complex conjugate of the denominator in the complex division, we will use the complex conjugate of $0 + j$.

This means we will have

$$\left(\frac{1+j0}{0+j}\right)\left(\frac{0-j}{0-j}\right)$$

You can see that we are only multiplying the original division by 1, but we are creating the value of 1 in a very special way.

We can multiply the two numerators separately and the two denominators separately:

$$(1+j0)(0-j)=(1\times 0)+(1\times -j)+(j0\times 0)+(j0\times -j)$$

This gives

$$-j+0+0+0=-j$$

Now if we do the same with the denominator, we get

$$(0+j)(0-j)=(0\times 0)+(0\times -j)+(j\times 0)+(j\times -j)$$

This gives

$$0+0+0+-j^2$$

However

$$j^2=-1$$

Therefore

$$-j^2=--1=1$$

This means that the division has now become

$$\frac{-j}{1}=-j$$

This proves that

$$\frac{1}{j} = -j$$

We could use the complex conjugate to prove that

$$j = -\frac{1}{j}$$

$$j = \frac{j}{1}$$

$$\left(\frac{j}{1}\right)\left(\frac{-j}{1}\right) = \frac{(j)(-j)}{(1)(-j)}$$

Here, we are using the complex conjugate of the numerator. Note we don't need to include the terms that are zero.

This means that we get

$$\frac{-j^2}{-j}$$

But we know $j^2 = -1$; therefore, we get

$$\frac{--1}{-j} = \frac{1}{-j} = -\frac{1}{j}$$

Using the complex conjugate, we can always say

$$(a + jb)(a - jb) = (a \times a) + (a \times -jb) + (jb \times a) + (jb \times -jb)$$

This gives

$$(a + jb)(a - jb) = a^2 + ajb - jba - j^2b^2$$

This gives

$$(a+jb)(a-jb)=a^2+b^2$$

This is sometimes called rationalizing a complex number, and we usually do that to rationalize the denominator of a complex division.

# An Example from Chapter 11

We can use the complex conjugate to rationalize this transfer function:

$$TF=\frac{2}{1-\frac{1}{(\omega CR)^2}-j\frac{1}{\omega CR}}$$

The complex conjugate of the denominator is

$$1-\frac{1}{(\omega CR)^2}+j\frac{1}{\omega CR}$$

If we multiply the top and bottom by 1 using the complex conjugate to create a term for 1, we get

$$\frac{(2)\left(1-\frac{1}{(\omega CR)^2}+j\frac{1}{\omega CR}\right)}{\left(1-\frac{1}{(\omega CR)^2}-j\frac{1}{\omega CR}\right)\left(1-\frac{1}{(\omega CR)^2}+j\frac{1}{\omega CR}\right)}$$

We know the denominator would produce $a^2 + b^2$ where

$$a=1-\frac{1}{(\omega CR)^2}$$

and

$$b = \frac{1}{\omega CR}$$

Therefore, the denominator becomes

$$\left(1 - \frac{1}{(\omega CR)^2}\right)^2 + \left(\frac{1}{\omega CR}\right)^2$$

The numerator becomes

$$2 - \frac{2}{(\omega CR)^2} + j\frac{2}{\omega CR}$$

Therefore, we get

$$TF = \frac{2 - \frac{2}{(\omega CR)^2} + j\frac{2}{\omega CR}}{\left(1 - \frac{1}{(\omega CR)^2}\right)^2 + \left(\frac{1}{\omega CR}\right)^2}$$

# Index




H. H. Ward, *Mastering Analog Electronics*, Maker Innovations Series,
https://doi.org/10.1007/979-8-8688-0245-4

## B

## C



## D

# E

## F

# G, H



# I



# J



# K



# L

## M



## N

# O

## T, U

## V



## W, X, Y

## Z

Printed in the United States
by Baker & Taylor Publisher Services